DISCRETE MATHEMATICS AND ITS APPLICATIONS

Series Editor KENNETH H. ROSEN

T0227815

Advanced Number Theory with Applications

Richard A. Mollin

University of Calgary
Alberta, Canada

CRC Press
Taylor & Francis Group
Boca Raton London New York

CRC Press is an imprint of the
Taylor & Francis Group, an **informa** business

A CHAPMAN & HALL BOOK

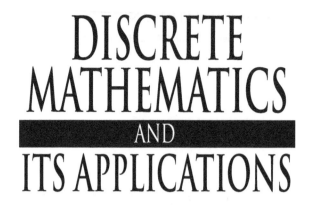

DISCRETE MATHEMATICS AND ITS APPLICATIONS

Series Editor

Kenneth H. Rosen, Ph.D.

Titles *(continued)*

Chapman & Hall/CRC
Taylor & Francis Group
6000 Broken Sound Parkway NW, Suite 300
Boca Raton, FL 33487-2742

First issued in paperback 2017

© 2010 by Taylor and Francis Group, LLC
Chapman & Hall/CRC is an imprint of Taylor & Francis Group, an Informa business

No claim to original U.S. Government works

ISBN-13: 978-1-4200-8328-6 (hbk)
ISBN-13: 978-1-138-11325-1 (pbk)

Library of Congress Cataloging-in-Publication Data

Mollin, Richard A., 1947-
    Advanced number theory with applications / Richard A. Mollin.
        p. cm. -- (Discrete mathematics its applications)
    Includes bibliographical references and index.
    ISBN 978-1-4200-8328-6 (hardcover : alk. paper)
    1. Number theory. I. Title.

QA241.M597 2009
512.7--dc22   2009026636

Visit the Taylor & Francis Web site at
http://www.taylorandfrancis.com

and the CRC Press Web site at
http://www.crcpress.com

For Kate Mollin

About the Cover

The surface on the cover was created using the equation for the lemniscate of Bernoulli in three dimensions, namely

$$f(x, y) = (x^2 + y^2)^2 - 2a^2(x^2 - y^2).$$

In two dimensions, the equation $(x^2 + y^2)^2 = 2a^2(x^2 - y^2)$ leads to the usual ∞ sign–see Biography 5.4 on page 207. The polar form is $r^2 = a^2 \cos(2\theta)$.

Contents

Preface

This book is designed as a second course in number theory at the senior undergraduate/junior graduate level to follow a course in elementary methods, such as that given in [68], the contents of which the reader is assumed to have knowledge. The material covered in the ten chapters of this book constitutes a course outline for one semester.

Chapter 1 begins with algebraic techniques including specialization to quadratic fields with applications to solutions of the Ramanujan–Nagell equations, factorization of Gaussian integers, Euclidean quadratic fields, and Gauss' proof of Fermat's Last Theorem (FLT) for $p = 3$. Applications of unique factorization are given in terms of both Euler's and Fermat's solution to Bachet's equation, concluding with a look at norm-Euclidean quadratic fields.

In Chapter 2 ideal theory is covered beginning with quadratic fields, and decomposition into prime ideals therein. Dedekind domains make up the second section, leading into Noetherian domains, and the unique factorization theorem for Dedekind domains. Principal Ideal Domains and Unique Factorization Domains are compared and contrasted. The section ends with the Chinese Remainder Theorem for ideals. The chapter concludes with an application to factoring using Pollard's cubic integer method, which serves as a preamble for the introduction of the number field sieve presented in the Appendix. Pollard's method is illustrated via factoring of the seventh Fermat number.

Chapter 3 is devoted to binary quadratic forms, starting with the basics on equivalence, discriminants, reduction, and class number. In the next section, composition is covered and linked to ideal theory. The form and ideal class groups are compared and contrasted, including an explicit formula for the relationship between the form class number and both the narrow and wide ideal class numbers. A proof of the finiteness of the ideal class number is achieved via the form class number, rather than the usual method of using Minkowski's Convex Body Theorem, which we cover in §4.3. Section 3.3 investigates the notion of ambiguous forms and ideals and the relationship between their classes. We show how this applies to representations of integers as a sum of two squares and to Markov triples. In Section 3.4, genus is introduced and the assigned values of generic characters are developed via Jacobi symbols. This is then applied to the principal genus, via a coset interpretation, using Dirichlet's Theorem on Primes in Arithmetic Progression, the proof of which is given in Chapter 7. This is a valuable vehicle for demonstrating the fact that two forms are in the same genus exactly when their cosets are equal. We tie the above together with the fact that the genus group is essentially the group of ambiguous forms. Section 3.5 uses the above to investigate representation problems. We begin with the algebraic interpretation of prime power representation as binary quadratic forms using the ideal class number. Numerous applications to representations of primes in the form $p = a^2 + Db^2$ are provided. The chapter ends with representations modulo a prime.

Chapter 4 develops Diophantine approximation techniques, starting with

Roth's celebrated result. We prove Liouville's Theorem, leading into an analysis of enumerable sets, including a proof that the set of all algebraic numbers is enumerable, followed by the countability of the rational numbers and the uncountability of the reals. Indeed, it follows from this that almost all reals are transcendental. The first section is completed with a proof of the fact that the n-th root of a rational integer is an algebraic integer of degree n, when that integer is not a certain power. Transcendence is covered in the second section with proofs that Liouville numbers, e, and π are all transcendental. Next the Lindemann–Weierstrass Theorem is established, allowing the statement of the more general Schanuel conjecture. The discussion is rounded out by a look at some renowned constants including those of Gel'fond, Gel'fond–Schneider, Proulet–Thue–Morse, Euler, Apéry, and Catalan. Section 4.3 introduces the geometry of numbers and its techniques with a goal of proving Minkowski's Convex Body Theorem that ends the chapter.

In Chapter 5, we extend the knowledge of arithmetic functions gained in a first course, by proving the Euler–Maclaurin summation formula, for which we introduce Bernoulli numbers, Bernoulli polynomials, and Fourier series. With this we are able to apply the formula to obtain Wallis' formula, Stirling's constant, Stirling's formula, and perhaps the slickest of applications, namely the accurate approximation of the Euler–Mascheroni constant. Average orders are the topic of the second section starting with a proof of Hermite's formula. This puts us into a position where we can derive the average order of the number of divisors function, the sum of divisors function, and Euler's totient $\phi(m)$. The third section concentrates upon the Riemann ζ-function. We apply the Euler–Maclaurin summation formula to obtain a formula for $\zeta(s)$. Then we discuss the Prime Number Theorem (PNT), Merten's Theorem, and various arithmetic function equivalences to the PNT. Then the Riemann hypothesis (RH) and its equivalent formulations are considered, after which we develop techniques to provide a rather straightforward proof of the functional equation for $\zeta(s)$ as a closing feature of the chapter.

In Chapter 6, we introduce p-adic analysis, commencing with solving modulo p^n for successively higher powers of a prime p. Hensel's Lemma is the featured result of the first section. The second section introduces valuations, including the p-adic versions. Then Cauchy sequences come into play giving rise to p-adic fields and domains. We have tools to prove that equivalent powers are valuations, which ends the section. We compare Archimedean and non-Archimedean valuations in the third section, featuring a proof of Ostrowski's Theorem. In the last section, we apply what we have learned to representation of p-adic numbers. This involves the proof that every rational number has a representation as a periodic power series in a given prime p to close the chapter.

Chapter 7 delves into Dirichlet, his characters, L-functions, and their zeros related to the RH. We see the implications of his theorem for primes in arithmetic progression, proved in the second section. In the third section we introduce Dirichlet density and applications such as Beatty's theorem. The chapter ends with Dirichlet density on primes in arithmetic progression modulo m which have density $1/\phi(m)$.

Chapter 8 comprises applications of the first seven chapters to Diophantine equations. We begin with an overview of Lucas–Lehmer theory, proving results promised earlier in the text such as solutions of the generalized Ramanujan–Nagell equations in the second section and Bachet's equation in the third section. The Fermat equation is the topic of the fourth section with Kummer's proof of FLT for regular primes. The chapter is rounded out with the ABC conjecture and Catalan's conjecture. We discuss the recent proof of the latter and its generalization, the still open Fermat–Catalan conjecture. More than a half-dozen consequences of the ABC conjecture are displayed and discussed, including the Thue–Siegel–Roth Theorem, Hall's conjecture, the Erdös–Mollin–Walsh conjecture, and the Granville–Langevin conjecture. We demonstrate how these follow from ABC.

Chapter 9 studies elliptic curves, launched by an introduction of the basics, illustrated and presented as a foundation. The second section defines torsion points, the Nagell–Lutz Theorem, Mazur's Theorem, Siegel's Theorem, and the notion of reduction. This sets the stage for Lenstra's elliptic curve factoring method and his primality testing method. We also look at the Goldwasser–Killian primality proving algorithm. The chapter closes with a description of the Menezes–Vanstone Elliptic Curve Cryptosystem as an application.

The last chapter is on modular forms. The modular group and modular forms are introduced as vehicles for much deeper considerations later in the chapter. Spaces and levels of modular forms are used as applications to elliptic curves including j-invariants and the Weierstrass $\wp$-function. The main text ends with Section 10.4 that looks, in detail, at the Shimura–Taniyama–Weil conjecture both in terms of L-functions and modular parametrizations. Modular elliptic curves are introduced as the steppingstone to the proof of FLT. Chapter 10 ends with Ribet's Theorem and a one-paragraph proof of FLT emanating from it, called the Frey–Serre–Ribet approach, a fitting conclusion and demonstration of the power of the theory.

An overview, without proofs, of sieve theory is relegated to the Appendix. We begin with a description of the goals of sieve theory and the effects its study has had on such open problems as the twin prime conjecture, the Goldbach conjecture, and Artin's conjecture, among others. We provide a description of the Eratosthenes sieve from the perspective of the Möbius function in order to lay the foundation for modern-day sieves. We begin with Brun's Theorem and his constant, including a discussion of how computation of Brun's constant led to the discovery of a flaw in the Pentium computer chip. Then we set the groundwork for presentation of Selberg's sieve by painting the picture of the basic sieve problem in terms of upper and lower limits on certain related functions. Selberg's sieve has many applications including the Brun–Titchmarsh Theorem, bounds for the twin prime conjecture, and the Goldbach conjecture. Then Linnik's large sieve is developed as a generalization of Brun's results and illustrated via applications to Artin's conjecture. Next is the Bombieri–Vinogradov Theorem and its applications to the Titchmarsh divisor problem. Then the classic result, Bombieri's asymptotic sieve, is presented via a hypothesis involving the generalized Mangoldt function. The most striking of the applications of the asymptotic sieve is the Friedlander–Iwaniec Theorem that there are infinitely many primes of the form $a^2 + b^4$. The aforementioned hypothesis involves the Elliot–Halberstram conjecture (EHC), so we are naturally led to the recent results by Goldston, Pintz, and Yildirim on gaps between primes. In particular, their result based upon the validity of the EHC is the satisfying conclusion that $\lim_{n\to\infty} \inf(p_{n+1} - p_n) \leq 16$, where p_n is the n-th prime. With these results as an illustration of the power of sieve theory, we turn our attention to the use of sieves in factoring by bringing out the big gun, the number field sieve and illustrate in detail its use in factoring of the ninth Fermat number.

The Bibliography has been set up in such a way that maximum information is imparted. This includes a page reference for each and every citing of a given item, so that no guesswork is involved as to where this reference is used. The index has more than 1,500 entries presented for maximum cross-referencing. Similarly, any reference, in text, to a theorem, definition, etc. is coupled with the page number on which it sits. These conventions ensure that the reader will find data with ease. There are nearly 50 mini-biographies of the mathematicians who helped to develop the results presented, in order to give a human face to the number theory and its applications. There are nearly 340 exercises with solutions of the odd-numbered exercises included at the end of the text, and a solutions manual for the even-numbered exercises available to instructors who adopt the text for a course. The website below is designed for the reader to access any updates and the e-mail address below is available for any comments.

◆ **Acknowledgments** First of all, I am deeply grateful to the Killam Foundation for providing the award allowing the completion of this project in a timely fashion. Also, I am grateful for the proofreading done by the following people. Thanks go to John Burke (U.S.A.) who took the time to effectively comment. Moreover, Keith Matthews (Australia) made valuable comments that helped polish the book. Also, thanks to John Robertson (U.S.A.) with whom I had lengthy electronic conversations over development of several sections of the book, especially Chapter 3 on binary quadratic forms. These interchanges had beneficial effects both for the book and our respective research programs. His insightful comments were most welcome. With Anitha Srinivasan (India), I similarly had lengthy electronic exchanges that led to creative, and even perspective-changing results. Her input was extremely valuable. My former student, Thomas Zaplachinski (Canada) who is now a working cryptographer in the field, gave the non-academic approach that was needed to round out the input received, and was deeply appreciated. Overall, this was an inspiring project, and one that is intended to be a service to students studying the most dynamic area of mathematics—number theory.

July 15, 2009

website: http://www.math.ucalgary.ca/~ramollin/

e-mail: ramollin@math.ucalgary.ca

About the Author

Richard Anthony Mollin is a professor in the Mathematics Department at the University of Calgary. Over the past twenty-three years, he has been awarded 6 Killam Resident Fellowships—a record number of these awards, see: *http://www.killamtrusts.ca/*. His 2009 Killam award provided the opportunity to complete this book, *Advanced Number Theory with Applications*. He has written over 190 publications including 11 books in algebra, number theory, and computational mathematics. He is a past member of the Canadian and American Mathematical Societies, the Mathematical Association of America and is a member of various editorial boards. He has been invited to lecture at numerous universities, conferences and scientific society meetings and has held several research grants from universities and governmental agencies. He is the founder of the Canadian Number Theory Association and hosted its first conference and a NATO Advanced Study Institute in Banff in 1988–see [60]–[61].

On a personal note—in the 1970s he owned a professional photography business, *Touch Me with Your Eyes*, and photographed many stars such as Paul Anka, David Bowie, Cher, Bob Dylan, Peter O'Toole, the Rolling Stones, and Donald Sutherland. His photographs were published in *The Toronto Globe and Mail* newspaper as well as *New Music Magazine* and elsewhere. Samples of his work can be viewed online at *http://math.ucalgary.ca/~ramollin/pixstars.html*.

His passion for mathematics is portrayed in his writings—enjoyed by mathematicians and the general public. He has interests in the arts, classical literature, computers, movies, and politics. He is a patron and a benefactor of The Alberta Ballet Company, Alberta Theatre Projects, The Calgary Opera, The Calgary Philharmonic Orchestra, and Decidedly Jazz Danceworks. His love for life comprises cooking, entertaining, fitness, health, photography, and travel, with no plans to slow down or retire in the foreseeable future.

About the Author

Chapter 1

Algebraic Number Theory and Quadratic Fields

> *I used to love mathematics for its own sake, and I still do, because it allows for no hypocrisy and no vagueness, my two bêtes noires.*
> **Henri Beyle Stendhal (1783–1842), French novelist**

In this introductory chapter, we introduce algebraic number theory with a concentration on quadratic fields. We begin with a general look at number fields. The reader should be familiar with the concepts in a course in number theory contained in [68] to which we will refer when needed.

1.1 Algebraic Number Fields

Algebraic number theory generalizes the notion of the ordinary or *rational integers*

$$\mathbb{Z} = \{\ldots, -2, -1, 0, 1, 2, \ldots\}.$$

To see how this is done, we consider the elements of $\mathbb{Z}$ as roots of linear monic polynomials, namely if $a \in \mathbb{Z}$, then a is a root of $f(x) = x - a$. Then we generalize as follows.

Definition 1.1 | **Algebraic Integers**

If $\alpha \in \mathbb{C}$ is a root of a monic, integral polynomial of degree d, namely a root of a polynomial of the form

$$f(x) = \sum_{j=0}^{d} a_j x^j = a_0 + a_1 x + \cdots + a_{d-1} x^{d-1} + x^d \in \mathbb{Z}[x],$$

which is irreducible over $\mathbb{Q}$, then α is called an algebraic integer of degree d.

Example 1.1 $a + b\sqrt{-1} = a + bi$, where $a, b \in \mathbb{Z}$, with $b \neq 0$ is an algebraic integer of degree 2 since it is a root of $x^2 - 2ax + a^2 + b^2$, but not a root of a linear, integral, monic polynomial since $b \neq 0$.

The following notion allows us to look at some distinguished types of algebraic integers.

Definition 1.2 | **Primitive Roots of Unity**

For $n \in \mathbb{N} = \{1, 2, 3, \ldots\}$ *(the natural numbers)*, ζ_n denotes a primitive n^{th} root of unity, which is a root of $x^n - 1$, but not a root of $x^d - 1$ for any natural number $d < n$.

Example 1.2 $\zeta_3 = (-1 + \sqrt{-3})/2$ is a primitive cube root of unity since it is a root of $x^3 - 1$, but clearly not a root of $x^2 - 1$ or $x - 1$.

A special kind of algebraic integer is given in the following.

Example 1.3 Numbers of the form

$$z_0 + z_1\zeta_n + z_2\zeta_n^2 + \cdots + z_{n-1}\zeta_n^{n-1}, \text{ for } z_j \in \mathbb{Z},$$

are called *cyclotomic integers* of order n.

Definition 1.2, in turn, is a special case of the following.

Definition 1.3 | **Units**

An element α in a commutative ring R with identity 1_R is called a *unit* in R when there is a $\beta \in R$ such that $\alpha\beta = 1_R$. The multiplicative group of units in R is denoted by $\mathfrak{U}_R$.

Example 1.4 In $\mathbb{Z}[\sqrt{2}] = R$, $1 + \sqrt{2}$ is a unit since

$$(1 + \sqrt{2})(-1 + \sqrt{2}) = 1_R = 1.$$

Definition 1.4 | **Algebraic Numbers and Number Fields**

An algebraic number, α, of degree $d \in \mathbb{N}$ is a root of a monic polynomial in $\mathbb{Q}[x]$ of degree d and not the root of any polynomial in $\mathbb{Q}[x]$ of degree less than d. In other words, an algebraic number is the root of an irreducible polynomial of degree d over $\mathbb{Q}$. An *algebraic number field*, or simply *number field*, is of the form $F = \mathbb{Q}(\alpha_1, \alpha_2, \ldots, \alpha_n) \subseteq \mathbb{C}$ for $n \in \mathbb{N}$ where α_j for $j = 1, 2, \ldots, n$ are algebraic numbers. Denote the subfield of $\mathbb{C}$ consisting of all algebraic numbers by $\overline{\mathbb{Q}}$, and the set of all algebraic integers in $\overline{\mathbb{Q}}$ by $\mathbb{A}$. An algebraic number of degree $d \in \mathbb{N}$ over a number field F is the root of an irreducible polynomial of degree d over F.

Remark 1.1 If F is a *simple extension*, namely of the form $\mathbb{Q}(\alpha)$, for an algebraic number α, then we may consider this as a vector space over $\mathbb{Q}$, in which case we may say that $\mathbb{Q}(\alpha)$ has dimension d over $\mathbb{Q}$ having basis $\{1, \alpha, \dots, \alpha^{d-1}\}$. (See [68, §2.1] and [68, Appendix A], where the background on these algebraic structures is presented. Also, see Exercise 1.4 on page 16 to see that all number fields are indeed simple.)

By Definition 1.4, $\mathbb{Q}$ is the smallest algebraic number field since it is of dimension 1 over itself, and the simple field extension $\mathbb{Q}(\alpha)$ is the smallest subfield of $\mathbb{C}$ containing both $\mathbb{Q}$ and α.

We now demonstrate that $\mathbb{A}$, as one would expect, has the proper structure in $\overline{\mathbb{Q}}$, which will lead us to a canonical subring of algebraic number fields.

Theorem 1.1 | **The Ring of All Algebraic Integers**

$\mathbb{A}$ *is a subring of* $\overline{\mathbb{Q}}$.

Proof. It suffices to prove that if $\alpha, \beta \in \mathbb{A}$, then both $\alpha + \beta \in \mathbb{A}$ and $\alpha\beta \in \mathbb{A}$. To this end we need the following.

Claim 1.1 *If* $\alpha \in \mathbb{A}$, *then* $\mathbb{Z}[\alpha] = \{f(\alpha) : f(x) \in \mathbb{Z}[x]\}$ *is a finitely generated* $\mathbb{Z}$*-module.*

Since $\alpha \in \mathbb{A}$, then there exist $a_j \in \mathbb{Z}$ for $j = 0, 1, \dots, d-1$ for some $d \geq 1$ such that
$$\alpha^d - a_{d-1}\alpha^{d-1} - \cdots - a_1\alpha - a_0 = 0.$$
Therefore,
$$\alpha^d = a_{d-1}\alpha^{d-1} + a_{d-2}\alpha^{d-2} + \cdots + a_1\alpha + a_0 \in \mathbb{Z}\alpha^{d-1} + \cdots + \mathbb{Z}\alpha + \mathbb{Z},$$
and
$$\alpha^{d+1} = a_{d-1}\alpha^d + a_{d-2}\alpha^{d-1} + \cdots + a_1\alpha^2 + a_0\alpha \in \mathbb{Z}\alpha^d + \mathbb{Z}\alpha^{d-1} + \cdots + \mathbb{Z}\alpha^2 + \mathbb{Z}\alpha$$
$$\subseteq \mathbb{Z}\alpha^{d-1} + \mathbb{Z}\alpha^{d-2} + \cdots + \mathbb{Z}\alpha + \mathbb{Z}.$$
Continuing in this fashion we conclude, inductively, that
$$\alpha^c \in \mathbb{Z}\alpha^{d-1} + \mathbb{Z}\alpha^{d-2} + \cdots + \mathbb{Z}\alpha + \mathbb{Z},$$
for any $c \geq d$. However, clearly,
$$\alpha^c \in \mathbb{Z}\alpha^{d-1} + \mathbb{Z}\alpha^{d-2} + \cdots + \mathbb{Z}\alpha + \mathbb{Z},$$
for $c = 1, 2, \cdots, d-1$, so
$$\alpha^c \in \mathbb{Z}\alpha^{d-1} + \mathbb{Z}\alpha^{d-2} + \cdots + \mathbb{Z}\alpha + \mathbb{Z},$$

for any $c \geq 0$. Hence, $\mathbb{Z}[\alpha]$ is a finitely generated $\mathbb{Z}$-module. This completes Claim 1.1.

By Claim 1.1, both $\mathbb{Z}[\alpha]$ and $\mathbb{Z}[\beta]$ are finitely generated. Suppose that $a_1, a_2, \ldots, a_k$ are generators of $\mathbb{Z}[\alpha]$ and $b_1, b_2, \ldots, b_\ell$ are generators of $\mathbb{Z}[\beta]$. Then $\mathbb{Z}[\alpha, \beta]$ is the additive group generated by the $a_i b_j$ for $1 \leq i \leq k$ and $1 \leq j \leq \ell$. Thus, $\mathbb{Z}[\alpha, \beta]$ is finitely generated. Since $\alpha + \beta, \alpha\beta \in \mathbb{Z}[\alpha, \beta] \subseteq \mathbb{A}$, then we have secured the theorem. $\qquad\qquad\square$

Given an algebraic number field F, $F \cap \mathbb{A}$ is a ring in F, by Exercise 1.2 on page 16. This leads to the following.

Definition 1.5 $\boxed{\textbf{Rings of Integers}}$

If F is an algebraic number field, then $F \cap \mathbb{A}$ is called the *ring of (algebraic) integers* of F, denoted by $\mathfrak{O}_F$.

With Definition 1.5 in hand, we may now establish a simple consequence of Theorem 1.1.

Corollary 1.1 *The ring of integers of $\mathbb{Q}$ is $\mathbb{Z}$, namely $\mathfrak{O}_\mathbb{Q} = \mathbb{Q} \cap \mathbb{A} = \mathbb{Z}$.*

Proof. If $\alpha \in \mathbb{A} \cap \mathbb{Q}$, then $\alpha = a/b$ where $a, b \in \mathbb{Z}$ and $\gcd(a, b) = 1$, with $b \neq 0$. Since $\alpha \in \mathbb{A}$, there exists an $f(x) = a_0 + \sum_{j=1}^{d} a_j x^j \in \mathbb{Z}[x]$, with $a_d = 1$, such that $f(\alpha) = 0$. If $d = 1$, then we are done since $a_0 + \alpha \in \mathbb{Z}$ and $a_0 \in \mathbb{Z}$. If $d > 1$, then $a_0 + \sum_{j=1}^{d} a_j \alpha^j \in \mathbb{Z}$, so

$$\sum_{j=1}^{d} a_j \alpha^j = \sum_{j=1}^{d} \frac{a_j a^j b^{d-j}}{b^d} \in \mathbb{Z}.$$

Therefore, $b^d \mid \sum_{j=1}^{d} a_j a^j b^{d-j}$. Since $d > 1$, $b \mid \sum_{j=1}^{d-1} a_j a^j b^{d-j}$, so $b \mid a^d$. But $\gcd(a, b) = 1$, so $b = 1$ and $\alpha \in \mathbb{Z}$. $\qquad\qquad\square$

Corollary 1.2 *If F is an algebraic number field, then $\mathbb{Q} \cap \mathfrak{O}_F = \mathbb{Z}$.*

Proof. Since $\mathfrak{O}_F \subseteq \mathbb{A}$, then by Corollary 1.1, $\mathbb{Q} \cap \mathfrak{O}_F \subseteq \mathbb{Z}$. But clearly $\mathbb{Z} \subseteq \mathbb{Q} \cap \mathfrak{O}_F$, so we have equality. $\qquad\qquad\square$

Remark 1.2 Now we establish the rings of integers for quadratic fields. First, we show that a given quadratic field is determined by a unique squarefree integer. We note that if
$$f(x) = x^2 + ax + b \in \mathbb{Q}[x],$$
is irreducible, and $\alpha \in \mathbb{C}$ is a root of $f(x)$, then the smallest subfield of $\mathbb{C}$ containing both $\mathbb{Q}$ and α is given by adjoining α to $\mathbb{Q}$, denoted by $\mathbb{Q}(\alpha)$ so
$$\mathbb{Q}(\alpha) = \{x + y\alpha : x, y \in \mathbb{Q}\},$$
which is what we call a *quadratic field*.

Quadratic polynomials with the the same squarefree part of the discriminant give rise to the same quadratic field. To see this suppose that $f(x) = x^2 + bx + c$, $g(x) = x^2 + b_1 x + c_1 \in \mathbb{Q}[x]$ are irreducible, $\Delta = b^2 - 4c = m^2 D$, and $\Delta_1 = b_1^2 - 4c_1 = m_1^2 D$, where $m, m_1 \in \mathbb{Z}$ and D is squarefree. Then

$$\mathbb{Q}(\sqrt{\Delta}) = \mathbb{Q}(\sqrt{m^2 D}) = \mathbb{Q}(m\sqrt{D}) = \mathbb{Q}(\sqrt{D}) =$$

$$\mathbb{Q}(m_1\sqrt{D}) = \mathbb{Q}\left(\sqrt{m_1^2 D}\right) = \mathbb{Q}(\sqrt{\Delta_1}).$$

Thus, we need the following to clarify the situation on uniqueness of quadratic fields.

Theorem 1.2 | **Quadratic Fields Uniquely Determined**

If F is a quadratic field, there exists a unique squarefree integer D such that $F = \mathbb{Q}(\sqrt{D})$.

Proof. Suppose that $F = \mathbb{Q}(\alpha)$, where α is a root of the irreducible polynomial $x^2 + bx + c$. By the quadratic formula,

$$\alpha \in \left\{ \alpha_1 = \frac{-b + \sqrt{b^2 - 4c}}{2}, \ \alpha_2 = \frac{-b - \sqrt{b^2 - 4c}}{2} \right\}.$$

Since $\alpha_1 = -\alpha_2 - b$ with $b \in \mathbb{Q}$, then $\mathbb{Q}(\alpha_1) = \mathbb{Q}(\alpha_2) = \mathbb{Q}(\alpha)$. However,

$$\mathbb{Q}(\alpha_1) = \mathbb{Q}\left(\frac{-b + \sqrt{b^2 - 4c}}{2}\right) = \mathbb{Q}(\sqrt{b^2 - 4c}).$$

Let $a = b^2 - 4c = e/f \in \mathbb{Q}$. Then $a \neq d^2$ for any $d \in \mathbb{Q}$ since $x^2 + bx + c$ is irreducible in $\mathbb{Q}[x]$. Without loss of generality we may assume that $\gcd(e, f) = 1$ and f is positive. Let $ef = n^2 D$, where D is the squarefree part of ef. Hence, $D \neq 1$, and arguing as in Remark 1.2, $\mathbb{Q}(\sqrt{D}) = \mathbb{Q}(\sqrt{a})$, observing that $\mathbb{Q}(\sqrt{e/f}) = \mathbb{Q}(\sqrt{ef})$. This shows existence. It remains to prove uniqueness.

If D_1 is a squarefree integer such that $\mathbb{Q}(\sqrt{D}) = \mathbb{Q}(\sqrt{D_1})$, then

$$\sqrt{D} = u + v\sqrt{D_1}$$

with $u, v \in \mathbb{Q}$. By squaring, rearranging, and assuming that $uv \neq 0$, we get

$$\sqrt{D_1} = \frac{D - u^2 - Dv^2}{2uv} \in \mathbb{Q},$$

which contradicts that D_1 is squarefree. Thus, $uv = 0$. If $v = 0$, then $\sqrt{D} \in \mathbb{Q}$, contradicting the squarefreeness of D. Therefore, $u = 0$ and $D = v^2 D_1$, but again, D is squarefree, so $v^2 = 1$, which yields that $D = D_1$. $\square$

Now we are in a position to determine the ring of integers of an arbitrary quadratic field.

Theorem 1.3 | **Rings of Integers in Quadratic Fields**

Let F be a quadratic field and let D be the unique squarefree integer such that $F = \mathbb{Q}(\sqrt{D})$. Then

$$
\mathfrak{O}_F = \begin{cases} \mathbb{Z}\left[\frac{1+\sqrt{D}}{2}\right] & \text{if } D \equiv 1\,(\mathrm{mod}\,4), \\ \mathbb{Z}[\sqrt{D}] & \text{if } D \not\equiv 1\,(\mathrm{mod}\,4). \end{cases}
$$

Proof. Let

$$
\sigma = \begin{cases} 2 & \text{if } D \equiv 1\,(\mathrm{mod}\,4), \\ 1 & \text{if } D \not\equiv 1\,(\mathrm{mod}\,4). \end{cases}
$$

Then since $(1 + \sqrt{D})/\sigma$ is a root of

$$
x^2 - \frac{2x}{\sigma} + \frac{1-D}{\sigma^2},
$$

then

$$
\mathbb{Z} + \mathbb{Z}\left(\frac{\sigma - 1 + \sqrt{D}}{\sigma}\right) \subseteq \mathfrak{O}_F.
$$

It remains to prove the reverse inclusion.

Let $\alpha \in \mathfrak{O}_F \subseteq F$. Then $\alpha = a + b\sqrt{D}$ where $a, b \in \mathbb{Q}$. We may assume that $b \neq 0$ since otherwise we are done given that $\mathbb{Z} \subseteq \mathbb{Z} + \mathbb{Z}\left(\frac{\sigma-1+\sqrt{D}}{\sigma}\right)$. Since $\mathfrak{O}_F$ is a ring, then $\alpha' = (a - b\sqrt{D})$, $\alpha + \alpha' = 2a$, and $\alpha\alpha' = a^2 - Db^2$ are all in $\mathfrak{O}_F$. However, the latter two elements are also in $\mathbb{Q}$, and by Corollary 1.2, $\mathfrak{O}_F \cap \mathbb{Q} = \mathbb{Z}$, so

$$
2a, a^2 - Db^2 \in \mathbb{Z}. \tag{1.1}
$$

Case 1.1 $a \notin \mathbb{Z}$.

We must have $a = (2c + 1)/2$ for some $c \in \mathbb{Z}$. Therefore, by (1.1), $4(a^2 - Db^2) \in \mathbb{Z}$, which implies $4Db^2 \in \mathbb{Z}$. However, since D is squarefree, then $2b \in \mathbb{Z}$. (To see this, observe that if $2b = g/f$ where $g, f \in \mathbb{Z}$ with $\gcd(f, g) = 1$, and $f > 1$ is odd, then $4Dg^2 = f^2 h$ for some $h \in \mathbb{Z}$. Thus, since $\gcd(4g, f) = 1$, $f^2 \mid D$ contracting its squarefreeness.) If $b \in \mathbb{Z}$ then, by (1.1), $a \in \mathbb{Z}$, contradicting that $a = (2c + 1)/2$. Therefore, $b = (2k + 1)/2$ for some $k \in \mathbb{Z}$. Thus,

$$
a^2 - Db^2 = \frac{(2c+1)^2}{4} - \frac{D(2k+1)^2}{4} = c^2 + c - (k^2 + k)D + \frac{1-D}{4},
$$

which implies

$$
\frac{D-1}{4} = c^2 + c - (k^2 + k)D - a^2 + Db^2 \in \mathbb{Z},
$$

hence, $D \equiv 1 \pmod 4$ and:

$$\alpha = \frac{2c+1}{2} + \frac{(2k+1)\sqrt{D}}{2} = (c-k) + \frac{(2k+1)(1+\sqrt{D})}{2}$$

$$\in \mathbb{Z} + \mathbb{Z}\left(\frac{1+\sqrt{D}}{2}\right) = \mathbb{Z} + \mathbb{Z}\left(\frac{\sigma-1+\sqrt{D}}{\sigma}\right).$$

Case 1.2 $a \in \mathbb{Z}$.

In this instance, by (1.1), $Db^2 \in \mathbb{Z}$, and arguing as above, since D is square-free, $b \in \mathbb{Z}$. Hence,

$$\alpha = a + b\sqrt{D} \in \mathbb{Z} + \mathbb{Z}\sqrt{D} = \mathbb{Z} + \mathbb{Z}\left(\frac{\sigma-1+\sqrt{D}}{\sigma}\right),$$

which completes the reverse inclusion that secures the theorem. $\qquad\square$

Definition 1.6 | **Field Discriminants**

If D is the unique squarefree integer such that $F = \mathbb{Q}(\sqrt{D})$ is a quadratic field, then the discriminant of F is given by

$$\Delta_F = \begin{cases} D & \text{if } D \equiv 1 \pmod 4, \\ 4D & \text{if } D \not\equiv 1 \pmod 4. \end{cases}$$

Remark 1.3 Definition 1.6 follows from the fact that the minimal polynomial of F is $x^2 - x + (1-D)/4$ if $D \equiv 1 \pmod 4$ and $x^2 - D$ if $D \not\equiv 1 \pmod 4$.

Example 1.5 Suppose we have an irreducible quadratic polynomial

$$f(x) = ax^2 + bx + c \in \mathbb{Q}[x].$$

Then $\Delta = b^2 - 4ac$ is the discriminant of not only $f(x)$, but also the quadratic field $\mathbb{Q}(\sqrt{\Delta})$. By the quadratic formula, the roots of $f(x)$ are given by

$$\alpha = \frac{-b+\sqrt{\Delta}}{2a}, \text{ and } \alpha' = \frac{-b-\sqrt{\Delta}}{2a},$$

where α' is called the *algebraic conjugate* of α. By Exercise 1.1 on page 16, $\mathbb{Q}(\alpha) = \mathbb{Q}(\sqrt{\Delta})$. This is the simplest nontrivial number field, a quadratic field over $\mathbb{Q}$—see Remark 1.2 on page 4.

The reader will note that some easily verified properties of conjugates are given as follows.

(a) $(\alpha\beta)' = \alpha'\beta'$.

(b) $(\alpha \pm \beta)' = \alpha' \pm \beta'$.

(c) $(\alpha/\beta)' = \alpha'/\beta'$, where $\alpha/\beta = \delta \in \mathbb{Q}(\sqrt{\Delta})$.

Remark 1.4 If, in Theorem 1.3, $D < 0$, F is called a *complex* (or *imaginary*) quadratic field, and if $D > 0$, F is called a *real* quadratic field. Also, the group of units in a quadratic field forms an abelian group. For real quadratic fields we will learn about this group in Chapter 7 since it is more complicated than the complex case which we tackle now. The reader will recall the notion of groups and notation for a cyclic group, $\langle g \rangle$, generated by an element g—see [68, p. 300], for instance, and recall Definition 1.2 on page 2.

Theorem 1.4 | **Units in Complex Quadratic Fields**

If $F = \mathbb{Q}(\sqrt{D})$ is a complex quadratic field, then

$$\mathfrak{U}_F = \mathfrak{U}_{\mathfrak{O}_F} = \begin{cases} \langle \zeta_6 \rangle = \left\langle \frac{1+\sqrt{-3}}{2} \right\rangle & \text{if } D = -3, \\ \langle \zeta_4 \rangle = \langle \sqrt{-1} \rangle & \text{if } D = -1, \\ \langle \zeta_2 \rangle = \langle -1 \rangle & \text{otherwise.} \end{cases}$$

Proof. By Theorem 1.3 on page 6 we may write $u = a + b\sqrt{D} \in \mathfrak{U}_{\mathfrak{O}_F}$, with $2a, 2b \in \mathbb{Z}$. Hence, if $D \not\equiv 1 \,(\mathrm{mod}\ 4)$, then $a^2 - b^2 D = 1$, for some $a, b \in \mathbb{Z}$ since $D < 0$. If $D < -1$, then $a^2 - b^2 D > 1$ for $b \neq 0$. Thus, $b = 0$ for $D \not\equiv 1 \,(\mathrm{mod}\ 4)$ with $D < -1$. In other words, $\mathfrak{U}_{\mathfrak{O}_F} = \langle -1 \rangle = \langle \zeta_2 \rangle$, if $D \equiv 2, 3 \,(\mathrm{mod}\ 4)$ and $D < -1$.

Now we assume that $D \equiv 1 \,(\mathrm{mod}\ 4)$, so $a^2 - Db^2 = 4$ for $a, b \in \mathbb{Z}$. If $D < -4$, then for $b \neq 0$, $a^2 - Db^2 > 4$, a contradiction. Hence, for $D \equiv 1 \,(\mathrm{mod}\ 4)$, and $D < -4$,

$$\mathfrak{U}_{\mathfrak{O}_F} = \langle \zeta_2 \rangle.$$

It remains to consider the cases $D = -1, -3$. If $D = -1$, then by Theorem 1.3 on page 6, $\mathfrak{O}_{\mathbb{Z}[i]} = \mathbb{Z} + \mathbb{Z}[i]$, $a + bi$ is a unit in $\mathfrak{O}_F$ if and only if $a^2 + b^2 = 1$. The solutions are $(a, b) \in \{(0 \pm 1), (\pm 1, 0)\}$. In other words, $\mathfrak{U}_{\mathbb{Z}[i]} = \{\pm 1, \pm i\}$.

If $D = -3$, then $a^2 + 3b^2 = 4$, so either $a = b = 1$, or $b = 0$ and $a = 2$. Hence, the units are ± 1, $(1 \pm \sqrt{-3})/2$, and $(-1 \pm \sqrt{-3})/2$. However, $1 = \zeta_6^6$, $-1 = \zeta_6^3$, $(1 + \sqrt{-3})/2 = \zeta_6$, $(1 - \sqrt{-3})/2 = \zeta_6^5$, $(-1 + \sqrt{-3})/2 = \zeta_6^2$, and $(-1 - \sqrt{-3})/2 = \zeta_6^4$. Hence, $\mathfrak{U}_{\mathfrak{O}_{\mathbb{Q}(\sqrt{-3})}} = \langle \zeta_6 \rangle$, as required.   □

The above development leads to the following notions and allows us to discuss divisibility in $\mathfrak{O}_F$, which is not closed under division.

Definition 1.7 | **Division in $\mathfrak{O}_F$**

If F is a number field and $\alpha, \beta \in \mathfrak{O}_F$, α is said to *divide* β if there exists a $\delta \in \mathfrak{O}_F$ such that $\beta = \alpha\delta$, denoted by $\alpha \mid \beta$ in $\mathfrak{O}_F$. If no such δ exists, we say that α *does not divide* β, denoted by $\alpha \nmid \beta$, in $\mathfrak{O}_F$. If $\alpha \mid \beta_1$ and $\alpha \mid \beta_2$ for $\beta_1, \beta_2 \in \mathfrak{O}_F$, α is said to be a *common divisor* of β_1 and β_2 in $\mathfrak{O}_F$.

Example 1.6 In $\mathbb{Z}[\sqrt{10}] = \mathfrak{O}_F$ where $F = \mathbb{Q}(\sqrt{10})$, by Theorem 1.3 on page 6, then $(4 + \sqrt{10})(4 - \sqrt{10}) = 6 = 6 + 0\sqrt{10}$, so $\alpha = (4 + \sqrt{10}) \mid 6 = \beta$ in $\mathbb{Z}[\sqrt{10}]$.

Now we look at a new perspective, namely elements *over* an integral domain—see [68, Remark 2.6, p. 81] for the basics on integral domains.

Definition 1.8 | **Elements Algebraic and Integral Over a Domain**

If $R \subseteq S$ where R and S are integral domains, then $\alpha \in S$ is said to be *integral over* R if there exists an

$$f(x) = x^d + r_{d-1}\gamma^{d-1} + \cdots + r_1 x + r_0 \in R[x]$$

such that $f(\alpha) = 0$. If R is a field and α is integral over R, then α is said to be *algebraic* over R. Also, if every nonconstant polynomial $f(x) \in R[x]$ has a root in R, then R is said to be *algebraically closed*. Moreover, any extension field that is algebraic over R and is algebraically closed is called an *algebraic closure* of R, and it may be shown that an algebraic closure is unique up to isomorphism.

Remark 1.5 It is trivially true that every element of R is integral over R since $\alpha \in R$ satisfies $f(\alpha) = 0$ for $f(x) = x - \alpha \in R[x]$. Note, as well, that in view of Definition 1.4 on page 2, and Definition 1.8, we may now restate the notion of an *algebraic number* as a complex number that is algebraic over $\mathbb{Q}$. Moreover, in view of Definition 1.1 on page 1 and Definition 1.8, we see that an *algebraic integer* is a complex number that is integral over $\mathbb{Z}$.

Given an element α that is algebraic over a number field F, Definition 1.8 tells us that there is a monic polynomial $f(x) \in F[x]$ with $f(\alpha) = 0$. We may assume that f has minimal degree. Hence, f must be irreducible, since otherwise, α would be the root of a polynomial of lower degree. Thus chosen, f is called *the minimal polynomial of α over F*. It turns out this polynomial is also unique—see Theorem 1.6 on the next page.

We now want to demonstrate that algebraic integers are sufficient to characterize algebraic number fields. First we need the following crucial result.

Lemma 1.1 | **Algebraic Numbers as Quotients of Integers**

Every algebraic number is of the form α/ℓ where α is an algebraic integer and $\ell \in \mathbb{Z}$ is nonzero.

Proof. By Definition 1.4 on page 2, if γ is an algebraic number, there exist $a_j \in \mathbb{Q}$ for $j = 0, 1, 2, \ldots, d - 1$ such that γ is a root of

$$f(x) = a_0 + a_1 x + a_2 x^2 + \cdots + a_{d-1}x^{d-1} + x^d.$$

Since $a_0 + a_1\gamma + a_2\gamma^2 + \cdots + a_{d-1}\gamma^{d-1} + \gamma^d = 0$, we may form the least common multiple, ℓ, of the denominators of the a_j for $j = 0, 1, \ldots, d$. Then

$$(\ell\gamma)^d + (\ell a_{d-1})(\ell\gamma)^{d-1} + \cdots + (\ell^{d-1}a_1)(\ell\gamma) + \ell^d a_0 = 0.$$

Thus $\ell\gamma$ is the root of a monic integral polynomial, so $\ell\gamma$ is an algebraic integer, say, α. Hence, $\gamma = \alpha/\ell$, with $\alpha \in \mathbb{A}$ and $\ell \in \mathbb{Z}$. $\qquad\square$

Theorem 1.5 | **Number Fields—Algebraic Integer Extensions**

    If F is an algebraic number field, then there is an algebraic integer α such that $F = \mathbb{Q}(\alpha)$. Also, $\beta \in F$ if and only if there are unique $q_j \in \mathbb{Q}$ for $j = 0, 1, \ldots, n-1$, such that

$$\beta = q_0 + q_1\alpha + \cdots + q_{n-1}\alpha^{n-1},$$

where $n = |F : \mathbb{Q}|$.

Proof. By Exercise 1.4 on page 16, $F = \mathbb{Q}(\gamma)$ for some algebraic number γ, and by Lemma 1.1, $\mathbb{Q}(\gamma) = \mathbb{Q}(\alpha/\ell) = \mathbb{Q}(\alpha)$ for some $\alpha \in \mathbb{A}$. The second statement follows from the first statement in conjunction with Claim 1.1 on page 3 and Definition 1.8 on the previous page. $\qquad\square$

Example 1.7 Let $E = \mathbb{Q}(\sqrt{2}, i)$, where $i = \zeta_4 = \sqrt{-1}$ is a primitive fourth root of unity. Then by Exercise 1.6 on page 17 ,

$$\mathbb{Q}(i, \sqrt{2}) = \mathbb{Q}\left(\frac{\sqrt{2}}{2}(1+i)\right),$$

and

$$\zeta_8 = \frac{\sqrt{2}}{2}(1+i),$$

where ζ_8 is a primitive eighth root of unity.

Theorem 1.6 | **Minimal Polynomials are Unique**

    A number $\alpha \in \mathbb{C}$ is an algebraic number of degree $d \in \mathbb{N}$ over a number field F if and only if α is the root of an unique irreducible monic polynomial, denoted by $m_{\alpha,F}(x) \in F[x]$.
    Any $h(x) \in F[x]$ such that $h(\alpha) = 0$ must be divisible by $m_{\alpha,F}(x)$ in $F[x]$.

Proof. If α is an algebraic number of degree d over F, then by Definition 1.4 on page 2, we may let $f(x) \in F[x]$ be a monic polynomial of minimal degree with $f(\alpha) = 0$, and let $h(x) \in F[x]$ be any other monic polynomial of minimal degree with $h(\alpha) = 0$. Then by the Euclidean algorithm for polynomials (see [68, Theorem A.11, p. 302]), there exist $q(x), r(x) \in F[x]$ such that

$$h(x) = q(x)f(x) + r(x),$$

where

$$0 \le \deg(r) < \deg(f) \text{ or } r(x) = 0, \text{ the zero polynomial.}$$

However $f(\alpha) = 0$ so $h(\alpha) = 0 = f(\alpha)$, so $r(\alpha) = 0$, contradicting the minimality of f unless $r(x) = 0$ for all x. Hence, $f(x) \mid h(x)$. The same argument can be used to show that $h(x) \mid f(x)$. Hence, $h(x) = cf(x)$ for some $c \in F$. However, f and h are monic, so $c = 1$ and $h = f$. This proves that $f(x) = m_{\alpha, F}(x)$ is the unique monic polynomial of α over F. The converse of the first statement follows a fortiori.

To prove the second statement, assume that $h(x) \in F[x]$ such that $h(\alpha) = 0$ and use the Euclidean algorithm for polynomials as above to conclude that $m_{\alpha, F}(x) \mid h(x)$ by letting $m_{\alpha, F}(x) = f(x)$ in the above argument. □

Corollary 1.3 *An irreducible polynomial over an algebraic number field has no repeated roots in* $\mathbb{C}$. *In particular, all the roots of* $m_{\alpha, F}(x)$ *are distinct.*

Proof. If F is a number field and $f(x) \in F[x]$ is irreducible with a repeated root α, then
$$f(x) = c(x - \alpha)^2 g(x),$$
for some $c \in F$ and $g(x) \in \mathbb{C}[x]$. By Theorem 1.6, $m_{\alpha, F}(x) \mid f(x)$ so $f(x) = a m_{\alpha, F}(x)$ for some $a \in F$, since f is irreducible. However,
$$f'(x) = 2c(x - \alpha)g(x) + c(x - \alpha)^2 g'(x),$$
where f' is the derivative of f. Hence, $f'(\alpha) = 0$, so by Theorem 1.6, again
$$m_{\alpha, F}(x) \mid f'(x),$$
contradicting the minimality of $m_{\alpha, F}(x)$ since $\deg(f') < \deg(f)$. □

Corollary 1.4 *If* $\alpha \in \mathbb{A}$, *then* $m_{\alpha, \mathbb{Q}}(x) \in \mathbb{Z}[x]$.

Proof. This follows from Definition 1.1 on page 1 and Theorem 1.6. □

Example 1.8 Returning to Example 1.7 on the preceding page, we see that if $F = \mathbb{Q}(i)$ and $\alpha = \zeta_8$, then
$$m_{\alpha, F}(x) = x^2 - i$$
is the minimal polynomial of α over F. Moreover, the minimal polynomial of α over $\mathbb{Q}$ is given by
$$m_{\alpha, \mathbb{Q}}(x) = \frac{x^8 - 1}{x^4 - 1} = x^4 + 1,$$
which is an example of the following type of distinguished polynomial.

Definition 1.9 $\boxed{\textbf{Cyclotomic Polynomials}}$

If $n \in \mathbb{N}$, then the nth *cyclotomic polynomial* is given by

$$\Phi_n(x) = \prod_{\substack{\gcd(n,j)=1 \\ 1\leq j\leq n}} (x - \zeta_n^j),$$

where ζ_n is given by Definition 1.2 on page 2. The degree of $\Phi_n(x)$ is $\phi(n)$ where $\phi(n)$ is the Euler totient—see [68].

Remark 1.6 The reader may think of the term *cyclotomic* as "circle dividing," since the nth roots of unity divide the unit circle into n equal arcs. The cyclotomic polynomial also played a role in Gauss's theory of constructible regular polygons.

Note that since the roots of the nth cyclotomic polynomial are precisely the primitive n^{th} roots of unity, then the degree of $\Phi_n(x)$ is necessarily $\phi(n)$. We now demonstrate the irreducibility of the cyclotomic polynomial.

Theorem 1.7 | **Irreducibility of the Cyclotomic Polynomial**

For $n \in \mathbb{N}$,

$$\Phi_n(x) = m_{\zeta_n,\mathbb{Q}}(x),$$

so $\Phi_n(x)$ is irreducible in $\mathbb{Z}[x]$.

Proof. We may let $\Phi_n(x) = m_{\zeta_n,\mathbb{Q}}(x)g(x)$ for some $g(x) \in \mathbb{Z}[x]$ by Theorem 1.6 on page 10.

Claim 1.2 $m_{\zeta_n,\mathbb{Q}}(\zeta_n^p) = 0$ for any prime $p \nmid n$.

If $m_{\zeta_n,\mathbb{Q}}(\zeta_n^p) \neq 0$, then $g(\zeta_n^p) = 0$, so ζ_n is a root of $g(x^p)$. By Theorem 1.6 again,

$$g(x^p) = m_{\zeta_n,\mathbb{Q}}(x)h(x)$$

for some $h(x) \in \mathbb{Z}[x]$. Let

$$f(x) = \sum_j a_j x^j \in \mathbb{Z}[x]$$

have image

$$\overline{f}(x) = \sum_j \overline{a_j} x^j$$

under the natural map

$$\mathbb{Z}[x] \mapsto (\mathbb{Z}/p\mathbb{Z})[x].$$

Thus,

$$\overline{g}(x^p) = \overline{m}_{\zeta_n,\mathbb{Q}}(x)\overline{h}(x).$$

However, $\bar{g}(x^p) = \bar{g}^p(x)$ since char$(\mathbb{Z}/p\mathbb{Z}) = p$. Therefore,

$$0 = \bar{g}(\zeta_n^p) = (\bar{g}(\zeta_n))^p = \bar{g}(\zeta_n).$$

Since $\Phi_n(x) \mid (x^n - 1)$, then

$$x^n - 1 = \Phi_n(x)k(x) = m_{\zeta_n,\mathbb{Q}}(x)g(x)k(x),$$

for some $k(x) \in \mathbb{Z}[x]$. Therefore, in $\mathbb{Z}/p\mathbb{Z}[x]$,

$$x^n - \bar{1} = \overline{x^n - 1} = \overline{m}_{\zeta_n,\mathbb{Q}}(x)\bar{g}(x)\bar{k}(x).$$

Since $\bar{g}$ and $\overline{m}_{\zeta_n,\mathbb{Q}}$ have a common root ζ_n, then $x^n - \bar{1}$ has a repeated root. However, this is impossible by irreducibility criteria for polynomials over finite fields, since $p \nmid n$, (see [68, Corollary A.2, p. 301], for instance, where we see that $x^n - \bar{1}$ is irreducible if and only if $\gcd(x^n - \bar{1}, x^{p^i} - x) = 1$ for all natural numbers $i \leq \lfloor n/2 \rfloor$). We have established Claim 1.2, namely that ζ_n^p is a root of $m_{\zeta_n,\mathbb{Q}}(x)$ for any prime $p \nmid n$.

Repeated application of the above argument shows that y^p is a root of $m_{\zeta_n,\mathbb{Q}}(x)$ whenever y is a root. Hence, ζ_n^j is a root of $m_{\zeta_n,\mathbb{Q}}(x)$ for all j relatively prime to n such that $1 \leq j < n$. Thus, $\deg(m_{\zeta_n,\mathbb{Q}}) \geq \phi(n)$. However, $m_{\zeta_n,\mathbb{Q}}(x) \mid \Phi_n(x)$, so $m_{\zeta_n,\mathbb{Q}}(x) = \Phi_n(x)$, as required. $\square$

At this juncture, we look at general properties of units in rings of integers, in keeping with one of the themes of this section.

Proposition 1.1 *Let $\alpha \in \mathbb{A}$. Then the following are equivalent.*

(a) α *is a unit.*

(b) $\alpha \mid 1$ *in* $\mathbb{A}$.

(c) *If* $F = \mathbb{Q}(\alpha)$, *then* $m_{\alpha,F}(0) = \pm 1$.

Proof. The equivalence of (a) and (b) comes from Definition 1.3 on page 2. Now assume that α is a unit. Then, by Exercise 1.5 on page 17, $m_{\alpha,F}(0) = (-1)^d \prod_{j=1}^d \alpha_j = \pm 1$ if and only $\alpha \in \mathfrak{U}_F$, so (a) and (c) are equivalent. $\square$

We have now developed sufficient algebraic number theory in quadratic fields to provide a solution to a Diophantine problem that we did not have the tools to do in a first course — see [68, closing paragraph, p. 272].

Definition 1.10 | **Generalized Ramanujan–Nagell Equations**

The Diophantine Equation

$$x^2 - D = p^n, \text{ for } D < 0, n \in \mathbb{N}, \text{ and } p \text{ prime} \qquad (1.2)$$

is called the *generalized Ramanujan–Nagell equation*. This is a generalization of the equation $x^2 + 7 = 2^n$ studied by Ramanujan—see [68, Biography 7.1, p. 273].

Theorem 1.8 | **Solutions of the Ramanujan–Nagell Equations**

The only solutions of
$$x^2 + 7 = 2^n \tag{1.3}$$

with $x > 0$ are $(x, n) \in \{(1, 3), (3, 4), (5, 5), (11, 7), (181, 15)\}$.

Proof. If n is even, then
$$(2^{n/2})^2 - x^2 = (2^{n/2} - x)(2^{n/2} + x) = 7,$$

which implies that $2^{n/2} \pm x = 7$ so $2^{n/2} \mp x = 1$, for which only $n = 4$ and $x = 3$ provide a solution. Now assume that $n - 2 = m$ for odd $m \in \mathbb{N}$ and since clearly $(x, n) = (1, 3)$ is a solution, we may assume that $m > 1$.

By Theorem 1.3 on page 6, since $-7 \equiv 1 \pmod 4$, then
$$\mathfrak{O}_{\mathbb{Q}(\sqrt{-7})} = \mathbb{Z}[(1 + \sqrt{-7})/2],$$

so since $x^2 + 7 = 2^n$, then
$$\left(\frac{x + \sqrt{-7}}{2}\right)\left(\frac{x - \sqrt{-7}}{2}\right) = 2^m.$$

Therefore, there exist $a, b \in \mathbb{Z}$ such that $(a^2 + 7b^2)/4 = 2$, or $a^2 + 7b^2 = 8$, where we may assume, without loss of generality, that $a > 0$. Thus, only $a = 1$ and $b = \pm 1$ work. Hence,
$$\left(\frac{x + \sqrt{-7}}{2}\right)\left(\frac{x - \sqrt{-7}}{2}\right) = \left(\frac{1 + \sqrt{-7}}{2}\right)^m \left(\frac{1 - \sqrt{-7}}{2}\right)^m. \tag{1.4}$$

Now let
$$\alpha = \frac{1 + \sqrt{-7}}{2} \text{ and } \beta = \frac{1 - \sqrt{-7}}{2},$$

so $\alpha + \beta = 1$ and $\alpha\beta = 2$.

Since there are no factorizations for the right-hand side of (1.4) up to units, we must have
$$\left(\frac{x \pm \sqrt{-7}}{2}\right) = \pm\alpha^m \text{ or } \pm\beta^m.$$

Using (1.4) we see that no matter which of the four possible selections is made for $(x + y\sqrt{-7})/2$, we have
$$\pm\sqrt{-7} = \alpha^m - \beta^m.$$

We show that the plus sign cannot occur. If the plus sign occurs, then
$$\alpha - \beta = \left(\frac{1 + \sqrt{-7}}{2}\right) - \left(\frac{1 - \sqrt{-7}}{2}\right) = \sqrt{-7} = \alpha^m - \beta^m. \tag{1.5}$$

Therefore, since $\alpha\beta = 2$, then $\alpha^2 = (1-\beta)^2 \equiv 1 \pmod{\beta^2}$, where the congruence, here and in what follows, takes place in $\mathbb{Z}[(1+\sqrt{-7})/2]$. Thus, $\alpha^m \equiv \alpha(\alpha^2)^{(m-1)/2} \equiv \alpha \pmod{\beta^2}$. Therefore, by (1.5),

$$\alpha \equiv \alpha^m - \beta^m + \beta \equiv \alpha + \beta \pmod{\beta^2},$$

so $\beta \equiv 0 \pmod{\beta^2}$, namely, $\beta \mid 1_{\mathfrak{O}_{\mathbb{Q}(\sqrt{-7})}}$, a contradiction, since β is not a unit. We have shown that $-\sqrt{-7} = \alpha^m - \beta^m$. Hence,

$$-1 = \frac{\alpha^m - \beta^m}{\sqrt{-7}} = \frac{\left(\frac{1+\sqrt{-7}}{2}\right)^m - \left(\frac{1-\sqrt{-7}}{2}\right)^m}{\sqrt{-7}}. \tag{1.6}$$

Now we expand (1.6) by using the Binomial Theorem (see [68, Theorem 1.6, p.9]) and once done, (1.6) equals,

$$\frac{\sum_{j=0}^{m}\binom{m}{j}(\sqrt{-7})^{j-1} - \sum_{j=0}^{m}\binom{m}{j}(-1)^j(\sqrt{-7})^{j-1}}{2^m} =$$

$$\frac{\sum_{j=0}^{m}\binom{m}{j}(\sqrt{-7})^{j-1}[1-(-1)^j]}{2^m} = \frac{\sum_{j=1}^{(m+1)/2}\binom{m}{2j-1}(7)^{j-1}}{2^{m-1}}.$$

Hence,

$$-2^{m-1} = \sum_{k=1}^{(m+1)/2}\binom{m}{2k-1}(7)^{k-1}. \tag{1.7}$$

From (1.7), we glean that $-2^{m-1} \equiv m \pmod{7}$, and this has solutions if and only if $m \equiv 3, 5, 13 \pmod{42}$. In other words, this occurs if and only if $n \equiv 5, 7, 15 \pmod{42}$, which are exactly the values for which we are searching. However, we must ensure that none of these *distinct* solutions are congruent modulo 42, our last remaining task.

If we have two distinct solutions m_1 and m_2 with $m_1 \equiv m_2 \pmod{42}$ and 7^ℓ for $\ell \in \mathbb{N}$ is the largest power of 7 dividing $m_1 - m_2$, then

$$\alpha^{m_1} = \alpha^{m_2}\alpha^{m_1-m_2} = \alpha^{m_2}\left(\frac{1}{2}\right)^{m_1-m_2}\left(1+\sqrt{-7}\right)^{m_1-m_2}, \tag{1.8}$$

where

$$\left(\frac{1}{2}\right)^{m_1-m_2} = \left[\left(\frac{1}{2}\right)^6\right]^{(m_1-m_2)/6} \equiv 1 \pmod{7^{\ell+1}}.$$

Now by an easy iterative argument, this leads to the congruence,

$$\alpha^{m_1-m_2} \equiv 1 + (m_1 - m_2)\sqrt{-7} \pmod{7^{\ell+1}}. \tag{1.9}$$

However, using the Binomial Theorem as above, we have

$$\alpha^{m_2} \equiv \frac{1 + m_2\sqrt{-7}}{2^{m_2}} \pmod{7}. \tag{1.10}$$

Substituting (1.10) and (1.9) into (1.8) yields the congruence

$$\alpha^{m_1} \equiv \alpha^{m_2} + \frac{m_1 - m_2}{2m_2}\sqrt{-7} \pmod{7^{\ell+1}}.$$

By a similar argument,

$$\beta^{m_1} \equiv \beta^{m_2} - \frac{m_1 - m_2}{2m_2}\sqrt{-7} \pmod{7^{\ell+1}}.$$

Hence,

$$\alpha^{m_1} - \beta^{m_1} \equiv \alpha^{m_2} - \beta^{m_2} + \frac{m_1 - m_2}{2^{m_2-1}}\sqrt{-7} \pmod{7^{\ell+1}}.$$

We also know, by the same argument as that used on m above, that

$$\alpha^{m_1} - \beta^{m_1} = \alpha^{m_2} - \beta^{m_2},$$

so $(m_1 - m_2)\sqrt{-7} \equiv 0 \pmod{7^{\ell+1}}$. Since $m_1, m_2 \in \mathbb{Z}$, then $m_1 \equiv m_2 \pmod{7^{\ell+1}}$, which contradicts the fact that ℓ is the *largest* power of 7 dividing such a difference. Hence, ℓ cannot exist, so $m_1 = m_2$.   □

Later, when we have developed more algebraic number theory such as ideal theory, we will be able to prove results for the *generalized* Ramanujan-Nagell equation—see §8.2. For now we have exploited the most out of our development thus far, so this is a suitable juncture to end this section.

In the following section, we will concentrate upon a special type of quadratic field called *Gaussian*, and we will look at it in detail as a mechanism for developing more general concepts.

Exercises

1.1. Let $\mathbb{Q}(\alpha)$ be an algebraic number field. Prove that $\mathbb{Q}(\alpha) = \mathbb{Q}(a\alpha + b)$ for any $a, b \in \mathbb{Q}$ with $a \neq 0$.

1.2. Let R be a ring and let $\{R_j : j \in \mathfrak{I}\}$ for some indexing set $\mathfrak{I}$ be any set of subrings of R. Prove that $\cap_{j \in \mathfrak{I}} R_j$ is a subring of R. Also, show that if $R_1 \subseteq R_2 \subseteq \cdots \subseteq R_j \subseteq \cdots$, then $\cup_{j \in \mathfrak{I}} R_j$ is a subring of R.

1.3. Let p be a prime and let ζ_p be a primitive p^{th} root of unity. Prove that $m_{\zeta_p, \mathbb{Q}}(x) = x^{p-1} + x^{p-2} + \cdots + x + 1$.

☆ 1.4. Prove that if an algebraic number field F is of the form

$$F = \mathbb{Q}(\alpha_1, \alpha_2, \ldots, \alpha_n)$$

for $n \in \mathbb{N}$ where α_j for $j = 1, 2, \ldots, n$ are algebraic numbers, then there is an algebraic number γ such that $F = \mathbb{Q}(\gamma)$. (Hence, all algebraic number fields are simple extensions of $\mathbb{Q}$.)

(*Hint: It suffices to prove this for* $n = 2$ *with* $\alpha_1 = \alpha$ *and* $\alpha_2 = \beta$. *Let*

$$m_{\alpha,\mathbb{Q}}(x) = \prod_{j=1}^{d_\alpha} (x - \alpha_j),$$

where the α_j *are the conjugates of* α *over* $\mathbb{Q}$, *and let*

$$m_{\beta,\mathbb{Q}}(x) = \prod_{j=1}^{d_\beta} (x - \beta_j),$$

where the β_j *are the conjugates of* $\beta_1 = \beta$ *over* $\mathbb{Q}$. *Select* $q \in \mathbb{Q}$ *with* $q \neq (\alpha - \alpha_k)/(\beta_j - \beta)$ *for any* $k = 1, 2, \ldots, d_\alpha$ *and any* $j = 1, 2, \ldots, d_\beta$, *and let*

$$\gamma = \alpha + q\beta$$

and

$$f(x) = m_{\alpha,\mathbb{Q}}(\gamma - qx).$$

Prove that β *is the only common root of* $f(x)$ *and* $m_{\beta,\mathbb{Q}}(x)$. *Show that this implies* $\mathbb{Q}(\alpha, \beta) \subseteq \mathbb{Q}(\gamma)$. *The reverse inclusion is clear.*)

1.5. Let F be an algebraic number field. Prove that if $\alpha \in \mathfrak{U}_F$, then $\alpha_j \in \mathfrak{U}_F$ for all $j = 1, 2, \ldots, d$ where

$$m_{\alpha,F}(x) = x^d + a_{d-1}x^{d-1} + \cdots + a_1 x + a_0,$$

for some $d \in \mathbb{N}$ is the minimal polynomial of α over F, and α_j are the roots of $m_{\alpha,F}(x)$. Conclude that if F is an algebraic number field, then

$$\alpha \in \mathfrak{U}_F \text{ if and only if } \prod_{j=1}^{d} \alpha_j = \pm 1.$$

1.6. Referring to Example 1.7 on page 10, prove that

$$\mathbb{Q}(i, \sqrt{2}) = \mathbb{Q}\left(\frac{\sqrt{2}}{2}(1 + i) \right),$$

and that if ζ_8 is a primitive eighth root of unity, then it is an odd power of $\sqrt{2}(1 + i)/2$.

1.7. Prove that

$$x^n - 1 = \prod_{d \mid n} \Phi_d(x),$$

where $\Phi_d(x)$ is the cyclotomic polynomial given in Definition 1.9 on page 11.

1.2   The Gaussian Field

> *One may say that mathematics talks about things which are of no concern to man. Mathematics has the inhuman quality of starlight, brilliant and sharp, but cold. But it seems an irony of creation that man's mind knows how to handle things the better the farther removed they are from the center of his existence. Thus, we are cleverest where knowledge matters least: in mathematics, especially number theory–see* [102].
>
> **Hermann Weyl, German mathematician**— see Biography 1.1 on page 31

The ring of Gaussian integers $\mathbb{Z}[i]$ in the Gaussian number field, $\mathbb{Q}(\sqrt{-1}) = \mathbb{Q}(i)$, exhibits properties of the algebraic integers such as the greatest common divisor, prime elements, relative primality, and unique factorization, which allow us a pedagogical means of introducing such concepts with minimal abstraction for later elucidation. (Note that by Theorem 1.3 on page 6, we know that $\mathbb{Z}[i]$ must be the ring of integers of $\mathbb{Q}(i)$.) Indeed, the study in this section may be viewed as a link to the general theory of algebraic numbers to which we were introduced in §1.1. For the following we need to recall Example 1.5 on page 7.

Definition 1.11 | **Quadratic Conjugates and Norms**

If $F = \mathbb{Q}(\sqrt{D})$ is a quadratic number field and $\alpha \in F$, then

$$N_F(\alpha) = \alpha\alpha'$$

is called the *norm* of α from F to $\mathbb{Q}$, where α' is the conjugate of α.

Remark 1.7   Definition 1.11 is a precursor to the more general notion of a "norm" that we will study later in the text. We will see that the norm is the product of all the "conjugates" of α from a given number field. Exercise 1.8 on page 28 tells us that the norm is multiplicative, and is equal to zero if and only if the preimage is zero, $N_F(\alpha) \in \mathbb{Q}$ for any algebraic number α, and $N_F(\alpha) \in \mathbb{Z}$ for any algebraic integer α. In particular, if $\alpha + bi \in \mathbb{Z}[i]$, then

$$N_F(\alpha) = a^2 + b^2 \geq 0.$$

Furthermore, for elements $\alpha, \beta \in \mathfrak{O}_F$, if $\alpha \mid \beta$ in $\mathfrak{O}_F$, then $N(\alpha) \mid N(\beta)$ in $\mathbb{Z}$.

Now we illustrate Definition 1.7 on page 8 for Gaussian integers, which displays the divisibility within $\mathbb{Z}[i]$.

Example 1.9   Since $5 = (2+i)(2-i)$, then we see that $\alpha = (2+i) \mid \beta = 5 = 5+0i$. Also, $\pm 1 \mid \beta$ and $\pm i \mid \beta$ for any $\beta \in \mathfrak{O}_F = \mathbb{Z}[i]$ by Theorem 1.4 on page 8 since $\pm 1, \pm i$ are the units of $\mathbb{Z}[i]$.

Definition 1.12 | Associates |

If F is an algebraic number field, $\alpha \in \mathfrak{O}_F$, and $u \in \mathfrak{U}_{\mathfrak{O}_F}$, then $u\alpha$ is called an *associate* of α. If α and β are associates, we denote this fact by $\alpha \sim \beta$ where the underlying $\mathfrak{O}_F$ will be assumed in context.

In order to introduce another concept that mimics one notion of "prime" number encountered in $\mathbb{Z}$, we introduce the following based on units. The other notion of "prime" is given in Exercise 1.28 on page 29. We will see the distinction between the two comes into focus in §1.3 — in particular, see Remark 1.16 on page 40.

Definition 1.13 | Gaussian Primes |

If $\alpha \neq 0$, and α is not a unit such that α is divisible only by units and associates in $\mathfrak{O}_F$, then α is called a *Gaussian prime* in $\mathfrak{O}_F$.

Example 1.10 In Example 1.9 on the preceding page, $2 \pm i$ are Gaussian primes since any divisor $a + bi$ of $2 \pm i$ must satisfy that $N_F(a + bi) = (a^2 + b^2) \mid 5$, by part (e) of Exercise 1.8 on page 28. Therefore, $a + bi$ is a unit or an associate of $2 \pm i$ given that the only solutions to $a^2 + b^2 = 5$ for $1 \leq |a| < |b|$ are $a = \pm 1$ and $b = \pm 2$.

Example 1.11 If F is a quadratic number field, then $\beta \mid \alpha$ in $\mathfrak{O}_F$ if and only if $\beta' \mid \alpha'$, where α' is the conjugate of α—see properties (a)–(c) given in Example 1.5 on page 7. Thus, $\alpha \in \mathbb{Z}[i]$ is a Gaussian prime if and only if α' is a Gaussian prime.

In [68, Section 1.2], we studied the greatest common divisor for rational integers. We now elevate this to the Gaussian integers. As with the rational integers, to do this we need the notion of a Euclidean algorithm, albeit in the case of $\mathbb{Z}[i]$, employing norms as follows. As with the rational integers, we first develop a division algorithm that is then repeatedly applied to yield the Euclidean algorithm. In the ensuing proof, we also use the floor function as studied in [68, Section 2.5], to define the *nearest integer function*, $Ne(x) = \lfloor x + 1/2 \rfloor$, which is the integer closest to $x \in \mathbb{R}$.

Theorem 1.9 | Division Algorithm for Gaussian Integers |

Let $\alpha, \beta \in \mathbb{Z}[i]$ with $\beta \neq 0$. Then there exists $\sigma, \delta \in \mathbb{Z}[i]$ such that

$$\alpha = \beta\sigma + \delta,$$

where $0 \leq N_F(\delta) < N_F(\beta)$.

Proof. Let $\alpha/\beta = c + di \in \mathbb{C}$. Set $f = \lfloor c + 1/2 \rfloor = Ne(c)$, and $g = \lfloor d + 1/2 \rfloor = Ne(d)$. Hence, there are $k, \ell \in \mathbb{R}$ such that

$$|k| \leq 1/2 \text{ and } |\ell| \leq 1/2 \tag{1.11}$$

with

$$c + di = (f + k) + (g + \ell)i. \tag{1.12}$$

Set

$$\sigma = f + gi \text{ and } \delta = \alpha - \beta\sigma. \tag{1.13}$$

Then it remains to show that $0 \leq N_F(\delta) < N_F(\beta)$. From Remark 1.7 on page 18, we know that $N_F(\delta) \geq 0$. Now we show that $N_F(\delta) < N_F(\beta)$.

By part (b) of Exercise 1.8 on page 28 (the multiplicativity of the norm), we have that

$$N_F(\delta) = N_F(\alpha - \beta\sigma) = N_F((\alpha/\beta - \sigma)\beta)$$

$$= N_F(\alpha/\beta - \sigma)N_F(\beta) = N_F(c + di - \sigma)N_F(\beta).$$

However, from (1.12)–(1.13), we get

$$c + di - \sigma = c + di - (f + gi) = (c - f) + (d - g)i = k + \ell i.$$

Therefore, by (1.11),

$$N_F(\delta) = N_F(k + \ell i)N_F(\beta) =$$

$$(k^2 + \ell^2)N_F(\beta) \leq ((1/2)^2 + (1/2)^2)N_F(\beta) \leq N_F(\beta)/2 < N_F(\beta),$$

as required. $\square$

Remark 1.8 The σ in Theorem 1.9 is called a *quotient* and the δ is called a *remainder* of the division. This follows the notions set up for the division algorithm in $\mathbb{Z}$.

Remark 1.9 Although Theorem 1.9 gives us a criterion for the existence of an algorithm for division in $\mathbb{Z}[i]$, there is no uniqueness attached to it. In other words, we may have many such representations as the following illustration demonstrates.

Example 1.12 Let $\alpha = 10 + i$ and $\beta = 2 + 5i$, then we may find $\sigma, \delta \in \mathbb{Z}[i]$ using the techniques established in the proof of Theorem 1.9. We have

$$c + di = \frac{\alpha}{\beta} = \frac{10 + i}{2 + 5i} = \frac{(10 + i)(2 - 5i)}{(2 + 5i)(2 - 5i)} = \frac{25}{29} - \frac{48}{29}i,$$

so

$$f = \left\lfloor c + \frac{1}{2} \right\rfloor = \left\lfloor \frac{25}{29} + \frac{1}{2} \right\rfloor = 1 \text{ and } g = \left\lfloor d + \frac{1}{2} \right\rfloor = \left\lfloor -\frac{48}{29} + \frac{1}{2} \right\rfloor = -2.$$

Therefore,

$$\sigma = 1 - 2i \text{ and } \delta = \alpha - \beta\sigma = 10 + i - (2 + 5i)(1 - 2i) = -2.$$

Moreover, we verify that $N_F(\delta) = N_F(-2) = 4 < N_F(\beta) = N_F(2 + 5i) = 29$ with

$$\alpha = 10 + i = (2 + 5i)(1 - 2i) - 2 = \beta\sigma + \delta. \tag{1.14}$$

However, these choices are not unique since we need not follow the techniques of Theorem 1.9. For instance, if we choose $\sigma = 1 - i$ and $\delta = 3 - 2i$, then

$$\alpha = 10 + i = (2 + 5i)(1 - i) + 3 - 2i = \beta\sigma + \delta, \tag{1.15}$$

where $N_F(\delta) = 13 < 29 = N_F(\beta)$. Thus, by *(1.14)–(1.15)*, we see that, when employing the division algorithm for Gaussian integers, the quotient and remainder are not unique. See Exercises 1.12–1.15 on page 28.

We are now in a position to exhibit the notion of *greatest common divisor* that we studied for the rational integers in [68, Section 1.2]. (Also, see Remark 1.13 on page 33.)

Definition 1.14 | GCD for Algebraic Integers

If F is a number field, and $\alpha, \beta \in \mathfrak{O}_F$, not both zero, then a *greatest common divisor* (gcd) of α and β is a $\gamma \in \mathfrak{O}_F$ such that both of the following are satisfied.

(a) $\gamma \mid \alpha$ and $\gamma \mid \beta$, namely γ is a common divisor of α and β.

(b) Suppose that $\delta \in \mathfrak{O}_F$ where $\delta \mid \alpha$, and $\delta \mid \beta$. Then $\delta \mid \gamma$, namely any common divisor of α and β divides γ.

The first thing we need to know is that every pair of Gaussian integers indeed has a gcd.

Theorem 1.10 | Gaussian GCDs Always Exist

If $\alpha, \beta \in \mathbb{Z}[i] = \mathfrak{O}_F$, where at least one of α or β is not zero, then there exists a gcd $\gamma \in \mathbb{Z}[i]$ of α and β which is unique.

Proof. Given fixed $\alpha, \beta \in \mathbb{Z}[i]$, not both zero, set

$$S = \{N_F(\sigma\alpha + \rho\beta) > 0 : \sigma, \rho \in \mathbb{Z}[i]\},$$

with $S \neq \varnothing$ since

$$N_F(\alpha) = N_F(1 \cdot \alpha + 0 \cdot \beta), \text{ and } N_F(\beta) = N_F(0 \cdot \alpha + 1 \cdot \beta) \tag{1.16}$$

are both in S, at least one of which is not zero, and by Remark 1.7 on page 18, nonnegative. Thus, we may employ the well-ordering principle studied in [68, Section 1.1, p. 11] to get the existence of an element $\gamma_0 = \sigma_0\alpha + \rho_0\beta \in S$, for which its norm is the least value in S, namely

$$N_F(\gamma_0) \leq N_F(\sigma\alpha + \rho\beta) \text{ for all } \sigma, \rho \in \mathbb{Z}[i].$$

Claim 1.3 γ_0 *is a greatest common divisor of α and β.*

Let $\tau \in \mathbb{Z}[i]$ with $\tau \mid \alpha$ and $\tau \mid \beta$. Thus, there exists $\delta_1, \delta_2 \in \mathbb{Z}[i]$ such that $\alpha = \tau \delta_1$ and $\beta = \tau \delta_2$. Hence,

$$\gamma_0 = \sigma_0 \alpha + \rho_0 \beta = \sigma_0 \tau \delta_1 + \rho_0 \tau \delta_2 = \tau(\sigma_0 \delta_1 + \rho_0 \delta_2), \qquad (1.17)$$

so $\tau \mid \gamma_0$. It remains to show that γ_0 divides both α and β.

Let $\kappa = \lambda_1 \alpha + \lambda_2 \beta$ such that $N_F(\kappa) \in \mathcal{S}$. Thus, by Theorem 1.9 on page 19, there exist $\mu, \nu \in \mathbb{Z}[i]$ such that

$$\kappa = \gamma_0 \mu + \nu \qquad (1.18)$$

with

$$0 \leq N_F(\nu) < N_F(\gamma_0). \qquad (1.19)$$

Also, by (1.17)–(1.18),

$$\nu = \kappa - \gamma_0 \mu = \lambda_1 \alpha + \lambda_2 \beta - (\sigma_0 \alpha + \rho_0 \beta)\mu = (\lambda_1 - \sigma_0 \mu)\alpha + (\lambda_2 - \rho_0 \mu)\beta,$$

so $N_F(\nu) \in \mathcal{S}$. However, by (1.19), this contradicts the minimality of $N_F(\gamma_0)$ in $\mathcal{S}$, unless $\nu = 0$, by part (c) of Exercise 1.8 on page 28. We have shown that γ_0 divides every element whose norm is in $\mathcal{S}$ so, in particular, by (1.16), it divides α and β, which secures claim 1.3 via Definition 1.14. Hence, we have the result.□

Remark 1.10 By Exercise 1.17 on page 29, we know that γ is a gcd of α and β in $\mathbb{Z}[i]$ if and only if all of its associates are also gcds. Therefore, we may ascribe "uniqueness" to the gcd of two elements by saying that we do not distinguish between associates when discussing their gcd. Another way of saying this is that gcds are "unique up to associates." In other words, the gcd, γ, of any two elements in $\mathbb{Z}[i]$ is unique in the sense that $\gamma \sim \delta$ for any greatest common divisor δ. In this sense, they are in the same "class." Essentially this is what we do in the ordinary integers $\mathbb{Z}$, since we allow only for a gcd to be positive given that the only units in $\mathbb{Z}$ are ± 1; this eliminates -1 as a choice, the only possible associate of a positive gcd in $\mathbb{Z}$. See Definition 1.20 on page 37.

Now we are in a position to state a generalization of another concept from the ordinary integers to the Gaussian integers.

Definition 1.15 | **Relatively Prime Algebraic Integers**

Two algebraic integers α and β are said to be *relatively prime* if 1 is a gcd of α and β. Equivalently, α and β are relatively prime if the only gcd of α and β is 1 up to associates, namely γ is a gcd of α and β if and only if $\gamma \sim 1$.

By Remark 1.10, 1 is a gcd of two Gaussian integers if and only if $\pm 1, \pm i$ are gcds of them. Now we are in a position to present a Euclidean algorithm as promised earlier.

Theorem 1.11 | A Euclidean Algorithm for Gaussian Integers

Let $\alpha = \alpha_0, \beta = \beta_0 \in \mathbb{Z}[i] = \mathfrak{O}_F$ be nonzero where $\beta \nmid \alpha$. By applying Theorem 1.9 on page 19 successively, the following sequence is obtained

$$\alpha_j = \beta_j \delta_j + \gamma_j \text{ with } N_F(\gamma_j) < N_F(\beta_j) \text{ for } j = 0, 1, \ldots, n$$

and $n \in \mathbb{N}$ is the least value such that $\gamma_n = 0$. The value δ_j is the quotient of the division of α_j by β_j; γ_j is its remainder; and γ_{n-1} is a greatest common divisor of α and β.

Proof. Applying Theorem 1.9 to α_0 and β_0 we get

$$\alpha_0 = \beta_0 \delta_0 + \gamma_0 \text{ with } N_F(\gamma_0) < N_F(\beta_0). \tag{1.20}$$

Then by repeated application, we get for $j \in \mathbb{N}$,

$$\alpha_j = \beta_j \delta_j + \gamma_j \text{ with } 0 \leq N_F(\gamma_j) < N_F(\beta_j), \tag{1.21}$$

where $\alpha_j = \beta_{j-1}$ and $\beta_j = \gamma_{j-1}$. Thus, for a given $j \in \mathbb{N}$,

$$0 \leq N_F(\gamma_{j-1}) \leq N_F(\beta_{j-1}) = N_F(\gamma_{j-2}) < N_F(\beta_{j-2}) < \cdots < N_F(\beta_0),$$

so by induction,

$$0 \leq N_F(\gamma_{j-1}) \leq N_F(\beta_0) - j,$$

which tells us that $N_F(\gamma_n) = 0$ for some $0 < n < N_F(\beta_0)$. Note that $n > 0$ since we assumed that α is not divisible by β.

Since $\alpha_n = \beta_n \delta_n + \gamma_n$, then

$$\gamma_{n-1} = \beta_n \mid \alpha_n = \beta_{n-1} = \gamma_{n-2},$$

and similarly, $\gamma_{n-2} \mid \gamma_{n-3}$. Continuing in this fashion, we see that $\gamma_{n-j} \mid \gamma_{n-j-1}$ for each natural number $j < n$, so

$$\gamma_{n-1} \mid \gamma_1 \mid \gamma_0 = \beta_1. \tag{1.22}$$

Thus, by Equation (1.21) with $j = 1$,

$$\gamma_{n-1} \mid \alpha_1 = \beta_0. \tag{1.23}$$

Therefore, by Equations (1.20), (1.22)–(1.23), $\gamma_{n-1} \mid \alpha_0$. Thus, γ_{n-1} is a common divisor of α and β. If σ is a common divisor of α and β, then by (1.20), $\sigma \mid \gamma_0$. However, by Equation (1.21) with $j = 1$,

$$\beta = \beta_0 = \alpha_1 = \beta_1 \delta_1 + \gamma_1 = \gamma_0 \delta_1 + \gamma_1,$$

so $\sigma \mid \gamma_1$. Continuing in this fashion, we see that $\sigma \mid \gamma_j$ for all nonnegative $j < n$. By Definition 1.14 on page 21, γ_{n-1} is a gcd of α and β. $\square$

Example 1.13 If $\alpha = 211 + 99i$ and $\beta = 12 + 69i$, then we may follow the steps of the Euclidean algorithm to find a gcd of α and β.

$$\alpha_0 = 211 + 99i = (12 + 69i)(1 - 3i) - 8 + 66i = \beta_0\delta_0 + \gamma_0 \qquad (1.24)$$

$$\alpha_1 = \beta_0 = 12 + 69i = (-8 + 66i)\cdot 1 + 20 + 3i = \gamma_0\delta_1 + \gamma_1 = \beta_1\delta_1 + \gamma_1 \quad (1.25)$$

$$\alpha_2 = \beta_1 = -8 + 66i = (20 + 3i)(3i) + 1 + 6i = \gamma_1\delta_2 + \gamma_2 = \beta_2\delta_2 + \gamma_2 \quad (1.26)$$

$$\alpha_3 = \beta_2 = 20 + 3i = (1 + 6i)(1 - 3i) + 1 = \gamma_2\delta_3 + \gamma_3 = \beta_3\delta_3 + \gamma_3 \qquad (1.27)$$

$$\alpha_4 = \beta_3 = 1 + 6i = 1 \cdot 6i + 1 = \gamma_3\delta_4 + \gamma_4 = \beta_4\delta_4 + \gamma_4 \qquad (1.28)$$

$$\alpha_5 = \beta_4 = 1 = \gamma_4 \cdot 1 + 0 = \beta_5\delta_5 + \gamma_5.$$

Hence, $\gamma_n = \gamma_5 = 0$ and $\gamma_4 = 1$ is a gcd of α and β, so α and β are relatively prime.

Now we may illustrate Theorem 1.10 on page 21 by working backward in the above steps to get the gcd as a linear combination of α and β as follows. We begin with $\gamma_{n-1} = \gamma_4 = 1$ in terms of α_4 and α_5. Then successively work back to get γ_4 in terms of α_j and α_{j-1} for $j = 5, 4, 3, 2$ thereby getting it as a linear combination of α and β. From (1.28),

$$\gamma_{n-1} = \gamma_4 = 1 = (1 + 6i) \cdot 1 - 6i \cdot 1,$$

but by (1.27),

$$\gamma_3 = 1 = (20 + 3i) - (1 + 6i)(1 - 3i),$$

so

$$1 = (1 + 6i)\cdot 1 - 6i[(20 + 3i) - (1 + 6i)(1 - 3i)] = (1 + 6i)(19 + 6i) - 6i(20 + 3i).$$

From (1.26),

$$1 + 6i = -8 + 66i - (20 + 3i)(3i),$$

so

$$1 = [-8 + 66i - (20 + 3i)(3i)](19 + 6i) - 6i(20 + 3i) =$$
$$(-8 + 66i)(19 + 6i) + (20 + 3i)(18 - 63i).$$

From (1.25),

$$20 + 3i = 12 + 69i - (-8 + 66i),$$

so

$$1 = (-8 + 66i)(19 + 6i) + [12 + 69i - (-8 + 66i)](18 - 63i) =$$
$$(12 + 69i)(18 - 63i) + (-8 + 66i)(1 + 69i).$$

From (1.24),

$$-8 + 66i = 211 + 99i - (12 + 69i)(1 - 3i),$$

so

$$1 = (12 + 69i)(18 - 63i) + [211 + 99i - (12 + 69i)(1 - 3i)](1 + 69i) =$$

$$(12 + 69i)(-190 - 129i) + (211 + 99i)(1 + 69i).$$

Hence,

$$\gamma_{n-1} = \gamma_4 = 1 = (1 + 69i)\alpha - (190 + 129i)\beta,$$

an expression of our gcd as a linear combination of α and β.

Now we describe a means of ascribing parity to Gaussian integers.

Definition 1.16 | Odd and Even Gaussian Integers

If $\alpha \in \mathbb{Z}[i]$, then α is said to be *odd* if $(1 + i) \nmid \alpha$, and α is said to be *even* if $(1 + i) \mid \alpha$.

Remark 1.11 The notion of parity for Gaussian integers is based upon the fact that if $(1 + i) \mid \alpha$, then $N_F(1 + i) = 2 \mid N_F(\alpha)$—see part (e) of Exercise 1.8 on page 28.

Now we show how factorizations unfold in the Gaussian integers. There is a methodology to ensure uniqueness of factorizations in a stricter sense than the following, which is developed in Exercise 1.34 on page 30.

Theorem 1.12 | Unique Factorization for Gaussian Integers

Let α be a nonunit, nonzero Gaussian integer. Then

(a) α *may be written as a product of Gaussian primes, and*

(b) *The factorization is unique in the following sense. If for $m, n \in \mathbb{N}$,*

$$\alpha = \prod_{j=1}^{m} \alpha_j = \prod_{j=1}^{n} \beta_j, \text{ where the } \alpha_j, \beta_j \text{ are Gaussian primes, then}$$

$m = n$ *and, after possibly renumbering, the α_j and β_j are associates for $j = 1, 2, \ldots, n$.*

Proof. For the proof of both parts, we use induction. For part (a), since α is a nonzero, nonunit, then $N_F(\alpha) \geq 2$. If α is a Gaussian prime, then by Definition 1.12 on page 19, $\alpha = \beta \cdot u$ is the only factorization of α into a product of primes, where $u \in \mathfrak{U}_F = \mathfrak{U}_{\mathbb{Z}[i]}$ and β is an associate of α. Assume now the induction hypothesis, namely that any Gaussian integer, δ, with $2 \leq N_F(\delta) < N_F(\alpha)$, may be factored into a product of Gaussian primes. By the above we may assume that α is not a prime, since otherwise we are done. Thus, $\alpha = \sigma_1 \sigma_2$ for $\sigma_j \in \mathbb{Z}[i]$, and $2 \leq N_F(\sigma_j) < N_F(\alpha)$ for $j = 1, 2$. By the induction hypothesis, σ_j may be factored into a product of primes for each of $j = 1, 2$. This is part (a).

If α is a Gaussian prime, then by Definition 1.12 on page 19, $\alpha = \beta \cdot u$ is the only factorization of α into a product of primes, where $u \in \mathfrak{U}_{\mathbb{Z}[i]}$ and β is an associate of α, a unique factorization in the sense of (b), namely up to associates.

This is the induction step. Assume now the induction hypothesis, namely that any Gaussian integer, δ, with $2 \leq N_F(\delta) < N_F(\alpha)$, may be uniquely factored into a product of Gaussian primes, up to associates. Suppose that α is not prime and

$$\alpha = \prod_{j=1}^{m} \alpha_j = \prod_{j=1}^{n} \beta_j, \text{ where the } m, n \in \mathbb{N} \text{ and } \alpha_j, \beta_j \text{ are Gaussian primes.}$$

Therefore, $\alpha_1 \mid \prod_{j=1}^{n} \beta_j$, which by Exercise 1.28 on page 29, tells us that $\alpha_1 \mid \beta_j$ for some $j = 1, 2, \ldots, n$. Without loss of generality, we may assume that $j = 1$ since we may reorder the $\beta_1, \beta_2, \ldots, \beta_n$, if necessary, to ensure $\alpha_1 \mid \beta_1$. However, since β_1 is a Gaussian prime, then α_1 must be an associate of β_1, namely, $\beta_1 = u\alpha_1$ for some Gaussian unit u. Thus,

$$\alpha_1\alpha_2 \cdots \alpha_m = \beta_1\beta_2 \cdots \beta_n = u\alpha_1\beta_2 \cdots \beta_n$$

so dividing both sides by α_1, we get

$$\alpha_2\alpha_3 \cdots \alpha_m = u\beta_2\beta_3 \cdots \beta_n.$$

Since $N_F(\alpha_1) \geq 2$, then $1 \leq N_F(\alpha_2\alpha_3 \cdots \alpha_m) < N_F(\alpha)$, so by the induction hypothesis, we infer that $m - 1 = n - 1$ and after possibly reordering the terms, α_j is an associate of β_j for $j = 2, 3, \ldots, n$. This proves part (b) by induction.□

Example 1.14 The factorization, up to associates, of the Gaussian integer $-91 + 117i$ is given by

$$-91 + 117i = (1 + i)(2 + 3i)^2(1 - 2i)(3 + 2i),$$

where $(1 + i), (2 + 3i), (1 - 2i), (3 + 2i)$ are all Gaussian primes by Exercise 1.38 on page 31, since

$$N_F(1 + i) = 2, N_F(2 + 3i) = 13 = N_F(3 + 2i), N_F(1 - 2i) = 5.$$

See Exercises 1.35–1.36.

In Chapter 3, we will be looking at sums of squares as representations of natural numbers, which will be an extension of the elementary presentation we gave in [68, Chapter 6]. However, the Gaussian integers provide a segue to such representations and thus a desirable topic with which to close this section. As noted in Remark 1.7 on page 18, the norms of Gaussian integers naturally represent the corresponding rational integers as sums of two squares. Now we look at which natural numbers are so represented. The following was proved in [68, Theorem 6.1, p. 244] via fundamental techniques. The result presented here uses the Gaussian integers as a vehicle.

Theorem 1.13 | **Primes as Sums of Two Squares**

If $p \equiv 1 \pmod 4$ is prime in $\mathbb{Z}$, then there exist unique $a, b \in \mathbb{N}$ with $1 \leq b < a$ such that $p = a^2 + b^2$.

Proof. By Exercise 1.38 on page 31, p is not a Gaussian prime. Therefore, there exist $\alpha, \beta \in \mathfrak{D}_F = \mathbb{Z}[i]$ neither of which is a unit such that $p = u\alpha\beta$, where u is a unit. By taking norms we get $p^2 = N_F(u\alpha\beta) = N_F(u)N_F(\alpha)N_F(\beta)$, but $N_F(u) = 1$, $N_F(\alpha) > 1$, and $N_F(\beta) > 1$, so $N_F(\alpha) = N_F(\beta) = p$ is the only possibility. Suppose that $\alpha = a \pm bi$ and $\beta = c \pm di$. Since we may absorb any multiplication of a unit times u into the representation for α and β, then we may assume without loss of generality that $a, b, c, d \in \mathbb{N}$,

$$1 \leq b < a, \tag{1.29}$$

and

$$1 \leq d < c, \tag{1.30}$$

then

$$p = a^2 + b^2 \tag{1.31}$$

and

$$p = c^2 + d^2. \tag{1.32}$$

It remains to show uniqueness.

Multiplying (1.31) by d^2 and subtracting b^2 times (1.32) we get

$$a^2d^2 - b^2c^2 = (ad - bc)(ad + bc) = p(d^2 - b^2).$$

Thus, since $a, b, c, d < \sqrt{p}$, and $p \mid (ad - bc) \leq p - 1$ or $p \mid (ad + bc) < 2p$, then either

$$ad - bc = 0, \tag{1.33}$$

or

$$ad + bc = p. \tag{1.34}$$

If (1.33) holds, $ad = bc$ so since p is prime, $\gcd(a, b) = \gcd(c, d) = 1$. Since $a \mid bc$, then $a \mid c$. Thus, for some $f \in \mathbb{N}$, $c = af$, so $ad = bc = baf$, which means that $d = bf$. Hence, $p = c^2 + d^2 = a^2f^2 + b^2f^2 = f^2(a^2 + b^2) = f^2p$, forcing $f = 1$. Therefore, $c = a$ and $d = b$.

If (1.34) holds, then since

$$p^2 = (a^2 + b^2)(c^2 + d^2) = (ad + bc)^2 + (ac - bd)^2 \tag{1.35}$$

$$= p^2 + (ac - bd)^2,$$

so $ac - bd = 0$. Thus, $ac = bd$, and a similar argument to the above yields that $a = d$ and $b = c$. However, by (1.29)–(1.30), $c = b < a = d < c$, a contradiction. This is uniqueness so we have the entire result. $\square$

Remark 1.12 The prime-squared representation given in (1.35) is a special case of the more general result given in [68, Remark 1.6, p. 46], namely for $x, y, u, v, D \in \mathbb{Z}$,

$$(x^2 + Dy^2)(u^2 + Dv^2) = (xu \pm Dyv)^2 + D(xv \mp yu)^2.$$

Example 1.15 The representation $13 = 3^2 + 2^2$ is the unique up to order of the factors. Notice that

$$13 = (2 + 3i)(2 - 3i) = (3 + 2i)(3 - 2i)$$

so the representation as a product of primes is unique up to the order of the factors since $3 - 2i = (2 + 3i)(-i)$. Thus, in the notation of Theorem 1.13, $\beta = c + di = a - bi$, so that $c + di$ is the algebraic conjugate of $\alpha = a + bi$.

However 3 has no representation as a sum of two integer squares. In fact, as we proved in [68, Theorem 6.2, p. 245], $N = m^2 n \in \mathbb{N}$ where n is squarefree and is representable as a sum of two integer squares if and only if n is not divisible by any prime $p \equiv 3 \,(\mathrm{mod}\ 4)$. In [68, Theorem 6.3, p. 247], we also found the total number of *primitive* representations of a given $N = a^2 + b^2$, namely where $\gcd(a, b) = 1$. Furthermore, in [68, Chapter 6, Sections 6.2–6.4], we dealt with sums of three and four squares as well as sums of cubes.

Exercises

1.8. Given a quadratic number field F, and $\alpha, \beta \in F$, prove that

   (a) $N_F(\alpha) \in \mathbb{Q}$.

   (b) $N_F(\alpha\beta) = N_F(\alpha)N_F(\beta)$.

   (c) $N_F(\alpha) = 0$ if and only if $\alpha = 0$.

   (d) If $\alpha \in \mathfrak{O}_F$, then $N_F(\alpha) \in \mathbb{Z}$.

   (e) If $\alpha \mid \beta$ in $\mathfrak{O}_F$, then $N_F(\alpha) \mid N_F(\beta)$ in $\mathbb{Z}$.

1.9. Let F be an algebraic number field and let α be algebraic over F with minimal polynomial

$$m_{\alpha,F}(x) = x^d + a_{d-1}x^{d-1} + \cdots + a_1 x + a_0, \text{ where } d \in \mathbb{N},$$

   and α_j for $j = 1, 2, \ldots, d$ are all the roots of $m_{\alpha,\mathbb{Q}}(x)$. Prove that $\alpha_j \neq \alpha_k$ for any $j \neq k$.

1.10. Prove that if $\alpha \in \mathbb{A}$, then $\alpha_j \in \mathbb{A}$ for all $j = 1, 2, \ldots, d$, where α_j are the roots of $m_{\alpha,\mathbb{Q}}(x)$.

1.11. If $\alpha \in \overline{\mathbb{Q}}$, prove that $\alpha \in \mathbb{A}$ if and only if $m_{\alpha,\mathbb{Q}}(x) \in \mathbb{Z}[x]$.

1.12. For each of the following find a quotient and remainder for α/β using the division algorithm for Gaussian integers given on page 19.

   (a) $\alpha = 3 + i$, $\beta = 4 - 3i$.   (b) $\alpha = 3$, $\beta = 3 + 5i$.
   (c) $\alpha = 11 - i$, $\beta = 4$.   (d) $\alpha = 4 - 3i$, $\beta = 3$.

1.13. For each of the following find a quotient and remainder for α/β using the division algorithm for Gaussian integers.

   (a) $\alpha = 7$, $\beta = 3 - 3i$.   (b) $\alpha = 2 - i$, $\beta = 1 + 5i$.
   (c) $\alpha = 1 - i$, $\beta = 3 - i$.   (d) $\alpha = -3i$, $\beta = 3 + 6i$.

1.14. If $\beta = 2 + i$ find all $\alpha \in \mathbb{Z}[i]$ such that $\beta \mid \alpha$ in $\mathbb{Z}[i]$.

1.15. If $\beta = 4 + 5i$ find all $\alpha \in \mathbb{Z}[i]$ such that $\beta \mid \alpha$ in $\mathbb{Z}[i]$.

1.16. Let F be a number field and let $\gamma_1, \gamma_2 \in \mathfrak{O}_F$. Prove that

γ_1 and γ_2 are associates of one another if and only if $\gamma_1 \mid \gamma_2$ and $\gamma_2 \mid \gamma_1$.

1.17. Suppose that F is a number field with $\alpha, \beta \in \mathfrak{O}_F$ not both zero. Prove that γ is a greatest common divisor of α and β if and only if all associates of γ are greatest common divisors thereof. Conclude that any two gcds, γ_1, γ_2, of α and β must be associates.

1.18. Given a number field F with $\alpha \in \mathfrak{O}_F$ and $u \in \mathfrak{U}_F$, prove that 1 is a gcd of α and u.

1.19. Let F be a quadratic number field. Prove that if $\alpha \in \mathfrak{O}_F$, then $N_F(\alpha) = \pm 1$ if and only if $\alpha \in \mathfrak{U}_F$.

1.20. Prove that if $\alpha, \beta \in \mathbb{Z}[i]$ and $\gcd(N_F(\alpha), N_F(\beta)) = 1$, then α and β are relatively prime as Gaussian integers.

1.21. Let F be a quadratic number field. If $\alpha, \beta \in F$ with $\alpha \sim \beta$, prove that $|N_F(\alpha)| = |N_F(\beta)|$.

1.22. Is the converse of Exercise 1.21 true? If so, prove it. If not, provide a counterexample.

1.23. Is the converse to Exercise 1.20 true? If so, prove that it is and if not provide a counterexample.

In each of Exercises 1.24–1.27, use the Euclidean algorithm given in Theorem 1.11 on page 23 to find a gcd in $\mathbb{Z}[i]$ for each pair.

1.24. (a) $(1 + 5i, 7 + 9i)$ (b) $(111 + 7i, 71 + 9i)$

1.25. (a) $(12 + 9i, 2 + 69i)$ (b) $(2 + 8i, 21 + 9i)$

1.26. (a) $(111 + 7i, 7 + 9i)$ (b) $(1 + 7i, 7 + 4i)$

1.27. (a) $(17 + 7i, 71 + 4i)$ (b) $(1 + 77i, 55 + 4i)$

1.28. Let $\rho \in \mathbb{Z}[i]$ be a prime, and suppose that $\alpha_j \in \mathbb{Z}[i]$ for $j = 1, 2, \ldots, n \in \mathbb{N}$. Prove that if $\rho \mid \prod_{j=1}^{n} \alpha_j$, then $\rho \mid \alpha_j$ for some $j = 1, 2, \ldots, n$.

1.29. Prove that $\alpha, \beta \in \mathbb{Z}[i]$ are relatively prime if and only if their conjugates, α' and β', are relatively prime.

In Exercises 1.30–1.34, a primary *Gaussian integer is an element*

$$\alpha = a + bi \in \mathbb{Z}[i] \text{ such that } a \text{ is odd, } b \text{ is even, and } a + b \equiv 1 \pmod 4.$$

These are often used in establishing properties of what are called higher reciprocity laws. *See* [64, pp.290 ff], *for instance. In the following exercises, we employ the topic to establish properties of primary integers that are of interest in their own right.*

1.30. Prove that the only primary Gaussian unit is 1.

1.31. Prove that $a + bi$ is a primary Gaussian integer if and only if

$$a + bi \equiv 1 \pmod{2 + 2i} \text{ in } \mathbb{Z}[i].$$

1.32. Prove that any primary Gaussian integer must be odd.

1.33. Prove that, given any odd Gaussian integer, exactly one of its four associates is primary.

1.34. Suppose that α is a primary non-unit Gaussian integer. Prove that α can be *uniquely* factored into a product of primary Gaussian primes α_j with

$$\alpha = \prod_{j=1}^{n} \alpha_j \text{ where } N_F(\alpha_j) \le N_F(\alpha_{j+1}) \text{ for } n \in \mathbb{N} \text{ with } j = 1.2, \ldots, n-1.$$

(Note that this is in contrast to the general case, given in Theorem 1.12 on page 25, where an arbitrary, non-unit, Gaussian integer can be factored into a product of Gaussian primes "up to associates," since there exist more than one associate for a given Gaussian integer but only one primary associate for a given primary integer by Exercise 1.33.)

In Exercises 1.35–1.36, find a factorization of the Gaussian integer into Gaussian primes with positive real part and units equal to 1.

1.35. (a) $323 + 1895i$.

    (b) $420 - 65i$.

    (c) $9497 + 4112i$.

    (d) $-355 + 533i$.

1.36. (a) $-64 + 83i$.

    (b) $-271 - 178i$.

    (c) $561 - 62i$.

    (d) $212 - 137i$.

1.37. Prove that any prime $p \in \mathbb{Z}$ with $p \equiv 3 \pmod 4$ is a Gaussian prime.

1.38. Prove that if $\alpha \in \mathbb{Z}[i] = \mathfrak{O}_F$ and $N_F(\alpha) = p$ where p is prime in $\mathbb{Z}$, then α is a Gaussian prime but p is not a Gaussian prime and $p \equiv 1 \pmod 4$ or $p = 2$.

Biography 1.1 Hermann Klaus Hugo Weyl *was born on November 9, 1885 in Elmshorn, Schleswig–Holstein, Germany. He began his advanced education at the University of Munich, studying mathematics and physics. Later he continued these studies at the University of Göttingen. His supervisor there was David Hilbert, under whose direction he received his doctorate in 1908–see Biography 3.5 on page 127. His thesis was on singular integral equations that invited deep study of Fourier integral theory. At Göttingen, he took up his first teaching position which he held until 1913. There he wrote his* habilitation *thesis, which involved the spectral theory of singular Sturm–Liouville problems. He also published his first book in 1913, entitled* Idee der Riemannschen Fläche, *which gave a rigorous foundation to the geometric function theory previously developed by Riemann. This was accomplished by essentially bringing together analysis, geometry, and topology. The fact that the original text was reprinted in 1997 shows its impact on the progress of mathematics. Eventually he took up a chair in Zürich, Switzerland, where he gave lectures that formed the foundation for his second book* Raum–Zeit–Materie, *published in 1919. Later editions developed his gauge field theory. During this time he also made contributions to the theory of uniform distribution modulo 1, an important area of analytic number theory. In 1927-28, he taught a course on group theory and quantum mechanics. This lead to his third book* Gruppentheorie und Quantenmechanik *which was published in 1928. Essentially Weyl had laid the foundation for the first unified field theory for which the Maxwell electromagnetic field and gravitational field appear as geometrical properties of space-time. From 1930-33, he held the chair of mathematics at Göttingen to fill the vacancy created by Hilbert's retirement. However, the Nazi rise to power convinced him to accept a position at the newly created Institute for Advanced Study at Princeton in the U.S.A., where he remained until his retirement in 1951. During his years at Princeton, he published other influential books, perhaps the most important of which was* Symmetry *published in 1952. On December 8, 1955, while on a visit to Zürich, he collapsed and died after mailing thank you letters to those who had wished him a happy seventieth birthday. During his life he contributed to the geometric foundations of manifolds and physics, topological groups, Lie groups, representation theory, harmonic analysis, analytic number theory, and the foundations of mathematics itself. In regard to the latter he said: "The question for the ultimate foundations and the ultimate meaning of mathematics remains open; we do not know in which direction it will find its final solution nor even whether a final objective answer can be expected at all. "Mathematizing" may well be a creative activity of man, like language or music, of primary originality, whose historical decisions defy complete objective rationalization."*

1.3   Euclidean Quadratic Fields

> *Keeping an open mind is a virtue—but as the space engineer James Oberg*
> *once said, not so open that your brains fall out.*
>
> From **The Demon-Haunted World (1995)**
> **Carl Sagan (1934–1996), American astronomer and astrochemist**

In Theorem 1.9 on page 19, we proved that for $\alpha, \beta \in \mathfrak{O}_F = \mathbb{Z}[i]$, there are Gaussian integers σ, δ such that

$$\alpha = \beta\sigma + \delta, \text{ where } 0 \leq N_F(\delta) < N_F(\beta), \tag{1.36}$$

where σ is a quotient and δ is called a remainder. Condition (1.36) is a special instance of the following notion that is the topic of this section. The title of this section speaks to Euclidean "fields," but this slight abuse of language is a succinct way of saying the "ring of integers of a given quadratic number field."

Definition 1.17 | **Euclidean Functions and Domains** |

Let R be an integral domain. If there exists a function,

$$f : R \mapsto \mathbb{N} \cup \{0\},$$

which satisfies the following conditions,

(a) If $\alpha, \beta \in R$ with $\alpha\beta \neq 0$, then $f(\alpha) \leq f(\alpha\beta)$, and

(b) If $\alpha, \beta \in R$ with $\beta \neq 0$, there exist $\sigma, \delta \in R$, such that

$$\alpha = \beta\sigma + \delta, \text{ where } f(\delta) < f(\beta),$$

then f is called a *Euclidean function*, and R is called a *Euclidean domain with respect to* f.

Example 1.16 We show that the Gaussian integers provide an illustration of a Euclidean domain. Let $\alpha = a + bi \in R = \mathbb{Z}[i]$ and define

$$f(\alpha) = a^2 + b^2.$$

Then by Exercise 1.41 on page 45, $f(\alpha) \leq f(\alpha\beta)$ for any $\alpha\beta \neq 0$. This is condition (a) of Definition 1.17. To verify condition (b), let

$$\beta = c + di \in R.$$

Then

$$\alpha/\beta = \frac{a + bi}{c + di} = \frac{(a + bi)(c - di)}{c^2 + d^2} =$$

$$\frac{ac+bd}{c^2+d^2} + \frac{bc-ad}{c^2+d^2}i = u + vi \in \mathbb{C}.$$

Let $x, y \in \mathbb{Z}$ such that

$$|u - x| \le 1/2, \text{ and } |v - y| \le 1/2.$$

Then,

$$|\alpha/\beta - (x + yi)| = |(u - x) + (v - y)i| =$$
$$(u - x)^2 + (v - y)^2 \le 1/4 + 1/4 < 1. \tag{1.37}$$

Hence, if we let

$$\sigma = x + yi, \text{ and } \delta = \alpha - \beta\sigma,$$

then

$$f(\delta) = f(\alpha - \beta\sigma) = |\alpha - \beta\sigma| = |\beta||\alpha/\beta - \sigma| < |\beta| = f(\beta),$$

where the inequality follows from (1.37). Hence, condition (b) is satisfied as well. Therefore, R is Euclidean with respect to the norm function $f(\alpha) = N_F(\alpha)$ —see Definition 1.18 on the following page.

Remark 1.13 In Theorem 1.11 on page 23, we provided a Euclidean algorithm for Gaussian integers. Now we generalize this, in light of Example 1.16 to an arbitrary Euclidean domain, and the proof follows along the lines of Theorem 1.11. *Note that the following also extends the notion of a gcd from algebraic integers given in Definition 1.14 on page 21 to elements in a Euclidean domain, and so extends the notion of relative primality given in Definition 1.15 on page 22 to Euclidean domains as well.* See Exercise 1.39 on page 45.

Theorem 1.14 | **Euclidean Algorithm in Euclidean Domains**

Let R be a Euclidean domain with respect to f, and let $\alpha = \alpha_0, \beta = \beta_0 \in R$ with $\alpha_0\beta_0 \ne 0$ and $\beta_0 \nmid \alpha_0$. We can define $\alpha_j \in R$ and $\beta_j \in R$ for $j = 1, 2, \ldots, n$ recursively by

$$\alpha_j = \beta_j\delta_j + \gamma_j, \text{ with } f(\gamma_j) < f(\beta_j), \tag{1.38}$$

where $\alpha_j = \beta_{j-1}$ and $\beta_j = \gamma_{j-1}$. Also, if $n \in \mathbb{N}$ is the least value such that $\gamma_n = 0$, then γ_{n-1} is a gcd of α and β.

Proof. For each $j = 1, 2, \ldots, n$, (1.38), follows from condition (b) of Definition 1.17, given that we begin with $\alpha = \alpha_0$ and $\beta = \beta_0$ where

$$f(0) < f(\alpha_j) < f(\alpha_{j+1})$$

for each $j = 0, 1, \ldots, n$. Thus, there must exist a value $n \in \mathbb{N}$ such that $\gamma_n = 0$, observing that $n \ne 0$ since $\beta \nmid \alpha$. Since $\alpha_n = \beta_n\delta_n + \gamma_n$, then

$$\gamma_{n-1} = \beta_n \mid \alpha_n = \beta_{n-1} = \gamma_{n-2}.$$

Similarly, $\gamma_{n-2} \mid \gamma_{n-3}$. Continuing in this way, we see that

$$\gamma_{n-j} \mid \gamma_{n-j-1}$$

for all natural numbers $j < n$. Hence, $\gamma_{n-j} \mid \gamma_1 \mid \gamma_0 = \beta_1$, so since

$$\alpha_1 = \beta_1 \delta_1 + \gamma_1,$$

then $\gamma_{n-1} \mid \alpha_1 = \beta_0$. Also, since

$$\alpha_0 = \beta_0 \delta_0 + \gamma_0, \tag{1.39}$$

then $\gamma_{n-1} \mid \alpha_0$. We have shown that γ_{n-1} is a common divisor of α and β. It remains to show that it is divisible by any common divisor of the two. If σ is a common divisor of α and β, then $\sigma \mid \gamma_0$ by (1.39). Thus,

$$\beta = \beta_0 = \alpha_1 = \beta_1 \delta_1 + \gamma_1 = \gamma_0 \delta_1 + \gamma_1,$$

so $\sigma \mid \gamma_1$. Continuing in this way, $\sigma \mid \gamma_j$ for all natural numbers $j < n$. In particular, $\sigma \mid \gamma_{n-1}$. We have shown that any pair of elements in a Euclidean domain possesses a gcd and that such a gcd may be found by the Euclidean algorithm described above.   □

Example 1.16 on page 32 is a motivator for another aspect of Euclidean quadratic fields that is worthy of exploring, namely those that satisfy the following property.

Definition 1.18 | **Norm-Euclidean Quadratic Fields**

A quadratic number field $F = \mathbb{Q}(\sqrt{D})$ is said to be *norm-Euclidean* if given $\alpha, \beta \in \mathfrak{O}_F$ with $\beta \neq 0$, there exist $\sigma, \delta \in \mathfrak{O}_F$ such that

$$\alpha = \beta\sigma + \delta \text{ where } |N_F(\delta)| < |N_F(\beta)|.$$

Remark 1.14   Now we look to determine which quadratic fields are Euclidean. The reader should first solve Exercise 1.46 on page 45. Note that condition (c) in that exercise was established by G.R. Veldkamp in [97], who essentially wanted to show that condition (a) of Definition 1.17 on page 32 is redundant.

Theorem 1.15 | **Euclidean Complex Quadratic Fields**

   Let $\mathfrak{O}_F$ be the ring of integers of the quadratic number field $F = \mathbb{Q}(\sqrt{D})$ with $D < 0$. Then the following are equivalent.

(a) $\mathfrak{O}_F$ *is Euclidean.*

(b) $\mathfrak{O}_F$ *is norm-Euclidean.*

(c) $D \in \{-1, -2, -3, -7, -11\}$.

Proof. To show that (a) and (b) are equivalent, we first show that the norm function for quadratic fields given in Definition 1.17 on page 32 is indeed a Euclidean function according to Definition 1.17 on page 32. Part (a) of Definition 1.17 is satisfied since if $\alpha\beta \neq 0$, then

$$|N(\alpha\beta)| = |N(\alpha)||N(\beta)|$$

by part (b) of Exercise 1.8 on page 28, and

$$|N(\alpha)||N(\beta)| \geq |N(\beta)|.$$

Part (b) of Definition 1.17 is part of Definition 1.18.

Now we show that Euclidean complex quadratic fields are norm-Euclidean. Assume that $|D| > 11$ and $\mathfrak{O}_F$ is Euclidean with respect to f. Select $\beta \in \mathfrak{O}_F$ with $\beta \neq 0, \pm 1$ such that

$$f(\beta) = \min\{f(\alpha) : \alpha \in \mathfrak{O}_F, \alpha \neq 0, \pm 1\}. \tag{1.40}$$

Thus, by property (b) of Definition 1.17, for every $\alpha \in \mathfrak{O}_F$, there exists a $\sigma \in \mathfrak{O}_F$ such that $\alpha - \sigma\beta = 0, \pm 1$. In particular, if $\alpha = 2$, then $|\beta| \leq 3$, since

$$\sigma\beta = \alpha \text{ or } \sigma\beta = \alpha \pm 1. \tag{1.41}$$

However, if $|\beta| = 3$, this contradicts (1.40) since

$$f(\sigma\beta) = 3 > f(\alpha) = f(2),$$

using part (a) of Definition 1.17. Thus, $|\beta| \leq 2$ since either $\beta \mid \alpha$ or $\beta \mid (\alpha \pm 1)$. Hence, $N_F(\beta) \leq 4$. If $D \not\equiv 1 \pmod 4$, there exist $a, b \in \mathbb{Z}$ such that $\beta = a + b\sqrt{D}$ by Theorem 1.3 on page 6. So since

$$4 \geq N_F(\beta) = a^2 - Db^2 > a^2 + 11b^2,$$

we must have $b = 0$ and $|a| \leq 1$, namely $\beta = 0$ or $|\beta| = 1$ both of which contradict the choice of β. If $D \equiv 1 \pmod 4$, then by Theorem 1.3 again, there exist integers a, b of the same parity such that $\beta = (a + b\sqrt{D})/2$. Hence,

$$16 \geq a^2 - Db^2 \geq a^2 + 15b^2,$$

so $|b| = 0, 1$, respectively $|a| = 0, 1$. In the former case, $\beta = 0$ contradicting the choice of β and in the latter case,

$$\beta = (1 + \sqrt{-15})/2.$$

However, by (1.41), we must have $\alpha = 2 = \sigma\beta$ in this case, so there exist $x, y \in \mathbb{Z}$ of the same parity such that

$$2 = \left(\frac{x + y\sqrt{-15}}{2}\right)\left(\frac{1 + \sqrt{-15}}{2}\right) = \left(\frac{x - 15y + (x + y)\sqrt{-15}}{4}\right),$$

so $x = -y$ and $x - 15y = 8$. This implies $-16y = 8$, a contradiction. Hence (a) is equivalent to (b).

To show that (b) is equivalent to (c), we employ condition (c) of Exercise 1.46 on page 45. Assume that $\mathfrak{O}_F$ is Euclidean for $D < 0$. First we look at the case where $D \not\equiv 1 \pmod 4$. Then by Theorem 1.3, for a given $\rho = q + r\sqrt{D} \in F$, we must find

$$\sigma = a + b\sqrt{D} \in \mathbb{Z}[\sqrt{D}]$$

with

$$|(q-a)^2 - D(r-b)^2| < 1. \tag{1.42}$$

Let $\rho = \sqrt{D}/2$, then we must have, from (1.42), that

$$\left| a^2 - D\left(\frac{1}{2} - b\right)^2 \right| < 1,$$

which means that

$$\left(b - \frac{1}{2}\right)^2 |D| + a^2 < 1.$$

However, for any $b \in \mathbb{Z}$, $(b - 1/2)^2 \geq 1/4$, so

$$\frac{|D|}{4} < 1 - a^2 \leq 1,$$

namely $|D| < 4$ for which only the values $D = -1, -2$ hold.

Now assume that $D \equiv 1 \pmod 4$ and let $\rho = (1 + \sqrt{D})/4$. Then by (1.42),

$$\left| \left(\frac{1}{4} - \frac{a}{2}\right)^2 - D\left(\frac{1}{4} - \frac{b}{2}\right)^2 \right| < 1,$$

namely

$$\left(\frac{1}{4} - \frac{a}{2}\right)^2 + |D|\left(\frac{1}{4} - \frac{b}{2}\right)^2 < 1.$$

However, for any $x \in \mathbb{Z}$, $|1/4 - x/2| \geq 1/4$, so $1 + |D| < 16$, from which we infer that $D = -3, -7, -11$. We have shown that (b) implies (c).

It remains to verify that the values on our list actually are Euclidean, in order to prove that (c) implies (b). To do this, we recall the nearest integer function, Ne, described on page 19.

If $D = -1, -2$, then by taking $a = Ne(q), b = Ne(r)$, (1.42) holds since

$$|(q-a)^2 - D(r-b)^2| \leq \left| \left(\frac{1}{2}\right)^2 + 2\left(\frac{1}{2}\right)^2 \right| < 1.$$

It remains to consider $D \equiv 1 \pmod 4$. We let

$$b = Ne(2r), \text{ for which } |2r - b| \leq 1/2.$$

If we select $a \in \mathbb{Z}$ to be such that $|q - a - b/2| \leq 1/2$. Then, for $D = -3, -7, -11$,

$$\left| \left(q - a - \frac{b}{2} \right)^2 - D \left(r - \frac{b}{2} \right)^2 \right| \leq \left| \frac{1}{4} + \frac{11}{16} \right| = \frac{15}{16} < 1,$$

so (1.42) holds, as required. □

Remark 1.15 *The case for real quadratic fields is more complicated. We'll also address some of these fields in §1.4–see Theorem 1.21 on page 50.*

We now look at factorization in rings of integers of number fields. To do so we need to introduce some notions related to that of primes. Note that this more general definition refines the definition given for Gaussian integers in Definition 1.13 on page 19, which we will shortly show to be equivalent in the case of domains having a certain property shared with $\mathbb{Z}[i]$. — see Definition 1.20.

Definition 1.19 | **Irreducible and Prime Elements**

A nonzero, nonunit element α in an integral domain R is called *irreducible* if whenever there exist $\beta, \gamma \in R$ with $\alpha = \beta\gamma$, then one of β or γ is unit. If this property fails to hold for α then it is called *reducible*.

If $\alpha \in R$, then α is called *prime* if whenever $\alpha \mid \beta\gamma$ for $\beta, \gamma \in R$, then $\alpha \mid \beta$ or $\alpha \mid \gamma$.

Example 1.17 In the Gaussian integers

$$5 = (2 + i)(2 - i)$$

where $2 + i$ and $2 - i$ are irreducibles, and shortly we will see that they are also primes in the sense of Definition 1.19. Also, in $\mathbb{Z}[\sqrt{10}]$,

$$6 = 2 \cdot 3 = (4 + \sqrt{10})(4 - \sqrt{10}),$$

where each of the four factors is irreducible. In the latter case the two factorizations are distinct since 2 and 3 are not associates of $4 + \sqrt{10}$ or $4 - \sqrt{10}$—See Exercises 1.47–1.48 for proofs of the above facts. This nonuniqueness of factorization is at the core of fundamental aspects of algebraic number theory and motivates the following notion.

Definition 1.20 | **Unique Factorization**

If R is an integral domain in which every nonzero, nonunit element of R can be expressed as a finite product of irreducible elements of R, then R is called a *factorization domain*. A factorization domain R is called a *unique factorization domain* (UFD) if the following property holds:

Suppose that $\alpha \in R$ such that

$$\alpha = u\beta_1^{b_1} \beta_2^{b_2} \cdots \beta_n^{b_n}$$

where $b_j \in \mathbb{N}$, and the β_j are nonassociated irreducible elements of R for $1 \leq j \leq n$, and u is a unit of R. Suppose further that we have another factorization given by

$$\alpha = v\gamma_1^{a_1}\gamma_2^{a_2}\cdots\gamma_m^{a_m}, \text{ where } a_j \in \mathbb{N} \text{ and } v \text{ is a unit of } R.$$

Then $m = n$, the γ_j are nonassociated irreducible elements of R and (after possibly rearranging the β_j), $\beta_j \sim \gamma_j$ for $j = 1, 2, \ldots, n$.

The following links Definition 1.13 on page 19 and Definition 1.20.

Lemma 1.2 │ Primes are Irreducible │

If α is prime in an integral domain R, then α is irreducible.

Proof. Let α be a prime element in R. If $\alpha = \beta\gamma$ where $\beta, \gamma \in R$, then $\alpha \mid \beta$ or $\alpha \mid \gamma$. Without loss of generality, assume that $\alpha \mid \beta$. Therefore, there exists a $\delta \in R$ such that $\beta = \alpha\delta$, so $\alpha = \beta\gamma = \alpha\delta\gamma$. Since R is an integral domain, we may cancel the α from both sides to get that $1 = \delta\gamma$, so $\gamma \sim 1$. We have shown that α is irreducible. □

Theorem 1.16 │ Criterion for Unique Factorization Domains │

An integral domain R is a unique factorization domain if and only if every irreducible element in R is prime.

Proof. Assume that R is a unique factorization domain. Let $\alpha \in R$ be irreducible, and assume that $\alpha \mid \beta\gamma$. It remains to show that $\alpha \mid \beta$ or $\alpha \mid \gamma$. Since β and γ may be uniquely represented as

$$\beta = u\sigma_1^{a_1}\sigma_2^{a_2}\cdots\sigma_m^{a_m}$$

and

$$\gamma = v\delta_1^{b_1}\delta_2^{b_2}\cdots\delta_n^{b_n}$$

where $a_j, b_k \in \mathbb{N}$, u, v are units in R, σ_j for $j = 1, 2, \ldots, m$, respectively δ_k for $k = 1, 2, \ldots, n$, are nonassociated irreducibles, there exists a $\rho \in R$ such that

$$\rho\alpha = \beta\gamma = uv\prod_{j=1}^{m}\sigma_j^{a_j}\prod_{k=1}^{n}\delta_k^{a_k}.$$

Since α is irreducible, then $\alpha \sim \sigma_j$ for some $j = 1, 2, \ldots, m$, or $\alpha \sim \delta_k$ for some $k = 1, 2, \ldots, n$. In other words, $\alpha \mid \beta$ or $\alpha \mid \gamma$.

Conversely, suppose that every irreducible in R is prime. Let

$$u\prod_{j=1}^{m}\sigma_j^{a_j} = v\prod_{k=1}^{n}\delta_k^{b_k}, \tag{1.43}$$

where u, v are units in R and σ_j, δ_k are nonassociated irreducibles (primes) for $j = 1, 2, \ldots, m$, respectively $k = 1, 2, \ldots, n$. We will use induction on m to prove that $m = n$ and $\sigma_j \sim \delta_k$ for some j, k. If $m = 0$, then the result vacuously holds. Assume that $m \in \mathbb{N}$ and induction hypothesis is that unique factorization holds for all factorizations of nonassociated irreducibles of length less than m. Then if (1.43) holds,

$$\sigma_m \;\Big|\; v \prod_{k=1}^{n} \delta_k^{b_k},$$

so $\sigma_m \mid \delta_k$ for some k, since σ_m is prime, so $\sigma_m \nmid v$. By renumbering the δ_k if necessary, we may conclude that $\sigma_m \mid \delta_n$. But since both σ_m and δ_n are primes, then $\sigma_m = w\delta_n$ for some unit w in R. Thus,

$$uw\sigma_1^{a_1}\sigma_2^{a_2}\cdots\sigma_{m-1}^{a_{m-1}}\delta_n^{a_m} = v\delta_1^{b_1}\delta_2^{b_2}\cdots\delta_{n-1}^{b_{n-1}}\delta_n^{b_n}.$$

Without loss of generality assume that $a_m \geq b_n$. Then

$$uw\sigma_1^{a_1}\sigma_2^{a_2}\cdots\sigma_{m-1}^{a_{m-1}}\delta_n^{a_m-b_n} = v\delta_1^{b_1}\delta_2^{b_2}\cdots\delta_{n-1}^{b_{n-1}},$$

so if $a_m > b_n$, then

$$\delta_n \;\Big|\; \prod_{k=1}^{n-1} \delta_k,$$

and since δ_n is prime it must be an associate of δ_k for some $1 \leq k \leq n-1$, contradicting the fact that the δ_k are nonassociated for distinct k. Hence, $a_m = b_n$, and

$$uw\sigma_1^{a_1}\sigma_2^{a_2}\cdots\sigma_{m-1}^{a_{m-1}} = v\delta_1^{b_1}\delta_2^{b_2}\cdots\delta_{n-1}^{b_{n-1}},$$

so by the induction hypothesis $m - 1 = n - 1$ and the σ_j are associates of the δ_k in some order. This completes the induction and we have unique factorization. $\square$

Theorem 1.17 | **Euclidean Domains are UFDs**

Euclidean domains are unique factorization domains.

Proof. Let R be a Euclidean domain with respect to f, and let $\alpha \in R$ be nonzero. First, we establish the existence of factorizations into irreducible elements. By Exercise 1.43 on page 45, $f(\alpha) = f(1)$ if and only if $\alpha \in \mathcal{U}_R$. In this case α is vacuously a product of irreducible elements. Hence, we may use induction on $f(\alpha)$. By Exercise 1.42, $f(1) \leq f(\alpha)$. Assume that $\alpha \notin \mathcal{U}_R$, and that any $\beta \in R$ with $f(\beta) < f(\alpha)$ has a factorization into irreducible elements. If α is irreducible, we are done. Assume otherwise. Then

$$\alpha = \beta\gamma \text{ for } \beta, \gamma \in R \text{ and } \beta, \gamma \notin \mathcal{U}_R.$$

Thus, by property (a) of Euclidean domains given in Definition 1.17 on page 32, $f(\beta) \leq f(\alpha)$, and $f(\gamma) \leq f(\alpha)$. By part (b) of Exercise 1.44, $f(\gamma) \neq f(\alpha)$, and

$f(\beta) \neq f(\alpha)$, so by the induction hypothesis, both β and γ have factorizations into irreducibles. Thus, so does α, and existence is established. It remains to establish uniqueness.

Let $\alpha \mid \beta\gamma$ where α is irreducible. If $\alpha \nmid \beta$, then α and β are relatively prime—see Remark 1.13 on page 33. Therefore, by Exercise 1.39 on page 45, there are $\sigma, \delta \in R$ such that

$$1 = \alpha\sigma + \beta\delta.$$

Therefore,

$$\gamma = \alpha\sigma\gamma + \beta\delta\gamma.$$

Since $\alpha \mid \beta\gamma$, the latter implies that $\alpha \mid \gamma$. In other words, α is prime. Hence, all irreducibles are primes. By Theorem 1.16, we have secured the result. $\qquad\square$

Remark 1.16 Note that in Theorem 1.10 on page 21, we proved that gcd's always exist for the Gaussian integers. This is clear from Theorem 1.17 and Example 1.16 on page 32 since the Gaussian integers form a Euclidean domain. We cannot ensure the existence of gcds *without* unique factorization, which is guaranteed in Euclidean domains by Theorem 1.17. Indeed, the definition of a Gaussian prime given in Definition 1.13 on page 19, uses the fact that all irreducibles in the Gaussian integers are primes, a fact we now know from Example 1.16 on page 32 and Theorem 1.17 on the previous page.

We may speak about factorizations in domains that are not UFDs. However, we cannot speak about factorizations of elements in this regard; rather we must move to the level of ideals and this is to come later when we study ideal theory. This is part of the history of the development of algebraic number theory where Dedekind looked at factorizations in non-UFDs using ideal theory that we will study in Chapter 2.

Example 1.18 The Gaussian integers $2 \pm i$ are primes, which is equivalent to being irreducible in any UFD, as noted in Remark 1.16. However, the converse of Lemma 1.2 does not hold. For instance, by Example 1.17 on page 37, 2 is irreducible in $\mathbb{Z}[\sqrt{10}]$, but

2 is not prime, since $2 \mid (4 + \sqrt{10})(4 - \sqrt{10})$ without dividing either factor.

As shown in Example 1.17, $\mathbb{Z}[\sqrt{10}]$ is not a unique factorization domain. At the heart of this fact *in general* for the *nonexistence* of unique factorization in a factorization domain is the failure of irreducibles to be primes, as Theorem 1.16 on page 38 essentially validates.

The next topic is the introduction of another property, which cannot be guaranteed to exist, unless we are in a UFD. This mimics the notion studied for the rational integers in [68, Section 1.2].

Definition 1.21 | **Least Common Multiple in UFDs**

Let R be a UFD. A *least common multiple* of $\alpha, \beta \in R$ is an element $\delta \in R$ satisfying the two properties:

(a) $\alpha \mid \delta$ and $\beta \mid \delta$.

(b) If $\alpha \mid \sigma \in R$ and $\beta \mid \sigma$, then $\delta \mid \sigma$.

Example 1.19 By Exercise 1.49 on page 45, any two least common multiples of a given pair of elements in a UFD are associates. Thus, as with greatest common divisors, least common multiples are unique up to associates.

For instance,

$$\alpha = (2+i) \mid 5 = \delta \text{ and } \beta = (2-i) \mid 5.$$

Moreover, if $(2+i) \mid \sigma \in R$ and $(2-i) \mid \sigma$, then

$$\sigma = (2+i)\delta_1 = (2-i)\delta_2$$

for $\delta_1, \delta_2 \in R$. In particular, $(2+i) \mid \delta_2$, so $5 \mid \sigma$. Thus, $\delta = 5$ is a least common multiple of $2+i$ and $2-i$. Hence, $\pm 5i$ and ± 5 are all of their common multiples.

We conclude this section with an application to a famous result due to Fermat.

Remark 1.17 In what follows, we use the symbol $\gcd(x, y)$ for

$$x, y \in \mathbb{Z}[\zeta_3] = \mathbb{Z}[(-1 + \sqrt{-3})/2]$$

to mean the *unique* gcd *of elements up to associates* as dictated by Exercise 1.17 on page 29. Moreover, the congruences in the following proof all take place in $\mathbb{Z}[\zeta_3]$. See [68, Biography 1.7, p. 33].

Note that Fermat's Last Theorem (FLT) is the assertion that

$$x^n + y^n = z^n \tag{1.44}$$

has no solutions in positive integers x, y, z for $n \in \mathbb{N}$ with $n > 2$. For an overview and background, see [68, Biography 1.10, p.38]. Also, see Biography 5.5 of Wiles on page 225 for a synopsis of its solution.

Theorem 1.18 $\boxed{\textbf{Gauss' Proof of FLT for p} = 3}$

There are no solutions of

$$\alpha^3 + \beta^3 + \gamma^3 = 0$$

for nonzero $\alpha, \beta, \gamma \in \mathfrak{O}_F = \mathbb{Z}[\zeta_3]$, *where* $F = \mathbb{Q}(\zeta_3)$. *In particular, there are no solutions to*

$$x^3 + y^3 = z^3,$$

in nonzero rational integers x, y, z.

Proof. We assume that there are nonzero $\alpha, \beta, \gamma \in \mathfrak{O}_F$ such that

$$\alpha^3 + \beta^3 + \gamma^3 = 0,$$

and achieve a contradiction. Without loss of generality, we may assume that

$$\gcd(\alpha, \beta) = \gcd(\alpha, \gamma) = \gcd(\beta, \gamma) = 1.$$

Let

$$\lambda = 1 - \zeta_3.$$

Then since

$$N_F(\lambda) = \lambda\lambda' = 3,$$

we must have $\lambda \mid 3$. Also, by Theorem 1.17 on page 39 and Exercise 1.52 on page 46, λ is prime in $\mathfrak{O}_F$. We will achieve the desired contradiction by an infinite descent argument. This is not done directly, but rather we get a contradiction to the equation

$$\alpha^3 + \beta^3 + \lambda^{3n}\rho^3 = 0,$$

for any $n \in \mathbb{N}$ and $\rho \in \mathfrak{O}_F$. Thus, we first show that the latter equation holds. We require three claims.

Claim 1.4 *If $\lambda \nmid \delta \in \mathfrak{O}_F$, then $\delta \equiv \pm 1 \, (\mathrm{mod} \, \lambda)$.*

Let

$$\delta = a + b\zeta_3, \text{ where } a, b \in \mathbb{Z}.$$

Then $\delta = u + v\lambda$, where $u, v \in \mathbb{Z}$. If $\lambda | u$, then $\lambda \mid \delta$, a contradiction, so $\lambda \nmid u$. Since $\lambda | 3$, then $3 \nmid u$, so $u \equiv \pm 1 \, (\mathrm{mod} \, 3)$ in $\mathbb{Z}$. Thus, there is a $t \in \mathbb{Z}$ such that

$$\delta = \pm 1 + 3t + v\lambda.$$

But $\lambda | 3$, so there exists a $\sigma \in \mathfrak{O}_F$ such that

$$\delta = \pm 1 + t\sigma\lambda + v\lambda = \pm 1 + \lambda(t\sigma + v).$$

In other words, $\delta \equiv \pm 1 \, (\mathrm{mod} \, \lambda)$ as required.

Claim 1.5 *If $\lambda \nmid \delta \in \mathfrak{O}_F$, then $\delta^3 \equiv \pm 1 \, (\mathrm{mod} \, \lambda^4)$.*

Since $\lambda \nmid \delta$, then by Claim 1.4, $\delta \equiv \pm 1 \, (\mathrm{mod} \, \lambda)$. We may assume that

$$\delta \equiv 1 \quad (\mathrm{mod} \, \lambda)$$

since the other case is similar. Therefore, $\delta = 1 + \lambda\sigma$ for some $\sigma \in \mathfrak{O}_F$. Thus,

$$\delta^3 - 1 = (\delta - 1)(\delta - \zeta_3)(\delta - \zeta_3^2) = \lambda\sigma(\lambda\sigma + 1 - \zeta_3)(\lambda\sigma + 1 - \zeta_3^2) =$$

$$\lambda\sigma(\lambda\sigma + \lambda)(\lambda\sigma + \lambda(1 + \zeta_3)) = \lambda^3\sigma(\sigma + 1)(\sigma - \zeta_3^2), \qquad (1.45)$$

where the last equality follows via Exercise 1.54 on page 46, from the fact that

$$1 + \zeta_3 + \zeta_3^2 = 0. \tag{1.46}$$

Since $\zeta_3^2 - 1 = (\zeta_3 + 1)(\zeta_3 - 1) = (\zeta_3 + 1)\lambda$, then $\zeta_3^2 \equiv 1 \,(\text{mod } \lambda)$, so by (1.45) and Claim 1.4,

$$0 \equiv (\delta^3 - 1)\lambda^{-3} \equiv \sigma(\sigma + 1)(\sigma - \zeta_3^2) \equiv \sigma(\sigma + 1)(\sigma - 1) \equiv \sigma(\sigma^2 - 1) \;\; (\text{mod } \lambda).$$

Hence,

$$\delta^3 \equiv 1 \;\; (\text{mod } \lambda^4),$$

and we have Claim 1.5.

Claim 1.6 $\lambda \mid \alpha\beta\gamma$.

Suppose that $\lambda \nmid \alpha\beta\gamma$. Then by Claim 1.5,

$$0 = \alpha^3 + \beta^3 + \gamma^3 \equiv \pm 1 \pm 1 \pm 1 \;\; (\text{mod } \lambda^4),$$

from which it follows that $\lambda^4 \mid 1$ or $\lambda^4 \mid 3$. The former is impossible since λ is prime, and the second is impossible since

$$3 = (1 - \zeta_3)(1 - \zeta_3^2) = (1 - \zeta_3)^2(1 + \zeta_3) = \lambda^2(1 + \zeta_3),$$

and $1 + \zeta_3$ is a unit, so not divisible by λ^2. This contradiction establishes Claim 1.6.

By Claim 1.6, we may assume without loss of generality that $\lambda \mid \gamma$. However, by the gcd condition assumed at the outset of the proof, $\lambda \nmid \alpha$, and $\lambda \nmid \beta$. Let $n \in \mathbb{N}$ be the highest power of λ dividing γ. In other words, assume that $\gamma = \lambda^n \rho$, for some $\rho \in \mathfrak{O}_F$ with $\lambda \nmid \rho$. Thus, we have

$$\alpha^3 + \beta^3 + \lambda^{3n}\rho^3 = 0. \tag{1.47}$$

We now use Fermat's method of infinite descent (which we studied in [68, §7.4, p. 281]) to complete the proof. First we establish that $n > 1$. If $n = 1$, then by Claim 1.5,

$$-\lambda^3 \rho^3 = \alpha^3 + \beta^3 \equiv \pm 1 \pm 1 \;\; (\text{mod } \lambda^4).$$

The signs on the right cannot be the same since $\lambda \nmid 2$. Therefore,

$$-\lambda^3 \rho^3 \equiv 0 \;\; (\text{mod } \lambda^4),$$

forcing $\lambda \mid \rho$, a contradiction that shows $n > 1$. Given the above, the following claim, once proved, will yield the full result by descent.

Claim 1.7 *If Equation (1.47) holds for $n > 1$, then it holds for $n - 1$.*

Let

$$X = \frac{\beta + \alpha\zeta_3}{\lambda}, Y = \frac{\beta\zeta_3 + \alpha}{\lambda}, \text{ and } Z = \frac{(\beta + \alpha)\zeta_3^2}{\lambda}.$$

Observe that

$$X, Y, Z \in \mathfrak{O}_F$$

by Claim 1.5, Equation (1.47), and the fact that $\zeta_3 \equiv 1 \,(\text{mod } \lambda)$. Also, by Exercise 1.54 again,

$$X + Y + Z = 0,$$

and

$$XYZ = \frac{\beta^3 + \alpha^3}{\lambda^3} = \left(\frac{-\lambda^n \rho}{\lambda}\right)^3 = \lambda^{3n-3}\,(-\rho)^3,$$

so

$$\lambda^{3n-3} \mid XYZ, \text{ but } \lambda^{3n} \nmid XYZ,$$

since $\lambda \nmid \rho$. Also, since

$$\beta = -\zeta_3 X + \zeta_3^2 Y, \text{ and } \alpha = \zeta_3 Z - X,$$

then by the gcd condition assumed at the outset of the proof, we have

$$\gcd(X, Y) = \gcd(X, Z) = \gcd(Y, Z) = 1.$$

Hence, each of X, Y, and Z is an associate of a cube in $\mathfrak{O}_F$. Also, we may assume without loss of generality that $\lambda^{3n-3} \mid Z$. By unique factorization in $\mathfrak{O}_F$, we may let

$$X = u_1\xi^3, Y = u_2\eta^3, \text{ and } Z = u_3\lambda^{3n-3}\nu^3$$

for some $\xi, \eta, \nu \in \mathfrak{O}_F$, and $u_j \in \mathfrak{U}_F$ for $j = 1, 2, 3$. Therefore, we have

$$\xi^3 + u_4\eta^3 + u_5\lambda^{3n-3}\nu^3 = 0, \tag{1.48}$$

where $u_j = u_1^{-1}u_{j-2}$ for $j = 4, 5$. Therefore, $\xi^3 + u_4\eta^3 \equiv 0 \,(\text{mod } \lambda^3)$. By Claim 1.5

$$\xi^3 \equiv \pm 1 \quad (\text{mod } \lambda^4), \text{ and } \eta^3 \equiv \pm 1 \quad (\text{mod } \lambda^4).$$

Hence, $\pm 1 \pm u_4 \equiv 0 \,(\text{mod } \lambda^3)$. Since the only choices for u_4 are ± 1, $\pm\zeta_3$, and $\pm\zeta_3^2$, then the only values that satisfy the last congruence are $u_4 = \pm 1$, since

$$\lambda^3 \nmid (\pm 1 \pm \zeta_3), \text{ and } \lambda^3 \nmid (\pm 1 \pm \zeta_3^2).$$

If $u_4 = 1$, then Equation (1.48) provides a validation of Claim 1.7. If $u_4 = -1$, then replacing η by $-\eta$ provides a validation of the claim. This completes the proof. □

Remark 1.18 In §8.3 we will generalize the above proof, also due to Kummer, to prove that (1.44) fails to hold for any $xyz \neq 0$ when $n = p \geq 3$ is any so-called "regular" prime $p \nmid xyz$—see Remark 8.6 on page 291. However, this requires deeper tools.

Exercises

1.39. Let R be a Euclidean domain. Theorem 1.14 on page 33 shows that any two nonzero elements $\alpha, \beta \in R$ have a greatest common divisor. Prove that any such gcd may be written in the form

$$\gamma = \rho\alpha + \eta\beta$$

for some $\rho, \eta \in R$.

(*Hint: Mimic the proof of Theorem 1.10 on page 21.*)

1.40. Give an example of a ring in which there exist elements with no greatest common divisor.

1.41. Prove that in Definition 1.17 on page 32, condition (a) is equivalent to the following condition.

If $\alpha \mid \beta$ for any $\alpha, \beta \in R$ with $\alpha\beta \neq 0$, then $f(\alpha) \leq f(\beta)$.

1.42. Let R be a Euclidean domain with respect to f and multiplicative identity 1_R. Prove that $f(1_R) \leq f(\alpha)$ for all nonzero $\alpha \in R$.

1.43. Prove that in a Euclidean domain R with respect to f, and multiplicative identity 1_R, $f(\alpha) = f(1_R)$ for $\alpha \in R$ if and only if α is a unit in R.

1.44. Prove that if R is a Euclidean domain with respect to f, then for $\alpha, \beta \in R$, each of the following hold.

  (a) If $\alpha \sim \beta$, then $f(\alpha) = f(\beta)$.
  (b) If $\alpha \mid \beta$ and $f(\alpha) = f(\beta)$, then $\alpha \sim \beta$.
  (c) $\alpha \in \mathfrak{U}_R$ if and only if $f(\alpha) = f(1_R)$.
  (d) If $\alpha \neq 0$, then $f(\alpha) > f(0)$.

1.45. Prove that any real quadratic field has infinitely many units.

(*You may use the fact, established in* [68, Theorem 5.15, pp. 234–235], *that the Pell equations $x^2 - Dy^2 = 1$ has infinitely many solutions.*)

1.46. Let $F = \mathbb{Q}(\sqrt{D})$ be a quadratic number field. Prove that the condition in Definition 1.18 on page 34 is equivalent to the statement.

  (c) For any $\rho \in F$ there exists a $\sigma \in \mathfrak{O}_F$ such that $|N_F(\rho - \sigma)| < 1$.

1.47. In Example 1.17 on page 37, show that $2 + i$ and $2 - i$ are irreducible in $\mathbb{Z}[i]$.

1.48. In Example 1.17, show that $2, 3, 4 + \sqrt{10}, 4 - \sqrt{10}$ are irreducible in $\mathbb{Z}[\sqrt{10}]$, and that $2, 3$ are not associates of $4 + \sqrt{10}, 4 - \sqrt{10}$.

1.49. Prove that if δ_1 and δ_2 are least common multiples of $\alpha, \beta \in R$ where R is a UFD, then $\delta_1 \sim \delta_2$.

1.50. Let $\alpha \in \mathfrak{O}_F$ where $F = \mathbb{Q}(\sqrt{D})$ is a quadratic number field. Prove that if $N_F(\alpha) = \pm p$, where p is prime in $\mathbb{Z}$, then α is irreducible in $\mathfrak{O}_F$.

1.51. Is the converse of Exercise 1.50 true? If so prove it, and if not provide a counterexample.

1.52. Let F be a quadratic number field that is a UFD. Prove that if $\alpha \in \mathfrak{O}_F$ with $N_F(\alpha) = \pm p$, a prime in $\mathbb{Z}$, then α is a prime in $\mathfrak{O}_F$.

1.53. Is the converse of Exercise 1.52 true? If so prove it, and if not provide a counterexample.

1.54. If ζ_p is a primitive p-th root of unity for a prime p, prove that

$$\sum_{j=0}^{p-1} \zeta_p^j = 0.$$

Biography 1.2 Julius Wilhelm Richard Dedekind (1831–1916) *was born in Brunswick, Germany on October 6, 1831. There he attended school from the time he was seven. In 1848, he entered the Collegium Carolinum, an educational bridge between high school and university. He entered Göttingen at the age of 19, where he became Gauss' last student, and achieved his doctorate in 1852, the topic being Eulerian integrals. Although he taught in Göttingen and in Zürich, he moved to Brunswick in 1862 to teach at the Technische Hochschule, a technical high school. In that year he also was elected to the Göttingen Academy, one of many honours bestowed on him in his lifetime. He maintained this position until he retired in 1894. Dedekind's creation of ideals was published in 1879 under the title* Uber die Theorie der ganzen algebraischen Zahlen. *Hilbert extended Dedekind's ideal theory, which was later advanced further by Emmy Noether. Ultimately this led to the general notion of unique factorization of ideals into prime powers in what we now call Dedekind domains.*

Another of his major contributions was a definition of irrational numbers in terms of what we now call Dedekind cuts. He published this work in Stetigkeit und Irrationale Zahlen *in 1872. He never married, and lived with his sister Julie until she died in 1914. He died in Brunswick on February 12, 1916.*

1.4   Applications of Unique Factorization

> *...a mathematical proof, like a chess problem, to be aesthetically satisfying, must possess three qualities: inevitability, unexpectedness, and economy; that it should 'resemble a simple and clear-cut constellation, not a scattered cluster in the milky way.'*
>
> From page 447 of **Enigma (2001)**
> by **Robert Harris**—see **[38]**

In §1.3 we looked at some instances of unique factorization in quadratic fields. In particular, we applied unique factorization in

$$\mathbb{Z}[\zeta_3] = \mathbb{Z}[(-1 + \sqrt{-3})/2]$$

to present the Gauss' proof of the Fermat result, Theorem 1.18 on page 41. In fact, earlier, we tacitly used unique factorization from $\mathbb{Z}[\sqrt{-7}]$, in Theorem 1.8 on page 14, to provide solutions of the Ramanujan-Nagell equation. In this section we look at unique factorization in other quadratic rings of integers. We begin with $\mathbb{Z}[\sqrt{-2}]$ to find solutions of certain Bachet equations, those of the form

$$y^2 = x^3 + k \qquad (1.49)$$

where $k \in \mathbb{Z}$—see [68, Biography 7.2, p. 279].

Let us begin with a solution of (1.49) for $k = -2$ by Euler—see [68, Biography 1.17, p. 56]. As with the proof of Theorem 1.8, we use the notation $\gcd(x, y)$ in this section for x, y in the ring of integers of a given quadratic field, to mean the unique gcd up to associates.

Theorem 1.19 | **Euler's Solution of Bachet's Equation**

The only integer solutions of (1.49) *for* $k = -2$ *are* $x = 3$ *and* $y = \pm 5$.

Proof. First, we rule out the possibility that x is even or y is even. If x is even, then $y^2 \equiv -2 \,(\mathrm{mod}\ 4)$, and if y is even, then $x^3 \equiv 2 \,(\mathrm{mod}\ 4)$, both of which are impossible. Hence, both x and y are odd. We may factor in

$$\mathfrak{O}_F = \mathbb{Z}[\sqrt{-2}],$$

where $F = \mathbb{Q}(\sqrt{-2})$ as follows,

$$(y + \sqrt{-2})(y - \sqrt{-2}) = x^3.$$

First we show that $\gcd(y + \sqrt{-2}, y - \sqrt{-2}) = 1$. Suppose that

$$(a + b\sqrt{-2}) \,\big|\, \gcd(y + \sqrt{-2}, y - \sqrt{-2}) \text{ for } a, b \in \mathbb{Z},$$

then in particular,

$$N_F(a + b\sqrt{-2}) = (a^2 + 2b^2) \mid N_F(y + \sqrt{-2} - (y - \sqrt{-2})) = N_F(2\sqrt{-2}) = -8,$$

$$(a^2 + 2b^2) \mid N_F(y + \sqrt{-2} + y - \sqrt{-2}) = 4y^2,$$

and

$$(a^2 + 2b^2) \mid N_F(x^3) = x^6.$$

The first two equations show that $(a^2 + 2b^2) \mid 4$, since y is odd. Coupled with the third equation, this shows $(a^2 + 2b^2) \mid 1$, so $a + b\sqrt{-2}$ is a unit. Thus,

$$\gcd(y + \sqrt{-2}, y - \sqrt{-2}) = 1.$$

We may now invoke unique factorization to conclude that

$$y + \sqrt{-2} = \pm(c + d\sqrt{-2})^3, \text{ for some } c, d \in \mathbb{Z},$$

since ± 1 are the only units in $\mathbb{Z}[\sqrt{-2}]$ by Theorem 1.4 on page 8. By multiplying out the right-hand side and comparing coefficients, we get

$$y = \pm(c^3 - 6cd^2), \text{ and } 1 = \pm d(3c^2 - 2d^2).$$

The latter equation implies that $d = \pm 1$, so $1 = \pm(3c^2 - 2)$, the only possible solutions of which are $c = \pm 1$. Hence, putting these back into the equation for y, we get that $y = \pm(\pm 1 \pm 6)$, the only possible solutions for which are $y \in \{\pm 7, \pm 5\}$. However, $y = \pm 7$ implies that $51 = x^3$, which is impossible. Hence, $y = \pm 5$, and $x = 3$.   □

We will extend the above solution to much more general instances of (1.49) in Theorem 8.4 on page 282. However, we will need to develop deeper tools before we get there. For now, we exploit the unique factorization in the Gaussian integers to solve another instance of Bachet's equation purportedly solved by Fermat.

Theorem 1.20 | **Fermat's Solution of Bachet's Equation**

The only integer solutions of (1.49) *for* $k = -4$ *are*

$$(x, y) \in \{(5, \pm 11), (2, \pm 2)\}.$$

Proof. We work in the ring of Gaussian integers $F = \mathbb{Z}[i]$, which has unique factorization by Theorem 1.15 on page 34. We have the factorization

$$(2 + xi)(2 - xi) = y^3,$$

in F. We first show that $\gcd(2 + xi, 2 - xi) = 1$ in the case where x is odd. Any common divisor $a + bi$ in F must satisfy the property that

$$(a^2 + b^2) \mid N_F(4) = 16 = N_F(2 + ix + 2 - ix),$$

and

$$(a^2 + b^2)|4x^2 = N_F(2 + xi - (2 - xi)),$$

so $(a^2 + b^2)|4$. Thus, $a, b \in \{\pm 2, \pm 1, 0\}$. By part (e) of Exercise 1.8 on page 28,

$$(a^2 + b^2) \mid (x^2 + 4),$$

which is odd, so one of a or b must be 0. In other words, the only common divisors are units, so $2 + ix$ and $2 - ix$ are relatively prime. Thus, by unique factorization in $\mathbb{Z}[i]$ ensured by Theorem 1.15 and Theorem 1.17 on page 39,

$$2 + ix = (a + bi)^3 \tag{1.50}$$

for some $a, b \in \mathbb{Z}$. Therefore,

$$2 - ix = (a - ib)^3. \tag{1.51}$$

(Note that although uniqueness is up to units and associates, assume without loss of generality that (1.50)–(1.51) hold since $u(a + bi)^3$, where $u \in \{\pm 1, \pm i\}$, for instance, may be written as a cube in $\mathbb{Z}[i]$ since all units are cubes therein.) Adding (1.50)–(1.51) yields

$$4 = 2a(a^2 - 3b^2),$$

so $a|2$ forcing $a = \pm 1, \pm 2$. Of these, only

$$(a, b) \in \{(-1, \pm 1), (2, \pm 1)\}$$

are possible. Hence,

$$y^3 = ((a + bi)(a - bi))^3 = (a^2 + b^2)^3,$$

where $y = a^2 + b^2 \in \{2, 5\}$. Therefore, since x is odd, $x^2 + 4 = 125$, with $x = \pm 11$. Thus, the solution $(x, y) = (\pm 11, 5)$ is achieved. Now we assume that x is even. Set $x = 2X$ and $y = 2Y$. Then

$$X^2 + 1 = 2Y^3,$$

where X must be odd. In other words, for odd X,

$$(1 - Xi)(1 + Xi) = (1 + i)(1 - i)Y^3.$$

Since $\gcd(1 + iX, 1 - iX) = 1 - i$, then by unique factorization,

$$1 + iX = (1 + i)(a + bi)^3,$$

for some $a + bi \in \mathbb{Z}[i]$. By comparing the constant terms,

$$1 = a^3 - 3a^2b - 3ab^2 + b^3 = (a + b)(a^2 - 4ab + b^2),$$

from which it follows that $a + b = \pm 1$, and $a^2 - 4ab + b^2 = \pm 1$. Therefore, one of a or b is zero and the other is ± 1. Hence, $X = \pm 1$, and $x = \pm 2$, so $y = 2$. $\square$

We will look at Bachet's equation again in §8.3 once we have even more tools to tackle more general solutions to (1.49).

Remark 1.19 Now we explore some of the real quadratic fields which are norm-Euclidean, and so UFDs—see Definition 1.18 on page 34 and Theorem 1.17 on page 39. Unlike the case with complex quadratic fields, there are real quadratic fields that are Euclidean but not norm-Euclidean. For instance, see [15] where it is shown that $\mathfrak{O}_F$ for $F = \mathbb{Q}(\sqrt{69})$ is Euclidean but not norm-Euclidean.

The history of the resolution towards the complete list of real quadratic norm-Euclidean fields is due to the efforts of many researchers. In 1938, H. Heilbronn proved in [41] that there are only finitely many such fields–see Biography 1.3. That list was eventually determined to be

$$D \in \{2, 3, 5, 6, 7, 11, 13, 17, 19, 21, 29, 33, 37, 41, 57, 73\}.$$

This was due to the efforts of O. Perron [77], R. Remak [80], and N. Hofreiter [43], among others–see Biographies 1.4 and 1.5 on page 53 for instance. The final step was accomplished by H. Chatland and H. Davenport [14] in 1950–see Biography 1.6 on page 54. We will not give the full result here since it involves the geometry of numbers. The following partial result was proved by A. Oppenheim in 1934 — see [76].

Biography 1.3 Hans Arnold Heilbronn (1908–1975) *was born in Berlin, Germany on October 8, 1908. He entered the University of Berlin in 1926, but eventually moved to Göttingen, where he began to study number theory under the direction of Edmund Landau. He obtained his degree in 1933, when Hitler came to power. Heilbronn, who was Jewish, fled to England after being dismissed from his position at Göttingen. Eventually he was offered a position at the University of Bristol where he published, arguably, his most famous result, coauthored with Linfoot, on a conjecture of Gauss concerning complex quadratic fields of class number equal to 1, showing that there are at most ten of them. A short while thereafter he was offered the Brevan Fellowship in Trinity College, Cambridge, in May 1935. There he began his most long-standing collaboration with Davenport that lasted until Davenport died in 1969. For his lifetime achievements, he was elected as a Fellow of the Royal Society in 1951. In 1964, he moved to North America, and after a brief stay in the U.S.A., he moved to the University of Toronto in Canada becoming a Canadian citizen in 1970, and a member of the Royal Society of Canada in 1971. A heart attack in November of 1973 eventually led to complications and he died while undergoing an operation to fit a pacemaker on April 28, 1975.*

Theorem 1.21 | **Some Norm-Euclidean Real Quadratic Fields**

If $F = \mathbb{Q}(\sqrt{D})$ *is a quadratic number field and*

$$D \in \{2, 3, 5, 6, 7, 13, 17, 21, 29\},$$

then $\mathfrak{O}_F$ *is norm-Euclidean.*

Proof. First set

$$\epsilon = \begin{cases} 2 & \text{if } D \equiv 1 \,(\mathrm{mod}\ 4), \\ 1 & \text{if } D \equiv 2, 3 \,(\mathrm{mod}\ 4), \end{cases}$$

and observe that any

$$\sigma = r_1 + s\sqrt{D} \in F, \ r_1, s \in \mathbb{Q},$$

may be assumed without loss of generality to be in the form

$$\sigma = r_1 + (r_2/\epsilon)\sqrt{D},$$

since we may write $s = 2s/\epsilon = r_2/\epsilon$ when $D \equiv 1 \,(\mathrm{mod}\ 4)$.

Based upon Exercise 1.46 on page 45, we need to establish that for any

$$\rho = r_1 + (r_2/\epsilon)\sqrt{D} \in F, \text{ for } r_1, r_2 \in \mathbb{Q},$$

there exists a

$$\sigma = (x + y\sqrt{D})/\epsilon \in \mathfrak{O}_F, \text{ where } x, y \in \mathbb{Z}$$

(by Theorem 1.3 on page 6) such that

$$|N_F(\rho - \sigma)| = |(r_1 - x/\epsilon)^2 - (r_2 - y)^2 D/\epsilon^2| < 1. \tag{1.52}$$

We assume that Equation (1.52) fails for some $r_1, r_2 \in \mathbb{Q}$ and all $x, y \in \mathbb{Z}$. Then we show that for $D \leq 8\varepsilon^2$, the only values for which (1.52) does not fail are the ones on our list.

Claim 1.8 *We may assume without loss of generality that $0 \leq r_j \leq 1/2$, for $j = 1, 2$.*

First, for $j = 1, 2$ we set

$$z_j = \begin{cases} \lfloor r_j \rfloor & \text{if } 0 \leq r_j - \lfloor r_j \rfloor \leq 1/2, \\ \lfloor r_j \rfloor + 1 & \text{if } 1 > r_j - \lfloor r_j \rfloor > 1/2, \end{cases}$$

where $\lfloor r_j \rfloor$ is the floor of r_j, or greatest integer less than or equal to r_j. Since we are assuming that Equation (1.52) fails for all $x, y \in \mathbb{Z}$, then in particular it will fail for

$$x = \epsilon z_1 + \delta_1 x_1, \text{ and } y = z_2 + \delta_2 y_1,$$

for any integers x_1, y_1, where $\delta_j = 1$ if $z_j = \lfloor r_j \rfloor$ and $\delta_j = -1$ otherwise for $j = 1, 2$. Thus,

$$|(r_1 - x/\epsilon)^2 - (r_2 - y)^2 D/\epsilon^2|$$

becomes

$$|(s_1 - x_1/\epsilon)^2 - (s_2 - y_1)^2 D/\epsilon^2|,$$

for any $x_1, y_1 \in \mathbb{Z}$, where $0 \leq s_j = |r_j - z_j| \leq 1/2$, $j = 1, 2$. This completes the proof of Claim 1.8.

By Claim 1.8, for all $x, y \in \mathbb{Z}$, one of the following inequalities must hold for some $0 \leq r_j \leq 1/2$, $j = 1, 2$,

$$(r_1 - x/\epsilon)^2 \geq 1 + (r_2 - y)^2 D/\epsilon^2, \tag{1.53}$$

or

$$(r_2 - y)^2 D/\epsilon^2 \geq 1 + (r_1 - x/\epsilon)^2. \tag{1.54}$$

If $x = y = 0$, and (1.53) holds, then

$$\frac{1}{4} \geq r_1^2 \geq 1 + \frac{r_2^2 D}{\varepsilon^2} \geq 1,$$

a contradiction, so (1.54) must hold, namely

$$\frac{r_2^2 D}{\varepsilon^2} \geq 1 + r_1^2. \tag{1.55}$$

If $x = \varepsilon$, $y = 0$, and (1.53) holds, then $1 \geq (r_1 - 1)^2 \geq 1 + r_2^2 D/\varepsilon^2 \geq 1$, a contradiction, unless $r_1 = r_2 = 0$, which contradicts (1.55), so (1.54) must hold, namely

$$\frac{r_2^2 D}{\varepsilon^2} \geq 1 + (r_1 - 1)^2. \tag{1.56}$$

If $x = -\varepsilon$, $y = 0$, and (1.53) holds, then

$$(r_1 + 1)^2 \geq 1 + \frac{r_2^2 D}{\varepsilon^2} \geq 2 + (r_1 - 1)^2, \tag{1.57}$$

which implies that $r_1 \geq 1/2$, which in turn forces $r_1 = 1/2$ by Claim 1.8. Plugging this into (1.57), we get

$$\frac{9}{4} = (r_1 + 1)^2 \geq 1 + \frac{r_2^2 D}{\varepsilon^2} \geq 2 + (r_1 - 1)^2 = \frac{9}{4},$$

which forces

$$1 + \frac{r_2^2 D}{\varepsilon^2} = \frac{9}{4}.$$

Thus, $4r_2^2 D = 5\varepsilon^2$, so $\varepsilon = 2$, which forces $r_2 = 1$ contradicting Claim 1.8. We have shown that if $x = -\varepsilon$, $y = 0$, then (1.54) must hold, namely,

$$\frac{r_2^2 D}{\varepsilon^2} \geq 1 + (r_1 + 1)^2,$$

and by Claim 1.8, this implies that

$$\frac{D}{4\varepsilon^2} \geq \frac{r_2^2 D}{\varepsilon^2} \geq 2,$$

whence,

$$D \geq 8\varepsilon^2. \tag{1.58}$$

In view of (1.58), if $D < 8\varepsilon^2$, then $\mathfrak{O}_F$ is norm-Euclidean. If $\varepsilon = 2$, then $D < 32$, for which we get the values $D = 5, 13, 17, 21, 29$ and if $\varepsilon = 1$, then $D < 8$ for which we get $D = 2, 3, 6, 7$. This yields the values of D listed in the statement of the theorem. $\qquad\square$

Biography 1.4 Oskar Perron (1880–1975) *was born in Frankenthal, Pfalz, Germany on May 7, 1880. He studied at several universities including Göttingen. One of his best-known texts is on continued fractions, entitled* Die Lehre von den Kettenbrüchen, *published in 1913 with revisions in 1950 and 1954. In 1914 he was appointed as ordinary professor in Heidelberg. After World War I, he was appointed a chair at Munich where he continued to teach, even beyond retirement, until 1960. He not only contributed to number theory, but also to differential equations, matrices, and geometry. He published over 200 papers and books including his text on non-Euclidean geometry published when he was 82. He died on February 22, 1975 in Munich.*

Biography 1.5 Robert Remak (1888–1942) *studied for his doctorate at the University of Berlin under Frobenius. He received his degree in 1911 and this important work, which has his name attached to it, along with Weddeburn, Schmidt, and Krull, was on the decomposition of finite groups into products of irreducible factors. He taught at the University of Berlin until 1933 when Hitler's new laws got him dismissed. He was arrested in 1938 and sent to a concentration camp near Berlin. After eight weeks there he was released and his wife arranged for him to go to Amsterdam. There in 1942, he was arrested and sent to Auschwitz, Poland where he died in that year. He made important contributions to algebraic number theory and the geometry of numbers during his life.*

Exercises

1.55. Show that the only rational integer solutions of

$$y^2 = x^3 - 1$$

are $x = 1$ and $y = 0$ using unique factorization in $\mathbb{Z}[i]$.

In Exercises 1.56–1.63, assume that $F = \mathbb{Q}(\sqrt{D})$ is a quadratic number field where $\mathfrak{O}_F$ is a UFD.

1.56. Prove that any rational prime p is either a prime in $\mathfrak{O}_F$ or a product of two primes therein.

(*Hint: See Exercises 1.37–1.38 on page 31.*)

1.57. Prove that if α is a prime in $\mathfrak{O}_F$, then there is *exactly* one rational prime p such that $\alpha \mid p$.

1.58. Establish each of the following where p is an odd rational prime.

(a) $p \nmid D$ is a product of two primes α, β in $\mathfrak{O}_F$ if and only if the Legendre symbol $(\frac{D}{p}) = 1$. (See [68, §4.1, pp. 177–188]).

(b) If $p = \alpha\beta$, the product of two primes, then $\alpha \not\sim \beta$, but $\alpha \sim \beta'$, the latter being equivalent to $\alpha' \sim \beta$.

1.59. Prove that if α is a prime in $\mathfrak{O}_F$, but α is not a rational prime, then $|N_F(\alpha)| = p$ for some rational prime p.

1.60. Prove that if $D \equiv 3 \,(\mathrm{mod}\ 4)$, then $2 \sim \alpha^2$ where α is a prime in $\mathfrak{O}_F$.

1.61. If $D \equiv 5 \,(\mathrm{mod}\ 8)$, prove that 2 is a prime in $\mathfrak{O}_F$.

1.62. If $D \equiv 1 \,(\mathrm{mod}\ 8)$ show that 2 is the product of two nonassociated primes in $\mathfrak{O}_F$.

1.63. Prove that if $p \mid D$, then $p \sim \alpha^2$, where α is a prime in $\mathfrak{O}_F$.

1.64. By Theorem 1.21 on page 50,

$$\mathfrak{O}_F = \mathbb{Z}[(1 + \sqrt{21})/2]$$

is a UFD. With reference to Exercise 1.63, find a prime $\alpha \in \mathfrak{O}_F$ such that $3 \mid \alpha^2$, and find a $u \in \mathfrak{U}_F$ such that $3 = \alpha^2 u$.

Biography 1.6 Harold Davenport (1907–1969) *was born in Huncoat, Lancashire, England on October 30, 1907. He entered Manchester University in 1924. He graduated in 1927, then went to Trinity College, Cambridge. There he wrote his doctorate under the direction of Littlewood. His thesis topic was the distribution of quadratic residues by employing new methodology using character and exponential sums. In 1930, he won the Rayleigh prize and two years later was elected to a Trinity fellowship. Shortly thereafter he visited Hasse in Marburg, Germany where he also met Heilbronn, with whom he began a lengthy collaboration. In 1937, he accepted an offer from Mordell at the University of Manchester, where he interacted with Mahler, Erdös, and Segre. In 1940, he was elected as a member of the Royal Society, and won the Adams prize from the University of Cambridge. In the following year, he was appointed as chair of mathematics at the University College of North Wales in Bangor. In 1945, he moved to London, to succeed Jeffrey as Astor professor of mathematics in University College there. From 1957 to 1959, he was President of the London Mathematical Society and in the middle of this, in 1958, he returned to Cambridge as Rouse Ball professor of mathematics after Besicovitch retired. During his life he contributed to number theory including his work on Waring's problem where he showed that every sufficiently large natural number is the sum of sixteen fourth powers. He also wrote several texts which have become classics such as* The Higher Arithmetic *published in 1952 and many subsequent editions. Indeed his distinguished contribution to the theory of numbers was perhaps best honoured by his being awarded the Sylvestor Medal in 1967. He was a heavy smoker and finally succumbed to lung cancer on June 9, 1969 in Cambridge.*

Chapter 2

Ideals

> *It remains an old maxim of mine that when you have excluded the impossible, whatever remains, however improbable, must be the truth.*
> spoken by **Sherlock Holmes** in **The Adventure of Beryl Coronet**.
> **Sir Arthur Conan Doyle (1859–1930), Scottish-born writer of** detective fiction

2.1   The Arithmetic of Ideals in Quadratic Fields

We first mentioned the notion of an ideal on page 16 in reference to how we would need such a theory to delve deeper into some Diophantine analysis problems such as the generalized Ramanujan–Nagell equation. We also have some background in [68, Appendix A, pp. 303–305]. Now we have sufficient tools to introduce the concepts involved here.

Definition 2.1 | Ideals |

An *R-ideal* is a nonempty subset I of a commutative ring R with identity having the following properties.

(a) If $\alpha, \beta \in I$, then $\alpha + \beta \in I$.

(b) If $\alpha \in I$ and $r \in R$, then $r\alpha \in I$.

Remark 2.1 It is inductively clear that Definition 2.1 implies the following. If $\alpha_1, \alpha_2, \ldots, \alpha_n \in I$ for any $n \in \mathbb{N}$, then $r_1\alpha_1 + r_2\alpha_2 + \cdots + r_n\alpha_n \in I$ for any $r_1, r_2, \ldots, r_n \in R$. Moreover, if $1 \in I$, then $I = R$. Also, if we are given a set of elements $\{\alpha_1, \alpha_2, \ldots, \alpha_n\}$ in an integral domain R, then the set of all linear combinations of the α_j for $j = 1, 2, \ldots, n$

$$\left\{ \sum_{j=1}^{n} r_j \alpha_j : r_j \in R \text{ for } j = 1, 2, \ldots, n \right\}$$

is an ideal of R denoted by $(\alpha_1, \alpha_2, \ldots, \alpha_n)$. In particular, when $n = 1$, we have the following.

Definition 2.2 | **Principal and Proper Ideals**

If R is an integral domain and I is an R-ideal, then I is called a *principal* R-ideal if there exists an element $\alpha \in I$ such that $I = (\alpha)$, where α is called a *generator* of I. If $I \neq R$, then I is called a *proper* ideal.

Example 2.1 Let $n \in \mathbb{Z}$ and set

$$n\mathbb{Z} = \{nk : k \in \mathbb{Z}\},$$

which is an ideal in $\mathbb{Z}$ and $n\mathbb{Z} = (n) = (-n)$ is indeed a principal ideal. Moreover, it is a proper ideal for all $n \neq \pm 1$.

Example 2.1 is a segue to the question about how rings of integers behave in terms of intersection with $\mathbb{Z}$. This is answered in the next result which will also be valuable in §2.2–see Theorem 2.9 on page 73–but is also of interest in its own right since it employs minimal polynomials characterized in Theorem 1.6 on page 10.

Theorem 2.1 | **$\mathfrak{O}_F$-Ideals Intersecting $\mathbb{Z}$**

If F is a number field and I is a nonzero $\mathfrak{O}_F$-ideal, then $I \cap \mathbb{Z}$ contains a nonzero element of $\mathbb{Z}$.

Proof. Let $\alpha \in I$ where $\alpha \neq 0$ and consider

$$m_{\alpha, \mathbb{Q}}(x) = a_0 + a_1 x + \cdots + a_{d-1} x^{d-1} + x^d,$$

where $a_j \in \mathbb{Z}$ for all $j = 0, 1, \ldots, d-1$ by Corollary 1.4 on page 11. If $d = 1$, then $a_0 = -\alpha \neq 0$, and if $d > 1$, then $a_0 \neq 0$ since $m_{\alpha, \mathbb{Q}}(x)$ is irreducible in $\mathbb{Q}[x]$ by Corollary 1.4. Hence,

$$a_0 = -a_1 \alpha - \cdots - a_{d-1} \alpha^{d-1} - \alpha^d \in I,$$

as required.   □

For the following illustration and what follows, the reader has to be familiar with basic module theory. For those not so well versed or needing a reminder, see Exercise 2.1 on page 65.

Example 2.2 In $R = \mathbb{Z}[i]$, (2) and (3) are proper principal ideals. Moreover, the latter is an example of a special type of ideal that we now define and about which we will prove this assertion—see Example 2.3 on the following page.

Definition 2.3 | Prime Ideals |

If R is an integral domain, then a proper R-ideal $\mathcal{P}$ is called a *prime R-ideal* if it satisfies the property that whenever, $\alpha\beta \in \mathcal{P}$, for $\alpha, \beta \in R$, then either $\alpha \in \mathcal{P}$ or $\beta \in \mathcal{P}$.

In order to discuss any more features of ideal theory, we need to understand how multiplication of ideals comes into play.

Definition 2.4 | Products of ideals |

If R is an integral domain and I, J are R-ideals, then the product of I and J, denoted by IJ, is given by

$$IJ = \{r \in R : r = \sum_{j=1}^{n} \alpha_j \beta_j \text{ where } n \in \mathbb{N}, \text{and } \alpha_j \in I, \beta_j \in J \text{ for } 1 \le j \le n\}.$$

Theorem 2.2 | Criterion for Prime Ideals |

If R is an integral domain and I is a proper R-ideal, then I is a prime R-ideal if and only the following property is satisfied:

for any two R-ideals J, K, with $JK \subseteq I$, either $J \subseteq I$ or $K \subseteq I$.   (2.1)

Proof. Suppose that (2.1) holds. Then if $\alpha, \beta \in R$ such that $\alpha\beta \in I$, then certainly

$$(\alpha\beta) = (\alpha)(\beta) \subseteq I,$$

taking $J = (\alpha)$ and $K = (\beta)$ in (2.1), which therefore implies that $(\alpha) \subseteq I$ or $(\beta) \subseteq I$. Hence, $\alpha \in I$ or $\beta \in I$. We have shown that (2.1) implies I is prime.
    Conversely, suppose that I is a prime R-ideal. If (2.1) fails to hold, then there exist R-ideals J, K such that $JK \subseteq I$ but $K \not\subseteq I$ and $J \not\subseteq I$. Let $\alpha \in J$

with $\alpha \notin I$ and $\beta \in K$ with $\beta \notin I$, then $\alpha\beta \in I$ with neither of them being in I which contradicts Definition 2.3 on the previous page. Hence, (2.1) holds and the result is secured.   □

Now we prove a result that links the notion of prime element and prime ideal in the principal ideal case.

Theorem 2.3 $\boxed{\textbf{Principal Prime Ideals and Prime Elements}}$

If R is an integral domain and $\alpha \in R$ is a nonzero, nonunit element, then

$$(\alpha) \text{ is a prime } R\text{-ideal if and only if } \alpha \text{ is a prime in } R.$$

Proof. Suppose first that (α) is a prime R-ideal. Then for any $\beta, \gamma \in R$ such that $\alpha \mid \beta\gamma$,

$$\beta\gamma \in (\beta\gamma) \subseteq (\alpha).$$

Since (α) is a prime R-ideal, then $\beta \in (\alpha)$ or $\gamma \in (\alpha)$ by Definition 2.3 on the preceding page. In other words, $\alpha \mid \beta$ or $\alpha \mid \gamma$, namely α is a prime in R.

Conversely, suppose that α is prime in R. If $\beta, \gamma \in R$ such that $\beta\gamma \in (\alpha)$, then there exists an $r \in R$ with $\beta\gamma = \alpha r$. Since α is prime, then $\alpha \mid \beta$ or $\alpha \mid \gamma$. Suppose, without loss of generality, that $\alpha \mid \beta$. Thus, there is an $s \in R$ such that $\beta = \alpha s$, so $\beta \in (\alpha)$. We have shown that (α) is a prime R-ideal by Definition 2.3, which completes the proof.   □

Example 2.3 In Example 2.2 on the previous page, (2) and (3) were considered as principal ideals in the Gaussian integers. By Exercises 1.37–1.38 on page 31, 3 is a Gaussian prime, but 2 is not. Therefore, by Theorem 2.3, (3) is a prime ideal in the Gaussian integers but (2) is not.

Now that we may look at products of ideals, we may we look at the notion of division in ideals in order to link this with the element level and primes. Moreover, it will provide a segue for us to talk about explicit representation of ideal products in $\mathfrak{O}_F$ for quadratic fields F.

Definition 2.5 $\boxed{\textbf{Division of Ideals}}$

If R is an integral domain, then a nonzero R-ideal I is said to *divide an R-ideal* J if there is another R-ideal H such that $J = HI$.

The following shows that division of ideals implies containment.

Lemma 2.1 $\boxed{\textbf{To Divide is to Contain}}$

If R is an integral domain and I, J are R-ideals, with $I \mid J$, then $J \subseteq I$.

Proof. Since $I \mid J$, then by Definition 2.3, there is an R-ideal H such that $J = IH$. However, by Definition 2.1, $J = IH \subseteq I$.   □

Corollary 2.1 *Suppose that R is an integral domain and I is an R-ideal satisfying the property that whenever $I \mid JK$ for R-ideals J, K, we have $I \mid J$ or $I \mid K$. Then I is a prime R-ideal.*

Proof. Suppose that $I \mid JK$, then by Lemma 2.1, $JK \subseteq I$, and the property implies that either $J \subseteq I$ or $K \subseteq I$, so by Theorem 2.2 on page 57, I is a prime R-ideal.   □

Now we look at multiplication of ideals in quadratic fields. If the reader is in need of a reminder about the basics involved in modules and their transition to ideals in the rings of integers in quadratic fields, then see Exercises 2.1–2.4. In any case, see Exercise 2.8 on page 66.

Multiplication Formulas for Ideals in Quadratic Fields

Suppose that $F = \mathbb{Q}(\sqrt{D})$ is a quadratic number field, and $\mathfrak{O}_F$ is its ring of integers–see Theorem 1.3 on page 6. Let Δ_F be the field discriminant given in Definition 1.6 on page 7, and let

$$I_j = (a_j, (b_j + \sqrt{\Delta_F})/2),$$

for $j = 1, 2$ be $\mathfrak{O}_F$-ideals, then

$$I_1 I_2 = (g) \left(a_3, \frac{b_3 + \sqrt{\Delta_F}}{2} \right),$$

where

$$a_3 = \frac{a_1 a_2}{g^2} \text{ with } g = \gcd\left(a_1, a_2, \frac{b_1 + b_2}{2} \right),$$

and

$$b_3 \equiv \frac{1}{g} \left(\delta a_2 b_1 + \mu a_1 b_2 + \frac{\nu}{2} (b_1 b_2 + \Delta_F) \right) \pmod{2a_3},$$

where $\delta, \mu,$ and ν are determined by

$$\delta a_2 + \mu a_1 + \frac{\nu}{2}(b_1 + b_2) = g.$$

Note the above formulas are intended for our context, namely the ring of integers of a quadratic field $\mathfrak{O}_F$, called the *maximal order*. In an order contained in $\mathfrak{O}_F$ that is not maximal, the above does not work unless we restrict to *invertible* ideals. For the details on, and background for, orders in general, see either

[62, §1.5] or [65, §3.5]. Also, see Definition 2.14 on page 76 and Exercise 2.18 on page 86.

Example 2.4 Consider $\Delta_F = 40$, with

$$I_1 = (3, 1 + \sqrt{10}) \text{ and } I_2 = (3, -1 + \sqrt{10}),$$

so in the notation of the above description of formulas for multiplication, we have

$$a_1 = a_2 = 3, b_1 = 2 = -b_2, g = 3, \delta = 0 = \nu, \mu = 1, b_3 = 1 \text{ , and } a_3 = 1,$$

so

$$I_1 I_2 = (3, 1 + \sqrt{10})(3, -1 + \sqrt{10}) = (3). \tag{2.2}$$

Hence, the product of I_1 and I_2 is the principal ideal (3) in $\mathbb{Z}[\sqrt{10}] = \mathfrak{O}_F$, and by Theorem 2.3 on page 58, (3) is not a prime ideal in $\mathfrak{O}_F$ since (3) divides $I_1 I_2$ but does not divide either factor. To see this, note that if

$$(3) \mid (3, \pm 1 + \sqrt{10}),$$

then by Lemma 2.1 on page 58,

$$(3, \pm 1 + \sqrt{10}) \subseteq (3),$$

which is impossible since it is easy to show that $\pm 1 + \sqrt{10} \notin (3)$. Moreover, by Exercise 2.7 on page 66, I_1 and I_2 *are* prime $\mathfrak{O}_F$-deals.

Example 2.4 motivates a study of prime decomposition of ideals in quadratic fields. For instance, (2.2) is the decomposition of the ideal (3) in $\mathbb{Z}[\sqrt{10}] = \mathfrak{O}_F$ for $F = \mathbb{Q}(\sqrt{10})$ into the product of the two prime ideals I_1 and I_2. In what follows, we have a complete description. The notation $(*/p)$ in the following denotes the Legendre symbol–see [68, §4.1].

Theorem 2.4 $\boxed{\textbf{Prime Decomposition in Quadratic Fields}}$

If $\mathfrak{O}_F$ is the ring of integers of a quadratic field $F = \mathbb{Q}(\sqrt{D})$, and $p \in \mathbb{Z}$ is prime, then the following holds, where $\mathcal{P}_1, \mathcal{P}_2$, and $\mathcal{P}$ are distinct prime $\mathfrak{O}_F$-ideals with norm p–see Exercise 2.7.

$$(p) = p\mathfrak{O}_F = \begin{cases} \mathcal{P}_1 \mathcal{P}_2 & \text{if } p > 2, (D/p) = 1, \text{ or } p = 2, D \equiv 1 \pmod{8}, \\ \mathcal{P} & \text{if } p > 2, (D/p) = -1, \text{ or } p = 2, D \equiv 5 \pmod{8}, \\ \mathcal{P}^2 & \text{if } p > 2, p \mid D, \text{ or } p = 2, D \equiv 2, 3 \pmod{4}. \end{cases}$$

Proof. For the sake of simplicity of elucidation in the following Cases 2.1–2.3, we present only the instance where $\mathfrak{D}_F = \mathbb{Z}[\sqrt{D}]$ since the proof for $\mathfrak{D}_F = \mathbb{Z}[(1 + \sqrt{D})/2]$ is similar.

Case 2.1 $(D/p) = 1$ *for* $p > 2$.

The Legendre symbol equality tells us that there exists a $b \in \mathbb{Z}$ such that $b^2 \equiv D \, (\mathrm{mod}\ p)$. Also, since $p \nmid D$, then $p \nmid b$. Let

$$\mathcal{P}_1 = (p, b + \sqrt{D}) \text{ and } \mathcal{P}_2 = (p, -b + \sqrt{D}).$$

If $\mathcal{P}_1 = \mathcal{P}_2$, then

$$2b = b + \sqrt{D} - (-b + \sqrt{D}) \in \mathcal{P}_1,$$

so $p \mid 2b$ by the minimality of p as demonstrated in Exercises 2.2–2.4, namely

$$2b \in \mathcal{P}_1 \cap \mathbb{Z} = (p).$$

Thus, $\mathcal{P}_1$ and $\mathcal{P}_2$ are distinct $\mathfrak{D}_F$-ideals. By the multiplication formulas given on page 59, we have, in the notation of those formulas, $a_3 = 1$ and $g = p$, so

$$\mathcal{P}_1 \mathcal{P}_2 = (p).$$

Case 2.2 $(D/p) = -1$ *for* $p > 2$.

Let $\alpha\beta \in (p)$, where

$$\alpha = a_1 + b_1\sqrt{D}, \beta = a_2 + b_2\sqrt{D} \in \mathbb{Z}[\sqrt{D}].$$

Suppose that $\beta \notin (p)$. We have

$$\alpha\beta = a_1 a_2 + b_1 b_2 D + (a_2 b_1 + a_1 b_2)\sqrt{D} = p(x + y\sqrt{D}),$$

for some $x, y \in \mathbb{Z}$. Therefore,

$$a_1 a_2 + b_1 b_2 D = px, \tag{2.3}$$

and

$$a_2 b_1 + a_1 b_2 = py. \tag{2.4}$$

If $b_1 = 0$, then by (2.3), $p \mid a_1 a_2$. If $p \mid a_1$, then $\alpha = a_1 \in (p)$, so by Definition 2.3 on page 57, (p) is an $\mathfrak{D}_F$-prime ideal. If $p \mid a_2$, then $p \nmid b_2$ since $\beta \notin (p)$, so by (2.4) $p \mid a_1$ and we again have that $\alpha \in (p)$. Hence, we may assume that $b_1 \neq 0$. Similarly, we may assume that $a_1 \neq 0$.

Multiplying (2.4) by a_1 and subtracting b_1 times (2.3), we get

$$b_2(a_1^2 - b_1^2 D) = p(a_1 y - b_1 x).$$

If $p \mid (a_1^2 - b_1^2 D)$, then there exists a $z \in \mathbb{Z}$ such that $a_1^2 - b_1^2 D = pz$. Therefore,

$$-1 = \left(\frac{D}{p}\right) = \left(\frac{b_1^2 D}{p}\right) = \left(\frac{a_1^2 - pz}{p}\right) = \left(\frac{a_1^2}{p}\right) = 1,$$

a contradiction. Hence, $p \mid b_2$. By (2.4), this means that $p \mid a_2 b_1$. If $p \mid a_2$, then $p \mid (a_2 + b_2\sqrt{D})$, so $\beta \in (p)$, a contradiction to our initial assumption. Thus, $p \mid b_1$, so $p \mid (a_1 + b_1\sqrt{D})$, which means that $\alpha \in (p)$.

Case 2.3 $p > 2$ *and* $p \mid D$.

Let $\mathcal{P} = (p, \sqrt{D})$. Then by the multiplication formulas on page 59, with $a_3 = 1$ and $g = p$ in the notation there, $\mathcal{P}^2 = (p)$. This completes Case 2.3.

It remains to consider the three cases for $p = 2$.

Case 2.4 $p = 2$ *and* $D \equiv 1 \, (\mathrm{mod} \, 8)$.

Let
$$\mathcal{P} = \left(2, (1 + \sqrt{D})/2\right) \text{ and } \mathcal{P}_2 = \left(2, (-1 + \sqrt{D})/2\right).$$
Then by the multiplication formulas as used above with $a_3 = 1$ and $g = 2$, we have
$$\mathcal{P}_1 \mathcal{P}_2 = (2).$$
If $\mathcal{P}_1 = \mathcal{P}_2$, then $(1 + \sqrt{D})/2 + (-1 + \sqrt{D})/2 = \sqrt{D} \in \mathcal{P}_1$, which is not possible. Thus, $\mathcal{P}_1$ and $\mathcal{P}_2$ are distinct. This is Case 2.4.

Case 2.5 $p = 2$ *and* $D \equiv 5 \, (\mathrm{mod} \, 8)$.

Let $\alpha\beta \in (2)$, where
$$\alpha = (a_1 + b_1\sqrt{D})/2, \beta = (a_2 + b_2\sqrt{D})/2 \in \mathbb{Z}[(1 + \sqrt{D})/2],$$
with a_j and b_j of the same parity for $j = 1, 2$. Suppose that $\beta \notin (2)$. We have
$$\alpha\beta = \frac{a_1a_2 + b_1b_2D + (a_2b_1 + a_1b_2)\sqrt{D}}{4} = 2\left(\frac{x + y\sqrt{D}}{2}\right) = x + y\sqrt{D},$$
where $x, y \in \mathbb{Z}$ are of the same parity. Thus,
$$a_1a_2 + b_1b_2D = 4x, \tag{2.5}$$
and
$$a_2b_1 + a_1b_2 = 4y. \tag{2.6}$$
Multiplying (2.6) by a_1 and subtracting b_1 times (2.5), we get
$$b_2(a_1^2 - b_1^2D) = 4(ya_1 - xb_1).$$
If $a_1^2 - b_1^2D$ is even then either a_1 and b_1 are both odd or both even. In the former case, $1 \equiv a_1^2 \equiv b_1^2D \equiv 5 \, (\mathrm{mod} \, 8)$, a contradiction, so they are both even. Hence,
$$\alpha = 2\left(\frac{a_1/2 + (b_1/2)\sqrt{D}}{2}\right) \in (2),$$
so (2) is a prime $\mathfrak{O}_F$-ideal by Definition 2.3. If b_2 is even, then by (2.6), $2 \mid a_2b_1$. If $2 \mid a_2$, then
$$\beta = 2\left(\frac{a_2/2 + (b_2/2)\sqrt{D}}{2}\right) \in (2),$$

contradicting our initial assumption. Hence, b_1 is even and so a_1 is even since they must be of the same parity. As above, this implies that $\alpha \in (2)$. Thus, (2) is prime. This completes Case 2.5.

Case 2.6 $p = 2$ *and* $D \equiv 2 \,(\mathrm{mod}\ 4)$.

Let $\mathcal{P} = (2, \sqrt{D})$, which is an $\mathfrak{O}_F$-ideal by Exercise 2.7 on page 66. Moreover, $\mathcal{P}^2 = (2)$, by the multiplication formulas on page 59 with $a_3 = 1$ and $g = 2$.

Case 2.7 $p = 2$ *and* $D \equiv 3 \,(\mathrm{mod}\ 4)$.

Let $\mathcal{P} = (2, 1 + \sqrt{D})$, which is an $\mathfrak{O}_F$-ideal by Exercise 2.7. Moreover, as in Case 2.6, $\mathcal{P}^2 = (2)$.
This completes all cases. $\qquad\qquad\qquad\qquad\qquad\qquad\qquad\qquad\qquad\qquad\qquad$ □

Remark 2.2 Although we have not developed the full theory for ideals in general number fields, we will be able to talk about decomposition of ideals in quadratic fields. The following terminology will be suited to the more general case so we introduce it here—see [64]. Suppose that $F = \mathbb{Q}(\sqrt{D})$ is a quadratic number field, Δ_F is given as in Definition 1.6 on page 7, and (Δ_F/p) denotes the Kronecker symbol–see [68, pp. 199–200]. If $p \in \mathbb{Z}$ is a prime, then

$$(p) \text{ is said to } split \text{ in } F \text{ if and only if } \left(\frac{\Delta_F}{p}\right) = 1,$$

$$(p) \text{ is said to } ramify \text{ in } F \text{ if and only if } \left(\frac{\Delta_F}{p}\right) = 0,$$

and

$$(p) \text{ is said to be } inert \text{ in } F \text{ if and only if } \left(\frac{\Delta_F}{p}\right) = -1.$$

Note, as well, that from the proof of Theorem 2.4, when $(p) = \mathcal{P}_1 \mathcal{P}_2$, namely when (p) splits, then $\mathcal{P}_2$ is the *conjugate* of $\mathcal{P}_1$. This means that if

$$\mathcal{P}_1 = (p, b + \sqrt{D}),$$

then

$$\mathcal{P}_2 = (p, -b + \sqrt{D}).$$

Example 2.5 In Example 2.4 on page 60, with $\Delta_F = 40$, we saw that

$$(3) = I_1 I_2 = (3, 1 + \sqrt{10})(3, -1 + \sqrt{10}),$$

where

$$\left(\frac{\Delta_F}{3}\right) = \left(\frac{40}{3}\right) = 1,$$

so (3) splits in $\mathbb{Q}(\sqrt{10})$ into the two prime $\mathbb{Z}[\sqrt{10}]$-ideals I_1 and I_2.

In Examples 2.2 on page 57 and 2.3 on page 58, we saw that (2) is not a prime ideal in $\mathbb{Z}[i]$ and that (3) is a prime $\mathbb{Z}[i]$-ideal. Since

$$(2) = (1+i)^2,$$

where

$$\mathcal{P} = (2, 1+i) = (1+i) = (2, 1-i) = (1-i)$$

is a prime $\mathbb{Z}[i]$-ideal, then (2) is ramified in $F = \mathbb{Q}(i)$, where

$$\left(\frac{\Delta_F}{2}\right) = \left(\frac{-4}{2}\right) = 0.$$

Also, (3) is a prime ideal and we see that

$$\left(\frac{\Delta_F}{3}\right) = \left(\frac{-4}{3}\right) = -1,$$

so (3) is inert in F.

The following illustration shows that the converse of Lemma 2.1 on page 58 does not hold in general and that the multiplication formulas, on page 59, do not necessarily hold if we do not have the ring of integers of a quadratic field in which to work.

Example 2.6 If $R = \mathbb{Z}[\sqrt{5}]$, then

$$I = (2, 1+\sqrt{5})$$

is an R-ideal by Exercise 2.3 on page 66, and clearly $(2) = (2, 2\sqrt{5}) \subseteq I$. If $I \mid (2)$, then there exists an R-ideal J such that $(2) = IJ$. Thus, J has a representation

$$J = (a, b + c\sqrt{D})$$

with $a, c \in \mathbb{N}$, $b \in \mathbb{Z}$, $0 \le b < a$, such that $c \mid a$, $c \mid b$ and $ac \mid (b^2 - c^2 D)$. Moreover, $J \mid (2)$, so by Lemma 2.1, $(2) \subseteq J$, so there exist $x, y \in \mathbb{Z}$ such that

$$2 = ax + (b + c\sqrt{D})y.$$

Therefore, $y = 0$ and $a \mid 2$. If $a = 1$, then $I = (2)$, which means that

$$1 + \sqrt{5} \in (2),$$

a contradiction, so $a = 2$. If $b = 1$, then $c = 1$, so

$$I^2 = (2). \qquad (2.7)$$

However, by considering the multiplication of basis elements for I we see that

$$I^2 = (4, 2(1 + \sqrt{5}), 6 + 2\sqrt{5}) = (4, 2(1 + \sqrt{5})),$$

where the last equality follows since $6 + 2\sqrt{5}$ is a linear combination of the other basis elements so is redundant. Thus,

$$I^2 = (4, 2(1 + \sqrt{5})) = (2)(2, 1 + \sqrt{5}) = (2)I,$$

and combining this with *(2.7)*, we get $(2) = (2)I$, which implies

$$2(1 + \sqrt{5}) \in (2),$$

again a contradiction. We have shown both that although $(2) \subseteq I$, I does not divide (2), and that the multiplication formulas for ideals in R fail to hold. Note, that R is not the ring of integers of a quadratic field by Theorem 1.3 on page 6. (R *is what is known as an* order *in* $\mathfrak{O}_F = \mathbb{Z}[(1+\sqrt{5})/2]$ *for* $F = \mathbb{Q}(\sqrt{5})$, *and* I *is an example of an ideal in* R *which is not* invertible *in* R–*see* [62, *Chapter 1, pp. 23–30*]. *In an integral domain* R, *an* invertible R-*ideal is one for which there is an* R-*ideal* J *such that* $IJ = R$. *It can be shown that all ideals in the ring of integers of a quadratic field are* invertible, *which is why the multiplication formulas work there since they fail only for ideals that are* not *invertible.*)
 There are rings of integers for which the converse of Lemma 2.1 holds, called *Dedekind domains*, the topic of §2.2.

Exercises

2.1. Suppose that G is an additive abelian group, and that R is a commutative ring with identity 1_R which satisfy each of the following axioms:

   (a) For each $r \in R$ and $g, h \in G$, $r(g + h) = (rg) + (rh)$.
   (b) For each $r, s \in R$ and $g \in G$, $(r + s)g = (rg) + (sg)$.
   (c) For each $r, s \in R$ and $g \in G$, $r(sg) = (rs)g$.
   (d) For each $g \in G$, $1_R \cdot g = g$.

   Then G is a (two-sided) *module* over R, or for our purposes, simply an R-module. Prove that (in general) being a $\mathbb{Z}$-module is equivalent to being an additive abelian group.

2.2. Let $R = \mathbb{Z}[\omega_D]$, $D \in \mathbb{Z}$ not a perfect square, and $\omega_D = (\sigma - 1 + \sqrt{D})/\sigma$, with $\sigma = 1$ if $D \not\equiv 1 \pmod 4$ and $\sigma = 2$ otherwise. Then every $\mathbb{Z}$-submodule of R has a representation in the form $I = [a, b + c\omega_D]$ where $a, c \in \mathbb{N}$ and $b \in \mathbb{Z}$ with $0 \le b < a$. Moreover, a is the smallest natural number in I and c is the smallest natural number such that $b + c\omega_D \in I$ for any $b \in \mathbb{Z}$. (Note that when $c = 1$, I is called *primitive*.)

2.3. With reference to Exercise 2.2, prove that $I = (a, b + c\omega_D)$ is an R-ideal if and only if $c \mid a$, $c \mid b$, and $(\sigma b + c(\sigma - 1))^2 \equiv c^2 D \,(\mathrm{mod}\; \sigma^2 ac)$. (Note that we use the square brackets for $\mathbb{Z}$-modules and the round brackets for ideals.)

2.4. With reference to Exercise 2.2, prove that the $\mathbb{Z}$-module $[a, b + c\omega_D]$ for $a, c \in \mathbb{N}$, $b \in \mathbb{Z}$, is an R-ideal $(a, b + c\omega_D)$ if and only if $c \mid a$, $c \mid b$, and $(\sigma b + c(\sigma - 1))^2 \equiv c^2 D \,(\mathrm{mod}\; \sigma^2 ac)$. (Here a is the smallest natural number in I, called the *norm* of I.)

2.5. Let $[\alpha, \beta] = \alpha\mathbb{Z} + \beta\mathbb{Z}$ and $[\gamma, \delta] = \gamma\mathbb{Z} + \delta\mathbb{Z}$ be two $\mathbb{Z}$-modules, where $\alpha, \beta, \gamma, \delta \in R$, where R is given in Exercise 2.2. Prove that $[\alpha, \beta] = [\gamma, \delta]$ if and only if

$$\begin{pmatrix} \alpha \\ \beta \end{pmatrix} = X \begin{pmatrix} \gamma \\ \delta \end{pmatrix},$$

where $X \in \mathrm{GL}(2, \mathbb{Z})$, which is the *general linear group* of 2×2-matrices with entries from $\mathbb{Z}$, namely, those 2×2-matrices A such that $\det(A) = \pm 1$, also called *unimodular* matrices. (*Note that, in general,* $\mathrm{GL}(n, \mathbb{Z})$ *is the general linear group of* $n \times n$ *matrices with entries from* $\mathbb{Z}$.)

2.6. With reference to Exercise 2.2, prove that if $\alpha \in R$, and $I = (a, \alpha)$ is an R-ideal, then $I = (a, na \pm \alpha)$ for any $n \in \mathbb{Z}$.

2.7. Let F be a quadratic number field and let $\mathcal{P} = (p, (b + \sqrt{\Delta_F})/2)$ be an $\mathfrak{O}_F$-ideal where $p \in \mathbb{N}$ is prime. Prove that $\mathcal{P}$ is a prime $\mathfrak{O}_F$-ideal.

2.8. Verify the multiplication formulas on page 59.

2.2 Dedekind Domains

> *I can't cut this steak, he confided*
> *To the waiter who simply recited,*
> *Your prime cut of course*
> *Is as tough as a horse*
> *Since you can't take a prime and divide it.*
> From **Mathematical Conversation Starters (2002)**—see [**22, p. 221**]
> **John dePillis, American mathematician at U.C. Riverside**

In §1.3 we discussed unique factorization of elements in integral domains and looked at applications thereof in §1.4. In §2.1 we introduced the notion of prime ideals, and so the question of unique factorization of ideals in integral domains naturally arises. In particular, at the end of §2.1, we talked about the validity of the converse of Lemma 2.1 on page 58 in certain domains which is the topic of this section. In order to discuss this topic, we must prepare the stage with some essential topics. First of all there are types of ideals which are core to the theory, and to which we were introduced in [68, Definition A.21, p. 305].

Definition 2.6 | Maximal Ideals |

In an integral domain R, an ideal M is called *maximal* if it satisfies the property that whenever $M \subseteq I \subseteq R$, for any R-ideal I, then either $I = R$ or $I = M$.

The next concept is necessary to prove our first result about maximal ideals. Note first that if I, J are R-ideals, then $I + J$ is necessarily an R-ideal since for any $r \in R$, $r(\alpha + \beta) \in I + J$ by Definition 2.1 on page 55. We formalize this in the following.

Definition 2.7 | Sums of Ideals are Ideals |

If R is a commutative ring with identity, and I, J are R-ideals, then

$$I + J = \{\alpha + \beta : \alpha \in I, \beta \in I\},$$

is an ideal in R.

We use the above to prove our first result that we need to link maximality with primality.

Theorem 2.5 | Quotients of Prime Ideals are Integral Domains |

If R is an integral domain, then an R-ideal $\mathcal{P}$ is prime if and only if $R/\mathcal{P}$ is an integral domain.

Proof. We note that $R/\mathcal{P}$ is a commutative ring with identity $1_R + \mathcal{P}$ and additive identity $0 + \mathcal{P}$. If $0 + \mathcal{P} = 1_R + \mathcal{P}$, then $\mathcal{P} = R$, contradicting that $\mathcal{P}$ is prime. If

$$(\alpha + \mathcal{P})(\beta + \mathcal{P}) = \mathcal{P},$$

then $\alpha\beta + \mathcal{P} = \mathcal{P}$, so $\alpha\beta \in \mathcal{P}$. Since $\mathcal{P}$ is prime, then either $\alpha \in \mathcal{P}$ or $\beta \in \mathcal{P}$. In other words, either $\alpha + \mathcal{P} = \mathcal{P}$ or $\beta + \mathcal{P} = \mathcal{P}$. We have shown that $R/\mathcal{P}$ has no zero divisors, namely that it is an integral domain.

Conversely, if $R/\mathcal{P}$ is an integral domain, then $1_R + \mathcal{P} \neq \mathcal{P}$, so $1_R \notin \mathcal{P}$, consequently $\mathcal{P} \neq R$. Since $R/\mathcal{P}$ has no zero divisors, then $\alpha\beta \in \mathcal{P}$ implies that

$$\alpha\beta + \mathcal{P} = \mathcal{P},$$

namely

$$(\alpha + \mathcal{P})(\beta + \mathcal{P}) = \mathcal{P}.$$

Thus, either $\alpha + \mathcal{P} = \mathcal{P}$ or $\beta + \mathcal{P} = \mathcal{P}$. In other words, either $\alpha \in \mathcal{P}$ or $\beta \in \mathcal{P}$, so $\mathcal{P}$ is a prime R-ideal. $\square$

Now we link prime ideals with maximal ones.

Theorem 2.6 | **Maximal ideals are Prime**

If R is an integral domain, then every nonzero maximal R-ideal is prime.

Proof. If $M \neq (0)$ is a maximal R-ideal, and $M \mid (\alpha)(\beta)$ for some $\alpha, \beta \in R$, with M dividing neither factor, then by Definition 2.7 on the preceding page, $M + (\alpha)$ and $M + (\beta)$ are R-ideals, both of which properly contain M, so $M \neq R$. Hence, by the maximality of M, we have,

$$M + (\alpha) = R = M + (\beta).$$

Therefore,

$$M \subset R = R^2 = (M + (\alpha))(M + (\beta)) \subseteq M^2 + (\alpha)M + (\beta)M + (\alpha)(\beta)M \subseteq M,$$

a contradiction. We have shown that either $M \mid (\alpha)$ or $M \mid (\beta)$. Therefore, by Corollary 2.1 on page 58, M is prime. $\square$

The next result tells us when an ideal is maximal with respect to quotients in general.

Theorem 2.7 | **Fields and Maximal ideals**

If R is an integral domain, then M is a maximal R-ideal if and only if R/M is a field.

Proof. First we need the following fact.

Claim 2.1 *R is a field if and only if the only ideals in R are (0) and R.*

If R is a field and $I \neq (0)$ is an R-ideal, then there exists a nonzero element $\alpha \in I$. However, since R is a field, then there exists an inverse $\alpha^{-1} \in R$ of α. By Definition 2.1 on page 55, $\alpha \alpha^{-1} = 1_R \in I$, so $I = R$.

Conversely, suppose that the only R-ideals are (0) and R. If $\alpha \in R$ is nonzero, let $(\alpha) = \alpha R = I$. By hypothesis, $I = R$. Thus, there exist a $\beta \in R$ such that $\beta \alpha = 1_R$, so α is a unit. However, α was chosen as an arbitrary nonzero element in R, so R is a field. This is Claim 2.1.

Suppose that R/M is a field for a given R-ideal M. If $M \subseteq I \subseteq R$ for an R-ideal I, then I/M is an ideal of R/M, so by Claim 2.1, $I/M = (0)$ or $I/M = R/M$. In other words, either $I = M$ or $R = M$, namely M is maximal.

Conversely, if M is maximal, then by Theorem 2.6, either $M = (0)$ or M is prime. If $M = (0)$, then $R/(0) \cong R$ is a field by Claim 2.1 given that (0) is maximal so R has no proper ideals. If M is prime, then by Theorem 2.5 on page 67, R/M is an integral domain. Thus, it remains to show that all nonzero elements of R/M have multiplicative inverses, namely that if $\alpha + M \neq M$, then $\alpha + M$ has a multiplicative inverse in R/M. Given $\alpha + M \neq M$, then $\alpha \notin M$. Thus, M is properly contained in the ideal $(\alpha) + M$. Hence, $(\alpha) + M = R$. In other words,

$$1_R = m + r\alpha$$

for some $m \in M$ and $r \in R$. Therefore, $1_R - r\alpha = m \in M$, so

$$1_R + M = r\alpha + M = (r + M)(\alpha + M),$$

namely $r + M$ is a multiplicative inverse of $\alpha + M$ in R/M, so R/M is a field.□

Example 2.7 If $R = \mathbb{Z}/n\mathbb{Z}$, where $n \in \mathbb{N}$, a ring we studied in [68, pp. 79 ff], then $\mathbb{Z}/n\mathbb{Z}$ is a field if and only if n is prime. Hence, $n\mathbb{Z}$ is a maximal ideal in $\mathbb{Z}$ if and only if n is prime–see [68, Theorem 2.2, p. 81].

Example 2.8 Let F be a field, $r \in F$ is a fixed nonzero element, and

$$I = \{f(x) \in F[x] : f(r) = 0\}.$$

We now demonstrate that I is a maximal ideal in $F[x]$. First, we show that I is indeed an ideal in $F[x]$. If $g(x) \in F[x]$, then for any $f(x) \in I$, $g(r)f(r) = 0$, so $g(x)f(x) \in I$, and clearly $f(r) + h(r) = 0$ whenever $f(x), h(x) \in I$, which shows that I is an $F[x]$-ideal. In fact, $I = \ker(\phi)$, where ϕ is the natural map

$$\phi : F[x] \mapsto F[x]/I,$$

so I is maximal and

$$F \cong F[x]/I$$

–see [68, Example A.8. p. 305].

Remark 2.3 A few comments on the notion of *finite generation* are in order before we proceed. By Definition 1.4 on page 2 and Claim 1.1 on page 3 in the proof of Theorem 1.1, we know that for any number field F, $\mathfrak{O}_F$ is finitely generated as a $\mathbb{Z}$-module. Thus, any $\mathfrak{O}_F$-ideal I will have a representation as

$$I = (\alpha_1, \alpha_2, \ldots, \alpha_d) \text{ with } \alpha_j \in \mathfrak{O}_F \text{ for } j = 1, 2, \ldots, d,$$

and we say that I is *finitely generated*. In the instance where $d = 1$, we are in the case of Definition 2.2 on page 56, namely a principal ideal.

We also need the following notion in order to complement Definition 1.8 on page 9.

Definition 2.8 | **Integral Closure**

If $R \subseteq S$ where R and S are integral domains, then R is said to be *integrally closed* in S if each element of S that is integral over R is actually in R.

Example 2.9 The integral domain $\mathbb{Z}$ is integrally closed in $\mathbb{Q}$, but not in $\mathbb{C}$ since $\sqrt{-1} \in \mathbb{C}$ is integral over $\mathbb{Z}$. However, $\mathbb{Q}$ is an instance of the following notion that is also of interest to us here.

Definition 2.9 | **Field of Quotients**

If D is an integral domain, then the field F consisting of all elements of the form $\alpha\beta^{-1}$ for $\alpha, \beta \in D$ with $\beta \neq 0$ is called the *field of quotients* or simply the *quotient field* of D.

Example 2.10 If F is any field, then the quotient field of the polynomial domain $F[x]$ is the field $F(x)$ of rational functions in x. Moreover, the quotient field of $\mathbb{Z}$ is $\mathbb{Q}$. Indeed, the following result shows that the quotient field of $\mathfrak{O}_F$ for *any* number field F is F.

Theorem 2.8 | **Quotient Fields of Number Rings**

If F is a number field, then the quotient field of $\mathfrak{O}_F$ is F.

Proof. Let

$$K = \{\alpha\beta^{-1} : \alpha, \beta \in \mathfrak{O}_F, \beta \neq 0\},$$

which is the quotient field of $\mathfrak{O}_F$. Suppose that $\gamma = \alpha\beta^{-1} \in K$. Since $\mathfrak{O}_F \subseteq F$, then $\gamma \in F$, so $K \subseteq F$. Now if $\gamma \in F$, then by Lemma 1.1 on page 9, $\gamma = \alpha/\ell$ where $\alpha \in \mathbb{A}$ and $\ell \in \mathbb{Z}$. However, since

$$\alpha = \gamma\ell \in F \cap \mathbb{A} = \mathfrak{O}_F$$

by Definition 1.5 on page 4, then $\alpha \in \mathfrak{O}_F \subseteq F$, so $K \subseteq F$. Hence, $K = F$.   □

Remark 2.4 It can easily be shown that if D is an integral domain and F is its field of quotients, then there is an isomorphic copy of D in F — just consider $D_1 = \{\alpha \cdot 1^{-1} = \alpha \cdot 1 : \alpha \in D\} \subseteq F$. We merely identify D_1 with D and consider D as a subdomain of F.

Now we are in a position to define the main topic of this section—see Biography 1.2 on page 46.

Definition 2.10 | **Dedekind Domains**

A *Dedekind Domain* is an integral domain R satisfying the following properties.

(A) Every ideal of R is finitely generated.

(B) Every nonzero prime R-ideal is maximal.

(C) R is integrally closed in its quotient field F.

Remark 2.5 Condition (C) says that if $\alpha/\beta \in F$ is the root of some monic polynomial over R, then $\alpha/\beta \in R$, namely $\beta \mid \alpha$ in R.

The following is crucial in the sequel.

Definition 2.11 | **Ascending Chain Condition (ACC)**

An integral domain R is said to satisfy the *ascending chain condition* (ACC) if every chain of R-ideals

$$I_1 \subseteq I_2 \subseteq \cdots I_n \subseteq \cdots$$

terminates, meaning that there is a $n_0 \in \mathbb{N}$ such that $I_n = I_{n_0}$ for all $n \geq n_0$.

Remark 2.6 An equivalent way of stating the ACC is to say that R does not possess an *infinite strictly* ascending chain of ideals.

The above is a segue to the following important notion that will carry us forward towards our goals–see Biography 2.1 on page 73.

Definition 2.12 | **Noetherian Domains**

An integral domain R possessing the ACC is called a *Noetherian Domain*.

Lemma 2.2 | **Finite Generation and Noetherian Domains**

If R is an integral domain, then R is a Noetherian Domain if and only if every R-ideal is finitely generated.

Proof. Suppose that every R-ideal is finitely generated. Let

$$I_1 \subseteq I_2 \subseteq \cdots \subseteq I_n \subseteq \cdots$$

be an ascending chain of ideals. It follows from Exercise 1.2 on page 16 that

$$I = \cup_{i=1}^{\infty} I_j$$

is an R-ideal, and since any R-ideal is finitely generated, then there exist $\alpha_j \in R$ for $j = 1, 2, \ldots, d \in \mathbb{N}$ such that

$$I = (\alpha_1, \alpha_2, \ldots, \alpha_d).$$

Therefore, for each $j = 1, 2, \ldots, d$, there exists a k_j such that $\alpha_j \in I_{k_j}$. Let

$$n = \max\{k_1, k_2, \ldots, k_d\}.$$

Then since $I_n \subseteq I$ and $I_{k_j} \subseteq I_n$ since $k_j \leq n$ for each such j, then

$$(\alpha_1, \alpha_2, \ldots, \alpha_d) \subseteq I_n,$$

which implies that $I \subseteq I_n$. Hence,

$$I_n = \cup_{i=1}^{\infty} I_j$$

and so $I_n = I_j$ for each $j \geq n$. Since the chain terminates, R satisfies the ACC so is a Noetherian domain.

Conversely, suppose that R is a Noetherian domain. If I is an R-ideal that is not finitely generated, then $I \neq (0)$, so there exists $\alpha_1 \in I$ with $\alpha_1 \neq 0$, and $(\alpha_1) \subset I$. Since $I \neq (\alpha_1)$, given that the former is not finitely generated, then there exists $\alpha_2 \in I$ and $\alpha_2 \notin (\alpha_1)$, so we have

$$(\alpha_1) \subset (\alpha_1, \alpha_2) \subset I.$$

Continuing inductively in this fashion, we get the strictly ascending chain of ideals,

$$(\alpha_1) \subset (\alpha_1, \alpha_2) \subset \cdots \subset (\alpha_1, \alpha_2, \ldots, \alpha_n) \subset \cdots \subset I,$$

which contradicts that R is a Noetherian domain. Hence, every R-ideal is finitely generated. $\square$

Corollary 2.2 *If F is a number field, then $\mathfrak{O}_F$ is a Noetherian domain.*

Proof. This follows from Remark 2.3 on page 70 and Lemma 2.2. $\square$

Corollary 2.3 *Let R be a Noetherian domain. Then every nonempty subset of R-ideals contains a maximal element.*

Proof. Let $\mathfrak{T}$ be the set of ideals with the property that for every ideal I of $\mathfrak{T}$, there exists an ideal J of $\mathfrak{T}$ with $I \subset J$. If $\mathfrak{T} \neq \varnothing$, then by its definition we may construct an infinite strictly ascending chain of ideals in $\mathfrak{T}$, contradicting Lemma 2.2 on the preceding page. This is the result. $\square$

Immediate from Corollary 2.3 is the following result.

Corollary 2.4 *In a Noetherian domain R, every proper R-ideal is contained in a maximal R-ideal.*

Remark 2.7 Given Lemma 2.2, Condition (A) of Definition 2.10 may be replaced by the condition that R is a Noetherian domain.

Biography 2.1 Emmy Amalie Noether (1882–1935) *was born in Erlangen, Bavaria, Germany on March 23, 1882. She studied there in her early years and, in 1900, received certification to teach English and French in Bavarian girls' schools. However, she chose a more difficult route, for a woman of that time, namely to study mathematics at university. Women were required to get permission to attend a given course by the professor teaching it. She did this at the University of Erlangen from 1900 to 1902, and passed her matriculation examination in Nürnberg in 1903, after which she attended courses at the University of Göttingen from 1903 to 1904. By 1907, she was granted a doctorate from the University of Erlangen. By 1909, her published works gained her enough notoriety to receive an invitation to become a member of the* Deutsche Mathematiker-Vereiningung, *and in 1915, she was invited back to Göttingen by Hilbert and Klein. However, it took until 1919 for the university to, grudgingly, obtain her habilitation, and permit her to be on the faculty. In that year she proved a result in theoretical physics, now known as* Noether's Theorem, *praised by Albert Einstein as a penetrating result, which laid the foundations for many aspects of his general theory of relativity. After this, she worked in ideal theory, developing ring theory to be of core value in modern algebra. Her work* Idealtheorie in Ringbereichen, *published in 1921, helped cement this value. In 1924, B.L. van der Waerden published his work* Moderne Algebra, *the second volume of which largely consists of Noether's results. Her most successful collaboration was in 1927 with Helmut Hasse and Richard Brauer on noncommutative algebra. She was recognized for her mathematical achievements through invitations to address the International Mathematical Congress, the last at Zurich in 1932. Despite this, she was dismissed from her position at the University of Göttingen in 1933 due to the Nazi rise to power given that she was Jewish. She fled Germany in that year and joined the faculty at Bryn Mawr College in the U.S.A. She died at Bryn Mawr on April 14, 1935. She was buried in the Cloisters of the Thomas Great Hall on the Bryn Mawr campus.*

One of our main goals is the following result that leads us toward a unique factorization theory for ideals in rings of algebraic integers.

Theorem 2.9 | **Rings of Integers are Dedekind Domains**

If F is an algebraic number field, then $\mathfrak{O}_F$ is a Dedekind domain.

Proof. By Corollary 2.2 (in view of Remark 2.7), condition (A) of Definition 2.10 is satisfied. In order to verify condition (B), we require some results as follows.

Assume that there is a prime $\mathfrak{O}_F$-ideal $\mathcal{P} \neq (0)$ that is not maximal. Therefore, the set $\mathcal{S} \neq \varnothing$, where $\mathcal{S}$ is the set of all proper $\mathfrak{O}_F$-ideals that strictly contain $\mathcal{P}$. By Corollary 2.3, there is a maximal ideal $M \in \mathcal{S}$ such that $\mathcal{P} \subset M \subset \mathfrak{O}_F$. By Theorem 2.6 on page 68, M is a prime $\mathfrak{O}_F$-ideal. By Theorem 2.1 on page 56, there exists a nonzero $a \in \mathcal{P} \cap \mathbb{Z}$. By Exercise 1.2 on page 16, $\mathcal{P} \cap \mathbb{Z}$ is a $\mathbb{Z}$-ideal. Suppose that $ab \in \mathcal{P} \cap \mathbb{Z}$, where $a, b \in \mathbb{Z}$. Since $\mathcal{P}$ is a prime $\mathfrak{O}_F$-ideal, then $a \in \mathcal{P}$ or $b \in \mathcal{P}$ so $a \in \mathcal{P} \cap \mathbb{Z}$ or $b \in \mathcal{P} \cap \mathbb{Z}$, which means that $\mathcal{P} \cap \mathbb{Z}$ is a prime $\mathbb{Z}$-ideal. If $p \in \mathcal{P} \cap \mathbb{Z}$ is a rational prime, then

$$(p) \subseteq \mathcal{P} \cap \mathbb{Z}$$

and (p) is a maximal $\mathbb{Z}$-ideal by Theorem 2.7 on page 68 since $\mathbb{Z}/(p)$ is a field by Example 2.7 on page 69. Hence, since $\mathcal{P} \cap \mathbb{Z} \neq \mathbb{Z}$, then $(p) = \mathcal{P} \cap \mathbb{Z}$. However,

$$(p) = \mathcal{P} \cap \mathbb{Z} \subseteq M \cap \mathbb{Z} \subset \mathbb{Z},$$

where $1 \notin M$ so

$$(p) = \mathcal{P} \cap \mathbb{Z} = M \cap \mathbb{Z}.$$

Since $M \in \mathcal{S}$, then $\mathcal{P} \neq M$, so there exists an $\alpha \in M$ such that $\alpha \notin \mathcal{P}$. Consider

$$m_{\alpha, \mathbb{Q}}(x) = x^d + a_{d-1} x^{d-1} + \cdots + a_1 x + a_0 \in \mathbb{Z}[x] \text{ for some } d \in \mathbb{N}.$$

Then $m_{\alpha, \mathbb{Q}}(\alpha) \in \mathcal{P}$. Now define $\ell \in \mathbb{N}$ to be the least value for which there exist integers b_j such that

$$\alpha^\ell + b_{\ell-1} \alpha^{\ell-1} + \cdots + b_1 \alpha + b_0 \in \mathcal{P}, \tag{2.8}$$

for $j = 0, 1, \cdots, \ell - 1$. Since $\alpha \in M$, then by properties of ideals,

$$\alpha(\alpha^{\ell-1} + b_{\ell-1} \alpha^{\ell-2} + \cdots + b_1) \in M.$$

Also, since $m_{\alpha, \mathbb{Q}}(\alpha) \in \mathcal{P} \subset M$, then, again by properties of ideals,

$$m_{\alpha, \mathbb{Q}}(\alpha) - \sum_{j=1}^{\ell-1} \alpha^j b_j - \alpha^\ell = b_0 \in M, \tag{2.9}$$

so $b_0 \in M \cap \mathbb{Z} = \mathcal{P} \cap \mathbb{Z}$. If $\ell = 1$, then $\alpha \in \mathcal{P}$, a contradiction, so $\ell > 1$. Thus, by (2.8)–(2.9),

$$\alpha^\ell + b_{\ell-1} \alpha^{\ell-1} + \cdots + b_1 \alpha + b_0 - b_0 = \alpha(\alpha^{\ell-1} + b_{\ell-1} \alpha^{\ell-2} + \cdots + b_1) \in \mathcal{P}.$$

However, since $\mathcal{P}$ is prime and $\alpha \notin \mathcal{P}$, then

$$\alpha^{\ell-1} + b_{\ell-1} \alpha^{\ell-2} + \cdots + b_1 \in \mathcal{P},$$

contradicting the minimality of $\ell > 1$. We have shown $\mathcal{S} = \varnothing$, which establishes that condition (B) of Definition 2.10 holds.

It remains to show that condition (C) holds. By Theorem 2.8 on page 70, $\mathfrak{O}_F$ has quotient field F. Let $\alpha \in F$ be integral over $\mathfrak{O}_F$. Also, $\mathfrak{O}_F$ is integral over $\mathbb{Z}$ – see Remark 1.5 on page 9 – so α is an algebraic integer in F. However, by Definition 1.5 on page 4, $F \cap \mathbb{A} = \mathfrak{O}_F$, so $\alpha \in \mathfrak{O}_F$, which means that $\mathfrak{O}_F$ is integrally closed and we have condition (C) that establishes the entire result. $\square$

Now we aim at the main goal of this section, which is a unique factorization theorem for rings of integers. To this end, we first settle conditions for which the converse of Lemma 2.1 on page 58 holds. First, we require a more general notion of "ideal" in order to proceed.

Definition 2.13 | **Fractional Ideals**

Suppose that R is an integral domain with quotient field F. Then a nonempty subset I of F is called a *fractional R-ideal* if it satisfies the following three properties.

1. For any $\alpha, \beta \in I$, $\alpha + \beta \in I$.

2. For any $\alpha \in I$ and $r \in R$, $r\alpha \in I$.

3. There exists a nonzero $\gamma \in R$ such that $\gamma I \subseteq R$.

When $I \subseteq R$, we call I an *integral R-ideal* (which is the content of Definition 2.1 on page 55) to distinguish it from the more general fractional ideal.

Remark 2.8 It is immediate from Definition 2.13 that if I is a fractional R-ideal, then there exists a nonzero $\gamma \in R$ such that $\gamma I = J$ where J is an integral R-ideal. Hence, if R is Noetherian domain, then by Lemma 2.2 on page 71, there exist $\alpha_1, \alpha_2, \ldots, \alpha_d$ for some $d \in \mathbb{N}$ such that $J = (\alpha_1, \ldots, \alpha_d)$. Hence,

$$I = \frac{1}{\gamma}J = \left(\frac{\alpha_1}{\gamma}, \frac{\alpha_2}{\gamma}, \ldots, \frac{\alpha_d}{\gamma}\right)$$

is also finitely generated. Indeed, in a Noetherian domain, a fractional R-ideal is the same as a finitely-generated R-submodule of the quotient field of R.

Example 2.11 Let $R = \mathbb{Z}$, and $F = \mathbb{Q}$. Then the fractional R-ideals are the sets

$$I_q = \{q\mathbb{Z} : q \in \mathbb{Q}^+\}.$$

Since $q\mathbb{Z} = (-q)\mathbb{Z}$, we may restrict attention to the positive rationals $\mathbb{Q}^+$ without loss of generality. Also,

$$I_{q_1}I_{q_2} = q_1 q_2 \mathbb{Z} = I_{q_1 q_2}.$$

We have the isomorphism

$$\mathcal{S} = \{I_q : q \in \mathbb{Q}\} \cong \mathbb{Q}^+,$$

as multiplicative groups. The unit element of $\mathcal{S}$ is $\mathbb{Z}$ and the inverse element of $I_q \in \mathcal{S}$ is $(I_q)^{-1} = q^{-1}\mathbb{Z}$. (See Exercise 2.18 on page 86.)

Example 2.11 motivates the following.

Theorem 2.10 | Inverse Fractional Ideals |

If R is an integral domain with quotient field F, and I is a fractional R-ideal, then the set

$$I^{-1} = \{\alpha \in F : \alpha I \subseteq R\}$$

is a nonzero fractional R-ideal.

Proof. If $\alpha, \beta \in I^{-1}$, then $\alpha I \subseteq R$ and $\beta I \subseteq R$, so

$$(\alpha + \beta)I \subseteq \alpha I + \beta I \subseteq R,$$

so $\alpha + \beta \in I^{-1}$. If $\alpha \in I^{-1}$ and $r \in R$, $\alpha I \subseteq R$ so $r\alpha I \subseteq R$, which implies $r\alpha \in I^{-1}$. Lastly, let γ be a nonzero element of I. Then for any $\alpha \in I^{-1}$, $\alpha I \subseteq R$, so in particular, $\gamma\alpha \in R$. Hence, $\gamma I^{-1} \subseteq R$. This satisfies all three conditions in Definition 2.13. □

Definition 2.14 | Invertible Fractional Ideals |

In an integral domain R, a fractional R-ideal I is called invertible if

$$II^{-1} = R,$$

where I^{-1}, given in Theorem 2.10, is called the *inverse* of I.

Now we may return to Dedekind domains and the pertinence of the above to them.

Theorem 2.11 | Invertibility in Dedekind Domains |

If R is a Dedekind domain, then every nonzero integral R-ideal is invertible.

Proof. Since R is a Dedekind Domain, then every R-ideal I is finitely generated, so for $I \neq (0)$, there are $\alpha_j \in R$ for $1 \leq j \leq d$ such that $I = (\alpha_1, \alpha_2, \ldots, \alpha_d)$. If $d = 1$, then $I^{-1} = (\alpha_1^{-1})$ and $II^{-1} = R$. Now the result may be extrapolated by induction, and the result is established. □

Via the above, we are in a position to provide the promised converse of Lemma 2.1 on page 58.

Corollary 2.5 | To Divide is the Same as to Contain |

If R is a Dedekind domain, and I, J are R-ideals, then

$$I \mid J \text{ if and only if } J \subseteq I.$$

Proof. In view of Lemma 2.1, we need only prove one direction. Suppose that

$$J \subseteq I. \tag{2.10}$$

Now let $H = I^{-1}J$, in which case $J = IH$ where H is an R-ideal since by (2.10),

$$I^{-1}J \subseteq II^{-1} = R,$$

where the equality follows from Theorem 2.11. Thus, $I \mid J$, and we have secured the result. $\qquad\square$

As a consequence of Corollary 2.5, we see that a prime R-ideal $\mathcal{P}$ in a Dedekind domain R satisfies the same property as prime elements in $\mathbb{Z}$.

Corollary 2.6 *Suppose that R is a Dedekind domain. Then $\mathcal{P}$ is a prime R-ideal if it satisfies the property that for any R-ideals I, J,*

$$\mathcal{P} \mid IJ \text{ if and only if } \mathcal{P} \mid I \text{ or } \mathcal{P} \mid J.$$

Proof. By Corollary 2.5, $\mathcal{P} \mid IJ$ if and only if $IJ \subseteq \mathcal{P}$ and the latter holds, by (2.1), if and only if $I \subseteq \mathcal{P}$ or $J \subseteq \mathcal{P}$, so applying Corollary 2.5 to the latter we get the result. $\qquad\square$

Also, we have the following result that mimics the same law for nonzero elements of $\mathbb{Z}$.

Corollary 2.7 | **Cancellation Law for Ideals in Dedekind Domains**

Let R be a Dedekind domain. If I, J, L are R-ideals with $I \neq (0)$, and $IJ \subseteq IL$, then $J \subseteq L$.

Proof. If $IJ = IL$, then by Theorem 2.11,

$$J = RJ = I^{-1}IJ \subseteq I^{-1}IL = RL = L,$$

as required. $\qquad\square$

Now we are ready for the promised unique factorization result.

Theorem 2.12 | **Unique Factorization of Ideals**

Every proper nonzero ideal in a Dedekind domain R is uniquely representable as a product of prime ideals. In other words, any R-ideal has a unique expression (up to order of the factors) of the form

$$I = \mathcal{P}_1^{a_1} \mathcal{P}_2^{a_2} \ldots \mathcal{P}_n^{a_n},$$

where the $\mathcal{P}_j$ are the distinct prime R-ideals containing I, and $a_j \in \mathbb{N}$ for $j = 1, 2, \ldots, n$.

Proof. First we must show existence. In other words, we must show that every ideal is indeed representable as a product of primes. Let $\mathcal{S}$ be the set of all nonzero proper ideals that are not so representable. If $\mathcal{S} \neq \varnothing$, then by Corollary 2.3 on page 72, $\mathcal{S}$ has a maximal member M. Thus, M is not a prime R-ideal, but by Corollary 2.4, $M \subseteq \mathcal{P}$ where $\mathcal{P}$ is maximal, and so prime by Theorem 2.6 on page 68. Hence, $R \subseteq \mathcal{P}^{-1} \subseteq M^{-1}$, which implies that

$$ M \subseteq M\mathcal{P}^{-1} \subseteq MM^{-1} = R, $$

where the equality follows from Theorem 2.11 on page 76. We have shown that $M\mathcal{P}^{-1}$ is an integral R-ideal. If $\mathcal{P}^{-1}M = M$, then

$$ \mathcal{P}\mathcal{P}^{-1}M = \mathcal{P}M \subseteq \mathcal{P}, $$

where the latter inclusion comes from the fact that $\mathcal{P}$ is an ideal. Hence, $M = \mathcal{P}$ by the maximality of $\mathcal{P}$, a contradiction to $M \in \mathcal{S}$. Thus, $M \subset \mathcal{P}^{-1}M$, so $\mathcal{P}^{-1}M$ is an integral ideal not in $\mathcal{S}$ which means there are prime ideals $\mathcal{P}_j$ for $j = 1, 2, \ldots d \in \mathbb{N}$ such that

$$ \mathcal{P}^{-1}M = \mathcal{P}_1\mathcal{P}_2 \cdots \mathcal{P}_d, $$

which implies

$$ M = RM = \mathcal{P}\mathcal{P}^{-1}M = \mathcal{P}\mathcal{P}_1\mathcal{P}_2 \cdots \mathcal{P}_d, $$

contradicting that $M \in \mathcal{S}$. We have shown $\mathcal{S} = \varnothing$, thereby establishing existence. It remains to show uniqueness of representation.

Let $\mathcal{P}_j$ and $\mathcal{Q}_k$ be (not necessarily distinct) prime R-ideals such that

$$ \mathcal{P}_1 \cdots \mathcal{P}_r = \mathcal{Q}_1 \cdots \mathcal{Q}_s. \tag{2.11} $$

Hence,

$$ \mathcal{P}_1 \supseteq \mathcal{Q}_1 \cdots \mathcal{Q}_s, $$

so $\mathcal{Q}_j \subseteq \mathcal{P}_1$ for some $j = 1, 2, \ldots, s$. Without loss of generality, we may assume that $j = 1$, by rearranging the $\mathcal{Q}_j$ if necessary. However, by condition B of Definition 2.10, $\mathcal{P}_1 = \mathcal{Q}_1$. Multiplying both sides of (2.11) by $\mathcal{P}_1^{-1}$, we get

$$ \mathcal{P}_2 \cdots \mathcal{P}_r = \mathcal{Q}_2 \cdots \mathcal{Q}_s. $$

Continuing in this fashion, we see that by induction, $r = s$ and $\mathcal{P}_j = \mathcal{Q}_j$ for $1 \leq j \leq s = r$.   □

In view of Theorem 2.12, we have an immediate consequence that is the primary goal sought in this section.

Corollary 2.8 *If F is a number field, then every proper, nonzero $\mathfrak{O}_F$-ideal is uniquely representable as a product of prime ideals.*

Proof. By Theorem 2.9 on page 73, $\mathfrak{O}_F$ is a Dedekind domain, so the result is a special case of Theorem 2.12.   □

Example 2.12 In $R = \mathbb{Z}[\sqrt{10}]$ let us look at the unique factorization of the R-ideal (6) as a product of prime ideals. Note that

$$\mathcal{P} = (2, \sqrt{10}), \mathcal{Q} = (3, 1 + \sqrt{10}), \text{ and } \mathcal{Q}' = (3, 1 - \sqrt{10})$$

are prime ideals in $\mathbb{Z}[\sqrt{10}] = \mathfrak{O}_F$ for $F = \mathbb{Q}(\sqrt{10})$. The unique factorization of the principal ideal (6) is now apparent, as an exercise for the reader by employing the multiplication formulas on page 59:

$$(6) = \mathcal{P}^2 \mathcal{Q} \mathcal{Q}'.$$

We note that the *element* 6 in R does *not* have unique factorization since

$$6 = 2 \cdot 3 = (4 + \sqrt{10})(4 - \sqrt{10}),$$

where each factor is irreducible. Hence, unique factorization is restored at the ideal level by Dedekind's contribution of the theory of ideals.

The developments in this section allow us to now define gcd and lcm concepts for ideals that mimic those for rational integers.

Definition 2.15 | **A gcd and lcm for Ideals**

If R is a Dedekind domain, and I, J are R-ideals, then

$$\gcd(I, J) = I + J,$$

and

$$\text{lcm}(I, J) = I \cap J.$$

If $\gcd(I, J) = R$, then I and J are said to be *relatively prime*.

Remark 2.9 The notion of relative primality given in Definition 2.15 is the direct analogue for rational integers since $R = (1_R)$ is a principal ideal. This is of course what we mean in $\mathbb{Z}$ since the pair of integers can have no common divisors. Let us look at this directly.

If I, J are relatively prime, then

$$\gcd(I, J) = I + J = R.$$

If an R-ideal H divides both I and J, then by Corollary 2.5 on page 76, $I \subseteq H$ and $J \subseteq H$, so $I + J = R \subseteq H$, which means that $H = R$. Hence, the only R-ideal that can divide both I and J is $R = (1)$.

The next result is the exact analogue for rational integers of the one that we proved in [68, Theorem 1.13 (b), p.26].

Lemma 2.3 | Product of the Ideal-Theoretic gcd and lcm

If R is a Dedekind domain and I, J are R-ideals, then

$$\gcd(I, J) \cdot \mathrm{lcm}(I, J) = (I + J)(I \cap J) = IJ.$$

Proof. By the definition of an ideal, any elements of $I + J$ times any element of $I \cap J$ must be in I and J, so in IJ. Thus,

$$(I \cap J)(I + J) \subseteq IJ.$$

Conversely, any element of IJ is in both I and J, so in $I \cap J$, and trivially in $I + J$. Thus,

$$IJ \subseteq (I \cap J)(I + J),$$

from which the desired equality follows.   □

The following exploits our unique factorization result to provide an analogue of the same result for rational integers that we proved in [68, Theorem 1.17, p. 34].

Theorem 2.13 | Prime Factorizations of gcd and lcm of Ideals

Suppose that R is a Dedekind domain and I, J are R-ideals with prime factorizations given via Theorem 2.12 by

$$I = \prod_{j=1}^{r} \mathcal{P}_j^{a_j}, \text{ and } J = \prod_{j=1}^{r} \mathcal{P}_j^{b_j},$$

where $\mathcal{P}_j$ are prime R-ideals with integers $a_j, b_j \geq 0$. Then

$$\gcd(I, J) = \prod_{j=1}^{r} \mathcal{P}_j^{m_j}, \text{ and } \mathrm{lcm}(I, J) = \prod_{j=1}^{r} \mathcal{P}_j^{M_j},$$

where $m_j = \min(a_j, b_j)$ and $M_j = \max(a_j, b_j)$, for each $j = 1, \ldots, r$.

Proof. Since $\gcd(I, J) = I + J$, then

$$\gcd(I, J) = \prod_{j=1}^{r} \mathcal{P}_j^{a_j} + \prod_{j=1}^{r} \mathcal{P}_j^{b_j} = \prod_{j=1}^{r} \mathcal{P}_j^{m_j} \left(\prod_{j=1}^{r} \mathcal{P}_j^{a_j - m_j} + \prod_{j=1}^{r} \mathcal{P}_j^{b_j - m_j} \right).$$

However, for each j, one of $a_j - m_j$ or $b_j - m_j$ is zero, so the right hand sum is R since the two summands are relatively prime. In other words,

$$\gcd(I, J) = \prod_{j=1}^{r} \mathcal{P}_j^{m_j},$$

as required. Now, by Lemma 2.3 on the preceding page, $(I \cap J)(I + J) = IJ$, so

$$IJ = \prod_{j=1}^{r} \mathcal{P}_j^{a_j+b_j} = \prod_{j=1}^{r} \mathcal{P}_j^{m_j}(I \cap J) = (I + J)(I \cap J),$$

so

$$\mathrm{lcm}(I, J) = I \cap J = \prod_{j=1}^{r} \mathcal{P}_j^{a_j+b_j-m_j} = \prod_{j=1}^{r} \mathcal{P}_j^{M_j},$$

and we have the complete result. $\qquad\square$

Remark 2.10 Theorem 2.13 tells us that, when R is a Dedekind domain, $\mathrm{lcm}(I, J)$ is actually the largest ideal contained in both I and J and $\gcd(I, J)$ is the smallest ideal containing both I and J.

The following allows us to compare unique factorization of elements with that of ideals and show where Dedekind's contribution comes into play.

Definition 2.16 | **Irreducible Ideals, gcds and lcms**

If R is an integral domain, then an R-ideal I is called *irreducible* if it satisfies the property that whenever an R-ideal $J \mid I$, then $J = I$ or $J = R$.

Theorem 2.14 | **Irreducible = Prime in Dedekind Domains**

If R is a Dedekind domain, and I is an R ideal, then I is irreducible if and only if I is a prime R-ideal.

Proof. Let I be irreducible and let J, K be R-ideals such that $I \mid JK$. Since $\gcd(I, J) \mid I$, then $\gcd(I, J) = I$ or $\gcd(I, J) = R$. If $\gcd(I, J) = I$, then $I + J = I$, which means that

$$I = J = \gcd(I, J).$$

Now suppose that $I \nmid J$. Then $\gcd(I, J) = R$, so there exist $\alpha \in I$ and $\beta \in J$ such that $\alpha + \beta = 1_R$. Therefore, given an arbitrary $\gamma \in K$,

$$\gamma = \gamma\alpha + \gamma\beta.$$

Since $I \mid JK$, then by Corollary 2.5 on page 76, $JK \subseteq I$, so $\beta\gamma \in I$ since $\beta\gamma \in JK$. However, $\alpha\gamma \in I$ so $\gamma \in I$. This shows that $K \subseteq I$, so by Corollary 2.5, we have that $I \mid K$. Hence, by Theorem 2.2 on page 57, I is prime.

Conversely, suppose that I is prime. If $I = HJ$ for some nontrivial R-ideals H and J, then either $I|H$ or $I|J$. If $I|H$, there is an R-ideal L such that $H = IL$. Therefore,

$$I = HJ = ILJ.$$

By Corollary 2.7 on page 77, $(1) = R = LJ$. Hence, $J = (1) = R$, so I is irreducible. $\qquad\square$

The following is immediate from Theorem 2.14, and is the analogue of the definition of a rational prime.

Corollary 2.9 *If R is a Dedekind domain, then I is a prime R-ideal if and only if it satisfies the property that*

$$whenever\ J \mid I\ for\ a\ proper\ R\text{-}ideal\ J\ then\ I = J.$$

Remark 2.11 It follows from Lemma 1.2 and Theorem 1.16 on page 38 that the failure of unique factorization in an integral domain R is the failure of irreducible elements to be prime in R. However, since Theorem 2.14 tells us that irreducible *ideals* are the *same* as prime ideals in a Dedekind domain, then we have unique factorization restored at the ideal level via Theorem 2.12 on page 77. In particular, rings of integers $\mathfrak{O}_F$ of number fields F have unique factorization ideals since Theorem 2.9 on page 73 tells us that $\mathfrak{O}_F$ is a Dedekind domain. Thus, the magnitude of of Dedekind's contribution is brought to light by this fact.

We need the following concept that is intimately linked to the notion of a UFD, especially when we are dealing with Dedekind domains–see Definition 1.20 on page 37.

Definition 2.17 $\boxed{\textbf{Principal Ideal Domain (PID)}}$

An integral domain R in which all ideals are principal is called a *principal ideal domain*, or *PID* for convenience.

Theorem 2.15 $\boxed{\textbf{PIDs and Noetherian Domains}}$

If R is a PID, then R is a Noetherian domain.

Proof. If we have a nested sequence of R-ideals

$$(\alpha_1) \subseteq (\alpha_2) \subseteq \cdots (\alpha_j) \subseteq \cdots,$$

then it follows from Exercise 1.2 on page 16 that $\cup_{j=1}^{\infty}(\alpha_j)$ is an R-ideal. Thus, since R is a PID, there exists an $\alpha \in R$ such that

$$\cup_{j=1}^{\infty}(\alpha_j) = (\alpha),$$

so there exists an $n \in \mathbb{N}$ such that $\alpha \in (\alpha_n)$. Therefore,

$$(\alpha_j) = (\alpha_n) = (\alpha)$$

for all $j \geq n$. Thus, the ACC condition of Definition 2.11 on page 71 is satisfied and R is a Noetherian domain. $\qquad\square$

Theorem 2.16 $\boxed{\textbf{PIDs and UFDs}}$

If R is a PID then R is a UFD.

Proof. Let S be the set of all $\alpha \in R$ such that (α) is not a product of irreducible elements. If $S \neq \varnothing$, then by Corollary 2.3 on page 72, via Theorem 2.15, S has a maximal element (m). Thus, (m) is a proper ideal (since a unit is vacuously a product of irreducible elements by Definition 1.19 on page 37). Therefore, (m) is contained in a maximal R-ideal (M) for some $M \in R$ by Corollary 2.4 on page 73, again via Theorem 2.15. Thus, $M \mid m$ and $(M) \neq (m)$ by Theorem 2.6 on page 68. Since M is a product of irreducible elements, there exists an $\alpha \mid m$ such that α is irreducible. Therefore, $m = \alpha\beta$ for some $\beta \in R$. If β is a unit, then m is irreducible since associates of irreducibles are also irreducible, a contradiction. Hence, β is not a unit. If $(\beta) \notin S$, then β is a product of irreducibles, and so is m, a contradiction. Thus, $(\beta) \in S$. However, $\beta \mid m$, so $(m) \subseteq (\beta)$, by Corollary 2.5 on page 76. Also, $(m) \neq (\beta)$ since α is not a unit, given that it is irreducible. Hence, (m) is properly contained in $(\beta) \subseteq S$, a contradiction to the maximality of (m) in S, so $S = \varnothing$. This establishes that all nonzero elements are expressible as a product of irreducible elements.

We may complete the proof by showing that all irreducible elements are prime and invoke Theorem 1.16 on page 38. Suppose that $r \in R$ is an irreducible element and $r \mid \alpha\beta$, $\alpha, \beta \in R$ with r not dividing α. Then by the irreducibility of r, we must have that r and α are relatively prime, namely

$$R = (r) + (\alpha),$$

so there exist $s_1, s_2 \in R$ such that $1_R = rs_1 + \alpha s_2$. Therefore,

$$(\beta) = (rs_1\beta + \alpha s_2\beta) \subseteq (r),$$

since $r \mid \alpha\beta$ implies that $(r) \supseteq (\alpha\beta)$, so both $rs_1\beta \in (r)$ and $\alpha s_2\beta \in (r)$. In other words, $r \mid \beta$, so r is prime as required. $\square$

Now we look at PIDs and UFDs in the case of Dedekind domains, which will be of value when we study binary quadratic forms in §3.2.

Theorem 2.17 $\boxed{\textbf{UFDs are PIDs for Dedekind domains}}$

If R is a Dedekind domain, then R is a UFD if and only if R is a PID.

Proof. In view of Theorem 2.16 on the preceding page, we need only prove that R is a PID when it is a UFD. Let R be a UFD. If there exists an R-ideal that is not principal, then by Theorem 2.12 on page 77, there exists a prime R-ideal $\mathcal{P}$ that is not principal. Let S consist of the set of all R-ideals I such that $\mathcal{P}I$ is principal. By Exercise 2.11 on page 85, $S \neq \varnothing$. By Remark 2.7 and Corollary 2.3 on page 72, S has a maximal element M. Let

$$\mathcal{P}M = (\alpha).$$

If $\alpha = \beta\gamma$ where $\beta \in \mathcal{P}$ is irreducible, then $(\beta) = \mathcal{P}J$ where J is an R-ideal such that $J \mid M$, so $J \supseteq M$. By the maximality of M, we have $J = M$, so γ is a unit and α is irreducible. Since $\mathcal{P}$ is not principal, there is a nonzero $\delta \in \mathcal{P} - (\alpha)$, and since $M = (\alpha)$ would imply that $\mathcal{P} = R$, there is a nonzero $\sigma \in M - (\alpha)$. Thus,

$$\delta\sigma \in \mathcal{P}M \subseteq (\alpha),$$

so $\alpha \mid \delta\sigma$. However, α divides neither δ nor σ, so α is not prime. This contradicts Theorem 1.16 on page 38. $\qquad\square$

In view of Theorem 1.17 on page 39 and Theorem 2.17, it is now apparent why we introduced Euclidean domains in §1.3, where we were concerned with introducing the importance of the notion of unique factorization of algebraic integers.

We conclude this section with a result that is the analogue of [68, Theorem 1.22, p. 40]. The reader should be familiar with the basics on ring actions such as that covered in [68, pp. 303–305].

Theorem 2.18 | **Chinese Remainder Theorem for Ideals**

Let R be a commutative ring with identity and let $I_1, \ldots, I_r$ be pairwise relatively prime ideals in R. Then the natural map

$$\psi : R/ \cap_{j=1}^{r} I_j \mapsto R/I_1 \times \cdots \times R/I_r$$

is an isomorphism.

The above statement is equivalent to saying that if $\beta_1, \beta_2, \ldots, \beta_r \in R$, there exists a $\beta \in R$ such that $\beta - \beta_j \in I_j$ for each $j = 1, 2, \ldots, r$, where β is uniquely determined modulo $\cap_{j=1}^{r} I_j$. The latter means that

$$\text{any } \gamma \text{ satisfying } \gamma - \beta_j \in I_j \text{ for each such } j \text{ implies } \beta - \gamma \in \cap_{j=1}^{r}I_j. \quad (2.12)$$

Proof. Since $\psi(s) = 0$ if and only if $s \in \cap_{j=1}^{r}I_j$, then $\ker(\psi) = (0)$, since the I_j are pairwise relatively prime. It remains to show that ψ is a surjection. Let

$$\beta_1, \beta_2, \ldots, \beta_r \in R.$$

We must show that there is a $\beta \in R$ such that $\psi(\beta) = (\beta_1, \ldots, \beta_r)$. This is tantamount to saying: there is a $\beta \in R$ such that $\beta - \beta_k \in I_k$ for each k. Since $I_i + I_j = R$ for all $i \neq j$, then by induction

$$I_k + \cap_{j\neq k}I_j = R.$$

Thus, for each such k, there exists an $\alpha_k \in I_k$ and $r_k \in \cap_{j\neq k}I_j$ such that

$$\beta_k = \alpha_k + r_k \text{ with } \beta_k - r_k \in I_k \text{ and } r_k \in I_j \text{ for all } j \neq k.$$

Set

$$\beta = \sum_{j=1}^{r} r_j.$$

Then

$$\beta - \beta_k = \sum_{j \neq k} r_j + (r_k - \beta_k) \in I_k,$$

as required. □

Remark 2.12 In Theorem 2.18, we may use the notation

$$\gamma \equiv \beta_j \pmod{I_j},$$

to denote $\gamma - \beta_j \in I_j$. Then *(2.12)* becomes:

any γ satisfying $\gamma \equiv \beta_j \pmod{I_j}$ for $1 \leq j \leq r$ implies $\beta \equiv \gamma \pmod{\cap_{j=1}^{r} I_j}$.

For more on this concept see Exercises 8.32–8.39 on pages 292–293.

Exercises

2.9. Let R be a Dedekind domain. If I, J are R-ideals, prove that there exists an $\alpha \in I$ such that $\gcd((\alpha), IJ) = I$.

2.10. Let R be a Dedekind domain, and let I, J, H be R-ideals. Prove that $I(J + H) = IJ + IH$.

2.11. Let R be a Dedekind domain and I, J nonzero R-ideals. Prove that there is an R-ideal H, relatively prime to J, such that HI is principal.

2.12. Prove that, in a Noetherian domain, every ideal can be represented as the intersection of a finite number of irreducible ideals.

2.13. A commutative ring R with identity is said to satisfy the descending chain condition, denoted by DCC for convenience, on ideals if every sequence $I_1 \supseteq I_2 \supseteq \cdots \supseteq I_j \supseteq \cdots$ of R-ideals terminates. In other words, there exists an $n \in \mathbb{N}$ such that $I_j = I_n$ for all $j \geq n$. Prove that R satisfies the DCC if and only if every nonempty collection of ideals contains a minimal element. (Rings of the above type are called *Artinian rings*–see Biography 2.2 on page 87.)

2.14. Let R be an integral domain with quotient field F. Prove that every invertible fractional R-ideal is a finitely generated R-module.

2.15. Let R, S be commutative rings with identity such that $R \subseteq S$, and $s \in S$. Prove that if s is integral over R, then $R[s]$ is a finitely-generated R-module.

2.16. Let R be an integral domain with quotient field F. Prove that every nonzero finitely-generated submodule I of F is a fractional R-ideal.

2.17. Prove that in an integral domain R, the following are equivalent.

(a) Every nonzero fractional R-ideal is invertible.

(b) The set G of all fractional R-ideals forms a multiplicative group.

2.18. Prove that in an integral domain R, the following are equivalent.

(i) R is a Dedekind domain.

(ii) Every proper R-ideal is a unique product of a finite number of prime ideals (up to order of the factors), and each is invertible.

(iii) Every nonzero R-ideal is invertible.

(iv) Every fractional R-ideal is invertible.

(v) The set G of all fractional R-ideals forms a multiplicative abelian group.

(vi) R is an integrally closed, Noetherian domain, and every nonzero prime ideal is maximal.

(*Hint: Use Exercises 2.14–2.17.*)

2.19. Suppose that R is a Dedekind domain with quotient field F and I is an R-ideal. Also, we define $\operatorname{ord}_{\mathcal{P}}(I) = a$ where $a \geq 0$ is the largest power of the prime ideal $\mathcal{P}$ dividing I, namely $\mathcal{P}^a \mid I$ but $\mathcal{P}^{a+1}$ does not divide I. The value $\operatorname{ord}_{\mathcal{P}}(I)$ is called the *order of I with respect to $\mathcal{P}$*. Prove the following.

(a) For R-ideals I, J, $\operatorname{ord}_{\mathcal{P}}(IJ) = \operatorname{ord}_{\mathcal{P}}(I) + \operatorname{ord}_{\mathcal{P}}(J)$.

(b) For R-ideals I, J, $\operatorname{ord}_{\mathcal{P}}(I + J) = \min(\operatorname{ord}_{\mathcal{P}}(I), \operatorname{ord}_{\mathcal{P}}(J))$.

(c) For any R-ideal I, there exists an $\alpha \in F$ such that $\operatorname{ord}_{\mathcal{P}}((\alpha)) = \operatorname{ord}_{\mathcal{P}}(I)$ for any prime R-ideal $\mathcal{P} \mid I$.

2.20. Prove that every R-ideal in a Dedekind domain R can be generated by at most two elements.

(*Hint: Use Exercise 2.19.*)

Biography 2.2 Emil Artin (1898–1962) *was born on March 3, Vienna, Austria in 1898. He served in the Austrian army in World War I, after which he entered the University of Leipzig. In 1921 he obtained his doctorate, the thesis of which was on quadratic extensions of rational function fields over finite fields. In 1923, he had his Habilitation, allowing him to become Privatdozent at the University of Hamburg. In 1925, he was promoted to extraordinary professor at Hamburg. In that same year, he introduced the theory of braids, which is studied today by algebraists and topologists. In 1928, he worked on rings with minimum condition, the topic of Exercise 2.13, which are now called* Artinian rings. *In 1937, Hitler enacted the New Official's Law, which enabled a mechanism for removing not only Jewish teachers from university positions but also those related by marriage. Since Artin's wife was Jewish, although he was not, he was dismissed. In 1937, he emigrated to the U.S.A. and taught at several universities there, including eight years at Bloomingdale at Indiana University during 1938–1946, as well as Princeton from 1946 to 1958. During this time, in 1955, he produced what was, arguably, the catalyst for the later classification of finite simple groups, by proving that the only (then-known) coincidences in orders of finite simple groups were those given by Dickson in his* Linear Groups. *In 1958, he returned to Germany where he was appointed again to the University of Hamburg. Artin's name is attached not only to the aforementioned rings, but also to the reciprocity law that he discovered as a generalization of Gauss's quadratic reciprocity law. One of the tools that he developed to do this is what we now call Artin L-functions. He also has the distinction of solving one of Hilbert's famous list of twenty-three problems posed in 1900.*

He was an outstanding and respected teacher. In fact, many of his Ph.D. students such as Serge Lang, John Tate, and Max Zorn went on to major accomplishments. He also had an interest in astronomy, biology, chemistry, and music. He was indeed an accomplished musician in his own right playing the flute, harpsichord, and clavichord. He died in Hamburg on December 20, 1962.

2.3 Application to Factoring

> *If you want a helping hand, you'll find one at the end of your arm.*
> **Audrey Hepburn (originally Edda Van Heemstra), (1929–1993)**
> **Belgian Actress**

In [68, §4.3, pp. 201–208], we saw the importance of factoring methods, especially in terms of the security of certain cryptosystems.

In this section, we will look at factoring using certain cubic integers, namely the integers from

$$\mathfrak{O}_F = \mathbb{Z}[\sqrt[3]{-2}] = \mathbb{Z}[\sqrt[3]{2}]$$

(since $\sqrt[3]{-2} = -\sqrt[3]{2}$), which is the ring of integers of

$$F = \mathbb{Q}(\sqrt[3]{-2}) = \mathbb{Q}(\sqrt[3]{2}),$$

by Exercise 2.21 on page 96). In this section, we will show how we may employ these cubic integers in $\mathbb{Z}[\sqrt[3]{-2}]$ to factor integers in $\mathbb{Z}$. In order to do this we need to introduce some more general aspects of number fields upon which we have only touched. In Definition 1.11 on page 18, we introduced the notion of the norm of an element in a quadratic field. We need to generalize this in order to apply the notion needed for cubic integers, and other number fields later on. In order to do this, we need to motivate another important concept related to a number field. This is motivated by our quadratic case. For instance, if $F = \mathbb{Q}(i)$ is the Gaussian field, then there are exactly two monomorphisms

$$\theta_1(x + yi) = x + yi \text{ and } \theta_2(x + yi) = x - yi \ (x, y \in \mathbb{Q})$$

from F into $\mathbb{C}$, the complex field. Since the degree of the Gaussian field over $\mathbb{Q}$ is $|F : \mathbb{Q}| = 2$, one might expect that the number of such monomorphisms is $|F : \mathbb{Q}|$ for a general number field F, and this is indeed the case. The reader should be familiar with the aforementioned notation for field degree, as well as polynomial degree and background material that is, for instance, contained in [68, Appendix A, pp. 298–306].

Theorem 2.19 | **Monomorphisms of a Number Field**

If F is a number field with degree $|F : \mathbb{Q}| = n$, then there exist exactly n monomorphisms

$$\theta_j : F \to \mathbb{C},$$

for $j = 1, 2, \ldots, n$.

Proof. By Theorem 1.5 on page 10, there is an algebraic integer α such that $F = \mathbb{Q}(\alpha)$. Let $m_{\alpha,\mathbb{Q}}(x)$ be the minimal polynomial of α over $\mathbb{Q}$. It follows from Corollary 1.3 on page 11 that

$$\deg(m_{\alpha,\mathbb{Q}}) = |\mathbb{Q}(\alpha) : \mathbb{Q}| = n.$$

Since $m_{\alpha,\mathbb{Q}}(x)$ has n distinct roots, say $\alpha = \alpha_1, \alpha_2, \ldots, \alpha_n$,

$$m_{\alpha,\mathbb{Q}}(x) = (x - \alpha_1)(x - \alpha_2) \cdots (x - \alpha_n).$$

By Theorem 1.5 on page 10, each element $\beta \in F$ can be expressed uniquely in the form

$$\beta = q_0 + q_1\alpha + \cdots + q_{n-1}\alpha^{n-1}$$

where $q_0, q_1, \ldots, q_{n-1} \in \mathbb{Q}$, so for $j = 1, 2, \ldots, n$ we define

$$\theta_j : F \to \mathbb{C},$$

by

$$\theta_j(\beta) = \theta_j(q_0 + q_1\alpha + \cdots + q_{n-1}\alpha^{n-1}) = q_0 + q_1\alpha_j + \cdots + q_{n-1}\alpha_j^{n-1}.$$

Claim 2.2 *For $j = 1, 2, \ldots, n$, θ_j is a field homomorphism.*

Let $\beta, \gamma \in F$. Then for $q_i, r_i \in \mathbb{Q}, (1 \leq i \leq n-1)$

$$\beta = q_0 + q_1\alpha + \cdots + q_{n-1}\alpha^{n-1} \text{ and } \gamma = r_0 + r_1\alpha + \cdots + r_{n-1}\alpha^{n-1}. \quad (2.13)$$

Therefore,

$$\beta + \gamma = (q_0 + r_0) + (q_1 + r_1)\alpha + \cdots + (q_{n-1} + r_{n-1})\alpha^{n-1},$$

from which we get, for $1 \leq j \leq n$,

$$\theta_j(\beta + \gamma) = (q_0 + r_0) + (q_1 + r_1)\alpha_j \cdots + (q_{n-1} + r_{n-1})\alpha_j^{n-1} =$$

$$(q_0 + q_1\alpha_j + \cdots + q_{n-1}\alpha_j^{n-1}) + (r_0 + r_1\alpha_j + \cdots + r_{n-1}\alpha_j^{n-1}) =$$

$$\theta_j(\beta) + \theta_j(\gamma),$$

so θ_j is additive. It remains to show the θ_j are multiplicative.

In view of (2.13), let

$$f(x) = q_0 + q_1x + \cdots + q_{n-1}x^{n-1} \text{ and } g(x) = r_0 + r_1x + \cdots + r_{n-1}x^{n-1},$$

and use the Euclidean algorithm for polynomials (which we had occasion to use in the proof of Theorem 1.6 on page 10), to establish that there exist $q(x), r(x) \in \mathbb{Q}[x]$ such that

$$f(x)g(x) = m_{\alpha,\mathbb{Q}}(x)q(x) + r(x),$$

where $\deg(r(x)) < \deg(m_{\alpha,\mathbb{Q}}(x)) = n$. Since $f(\alpha) = \beta$ and $g(\alpha) = \gamma$ while $m_{\alpha,\mathbb{Q}}(\alpha) = 0$, then

$$\beta\gamma = f(\alpha)g(\alpha) = m_{\alpha,\mathbb{Q}}(\alpha)q(\alpha) + r(\alpha) = r(\alpha).$$

Thus,

$$\theta_j(\beta\gamma) = \theta_j(r(\alpha)) = r(\alpha_j) = m_{\alpha,\mathbb{Q}}(\alpha_j)q(\alpha_j) + r(\alpha_j) = f(\alpha_j)g(\alpha_j) = \theta_j(\beta)\theta_j(\gamma),$$

so θ_j is multiplicative and we have established Claim 2.2.

Claim 2.3 *For $j = 1, 2, \ldots, n$, θ_j is a monomorphism.*

Suppose that

$$\beta = q_0 + q_1\alpha + \cdots + q_{n-1}\alpha^{n-1} \in F, \gamma = r_0 + r_1\alpha + \cdots + r_{n-1}\alpha^{n-1} \in F,$$

and

$$\theta_j(\beta) = \theta_j(\gamma),$$

so

$$q_0 + q_1\alpha_j + \cdots + q_{n-1}\alpha_j^{n-1} = r_0 + r_1\alpha_j + \cdots + r_{n-1}\alpha_j^{n-1},$$

which means that α_j is a root of

$$h(x) = (q_0 - r_0) + (q_1 - r_1)x + \cdots + (q_{n-1} - r_{n-1})x^{n-1},$$

where $\deg(h(x)) < n$. Since $\deg(h(x)) > 0$ would contradict that

$$m_{\alpha,\mathbb{Q}}(x) = m_{\alpha_j,\mathbb{Q}}(x)$$

is the minimal polynomial of α_j, then $\deg(h(x)) = 0$, so $q_i - r_i = 0$ for $i = 0, 1, \ldots, n - 1$. This means that $q_i = r_i$ for each such i so

$$\beta = q_0 + q_1\alpha + \cdots + q_{n-1}\alpha^{n-1} = r_0 + r_1\alpha + \cdots + r_{n-1}\alpha^{n-1} = \gamma,$$

which secures Claim 2.3.

It remains to show that there are no other monomorphisms of F into $\mathbb{C}$. Let

$$\sigma : F \to \mathbb{C}$$

be a monomorphism. Then

$$m_{\alpha,\mathbb{Q}}(\sigma(\alpha)) = \sigma(m_{\alpha,\mathbb{Q}}(\alpha)) = \sigma(0) = 0,$$

which implies that

$$\sigma(\alpha) = \alpha_j$$

for some $j = 1, 2, \ldots, n$, since these are the only roots of the minimal polynomial. Hence,

$$\sigma(\alpha) = \theta_j(\alpha),$$

so

$$\sigma(q_0 + q_1\alpha + \cdots + q_{n-1}\alpha^{n-1}) = q_0 + q_1\alpha_j + \cdots + q_{n-1}\alpha_j^{n-1} =$$

$$\theta_j(q_0 + q_1\alpha + \cdots + q_{n-1}\alpha^{n-1}),$$

for all $q_j \in \mathbb{Q}$. We have shown that $\sigma = \theta_j$ for some $j = 1, 2, \ldots, n$, which secures the result. $\square$

Theorem 2.19 motivates the following. The reader should solve Exercise 2.22 on page 96 in preparation.

Definition 2.18 | **Conjugates of an Element and a Field**

If $\alpha \in \mathbb{C}$ and F is a number field such that α is algebraic over F, then the *conjugates* of α over F, also called the *F-conjugates* of α, are the roots of $m_{\alpha,F}(x)$ in $\mathbb{C}$. If $F = \mathbb{Q}(\alpha)$, then the fields $\mathbb{Q}(\alpha_j)$ are called the *conjugate fields* of F.

We are now in a position to provide the promised generalization of the notion of norm and related notions.

Definition 2.19 | **Norm and Trace of Elements**

If F is an algebraic number field, $|F : \mathbb{Q}| = n$, and $\alpha \in F$, let $\alpha = \alpha_1, \alpha_2, \ldots, \alpha_n$ be the F-conjugates of α. Then the *norm* of α is

$$N_F(\alpha) = \alpha_1 \alpha_2 \cdots \alpha_n,$$

and the *trace* of α is

$$T_F(\alpha) = \alpha_1 + \alpha_2 + \cdots + \alpha_n.$$

Remark 2.13 From Exercise 2.25 on page 96, we see that $N_F(\alpha), T_F(\alpha) \in \mathbb{Q}$ and

$$\prod_{j=1}^{n}(x - \alpha_j) \in \mathbb{Q}[x].$$

This polynomial is distinguished as follows.

Definition 2.20 | **Field Polynomials over F**

If $\alpha \in F$ where F is a number field, and $\alpha = \alpha_1, \alpha_2, \ldots, \alpha_n$ are the F-conjugates of α, then the *field polynomial of α over F* is given by

$$f_F(\alpha) = (x - \alpha)(x - \alpha_2) \cdots (x - \alpha_n).$$

We now look at a motivating example.

Example 2.13 We look at how to factor the fifth Fermat number

$$F_5 = 2^{32} + 1.$$

For convenience, set $\alpha = \sqrt[3]{-2}$. First, notice that

$$2F_5 = x^3 + 2, \text{ where } x = 2^{11},$$

and that

$$N_F(x - \alpha) = x^3 + 2, \text{ with } x - \alpha \in \mathbb{Z}[\alpha].$$

In fact, by Exercise 2.27, any $\beta = a + b\alpha + c\alpha^2$ has norm

$$N_F(\beta) = a^3 - 2b^3 + 4c^3 + 6abc. \tag{2.14}$$

By Exercise 2.26, there is a prime $\beta \in \mathbb{Z}[\alpha]$ such that $\beta \mid (x - \alpha)$, so by Exercise 2.29,

$$N_F(\beta) \mid N_F(x - \alpha) = x^3 + 2.$$

Hence, we may be able to find a nontrivial factorization of F_5 via norms of certain elements of $\mathbb{Z}[\alpha]$. We do this as follows.

Consider elements of the form $a + b\alpha \in \mathbb{Z}[\alpha]$, for convenience, and sieve over values of a and b, testing for

$$\gcd(N_F(a + b\alpha), F_5) = \gcd(a^3 - 2b^3, F_5) > 1.$$

For convenience, we let a run over the values $1, 2, \ldots, 100$, and b run over the values $b = 1, 2, \ldots 20$. Formal reasons for this approach will be given later. We fix each value of a, and let b run over its range of values. The runs for $1 \leq a \leq 15$ and $1 \leq b \leq 20$ yield

$$\gcd(a^3 - 2b^3, F_5) = 1.$$

However, at $a = 16$, $b = 5$, we get

$$\gcd(16^3 - 2 \cdot 5^3, F_5) = 641.$$

In fact,

$$F_5 = 641 \cdot 6700417.$$

We may factor $16 + 5\alpha$ as follows.

$$16 + 5\alpha = (1 + \alpha)(-1 + \alpha)(\alpha)(-9 + 2\alpha - \alpha^2),$$

where $1 + \alpha$ is a unit with norm -1; $-1 + \alpha$ has norm -3; α has norm -2; and $\beta = -9 + 2\alpha - \alpha^2$ has norm -641. This accounts for

$$16^3 - 2 \cdot 5^3 = 2 \cdot 3 \cdot 641,$$

and shows that β is the predicted prime divisor of $x - \alpha$, which gives us the nontrivial factor of F_5.

The method in Example 2.13 works well largely because of the small value of F_5. However, it may not be feasible for larger values to check all of the gcd conditions over a much larger range. The following method of Pollard, which he introduced in 1991 in [78], uses the above notions of factorizations in $\mathbb{Z}[\alpha]$ to factor F_7, which was first accomplished in 1970.

An important role in factorization is played by the following notion, which we will need as part of the algorithm to be described.

Definition 2.21 $\boxed{\text{Smooth Integers}}$

A rational integer z is said to be smooth with respect to $y \in \mathbb{Z}$, or simply y-smooth, if all prime factors of z are less than or equal to y.

As in the above case, suppose that $n \in \mathbb{N}$ with

$$2n = m^3 + 2.$$

For instance,

$$2F_7 = m^3 + 2$$

where $m = 2^{43}$. Pollard's idea to factor $n = F_7$ involves B-smooth numbers of the form $a + bm$, for some suitable B that will be the number of primes in a prescribed set defined in the algorithm below. Also, $a + b\alpha$ will be B-smooth meaning that its *norm* is B-smooth in the sense of Definition 2.21. Thus, if we get a factorization of $a + b\alpha$ in $\mathbb{Z}[\alpha]$, we also get a corresponding factorization of $a + bm$ modulo F_7. To see this, one must understand a notion that we will generalize when we discuss the number field sieve in Appendix A. We let

$$\psi : \mathbb{Z}[\alpha] \mapsto \mathbb{Z}/n\mathbb{Z}$$

be a ring homomorphism such that $\psi(\alpha) = m$. Thus, in $\mathbb{Z}/n\mathbb{Z}$,

$$x^3 = -2 = -(1+1), \text{ where } 1 \text{ is the identity of } \mathbb{Z}/n\mathbb{Z}.$$

Hence, ψ is that unique map which is defined element-wise by the following.

$$\psi\left(\sum_{j=0}^{2} z_j \alpha^j\right) = \sum_{j=0}^{2} z_j m^j \in \mathbb{Z}/n\mathbb{Z}, \text{ where } z_j \in \mathbb{Z}.$$

The role of this map ψ in attempting to factor a number n is given by the following.

Suppose that we have a set $\mathcal{S}$ of polynomials

$$g(x) = \sum_{j=0}^{2} z_j x^j \in \mathbb{Z}[x]$$

such that

$$\prod_{g \in \mathcal{S}} g(\alpha) = \beta^2$$

where $\beta \in \mathbb{Z}[\alpha]$, and

$$\prod_{g \in \mathcal{S}} g(m) = y^2,$$

where $y \in \mathbb{Z}$. Then if $\psi(\beta) = x \in \mathbb{Z}$, we have

$$x^2 \equiv \psi(\beta)^2 \equiv \psi(\beta^2) \equiv \psi\left(\prod_{g \in \mathcal{S}} g(\alpha)\right) \equiv \prod_{g \in \mathcal{S}} g(m) \equiv y^2 \pmod{n}.$$

In other words, this method finds a pair of integers x, y such that

$$x^2 - y^2 \equiv (x - y)(x + y) \equiv 0 \pmod{n},$$

so we may have a nontrivial factor of n by looking at $\gcd(x - y, n)$.

We now describe the algorithm, but give a simplified version of it, since this is meant to be a simple introduction to the ideas behind the number field sieve, which we will present in detail in Appendix A. The following is adapted from [64].

We use a very small value of n as an example for the sake of simplicity, namely $n = 23329$. Note that $2n = 36^3 + 2 = m^3 + 2$. We will also make suitable references in the algorithm in terms of how Pollard factored $n = F_7$.

◆ Pollard's Algorithm

Step 1. Compute a *factor base.*

The term "factor base" means the choice of a suitable set of rational primes over which we may factor a set of integers. In the case of cubic integers in $\mathbb{Z}[\alpha] = \mathbb{Z}[\sqrt[3]{-2}]$, we take for $n = 23329$ only the first eleven primes, those up to and including 41 (or for $n = F_7$, Pollard chose the first five hundred rational primes) as $\mathcal{FB}_1$, the first part of the factor base, and for the second part, $\mathcal{FB}_2$, we take those primes of $\mathbb{Z}[\alpha]$ with norms $\pm p$, where $p \in \mathcal{FB}_1$. (The reasons behind the choice of the number of primes in $\mathcal{FB}_1$ are largely empirical.) Also, we include the units $-1, 1 + \alpha$, and $1/(1 + \alpha) = -1 + \alpha - \alpha^2$ in $\mathcal{FB}_2$. Here, we have discarded the $\mathbb{Z}[\alpha]$-primes of norm p^2 or p^3, since these cannot divide our n, given that they cannot divide the $a + b\alpha$, with the assumptions we are making.

Step 2. Run the sieve.

In this instance, the sieve involves finding numbers $a + bm$ that are composed of some primes from $\mathcal{FB}_1$. For $n = 23329$, we sieve over values of a from -5 to 5 and values of b from 1 to 10 (or for $n = F_7$, Pollard chose values of a from -4800 to 4800, and values of b from 1 to 2000). Save only coprime pairs (a, b).

Step 3. Look for smooth values of the norm, and obtain factorizations of $a + bx$ and $a + b\alpha$.

Here, smooth values of the norm means that $N = N_F(a + b\alpha) = a^3 - 2b^3$ is not divisible by any primes bigger than those in $\mathcal{FB}_1$. For those (a, b) pairs, factor $a + bm$ by trial division, and eliminate unsuccessful trials. Factor $a + b\alpha$ by computing the norm $N_F(a + b\alpha)$ and using trial division. When a prime p is found, then divide out a $\mathbb{Z}[\alpha]$-prime of norm $\pm p$ from $a + b\alpha$. This will involve getting primes in the factorization of the form $a + b\alpha + c\alpha^2$ where $c \neq 0$. Units may also come into play in the factorizations, and a table of values of $(1 + \alpha)^j$ is kept for such purposes with $j = -2, \cdots, 2$ for $n = 23329$ (or for F_7, one should choose to keep a record of units for $j = -8, -7, \ldots, 8$). Some data extracted for the run on $n = 23329$ is given as follows.

Table 2.1

$a + b\alpha + c\alpha^2$	N	factorization of $a + b\alpha + c\alpha^2$
$5 + \alpha$	$3 \cdot 41$	$(-1+\alpha)(-1-2\alpha-2\alpha^2)$
$4 + 10\alpha$	$-2^4 \cdot 11^2$	$-(3+2\alpha)^2\alpha^4(-1+\alpha-\alpha^2)^2$
$-1 + \alpha$	-3	$-1 + \alpha$
$-1 - 2\alpha - 2\alpha^2$	-41	$-1 - 2\alpha - 2\alpha^2$
$3 + 2\alpha$	11	$3 + 2\alpha$
α	-2	α
$-1 + \alpha - \alpha^2$	-1	*unit*

Table 2.2

$a + bm + cm^2$	factorization of $a + bm + cm^2$
$5 + m$	41
$4 + 10m$	$2^2 \cdot 7 \cdot 13$
$-1 + m$	$5 \cdot 7$
$-1 - 2m - 2m^2$	$-5 \cdot 13 \cdot 41$
$3 + 2m$	$3 \cdot 5^2$
m	$2^2 \cdot 3^2$
$-1 + m - m^2$	$-13 \cdot 97$

Step 4. Complete the factorization.

By selecting -1 times the first four rows in the third column of Table 2.1, we get a square in $\mathbb{Z}[\alpha]$:

$$\beta^2 = (-1+\alpha)^2(-1-2\alpha-2\alpha^2)^2(3+2\alpha)^2\alpha^4(-1+\alpha-\alpha^2)^2, \qquad (2.15)$$

and correspondingly, since β^2 is also -1 times the first four rows in the first column of Table 2.1, we get:

$$\beta^2 = (5+\alpha)(-4-10\alpha)(-1+\alpha)(-1-2\alpha-2\alpha^2). \qquad (2.16)$$

Then we get a square in $\mathbb{Z}$ from Table 2.2 by applying ψ to (2.16):

$$\psi(\beta^2) = (5+m)(-4-10m)(-1+m)(-1-2m-2m^2) = 2^2 \cdot 5^2 \cdot 7^2 \cdot 13^2 \cdot 41^2 = y^2.$$

Also, by applying ψ to β via (2.15), we get:

$$\psi(\beta) = (-1+m)(-1-2m-2m^2)(3+2m)m^2(-1+m-m^2) \equiv 9348 \pmod{23329},$$

so by setting $x = \psi(\beta)$, we have $x^2 = \psi^2(\beta) = \psi(\beta^2) \equiv y^2 \pmod{n}$. Since $y = 2 \cdot 5 \cdot 7 \cdot 13 \cdot 41 \equiv 13981 \pmod{23329}$, then $y - x \equiv 4633 \pmod{23329}$. However, $\gcd(4633, 23329) = 41$. In fact $23329 = 41 \cdot 569$.

Pollard used the algorithm in a similar fashion to find integers X and Y for the more serious factorization $\gcd(X - Y, F_7) = 59649589127497217$. Hence, we have a factorization of F_7 as follows.

$$F_7 = 59649589127497217 \cdot 5704689200685129054721.$$

Essentially, the ideas for factoring using cubic integers above is akin to the notion of the strategy used in the quadratic sieve method. There, we try to generate sufficiently many smooth quadratic residues of n close to $\sqrt{n}$. In the cubic case, we try to factor numbers that are close to perfect cubes. In Appendix A, we will extend these ideas to show how F_9 was factored using the number field sieve, and $\mathbb{Z}[\sqrt[5]{2}]$.

Exercises

2.21. Prove that $\mathbb{Z}[\sqrt[3]{-2}]$ is the ring of integers of $\mathbb{Q}(\sqrt[3]{-2})$.

2.22. Let F be a number field and let $\alpha \in \mathbb{A}$ such that $F = \mathbb{Q}(\alpha)$. Prove that if α_j for $j = 1, 2, \ldots, n$ are all the F-conjugates of α, then all the fields $\mathbb{Q}(\alpha_j)$ are isomorphic for $j = 1, 2, \ldots, n$.

2.23. Prove that if $F \subseteq K \subseteq E \subseteq \mathbb{C}$, where F, K, E are fields, then $|E : F| = |E : K| \cdot |K : F|$, where any of the degrees may be infinite.

2.24. Suppose that $F = \mathbb{Q}(\alpha)$ where $\alpha \in \mathbb{A}$, $\beta \in F$, and $\beta = \beta_1, \beta_2, \ldots, \beta_n$ are the F-conjugates of β. Prove that if $m_{\beta,\mathbb{Q}}(x) = x^d + q_{d-1}x^{d-1} + \cdots + q_1 x + q_0$ is the minimal polynomial of β over $\mathbb{Q}$, then

$$\prod_{j=1}^{n}(x - \beta_j) = m_{\beta,\mathbb{Q}}(x)^{n/d} \in \mathbb{Q}[x].$$

2.25. If F is a number field and $\beta \in F$ prove that $N_F(\beta) \in \mathbb{Q}$ and $T_F(\beta) \in \mathbb{Q}$, and if $\beta \in \mathfrak{O}_F$, then $N_F(\beta) \in \mathbb{Z}$ and $T_F(\beta) \in \mathbb{Z}$.

Conclude that if $f_F(x) = \prod_{j=1}^{n}(x - \beta_j)$ is the field polynomial of β, then $-T_F(\beta)$ is the coefficient of x^{n-1} and $\pm N_F(\beta)$ is the constant term.

2.26. Prove that every nonzero ideal in a Dedekind domain R must contain a prime element.

2.27. Prove that (2.14) holds in Example 2.13 on page 91.

2.28. Prove that the norm given in Definition 2.19 on page 91 is multiplicative and the trace is additive. In other words, for any $\alpha, \beta \in \mathfrak{O}_F$, $N_F(\alpha\beta) = N_F(\alpha)N_F(\beta)$, and $T_F(\alpha + \beta) = T_F(\alpha) + T_F(\beta)$.

2.29. Prove that if $\beta \mid \gamma$ for $\beta, \gamma \in \mathfrak{O}_F$, where F is a number field, then $N_F(\beta)$ divides $N_F(\gamma)$.

2.30. Use Pollard's method to factor F_6.

In Exercises 2.31–2.33, use the gcd method described before Pollard's method to find an odd factor of the given integer.

2.31. $5^{77} - 1$.

2.32. $7^{149} + 1$. (*Hint: Use $\mathbb{Z}[\sqrt[3]{-7}]$.*)

2.33. $3^{239} - 1$. (*Hint: Use $\mathbb{Z}[\sqrt[3]{3}]$.*)

Chapter 3

Binary Quadratic Forms

> *What is it indeed that gives us the feeling of elegance in a solution, in demonstration? It is the harmony of the diverse parts, their symmetry, their happy balance; in a word it is all that introduces order, all that gives unity, that permits us to see clearly and to comprehend at once both the ensemble and the details.*
>
> **Henri Jules Poincaré (1854–1912)**
> **French mathematician–see Biography 3.8 on page 147**

This chapter requires that the reader have a basic understanding of the fundamental background material on abstract algebra such as to be found, for instance, in [68, Appendix A]. We take an algebraic approach to binary quadratic forms that is straightforward and unmasks some of the otherwise difficult-to-interpret underpinnings of the theory.

3.1 Basics

Lagrange was the first to introduce the theory of quadratic forms, later expanded by Legendre, and greatly magnified even later by Gauss (see [68, Biography 2.7, p. 114], [68, Biography 4.1, p. 181], and [68, Biography 1.7, p. 33]). An *integral binary quadratic form* is given by

$$f(x,y) = ax^2 + bxy + cy^2 \text{ with } a, b, c \in \mathbb{Z}. \tag{3.1}$$

For simplicity, we may suppress the variables, and denote f by (a, b, c). The value a is called the *leading coefficient*, the value b is called the *middle coefficient*, and c is called the *last coefficient*. If $\gcd(a, b, c) = 1$, then we say that $f(x,y)$ is a *primitive* form.

The aforementioned three great mathematicians looked at the *representation problem*: Given a binary quadratic form (3.1), which $n \in \mathbb{Z}$ are represented by $f(x, y)$? In other words, for which n do there exist integers x, y such that $f(x, y) = n$? If $\gcd(x, y) = 1$, then we say that n is *properly represented* by $f(x, y)$. For instance, when studying criteria for the representation of a natural number n as sums of two squares, such as in Theorem 1.13 on page 26, or [68, Section 6.1, pp. 243–251], a simple answer can be given. When looking at norm-forms $x^2 + Dy^2 = n$, where $D \in \mathbb{Z}$, such as in [18] or [68, Section 7.1, pp. 265–273], the problem can be given a relatively simple answer for certain n and D. In general, there is no simple complete answer. Moreover, an even more general and difficult problem arises, namely when can an integer be represented by a binary quadratic form from a given set of such forms? The theory of binary quadratic forms deals with this question via the following notion. In the balance of our discussion, we use the term *form* to mean *binary quadratic form*.

Definition 3.1 | **Equivalent Binary Quadratic Forms**

Two forms $f(x, y)$ and $g(x, y)$ are said to be *equivalent* if there exist integers p, q, r, s, such that

$$f(x, y) = g(px + qy, rx + sy) \text{ and } ps - qr = \pm 1. \tag{3.2}$$

For simplicity, we may denote equivalence of f and g by $f \sim g$. If $ps - qr = 1$, then f and g are said to be *properly equivalent*, and if $ps - qr = -1$, they are said to be *improperly equivalent*. Two forms f and g are said to be in the same *equivalence class* or simply in the *same class*, if f is properly equivalent to g.

Remark 3.1 Definition 3.1 says that equivalent forms represent the same integers, and the same is true for proper representation – see Exercise 3.1 on page 103. Moreover, since

$$\det \begin{pmatrix} p & q \\ r & s \end{pmatrix} = ps - qr = \pm 1,$$

this means that

$$\begin{pmatrix} p & q \\ r & s \end{pmatrix} \in \mathrm{GL}(2, \mathbb{Z}),$$

– see Exercise 2.5 on page 66. Note, as well, that proper equivalence means that $ps - qr = 1$ so

$$\begin{pmatrix} p & q \\ r & s \end{pmatrix} \in \mathrm{SL}(2, \mathbb{Z}),$$

the subgroup of $\mathrm{GL}(2, \mathbb{Z})$ with elements having determinant 1. Properly equivalent forms are said to be related by a *unimodular transformation*, namely $X = px + qy$ and $Y = rx + sy$ with $ps - qr = 1$. Note as well, by Exercise 3.3 on page 103, proper equivalence of forms is an equivalence relation.

The notion of proper and improper equivalence is due to Gauss. Lagrange initiated the idea of equivalence, although he did not use the term. He merely said that one could be "transformed into another of the same kind," but did not make the distinction between the two kinds. Similarly Legendre did not recognize proper equivalence. However, there is a very nice relationship between proper representation and proper equivalence, since as Exercise 3.2 on page 103 shows, the form $f(x, y)$ properly represents $n \in \mathbb{Z}$ if and only if $f(x, y)$ is properly equivalent to the form $nx^2 + bxy + cy^2$ for some $b, c \in \mathbb{Z}$.

Example 3.1 For $f(x, y) = x^2 + 7y^2$, $n = 29 = 1 + 7 \cdot 2^2 = f(1, 2)$, $f(x, y)$ is properly equivalent to $g(x, y) = 29x^2 + 86xy + 64y^2$ since $f(x, y) = g(3x - y, -2x + y)$, where $p = 3, q = -1, r = -2, s = 1$. With reference to Remark 3.1 on the facing page, $X = 3x - y$, $Y = -2x + y$ represents a unimodular transformation.

The following notion is central to the discussion and links equivalent forms in another way.

Definition 3.2 | **Discriminants of Forms**

The *discriminant* of the form $f(x, y) = ax^2 + bxy + cy^2$ is given by

$$D = b^2 - 4ac.$$

If $D > 0$, then f is called an *indefinite* form. If $D < 0$ and $a < 0$, then f is called a *negative definite* form, and if $D < 0$ and $a > 0$, then f is called a *positive definite* form.

Remark 3.2 By Exercise 3.7 on page 103, if forms f and g have discriminants D and D_1, respectively, and $f(x, y) = g(px+qy, rx+sy)$, then $D = (ps-qr)^2 D_1$. Thus, equivalent forms have the same discriminant. However, forms with the same discriminant are not necessarily equivalent — see Exercise 3.8 on page 104. Furthermore, if $f(x, y) = ax^2 + bxy + cy^2$, then by completing the square, we get

$$4af(x, y) = (2ax + by)^2 - Dy^2,$$

so when $D > 0$, the form $f(x, y)$ represents both positive and negative integers. This is the justification for calling such forms "indefinite." If $D < 0$ and $a < 0$, then $f(x, y)$ represents only negative integers, thus the reason they are called "negative definite," and if $a > 0$, then they represent only positive integers, whence the term "positive definite." Since we may change a negative definite form into a positive definite one by changing the signs of all the coefficients, it is sufficient to consider only positive definite forms when $D < 0$. *We will, therefore, not consider negative definite forms in any discussion hereafter.*

Congruence properties of the discriminant of a form may provide us with information on representation. For instance, Exercise 3.9 on page 104 tells us that congruence properties modulo 4 determine when an integer may be represented by forms with discriminant $D \equiv 0,1\,(\mathrm{mod}\ 4)$. Furthermore, what this tells us is that we can take the equation $D = b^2 - 4ac$ and let $a = 1$ and $b = 0$ or 1 according as $D \equiv 0$ or $1\,(\mathrm{mod}\ 4)$, so then $c = -D/4$ or $-(D-1)/4$, respectively. Thus, we get a distinguished form of discriminant D given as follows.

Definition 3.3 | **Principal Forms**

If $D \equiv 0,1\,(\mathrm{mod}\ 4)$, then $(1,0,-D/4)$ or $(1,1,-(D-1)/4)$, respectively, are called *principal forms* of discriminant D.

Remark 3.3 Via Exercise 3.10, we see that if $D = -4m$, we get the form x^2+my^2. As we shall see, these forms are particularly important in the historical development of the representation problem. Indeed, entire books, such as [18] are devoted to discussing this issue. There is a general notion that allows us to look at canonical forms for more illumination of the topic. This is given in the following which is due to Lagrange.

Definition 3.4 | **Reduced Forms**

A primitive form $f(x,y) = ax^2 + bxy + cy^2$, of discriminant D, is said to be *reduced* if

(a) When $D < 0$ and $a > 0$, then

$$|b| \le a \le c, \text{ and if either } |b| = a \text{ or } a = c, \text{ then } b \ge 0. \qquad (3.3)$$

(b) When $D > 0$, then

$$0 < b < \sqrt{D} \text{ and } \sqrt{D} - b < 2|a| < \sqrt{D} + b. \qquad (3.4)$$

Note that since f is positive definite in part (a) of Definition 3.4, then by Definition 3.2 on the preceding page, both a and c are positive.

With the notion of reduction in hand, we have the following result, which provides us with a unique canonical representative for equivalence classes of positive definite forms.

Theorem 3.1 | **Positive Definite and Reduced Forms**

Every positive definite form is properly equivalent to a unique reduced form.

Proof. Let $f(x,y) = ax^2 + bxy + cy^2$ be a primitive positive definite form. Let n be the least positive integer represented by f. By Exercise 3.2 on page 103, there exist $B, C \in \mathbb{Z}$ such that $f \sim g$ properly, where $g(X,Y) = nX^2 + BXY + CY^2$. For any integer z, the transformation $X = x - zy$, $Y = y$ yields

$$g(X,Y) = nx^2 + (B - 2nz)xy + (nz^2 - Bz + C)y^2.$$

If we set $z = Ne\left(\frac{B}{2n}\right)$, the nearest integer to $B/(2n)$, then

$$-\frac{1}{2} < \frac{B}{2n} - z \le \frac{1}{2}, -n \le B - 2nz \le n, \text{ and } |B - 2nz| \le n.$$

Thus, if we set $b_1 = B - 2nz$ and $c_1 = nz^2 - Bz + C$, then

$$g(X, Y) = nx^2 + b_1 xy + c_1 y^2,$$

where $|b_1| \le n$. Thus, f is properly equivalent to g, g is positive definite, and $g(0, 1) = c_1$. Therefore, g represents c_1, which implies $c_1 \in \mathbb{N}$, and $c_1 \ge n$ by the minimality of n. We have shown that f is properly equivalent to a reduced form. The balance of the result will follow from the next result.

Claim 3.1 *Any two properly equivalent reduced forms must be identical.*

Suppose that the form $f(x, y) = ax^2 + bxy + cy^2$ is reduced and properly equivalent to the reduced form $g(x, y) = Ax^2 + Bxy + Cy^2$ via the transformation $f(x, y) = g(px + qy, rx + sy)$ with $ps - qr = 1$. We may assume without loss of generality that $a \ge A$. Also, a straightforward calculation shows that

$$A = ap^2 + bpr + cr^2,$$

$$B = 2apq + b(ps + qr) + 2crs, \tag{3.5}$$

$$C = aq^2 + bqs + cs^2.$$

Furthermore, we have that

$$|b| \le a \le c. \tag{3.6}$$

Using (3.6) we get,

$$A = ap^2 + bpr + cr^2 \ge ap^2 - |bpr| + cr^2$$

$$\ge ap^2 - |bpr| + ar^2 = a(p^2 + r^2) - |bpr|. \tag{3.7}$$

However, since $p^2 + r^2 \ge 2|pr|$, then (3.7) is greater than or equal to $2a|pr| - |bpr| \ge a|pr|$, where the latter inequality follows from (3.6) again. We have shown that

$$A \ge a|pr|. \tag{3.8}$$

However, by assumption $a \ge A$, so $|pr| \le 1$. If $|pr| = 0$, then

$$A = ap^2 + bpr + cr^2 \ge ap^2 + ar^2 = a(p^2 + r^2) \ge a,$$

from which it follows that $A = a$. On the other hand, if $|pr| = 1$, then by (3.8) $A \ge a$, so again we get $A = a$.

It remains to show that $B = b$ since, once shown, it follows from Exercise 3.7 on page 103 that $C = c$ since $B^2 - 4AC = b^2 - 4ac$.

Suppose that $c > C$. Then $c > a$ since $C \ge A = a$. If $|pr| = 1$, then by (3.7), using the fact that $cr^2 > ar^2$, we get that $A > a$, a contradiction. Hence, $|pr| = 0$. If $p = 0$, then again using (3.7), we conclude that $A > a$, so $r = 0$.

Since $ps - qr = 1$, then $ps = 1$. Moreover, since $|B| \leq A = a$ given that g is reduced, then from (3.6), we get $-a \leq |B| - |b| \leq a$. However, by (3.5), $B = 2apq + b$. It follows that $q = 0$ and $B = b$.

Lastly, suppose that $c < C$. By solving for a, b, c in terms of A, B, C we may reverse the roles of the variables and argue as above to the same conclusion that $B = b$. This completes the proof. $\square$

Remark 3.4 The above says that there is a unique representative for each equivalence class of positive definite binary quadratic forms. Furthermore, by Exercise 3.11 on page 104, when $D < 0$, the number h_D of classes of primitive positive definite forms of discriminant D is finite, and h_D is equal to the number of reduced primitive forms of discriminant D. (Note that we prove $h_D < \infty$ in general for field discriminants in Theorem 3.7 on page 116.)

The case for indefinite forms is not so straightforward. The uniqueness issue, in particular, is complicated since we may have many reduced forms equivalent to one another, and the determination as to which reduced forms are equivalent is more difficult. Yet, we resolve this issue in Theorem 3.5 on page 110.

We conclude this section with a result due to Landau–see Biography 3.1 on page 104. This result precisely delineates the negative discriminants $D = -4n$ for which $h_D = 1$ and the proof is essentially that of Landau [48].

Theorem 3.2 | When h$_{-4n}$ = 1 for n > 0 |

If $n \in \mathbb{N}$, then $h_{-4n} = 1$ if and only if $n \in \{1, 2, 3, 4, 7\}$.

Proof. Suppose that $h_{-4n} = 1$. $f(x, y) = x^2 + ny^2$ is clearly reduced since $a = 1$, $b = 0$, and $c = n \geq 1$ in Definition 3.4 on page 100. The result is clear for $n = 1$, so we assume that $n > 1$.

Case 3.1 *n is not a prime power.*

There exists a prime $p \mid n$ such that $p^d \| n$, for $d \in \mathbb{N}$, where $\|$ denotes *proper division*, also commonly called *exactly divides*, namely $p^d \mid n$, but $p^{d+1} \nmid n$ — see [68, Definition 1.3, p. 16] for the general notion. Let $a = \min(p^d, n/p^d)$ and $c = \max(p^d, n/p^d)$. Thus, $\gcd(a, c) = 1$, where $1 < a < c$, since n is not a prime power. Thus, $g(x, y) = ax^2 + cy^2$ is a reduced form of discriminant $-4ac = -4n$, so $h_{-4n} > 1$, given that $f(x, y)$ is also a reduced form of discriminant D, unequal to $g(x, y)$. This completes Case 3.1.

Case 3.2 *n = 2^ℓ where $\ell \in \mathbb{N}$.*

We need to show that $h_{-4n} > 1$ for $\ell \geq 3$. If $\ell = 3$, then $D = -32$ and the form $g(x, y) = 3x^2 + 2xy + 3y^2$ is a reduced form of discriminant $2^2 - 4 \cdot 3 \cdot 3 = -32$ not equal to $f(x, y)$, so we may assume that $\ell \geq 4$. Set

$$g(x, y) = 4x^2 + 4xy + (2^{\ell-2} + 1)y^2,$$

which is primitive since $\gcd(4, 4, 2^{\ell-2}+1) = 1$, and reduced since $4 < 2^{\ell-2}+1$. Moreover, the discriminant is

$$D = 4^2 - 4 \cdot 4 \cdot (2^{\ell-2}+1) = -16 \cdot 2^{\ell-2} = -2^{\ell+2} = -4n,$$

but $g \neq f$. This completes Case 3.2.

Case 3.3 $n = p^k$ *where* $p > 2$ *is prime and* $k \in \mathbb{N}$.

Suppose that $n+1$ is not a prime power. Then, as in Claim 3.1, we may write $n+1 = ac$, where $1 < a < c$ and $\gcd(a, c) = 1$. Thus, $g(x, y) = ax^2 + 2xy + cy^2$ is a reduced form of discriminant $2^2 - 4ac = 4 - 4(n+1) = -4n$, and $f \neq g$, so $h_{-4n} > 1$.

Lastly suppose that $n + 1 = 2^t$ where $t \in \mathbb{N}$, observing that $n + 1 = p^k + 1$ is even. If $t \geq 6$, then $g(x, y) = 8x^2 + 6xy + (2^{t-3}+1)y^2$ is reduced since $8 < 2^{t-3}+1$, and $\gcd(8, 6, 2^{t-3}+1) = 1$. Also, g has discriminant

$$D = 6^2 - 4 \cdot 8(2^{t-3}+1) = 4 - 4 \cdot 2^t = 4 - 4(n+1) = -4n,$$

and $f \neq g$, so $h_{-4n} > 1$. For $t \leq 5$ we have that $t \in \{1, 2, 3, 4, 5\}$ have the corresponding values $n \in \{1, 3, 7, 15, 31\}$. It remains to exclude $n = 15, 31$.

If $n = 15$, then n is not a prime power so this violates the hypothesis of Case 3.3. If $n = 31$, then the form $g(x, y) = 5x^2 + 4xy + 7y^2$ is reduced since $b = 4 < a = 5 < c = 7$, and is primitive since $\gcd(a, b, c) = 1$. Lastly, the discriminant is

$$D = 4^2 - 4 \cdot 5 \cdot 7 = -4 \cdot 31.$$

This completes Case 3.3, and we are done for this direction of the proof.

Now we assume that $n \in \{1, 2, 3, 4, 7\}$. That $h_{-4n} = 1$ is Exercise 3.13. $\quad\square$

Exercises

3.1. Prove that equivalent forms represent the same integers, and the same is true for proper representation.

3.2. Prove that the form $f(x, y)$ properly represents n if and only if $f(x, y)$ is properly equivalent to the form $nx^2 + Bxy + Cy^2$ for some $B, C \in \mathbb{Z}$.

3.3. Prove that proper equivalence of forms is an equivalence relation, namely that the properties of reflexivity, symmetry, and transitivity are satisfied.

3.4. Prove that improper equivalence is not an equivalence relation.

3.5. Prove that any form equivalent to a primitive form must itself be primitive.

3.6. Prove that if f represents $n \in \mathbb{Z}$, then there exists a $g \in \mathbb{N}$ such that $n = g^2 n_1$ and f properly represents n_1.

3.7. Suppose that $f \sim g$ where f is a form of discriminant D and g is a form of discriminant D_1, then $D = (ps - qr)^2 D_1 = D_1$, where $f(x, y) = g(px + qy, rx + sy)$.

3.8. Provide an example of forms with the same discriminant that are not equivalent.

3.9. Let $D \equiv 0, 1 \pmod 4$ and let n be an integer relatively prime to D. Prove that if n is properly represented by a primitive form of discriminant D, then D is a quadratic residue modulo $|n|$, and if n is even, then $D \equiv 1 \pmod 8$. Conversely, if n is odd and D is a quadratic residue modulo $|n|$, or n is even and D is a quadratic residue modulo $4|n|$, then $n \in \mathbb{Z}$ is properly represented by a primitive form of discriminant D.

3.10. Let $n \in \mathbb{Z}$ and $p > 2$ a prime not dividing n. Prove that p is represented by a primitive form of discriminant $-4n$ if and only if the Legendre symbol equality $(-n/p) = 1$ holds.

(*Hint: Use Exercise 3.9.*)

3.11. For a fixed integer $D < 0$, let h_D be the number of classes of primitive positive definite forms of discriminant D. Prove that h_D is finite and is equal to the number of reduced forms of discriminant D.

3.12. Let $n \in \mathbb{N}$ and $p > 2$ prime with $p \nmid n$. Prove that the Legendre symbol $(-n/p) = 1$ if and only if p is represented by one of the h_{-4n} reduced forms of discriminant $-4n$.

(*Hint: See Exercises 3.10–3.11 and Theorem 3.1 on page 100.*)

3.13. Prove that if $n \in \{1, 2, 3, 4, 7\}$, then $h_{-4n} = 1$.

Biography 3.1 Edmund Landau (1877–1938) *was born in Berlin, Germany on February 14, 1877. He studied mathematics at the University of Berlin, where his doctoral thesis, awarded in 1899, was supervised by Frobenius. Landau taught at the University of Berlin for the decade 1899–1909. In 1909, when he was appointed as ordinary professor at the University of Göttingen, he had amassed nearly seventy publications. His appointment at Göttingen was as a successor to Minkowski. Hilbert and Klein were also colleagues there–see Biography 3.5 on page 127. He became full professor there until the Nazis forced him out in 1933. On November 19, 1933, he was given permission to work at Groningen, Netherlands, where he remained until he retired on February 7, 1933. He returned to Berlin where he died of a heart attack on February 19, 1938.*

Landau's major contributions were in analytic number theory and the distribution of primes. For instance, his proof of the prime number theorem, published in 1903, was much more elementary than those given by Poussin and Hadamard–see [68, §1.9, pp. 65–72] for a detailed overview. He established more than 250 publications in number theory and wrote several books on number theory, which were influential.

3.2 Composition and the Form Class Group

> *The further mathematical theory is developed, the more harmoniously and uniformly does its construction proceed, and unsuspected relations are disclosed between hitherto separated branches of the science.*
>
> **David Hilbert–see Biography 3.5 on page 127**

Gauss is responsible for being the first to see the deep connections within genus theory (which we will study in §3.4) and composition, even though the seeds were there in the earlier work of Lagrange. However Gauss's definition of composition is difficult to use. Something close to Gauss' idea is given via Exercise 3.30 on page 145 in the positive definite case, where the product of two forms $f(x_1, y_1)$ and $g(x_2, y_2)$ of discriminant $-4n$ is equal to a form $F(X, Y)$ where X and Y are integral bilinear forms. We take an approach that is due to Dirichlet and is much easier. First we need to develop some new notions. The first result allows us to select a canonical form in each equivalence class. For ease of elucidation, we restrict our attention to discriminants that are field discriminants–see Definition 1.6 on page 7.

Lemma 3.1 | **Canonical Forms**

Let $F = \mathbb{Q}(\sqrt{\Delta_F})$ be a quadratic field of discriminant Δ_F and let $m \in \mathbb{Z}$. Then every proper equivalence class of forms of discriminant Δ_F contains a primitive form with positive leading coefficient that is relatively prime to m.

Proof. Let $f = (a, b, c) \in C_{\Delta_F}$ and set

$$P_{a,m,c} = \prod_p p$$

where the product ranges over all distinct primes p such that $p \mid a$, $p \mid c$ and $p \mid m$. Also set

$$P_{a,m} = \prod_q q$$

where the product ranges over all distinct primes q such that $q \mid a$, $q \mid m$, but $q \nmid c$, let

$$P_{c,m} = \prod_r r$$

where the product ranges over all distinct primes r such that $r \mid c$, $r \mid m$, but $r \nmid a$, and

$$S_m = \prod_s s$$

where the product ranges over all distinct primes s such that $s \mid m$ but $s \nmid P_{a,m,c}P_{a,m}P_{c,m}$. Then f represents

$$aP_{a,m}^2 + bP_{a,m}P_{c,m}S_m + c(P_{c,m}S_m)^2 = N. \tag{3.9}$$

Claim 3.2 $\gcd(N, m) = 1$.

Assume that a prime $t \mid N$ and $t \mid m$. Assume first that $t \mid a$. Then

$$t \mid P_{a,m,c}P_{a,m}$$

by the definition of the latter. If $t \mid P_{a,m}$, then by (3.9),

$$t \mid cP_{c,m}S_m.$$

However, $t \nmid P_{c,m}S_m$, so $t \mid c$. This contradicts the fact that $t \mid P_{a,m}$. Hence, $t \nmid P_{a,m}$, so $t \mid P_{a,m,c}$. It follows from (3.9) that

$$t \mid bP_{a,m}P_{c,m}S_m.$$

However, we have already shown that $t \nmid P_{a,m}$ and since $t \mid a$, then $t \nmid P_{c,m}$. Also, $t \mid P_{a,m,c}$, so $t \nmid S_m$, which implies that $t \mid b$. We have shown that $t \mid \gcd(a, b, c)$, contradicting that f is primitive. Hence, our initial assumption was false, namely we have shown that $t \nmid a$. Therefore,

$$t \mid P_{c,m}S_m$$

by the definition of the latter. However, by (3.9), this implies that $t \mid aP_{a,m}$, a contradiction to what we have already shown. This secures the claim.

By Exercise 3.2 on page 103, Claim 3.2 tells us that f is properly equivalent to the form

$$g(x, y) = Nx^2 + Bxy + Cy^2$$

for some $B, C \in \mathbb{Z}$. If $N > 0$, then we have our result.

If $N < 0$, then by setting $x_0 = Bm\ell + 1$ and $y_0 = -2N\ell m$ for some $\ell \in \mathbb{Z}$,

$$g(x_0, y_0) = Nx_0^2 + Bx_0y_0 + Cy_0^2$$

$$= N(Bm\ell + 1)^2 + B(Bm\ell + 1)(-2N\ell m) + C(2N\ell m)^2$$

$$= NB^2m^2\ell^2 + 2NBm\ell + N - 2NB^2m^2\ell^2 - 2NB\ell m + 4CN^2\ell^2m^2$$

$$= N(1 - m^2\ell^2(B^2 - 4NC)) = N(1 - m^2\ell^2\Delta_F) = Q,$$

where $Q > 0$ if $N < 0$.

Since f represents

$$Q = N(1 - m^2\ell^2\Delta_F)$$

and Q is relatively prime to m, given that N and $1 - m^2\ell^2\Delta_F$ are relatively prime to m, then Exercise 3.2 gives us the complete result.   □

Now we make a connection with ideals that greatly simplifies the presentation.

Theorem 3.3 | Ideals and Composition of Forms

Suppose that $\mathfrak{O}_F$ is the ring of integers of a quadratic field of discriminant Δ_F and

$$f(x,y) = ax^2 + bxy + cy^2$$

is a primitive form, with $a > 0$, of discriminant $\Delta_F = b^2 - 4ac$. Then

$$I = (a, (-b + \sqrt{\Delta_F})/2)$$

is an $\mathfrak{O}_F$-ideal.

Proof. Since $\Delta_F = b^2 - 4ac$, then $b^2 \equiv \Delta_F \pmod{4a}$, so by Exercise 2.4 on page 66, I is an $\mathfrak{O}_F$-ideal. $\qquad\square$

Note that in Theorem 3.3, we must exclude the case $a < 0$ since the norm of an ideal must be positive. This excludes the negative definite case, but in view of Remark 3.2 on page 99, there is no loss of generality. Moreover, in the indefinite case, with $a < 0$, we may circumvent this via the techniques given in the proof of Theorem 3.5 on page 110. In particular, see (3.13) on page 112.

Definition 3.5 | United Forms

Two primitive forms $f = (a_1, b_1, c_1)$ and $g = (a_2, b_2, c_2)$ of discriminant D are called *united* if $\gcd(a_1, a_2, (b_1 + b_2)/2) = 1$.

Note that in Definition 3.5, since $b_1^2 - 4a_1c_1 = b_2^2 - 4a_2c_2$, then b_1 and b_2 have the same parity so $(b_1 + b_2)/2 \in \mathbb{Z}$.

Theorem 3.4 | United Forms and Uniqueness

If $f = (a_1, b_1, c_1)$ and $g = (a_2, b_2, c_2)$ are united forms of discriminant D, where D is a field discriminant, then there exists a unique integer b_3 modulo $2a_1a_2$ such that $b_3 \equiv b_j \pmod{2a_j}$ for $j = 1, 2$ and $b_3^2 \equiv D \pmod{4a_1a_2}$.

Proof. This is an immediate consequence of the multiplication formulas for quadratic ideals on page 59. $\qquad\square$

Definition 3.6 | Dirichlet Composition

Suppose that $f = (a_1, b_1, c_1)$ and $g = (a_2, b_2, c_2)$ are primitive, united forms of discriminant Δ_F where Δ_F is a field discriminant, $a_3 = a_1a_2$, b_3 is the value given in Theorem 3.4, and

$$c_3 = \frac{b_3^2 - \Delta_F}{4a_3}.$$

Then the *Dirichlet composition* of f and g is the form

$$f \circ g = G = (a_3, b_3, c_3).$$

Remark 3.5 Note that
$$(a_3, (b_3 + \sqrt{\Delta_F})/2)$$
is an-$\mathfrak{O}_F$-ideal where $F = \mathbb{Q}(\sqrt{\Delta_F})$ by the multiplication formulas given on page 59. This shows the intimate connection between multiplication of quadratic ideals and composition of forms. Indeed, we need not restrict to field discriminants for this to work. We could expand the discussion to *nonmaximal orders in quadratic fields* but then the delineation becomes more complicated since we must rely on special conditions for invertibility of ideals and other considerations all of which are satisfied in the so-called *maximal order* $\mathfrak{O}_F$. See [62] for the more general approach.

The form G, in Definition 3.6, is a form of discriminant
$$b_3^2 - 4a_3c_3 = b_3^2 - 4a_3(b_3^2 - \Delta_F)/(4a_3) = b_3^2 - b_3^2 + \Delta_F = \Delta_F.$$

Also it is primitive since if a prime $p \mid \gcd(a_3, b_3, c_3)$, then $p \mid a_1$ or $p \mid a_2$. Without loss of generality suppose it divides a_1. Then since $p \mid b_3$, we must have that $p \mid b_1$ since $b_3 \equiv b_1 \pmod{2a_1}$ by Theorem 3.4 on the preceding page. However, since $p \mid c_3$ and $b_3^2 - 4a_3c_3 = D$, then $p^2 \mid \Delta_F$. However, Δ_F is a field discriminant so $p = 2$ and $\Delta_F \equiv 0 \pmod 4$ is the only possibility. By Definition 1.6 on page 7, $\Delta_F/4 \equiv 2, 3 \pmod 4$. If $\Delta_F/4 \equiv 2 \pmod 4$, then by Theorem 3.4, $b_3/2$ is even since
$$\left(\frac{b_3}{2}\right)^2 \equiv \frac{\Delta_F}{4} \pmod{a_1a_2},$$
given that $2 \mid a_1$. However, we have
$$\left(\frac{b_3}{2}\right)^2 - a_3c_3 = \frac{\Delta_F}{4}, \tag{3.10}$$
so since $2 \mid a_3$ and $2 \mid c_3$, then $\Delta_F/4 \equiv 0 \pmod 4$, a contradiction. Thus,
$$\Delta_F/4 \equiv 3 \pmod 4,$$
so by (3.10), $b_3/2$ is odd. However, (3.10) implies $\Delta/4 \equiv 1 \pmod 4$, a contradiction. We have shown that, indeed, G is a primitive form of discriminant Δ_F.

Remark 3.6 The *opposite* of
$$f = (a, b, c)$$
is
$$f^{-1} = (a, -b, c),$$
which is the *inverse* of f under Dirichlet composition. To see this we note that under the proper equivalence that sends (x, y) to $(-y, x)$, $f^{-1} \sim (c, b, a)$, for

which $\gcd(a, c, b) = 1$. This allows us to choose a united form in the class of f^{-1} by Definition 3.5 on page 107, so we may perform Dirichlet composition to get

$$f \circ f^{-1} = G = \left(ac, b, \frac{b^2 - \Delta_F}{4ac} \right) = (ac, b, 1).$$

Moreover, by Exercise 3.31 on page 145,

$$G \sim (1, 0, \Delta_F/4) \text{ when } \Delta_F \equiv 0 \pmod 4$$

and

$$G \sim (1, 1, (1 - \Delta_F)/4) \text{ when } \Delta_F \equiv 1 \pmod 4.$$

Thus, G is in the principal class by Corollary 3.1 on page 112.

We now need to introduce the ideal class group as a vehicle for defining the form class group since Theorem 3.3 on page 107 gives us the connection.

Definition 3.7 | Equivalence of Ideals |

Let $\mathfrak{O}_F$ be the ring of integers of a number field F. Then two $\mathfrak{O}_F$-ideals I, J are said to be in the same *equivalence class* if there exist nonzero $\alpha, \beta \in \mathfrak{O}_F$ such that $(\alpha)I = (\beta)J$ denoted by $I \sim J$.

Remark 3.7 By Theorem 2.9 on page 73 and Exercise 2.17 on page 85, we know that the set of all fractional $\mathfrak{O}_F$-ideals forms a multiplicative abelian group. If we denote this group by I_{Δ_F} and let P_{Δ_F} denote the group of principal ideals, then the quotient group

$$\frac{I_{\Delta_F}}{P_{\Delta_F}} = \mathbf{C}_{\mathfrak{O}_\mathbf{F}}$$

is called the *class group* of $\mathfrak{O}_F$. Also, the class of an $\mathfrak{O}_F$-ideal I is denoted by **I**. Thus a product of classes $\mathbf{IJ} = \mathbf{C}$ is the class belonging to any ideal $C = IJ$ formed by multiplying representatives $I \in \mathbf{I}$ and $J \in \mathbf{J}$. The identity element **1** is the *principal class*, namely all principal ideals $(\alpha) \sim (1)$, meaning $(\alpha) \in \mathbf{1}$. The existence of inverse classes $\mathbf{I}^{-1}$ for any class **I** is guaranteed by Exercise 2.18 and Theorem 2.9, namely $\mathbf{II}^{-1} = \mathbf{1}$. The commutative and multiplicative laws are clear, namely

$$\mathbf{IJ} = \mathbf{JI}, \text{ and } \mathbf{I}(\mathbf{JK}) = (\mathbf{IJ})\mathbf{K}, \text{ for } \mathfrak{O}_F\text{-ideals } \mathbf{I}, \mathbf{J}, \mathbf{K}.$$

Note as well, that the *conjugate* ideal I' for I, first mentioned in Remark 2.2 on page 63, satisfies

$$\mathbf{I}^{-1} = \mathbf{I}'$$

–see Exercise 3.19 on page 127. In what follows, we will need to refine this concept a bit in order to be able to include indefinite binary quadratic forms. We let $P_{\Delta_F}^+$ denote the group of principal ideals (α) where $N_F(\alpha) > 0$–see Definition 2.19 on page 91. Then we let

$$\frac{I_{\Delta_F}}{P_{\Delta_F}^+} = \mathbf{C}_{\mathfrak{O}_\mathbf{F}}^+$$

known as the *narrow ideal class group*, or sometimes called the *strict* ideal class group. Clearly, when F is a complex quadratic field, then $\mathbf{C}_{\mathfrak{D}_F} = \mathbf{C}^+_{\mathfrak{D}_F}$, since norms are necessarily positive in this case. In the real case we will learn more as we progress.

Note that in what follows, we use the symbol $\sim$ to denote both equivalence in the *ordinary* ideal class group $\mathbf{C}_{\mathfrak{D}_F}$ as well as equivalence of forms, but this will not lead to confusion when taken in context.

We use the symbol $\approx$ to denote *strict* equivalence in $\mathbf{C}^+_{\mathfrak{D}_F}$, i.e., $I \approx J$ in $\mathbf{C}^+_{\mathfrak{D}_F}$

when there exist $\alpha, \beta \in \mathfrak{D}_F$ such that

$$(\alpha)I = (\beta)J$$

where $N_F(\alpha\beta) > 0$. The next result shows that this is tantamount to form equivalence.

Theorem 3.5 | **Form and Ideal Class Groups** |

If C_{Δ_F} *denotes the set of classes of primitive forms of discriminant* Δ_F, *where* F *is a quadratic field, then* C_{Δ_F} *is a group with multiplication given by Dirichlet composition and*

$$C^+_{\mathfrak{D}_F} \cong C_{\Delta_F}.$$

Proof. Let $f = (a_1, b_1, c_1)$ and $g = (a_2, b_2, c_2)$, then by Exercises 3.2 on page 103 and 3.9 on page 104, $g \sim (a'_2, b'_2, c'_2)$ where $\gcd(a_1, a'_2) = 1$. Thus, Dirichlet composition is defined so we may assume the f and g to be united, without loss of generality. Let $F = (a_3, b_3, c_3)$ be given as in Definition 3.6 on page 107. Then we know that via the ideal correspondence given in Theorem 3.3 on page 107,

$$(a_1, (b_1 - \sqrt{\Delta_F})/2)(a_2, (b_2 - \sqrt{\Delta_F})/2) = (a_3, (b_3 - \sqrt{\Delta_F})/2), \qquad (3.11)$$

via the multiplication formulas on page 59. Thus, by Theorem 3.3 and (3.11), the Dirichlet composition of $f(x, y)$ and $g(x, y)$ corresponds to the product of the corresponding ideal classes, which shows that Dirichlet composition induces a well defined binary operation on C_{Δ_F}.

Note that in what follows, if we have strict equivalence of ideals given by

$$I = (a, (-b + \sqrt{\Delta_F})/2) \approx J = (a', (-b' + \sqrt{\Delta_F})/2), \qquad (3.12)$$

then we may replace I by $(aa')I$ and J by $(a^2)J$, so we may assume without loss of generality that $a = a'$. Via Theorem 3.3, we may define a mapping from $\mathbf{C}^+_{\mathfrak{D}_F}$ to C_{Δ_F} as follows

$$\tau : (a, (-b + \sqrt{\Delta_F})/2) \mapsto f = (a, b, c),$$

where

$$c = (b^2 - \Delta_F)/(4a).$$

Moreover, by the above,

$$\tau(IJ) = \tau(I)\tau(J)$$

since we have shown that ideal multiplication corresponds to form multiplication. To see that that τ is well defined, assume that $a' > 0$ and $b' \in \mathbb{Z}$ in (3.12). Thus, since there are $\delta, \gamma \in \mathfrak{D}_F$ such that $(\delta)I = (\gamma)J$ where $N_F(\delta\gamma) > 0$ then

$$N_F(\delta/\gamma)N(I) = N(J) = a,$$

so $N_F(\delta/\gamma) = 1$. By Exercise 3.21 on page 127, there is a $\sigma \in \mathfrak{D}_F$ such that $\delta/\gamma = \sigma/\sigma'$. If

$$m_{\sigma,\mathbb{Q}}(x) = ux^2 + vx + w$$

is the minimal polynomial of σ over $\mathbb{Q}$, then it is for σ' as well, so $\tau(\sigma) = \tau(\sigma') = (u, v, w)$. Hence,

$$\tau((\delta/\gamma)I) = \tau((\sigma/\sigma'))\tau(I) = \tau(I).$$

Hence, it suffices to prove that $\tau(I) = \tau(J)$ when $I = J$. By Exercise 2.5 on page 66, there exists

$$X = \begin{pmatrix} p & q \\ r & s \end{pmatrix} \in \mathrm{GL}(2, \mathbb{Z}),$$

such that

$$\begin{pmatrix} (-b + \sqrt{\Delta_F})/2 \\ a \end{pmatrix} = X \begin{pmatrix} (-b' + \sqrt{\Delta_F})/2 \\ a \end{pmatrix}.$$

Therefore,

$$p\left(\frac{-b' + \sqrt{\Delta_F}}{2}\right) + qa = \frac{-b + \sqrt{\Delta_F}}{2}$$

and

$$r\left(\frac{-b' + \sqrt{\Delta_F}}{2a}\right) + sa = a,$$

from which it follows that $r = 0$, $s = p = 1$, and $b = b' - 2qa$. Hence,

$$ax^2 + bxy + cy^2 = f(x, y) = g(x - qy, y) = a(x - qy)^2 + b'(x - qy)y + c'y^2,$$

so f and g are properly equivalent, namely they are in the same class in C_{Δ_F}, so τ is well defined. Now we establish the isomorphism.

First we show that τ is injective. Let

$$\tau(a, (-b + \sqrt{\Delta_F})/2) = f = (a, b, c) \sim \tau(a', (-b' + \sqrt{\Delta_F})/2) = g = (a', b', c')$$

in C_{Δ_F}. Since

$$(aa')(a, (-b + \sqrt{\Delta_F})/2) \approx (a^2)(a', (-b' + \sqrt{\Delta_F})/2)$$

as $\mathfrak{D}_F$-ideals, then we may assume that $a = a'$ without loss of generality since, if they are not equal, we may change the preimage to make it so as above. Now since

$$f\left(\frac{-b + \sqrt{\Delta_F}}{2a}, 1\right) = 0 = f\left(\frac{-b' + \sqrt{\Delta_F}}{2a}, 1\right),$$

then

$$\text{either}\quad \frac{-b+\sqrt{\Delta_F}}{2a} = \frac{-b'+\sqrt{\Delta_F}}{2a} \quad\text{or}\quad \frac{-b+\sqrt{\Delta_F}}{2a} = \frac{-b'-\sqrt{\Delta_F}}{2a},$$

given that these are the only two roots of $f(x,1) = ax^2 + bx + c = 0$. However, the latter is impossible by comparing coefficients so the former holds, from which we get that $b = b'$ so $c = c'$. Thus, τ is injective.

Lastly, we show that τ is surjective. Let

$$f(x,y) = ax^2 + bxy + cy^2$$

be a primitive form of discriminant Δ_F and let

$$\alpha = (-b + \sqrt{\Delta_F})/(2a).$$

Then $f(\alpha,1) = 0$, and $a\alpha \in \mathfrak{O}_F$. Define an $\mathfrak{O}_F$-ideal as follows. Set

$$I = \begin{cases} (a, a\alpha) & \text{if } a > 0, \\ (\sqrt{\Delta_F})(a, a\alpha) & \text{if } a < 0 \text{ and } \Delta_F > 0. \end{cases} \tag{3.13}$$

Therefore, $\tau(I) = (a,b,c)$ in the first instance is clear. In the second instance, we note that $I \approx (a, (-b + \sqrt{\Delta_F})/2)$ so

$$\tau(I) = \tau((a, (-b + \sqrt{\Delta_F})/2)) = (a,b,c).$$

Hence, τ is surjective and the isomorphism is established. □

Corollary 3.1 *The identity element of C_{Δ_F} is the class containing the principal form $(1,0,-\Delta_F/4)$ or $(1,1,(1-\Delta_F)/4)$ for $\Delta_F \equiv 0, 1 \,(\mathrm{mod}\ 4)$, respectively.*

Proof. Since

$$\tau(1, \sqrt{\Delta_F}/2) = (1,0,-\Delta_F/4) \quad\text{or}\quad \tau(1, (-1+\sqrt{\Delta_F})/2) = (1,1,(1-\Delta_F)/4)$$

depending on congruence modulo 4 of Δ_F, and the preimages are the identity elements in the principal class of $\mathbf{C}_{\mathfrak{O}_F}^+$, then the images are clearly the identity elements in the principal class of C_{Δ_F}. □

Remark 3.8 When F is a complex quadratic field, as noted in Remark 3.7 on page 109,

$$\mathbf{C}_{\mathfrak{O}_F} = \mathbf{C}_{\mathfrak{O}_F}^+,$$

so by Theorem 3.5 on page 110,

$$C_{\Delta_F} \cong \mathbf{C}_{\mathfrak{O}_F}.$$

However, *in the real case*, this is not always true. For instance, by Exercise 3.14, in the case where $\Delta_F = 12$, $C_{\Delta_F} \neq \{1\}$ and $\mathbf{C}_{\mathfrak{O}_F}$ has order 1. Yet by Theorem 3.5,

$$\mathbf{C}^+_{\mathfrak{O}_F} \cong C_{\Delta_F}.$$

Indeed, the case where the field F is real and has a unit of norm -1 or F is complex, then by Exercise 3.17 on page 117, $\mathbf{C}_{\mathfrak{O}_F} = \mathbf{C}^+_{\mathfrak{O}_F}$ always holds. When F is *real and has no such unit*, for instance as in the $\Delta_F = 12$ case, then by Exercise 3.16,

$$|\mathbf{C}^+_{\mathfrak{O}_F} : \mathbf{C}_{\mathfrak{O}_F}| = 2.$$

Note as well, if

$$h^+_{\mathfrak{O}_F} = |C^+_{\mathfrak{O}_F}|,$$

the *narrow ideal class number*, then by Theorem 3.5,

$$h^+_{\mathfrak{O}_F} = h_{\Delta_F},$$

the number of classes of forms of discriminant Δ_F. Also, if

$$h_{\mathfrak{O}_F} = |C_{\mathfrak{O}_F}|,$$

the ordinary or *wide class number*, by the above discussion, we have demonstrated the following.

Theorem 3.6 | Class Numbers of Forms and Ideals |

If Δ_F is the discriminant of a quadratic field F, then the class number of the form class group h_{Δ_F}, as well as that of both the wide ideal class group $h_{\mathfrak{O}_F}$ and the narrow ideal class $h^+_{\mathfrak{O}_F}$, is related by the following.

$$h_{\Delta_F} = h^+_{\mathfrak{O}_F} = \begin{bmatrix} h_{\mathfrak{O}_F} & \text{if } \Delta_F < 0, \\ h_{\mathfrak{O}_F} & \text{if } \Delta_F > 0 \text{ and there exists a } u \in \mathfrak{U}_F \\ & \text{with } N_F(u) = -1, \\ 2h_{\mathfrak{O}_F} & \text{if } \Delta_F > 0 \text{ and there is no } u \in \mathfrak{U}_F \\ & \text{with } N_F(u) = -1. \end{bmatrix}$$

We conclude this section with a verification that h_{Δ_F} is finite. To do this we first need the following result.

Lemma 3.2 | A Form of Reduction |

If Δ_F is the discriminant of a quadratic field F, then in each class of C_{Δ_F} there is a form $f = (a, b, c)$ such that

$$|b| \leq |a| \leq |c|.$$

Proof. Let the form $f = (a_1, b_1, c_1)$ be in an arbitrary class of C_{Δ_F}. We may select an integer a such that $|a|$ is the least value from the set of nonzero integers represented by forms in the class of f. Then there exist $p, r \in \mathbb{Z}$ such that

$$a = a_1 p^2 + b_1 pr + c_1 r^2. \tag{3.14}$$

If $g = \gcd(p, r)$, then a/g^2 is represented by f, contradicting the minimality of $|a|$ unless $g = 1$. Therefore, by the Euclidean algorithm, there exist integers q, s such that $ps - qr = 1$. Also,

$$f(px + qy, rx + sy) = a_1(px + qy)^2 + b_1(px + qy)(rx + sy) + c_1(rx + sy)^2 =$$

$$(a_1 p^2 + b_1 pr + c_1 r^2)x^2 + (a_1 2pq + b_1(ps + qr) + c_1 2rs)xy +$$

$$(a_1 q^2 + b_1 qs + c_1 s^2)y^2 =$$

$$ax^2 + Bxy + Cy^2,$$

where the coefficient for x^2 comes from (3.14),

$$B = (2pqa_1 + (ps + qr)b_1 + 2rsc_1),$$

and

$$C = a_1 q^2 + b_1 qs + c_1 s^2.$$

Set $g(x, y) = ax^2 + Bxy + Cy^2$ and we have $f \sim g$ in C_{Δ_F}. We may select an integer m such that

$$|2am + B| \leq |a|. \tag{3.15}$$

Thus,

$$g(x + my, y) = a(x + my)^2 + B(x + my)y + Cy^2 =$$

$$ax^2 + (2am + B)xy + (am^2 + Bm + C)y^2 =$$

$$ax^2 + bxy + cy^2,$$

with

$$b = 2am + B,$$

and

$$c = am^2 + Bm + C.$$

Set

$$h(x, y) = ax^2 + bxy + cy^2.$$

Then, since $\Delta_F = b^2 - 4ac$, given that $f \sim g \sim h$, then $c = 0$ implies that $\Delta_F = b^2$, a contradiction to the fact that Δ_F is a field discriminant. Hence, since $h(0, 1) = c$, then $|c| \geq |a|$ by the minimality of $|a|$. Thus, from (3.15), we have the result. $\qquad \square$

Corollary 3.2 *Any form of discriminant Δ_F is equivalent to a reduced form of the same discriminant.*

Proof. By Theorem 3.1 on page 100, we need only prove the result for $\Delta_F > 0$.

Claim 3.3 *We may assume that (a, b, c) satisfies $|a| \le |c|$ with*

$$\sqrt{\Delta_F} - 2|a| < b < \sqrt{\Delta_F}.$$

By Lemma 3.2, we may select a form (a, b, c) such that $|b| \le |a| \le |c|$. If $\sqrt{\Delta_F} - 2|a| > b$, then by setting

$$m = \left\lfloor \frac{\sqrt{\Delta_F}}{2c} + \frac{b|c|}{2c} + \varepsilon \right\rfloor,$$

where

$$\varepsilon = \begin{cases} 1 & \text{if } c < 0, \\ 0 & \text{if } c > 0 \end{cases}$$

we get

$$\sqrt{\Delta_F} - 2|c| < -b + 2cm < \sqrt{\Delta_F}.$$

We now show that

$$(a, b, c) \sim (c, -b + 2cm, a - bm + cm^2). \tag{3.16}$$

Via the map τ in Theorem 3.5 on page 110,

$$\tau : \left(a, \frac{-b + \sqrt{\Delta_F}}{2} \right) \mapsto (a, b, c),$$

and

$$\tau : \left(c, \frac{b - 2cm + \sqrt{\Delta_F}}{2} \right) \mapsto (c, -b + 2cm, a - bm + cm^2),$$

as $\mathfrak{O}_F$-ideals. However, by Exercise 2.6 on page 66

$$\left(c, \frac{b - 2cm + \sqrt{\Delta_F}}{2} \right) = \left(c, \frac{b + \sqrt{\Delta_F}}{2} \right),$$

so

$$\left(c, \frac{b - 2cm + \sqrt{\Delta_F}}{2} \right) \sim \left(\frac{b - \sqrt{\Delta_F}}{2c} \right) \left(c, \frac{b + \sqrt{\Delta_F}}{2} \right)$$

$$= \left(a, \frac{b - \sqrt{\Delta_F}}{2} \right) = \left(a, \frac{-b + \sqrt{\Delta_F}}{2} \right).$$

Since τ is a bijection, we have established (3.16).

If $|a - bm + cm^2| < |c|$, then we repeat the (finite) process, this time on

$$(c, -b + 2cm, a - bm + cm^2),$$

which must terminate in

$$(A, B, C) \sim (a, b, c)$$

with
$$|A| \leq |C| \text{ and } \sqrt{\Delta_F} - 2|A| < B < \sqrt{\Delta_F}.$$

This is Claim 3.3.

Therefore,
$$0 < \sqrt{\Delta_F} - b < 2|a| \leq 2|c| = \frac{|\Delta_F - b^2|}{2|a|} < \left| \sqrt{\Delta_F} + b \right|.$$

Hence, $b > 0$, so $b^2 < \Delta_F$ and $|2a|^2 \leq 4|ac| = \Delta_F - b^2 < \Delta_F$, so
$$2|a| < \sqrt{\Delta_F} < \sqrt{\Delta_F} + b,$$

from which it follows that (a, b, c) is reduced. $\square$

Theorem 3.7 $\boxed{\mathbf{h}_{\Delta_F} < \infty}$

If F is a quadratic field with discriminant Δ_F, then h_{Δ_F} is finite.

Proof. Note that by Exercise 3.11 on page 104, we need only consider the case where $\Delta_F > 0$. By Lemma 3.2 on page 113, for any class of C_{Δ_F}, there is a form $f = (a, b, c)$ in the class with
$$|ac| \geq b^2 = \Delta_F + 4ac > 4ac,$$

so $ac < 0$. Moreover,
$$4a^2 \leq 4|ac| = -4ac = \Delta_F - b^2 \leq \Delta_F.$$

Therefore,
$$|a| \leq \sqrt{\Delta_F}/2, \tag{3.17}$$

so by Lemma 3.2,
$$|b| \leq \sqrt{\Delta_F}/2. \tag{3.18}$$

Hence, by the bounds in (3.17)–(3.18), there can only be finitely many choices for the values a and b for a given discriminant Δ_F. Since
$$c = \frac{b^2 - \Delta_F}{4a},$$

we have established the result. $\square$

Corollary 3.3 $\boxed{\text{Positive Definite Forms and Reduction}}$

When $\Delta_F < 0$, then the number of inequivalent positive definite forms with discriminant Δ_F is the same as the number of reduced forms.

Proof. See Exercise 3.11.   □

Corollary 3.4 $\boxed{h_{\mathfrak{O}_F} < \infty}$

If Δ_F is the discriminant of a quadratic field F, then $h_{\mathfrak{O}_F}$ is finite.

Proof. This follows from Theorem 3.6 on page 113 and Theorem 3.7.   □

Exercises

3.14. Prove that when $\Delta_F = 12$ where $F = \mathbb{Q}(\sqrt{3})$, then the form $f = (-1, 0, 3)$
is not properly equivalent to the form $g = (1, 0, -3)$. This shows that
$C_{\Delta_F} \neq \{1\}$. Show, however, that $\mathbf{C}_{\mathfrak{O}_F} = \{1\}$.

(*Hint: See Remark 1.19 on page 50 and Theorem 2.17 on page 83.*)

In Exercises 3.15-3.17, assume that Δ_F is the discriminant of a quadratic
field F.

3.15. Let F be a real quadratic field and set

$$\alpha = \begin{cases} (1, 0, -\Delta_F/4) & \text{if } \Delta_F \equiv 0 \,(\mathrm{mod}\ 4), \\ (1, 1, (1 - \Delta_F)/4) & \text{if } \Delta_F \equiv 1 \,(\mathrm{mod}\ 4). \end{cases}$$

Prove that $\alpha \sim -\alpha$ in C_{Δ_F} if and only if $\mathfrak{O}_F$ has a unit u such that
$N_F(u) = -1$.

3.16. Let F be a real quadratic field. Assume that $\mathfrak{O}_F$ does *not* have a unit of
norm -1. Prove that
$$|\mathbf{C}^+_{\mathfrak{O}_F} : \mathbf{C}_{\mathfrak{O}_F}| = 2.$$

(*Hint: Use Exercise 3.15.*)

3.17. Prove that $\mathbf{C}^+_{\mathfrak{O}_F} = \mathbf{C}_{\mathfrak{O}_F}$ if F is either a complex quadratic field or F is a
real quadratic field such that $\mathfrak{O}_F$ has a unit u with $N_F(u) = -1$.

(*Hint: Use Exercise 3.15.*)

3.18. Let F be a number field and let $h_{\mathfrak{O}_F}$ be the (wide) class number of F.
Prove that if I is an integral $\mathfrak{O}_F$-ideal, then $I^{h_{\mathfrak{O}_F}} \sim 1$.

(*Hint: By Theorem 3.7 on the facing page, $|h_{\mathfrak{O}_F}| < \infty$.*)

3.3   Applications via Ambiguity

> *Seal up the mouth of outrage for a while,*
> *Till we can clear the ambiguities.*
>   from act five, scene 3, line 216 of *Romeo and Juliet* (1595)
>   **William Shakespeare**

In Remark 2.2 on page 63, we first mentioned the *conjugate* I' of an ideal I in $\mathbf{C}_{\mathfrak{D}_F}$ and we mentioned norms of ideals in Exercise 2.4 on page 66. These are important concepts that we now formalize.

Definition 3.8 | **Conjugates and Norms of Ideals**

Suppose that F is a quadratic field of discriminant Δ_F. If

$$I = \left(a, \frac{-b + \sqrt{\Delta_F}}{2} \right) \tag{3.19}$$

is an $\mathfrak{D}_F$-ideal, then

$$I' = \left(a, \frac{b + \sqrt{\Delta_F}}{2} \right)$$

is called the *conjugate ideal* of I. The representation of I given in (3.19) is called the *Hermite normal form* of I, and similarly for its conjugate–see Biography 3.4 on page 126. The value $a > 0$ is called the *norm* of I (and of I') denoted by

$$a = N(I) = N(I'),$$

the smallest positive integer in the ideal. Also,

$$N(IJ) = N(I)N(J) \text{ for } \mathfrak{D}_F\text{-ideals } I, J.$$

By Exercises 3.19–3.20 on page 127,

an ideal I has order at most 2 in $\mathbf{C}_{\mathfrak{D}_F}$ if and only if $\mathbf{I} \sim \mathbf{I'} \sim \mathbf{I}^{-1}$ in $\mathbf{C}_{\mathfrak{D}_F}$.

The elements of order 2 in both the form and ideal class groups are intimately linked and play an important role, including some interesting and valuable applications that we present in this section. First, we need the following which will be the gateway to linking forms and ideals in this context.

Definition 3.9 | **Ambiguous Ideals**

If F is a quadratic field of discriminant Δ_F and

$$I = (a, (-b + \sqrt{\Delta_F})/2)$$

is an $\mathfrak{D}_F$-ideal, then I is called *ambiguous* if

$$I = I' = (a, (b + \sqrt{\Delta_F})/2).$$

An *ambiguous class of ideals* in $C_{\mathfrak{D}_F}^+$ is one that contains an ideal I such that $I \approx I'$.

Definition 3.10 | Ambiguous Forms |

If F is a quadratic field of discriminant Δ_F and $f = (a, b, c)$ is a primitive form of discriminant $\Delta_F = b^2 - 4ac$, then f is said to be *ambiguous* if $a \mid b$. An *ambiguous class* in C_{Δ_F} is one that contains an ambiguous form.

Now we embark upon linking Definitions 3.9–3.10.

Lemma 3.3 *If F is a quadratic field of discriminat Δ_F and I is a primitive $\mathfrak{O}_F$-ideal, then*

$$I = I'$$

if and only if

$$N(I) \mid \Delta_F.$$

Proof. If $I = I'$, then

$$\left(N(I), \frac{-b + \sqrt{\Delta_F}}{2} \right) = \left(N(I), \frac{b + \sqrt{\Delta_F}}{2} \right),$$

so

$$\frac{b + \sqrt{\Delta_F}}{2} + \frac{-b + \sqrt{\Delta_F}}{2} = \sqrt{\Delta_F} \in I.$$

Thus, $(\Delta_F) \subseteq I$, so by Corollary 2.5 on page 76, $N(I) \mid \Delta_F$.

Conversely, suppose that $N(I) \mid \Delta_F$. Since Exercise 2.4 on page 66 tells us that

$$N(I) \mid N_F((b + \sqrt{\Delta_F})/2),$$

then $N(I) \mid b$, so we set $b = N(I)d$ for some $d \in \mathbb{Z}$. Then from Exercise 2.6,

$$I = \left(N(I), \frac{-b + \sqrt{\Delta_F}}{2} \right) = \left(N(I), \frac{-2dN(I) + b + \sqrt{\Delta_F}}{2} \right) =$$

$$\left(N(I), \frac{b + \sqrt{\Delta_F}}{2} \right) = I',$$

as required. $\square$

Corollary 3.5 *An $\mathfrak{O}_F$-ideal $I = (a, (-b + \sqrt{\Delta_F})/2)$ is ambiguous if and only if $a \mid b$.*

Proof. If $I = I'$, then

$$N(I) = a \mid \Delta_F = b^2 - 4ac,$$

by Lemma 3.3, so $a \mid b$ since Δ_F is either $4D$ or D where D is squarefree. Note that it is not possible that $a = 4$ since, when $\Delta_F \equiv 0 \pmod 4$, we must have that $D \equiv 2, 3 \pmod 4$. Conversely, if $a \mid b$, then $a = N(I) \mid \Delta_F$, so by Lemma 3.3, $I = I'$. $\square$

The next result gives us conditions on strict equivalence, namely equivalence in $C_{\mathfrak{O}_F}^+$, not explicit in the literature. The reader should be reminded of the distinction between strict ideal equivalence, denoted by $I \approx J$ and ordinary ideal equivalence, denoted by $I \sim J$, as discussed in Remark 3.7 on page 109.

Lemma 3.4 *If I is a primitive $\mathfrak{D}_F$-ideal of $C_{\mathfrak{D}_F}$, then the following are equivalent.*

(a) $I \approx I'$.

(b) There exists an $\mathfrak{D}_F$-ideal J such that $N(J) \mid \Delta_F$ and $I \sim J$.

(c) There exists a primitive $\mathfrak{D}_F$-ideal J such that $I \sim J$ and $J = J'$.

Proof. If $I \approx I'$, then there exist $\alpha, \beta \in \mathfrak{D}_F$ such that

$$(\alpha)I = (\beta)I'$$

where $N_F(\alpha) > 0$ and $N_F(\beta) > 0$. Thus, $N_F(\alpha/\beta) = 1$, so by Exercise 3.21 on page 127, we know there exists $\sigma \in \mathfrak{D}_F$ such that

$$\frac{\alpha}{\beta} = \frac{\sigma}{\sigma'},$$

so

$$(\sigma)I = (\sigma')I'. \tag{3.20}$$

Suppose now that $n \in \mathbb{N}$ is the largest value such that

$$(\sigma)I = (n)J,$$

where J is a primitive $\mathfrak{D}_F$-ideal. Then from (3.20), $J = J'$. Hence, from Lemma 3.3, $N(J) \mid \Delta_F$ and from (3.20), $I \sim J$. Thus, (a) implies (b).

If (b) holds, then (c) holds by Lemma 3.3. If (c) holds, then I is in an ambiguous class of $C_{\mathfrak{D}_F}$ having an ambiguous ideal J, so there exist $\alpha, \beta \in \mathfrak{D}_F$ such that

$$(\alpha)I = (\beta)J.$$

Hence, since $J = J'$, it follows that

$$(\beta\alpha')I' = (\beta'\alpha)I.$$

Since

$$N_F(\beta\alpha') = N_F(\beta'\alpha),$$

then $N(\beta\alpha'\beta'\alpha) > 0$, so by Remark 3.7 on page 109, $I \approx I'$. Thus, (c) implies (a) and this completes the logical circle. $\square$

Now we bring in forms and the connection to ambiguous ideals will materialize.

Theorem 3.8 $\boxed{\textbf{Forms of Order} \leq 2 \textbf{ Are in an Ambiguous Class}}$

Suppose that f is a binary quadratic form with discriminant Δ_F where F is a quadratic field, and C_{Δ_F} is the form class group. Then the following are equivalent.

(a) f has order 1 or 2 in C_{Δ_F}.

(b) $f \sim f^{-1}$.

(c) f is equivalent to an ambiguous form.

Proof. Suppose that $f = (a, b, c)$. If f has order at most 2, then

$$f \circ f \sim 1$$

so

$$f \sim f^{-1}.$$

Thus, (a) implies (b). If (b) holds, then by Theorem 3.5 on page 110,

$$I = (a, (-b + \sqrt{\Delta_F})/2) \approx (a, (b + \sqrt{\Delta_F})/2) = I',$$

so by Lemma 3.4, $I \sim J$ where J is ambiguous. Thus,

either $I \approx J$ or $I \approx (\sqrt{\Delta_F})J$, where $(\sqrt{\Delta_F})J$ is also ambiguous.

Hence, f is equivalent to an ambiguous form. If (c) holds, then by the multiplication formulas for ideals on page 59, and the correspondence via Theorem 3.5, $f^2 \sim 1$, so (c) imples (a). This establishes the equivalence of (a), (b) and (c). □

To show that the concept of ambiguity has even more formidable relationships, we state two of them as closing features to highlight the connections.
We need the following concept.

Definition 3.11 | Radicands of Quadratic Fields |

If Δ_F is the discriminant of a quadratic field F, then the *radicand* is given by

$$D_F = \begin{cases} \Delta_F/4 & \text{if } \Delta_F \equiv 0\,(\text{mod } 4), \\ \Delta_F & \text{if } \Delta_F \equiv 1\,(\text{mod } 4). \end{cases}$$

It was an outstanding problem to give criteria for a sum of two squares under the following situation. We quote from the well-written paper [82]: "An apparently open problem is to characterize those D that are a sum of two relatively prime squares but for which $x^2 - Dy^2$ does not represent -1. Such D include $34, 146, 178, 194, 205, 221, 305, 377, 386,$ and 410." This is accomplished as follows. We state the result for quadratic field radicands, although in [70] it is proved for arbitrary nonsquare integers.

Theorem 3.9 | Ambiguity and Sums of Squares |

Let Δ_F be the discriminant of a quadratic field F. Then the following are equivalent.

(a) There is an element of order 2 in $C_{\mathfrak{D}_F}$ that is not the image of an ambiguous class under the natural mapping $\rho : C^+_{\mathfrak{D}_F} \mapsto C_{\mathfrak{D}_F}$.

(b) D_F is a sum of two (relatively prime) squares and there is *no* unit $u \in \mathfrak{D}_F$ such that $N_F(u) = -1$.

Proof. If (a) holds, then by Lemma 3.4 on page 120, there is an $\mathfrak{D}_F$-ideal I such that $I \not\approx I'$ and $\rho(I)$ is an element of order 2 in $C_{\mathfrak{D}_F}$. Therefore,

$$C_{\mathfrak{D}_F} \not\cong C^+_{\mathfrak{D}_F},$$

so by Theorem 3.6 on page 113, $\Delta_F > 0$ and there is no unit $u \in \mathfrak{D}_F$ such that $N_F(u) = -1$. Thus, we need only show that there is no prime $p \mid \Delta_F$ with $p \equiv 3$ (mod 4) since, once established, D_F is a sum of two squares–see Example 1.15 on page 28, for instance. If such a prime p exists, then there exists

$$\gamma = (x + y\sqrt{\Delta_F})/(2z) \in \mathbb{Q}(\sqrt{\Delta_F})$$

such that

$$I = (\gamma)I' \text{ where } N_F(\gamma) = -1,$$

since $I \sim I'$, but $I \not\approx I'$. We may assume without loss of generality that $\gcd(z, p) = 1$ given that D_F is squarefree. Since

$$\frac{x^2 - y^2 \Delta_F}{4z^2} = -1 \text{ then } x^2 \equiv -4z^2 \pmod{p}, \text{ so } (x \cdot (2z)^{-1})^2 \equiv -1 \pmod{p}.$$

However, this is a contradiction since we know -1 is not a quadratic residue of such primes. We have shown that (a) implies (b). Now assume that (b) holds. Thus, there are $a, b \in \mathbb{Z}$,

$$D_F = 4a^2 + b^2, \text{ for some } a, b \in \mathbb{N}. \tag{3.21}$$

By (3.21), and Exercise 2.4 on page 66,

$$I = (a, (-b + \sqrt{\Delta_F})/2)$$

is an $\mathfrak{D}_F$-ideal. Also, from the multiplication formulas on page 59, it follows that

$$II' = (a) \tag{3.22}$$

and $I^2 = ((b + \sqrt{\Delta_F})/2)$. Assume that $I \approx I'$, so by Lemma 3.4, $I \sim J = J'$. Therefore, $I'J \sim (1)$ so there exists $\alpha \in \mathfrak{D}_F$ such that $(\alpha) = I'J$. Hence,

$$(\alpha)I = II'J = (a)J \tag{3.23}$$

where the last equality comes from (3.22). Taking conjugates in (3.23), we get $(\alpha')I' = (a)J' = (a)J = (\alpha)I$.

Thus, $(\alpha')I'I = (\alpha)I^2$, which implies that $(\alpha')(a) = (\alpha)((b + \sqrt{\Delta_F})/2)$. Hence,

$$\frac{\alpha'}{\alpha} = u\left(\frac{b + \sqrt{\Delta_F}}{2a}\right), \tag{3.24}$$

for some unit $u \in \mathfrak{O}_F$. However, by the hypothesis in (b), $N_F(u) = 1$, so (3.24) implies

$$1 = N_F\left(\frac{\alpha'}{\alpha}\right) = N_F\left(\frac{b + \sqrt{\Delta_F}}{2a}\right) = \frac{b^2 - \Delta_F}{4a^2} = -1,$$

a contradiction that proves $I \not\approx I'$, so I is not the image under ρ of an ambiguous class. Hence, we have (a) holds, so the result is secured. □

We conclude the section and thus the chapter with a look at applications of ambiguity to another concept–see Biography 3.3 on page 126.

Definition 3.12 | **Markov Triples**

A *Markov triple* is a triple of positive integers (a, b, c) satisfying the *Markov equation* $a^2 + b^2 + c^2 = 3abc$, and a, b, c are called *Markov numbers*.

Conjecture 3.1 | **The Markov Conjecture**

If (a_1, b_1, c) and (a_2, b_2, c) are Markov triples with $a_j \le b_j \le c$ for $j = 1, 2$, then $a_1 = a_2$ and $b_1 = b_2$, in which case c is said to be *unique*. In other words, the maximal element of a Markov triple uniquely determines the triple.

Markov came across this topic in 1879 when he was looking for the minimum positive value represented by real indefinite binary quadratic forms. He looked at the equation $x^2 + y^2 + z^2 = 3xyz$ and sought integral solutions x, y, z. We will adapt this quest to suit our setup drawing upon some recent results in the literature.

The following is essentially the approach taken from [4]. Note that below the discriminants may not be field discriminants. Let c be a Markov number and consider the Diophantine equation

$$x^2 + y^2 - 3cxy = -c^2. \tag{3.25}$$

Let

$$\omega = \frac{-3c + \sqrt{9c^2 - 4}}{2},$$

and set $F = \mathbb{Q}(\omega)$. Then there is a one-to-one correspondence between the solutions of (3.25) and elements $\beta = x + y\omega \in \mathbb{Z}[\omega]$ with

$$N_F(\beta) = \left(\frac{2x - 3cy + y\sqrt{9c^2 - 4}}{2}\right)\left(\frac{2x - 3cy - y\sqrt{9c^2 - 4}}{2}\right) = -c^2. \tag{3.26}$$

If we look at the group of automorphisms, acting on (3.25) that fix c, given by

$$\sigma : (x, y, z) \mapsto (y, x, c),$$

$$\rho : (x, y, c) \mapsto (-x, -y, c),$$

and

$$\phi : (x, y, z) \mapsto (y, 3yc - x, c),$$

then it follows from what Markov proved in 1879, that σ, ρ, ϕ essentially generate all solutions to (3.25). Now we show how this relates to $R = \mathbb{Z}[\omega]$ and put it into our context. First, ρ corresponds to multiplication by -1 in R, ϕ corresponds to multiplication by ω in R, and σ is a permutation that corresponds to taking a conjugate followed by multiplication by ω. Furthemore, ω is the smallest positive unit in R when $m \neq 1$. Therefore, there is a one-to-one correspondence between the solutions of (3.25) and pairs of principal ideals $I = (\beta)$, and $I' = (\beta')$ generated by elements $\beta, \beta' \in R$ satisfying (3.26). We have just proved the following.

Theorem 3.10 *An integer c is the maximal element of exactly one Markov triple if and only if there exists exactly one pair of primitive, principal ideals in $\mathfrak{O}_F$, $\{(\beta), (\beta')\}$ where $N_F(\beta) = -c^2$.*

Biography 3.2 Eduard Kummer (1810–1893) *was born on January 29, 1810 in Sorau, Brandenburg, Prussia (now Germany). He entered the University of Halle in 1828. By 1833, he was appointed to a teaching post at the Gymnasium in Liegniz which he held for 10 years. In 1836, he published an important paper in* Crelle's Journal *on hypergeometric series, which led to his correspondence with Jacobi and Dirichlet, who were impressed with his talent. Indeed, upon Dirichlet's recommendation, Kummer was elected to the Berlin academy in 1839, and was Secretary of the Mathematics Section of the Academy from 1863 to 1878. In 1842, with the support of Dirichlet and Jacobi, Kummer was appointed to a full professor at the University of Breslau, now Wroclaw, in Poland. In 1843, Kummer was aware that his attempts to prove Fermat's Last Theorem were flawed due to the lack of unique factorization in general –see the discussions of the topic in Chapter 1. He introduced his "ideal numbers" that was the basis for the concept of an ideal thus allowing the development of ring theory, and a substantial amount of abstract algebra later. In 1855, Dirichlet left Berlin to succeed Gauss at Göttingen, and recommended to Berlin that they offer the vacant chair to Kummer, which they did. In 1857, the Paris Academy of Sciences awarded Kummer the Grand Prize for his work. In 1863, the Royal Society of London elected him as a Fellow. He died in Berlin on May 14, 1893.*

Example 3.2 If $\Delta_F = 221$, we have $(\beta) = (14 + \sqrt{221})$ and $(\beta') = (14 - \sqrt{221})$ with $c = 5$ and $N_F(\beta) = -25$. Here $(a, b, c) = (1, 2, 5)$ is the Markov triple, $14 = (3cb - 2a)/2$, and $9c^2 - 4 = 221$.

Example 3.3 If $\Delta_F = 776 = 2^3 \cdot 97$ and $c = 194$, then $(\beta) = (3778 + 13\sqrt{84680})$ and $(3778 - 13\sqrt{84680})$ where $N_F(\beta) = -194^2$. In this case, the Markov triple is $(5, 13, 194)$, $3778 = (3cb - 2a)/2$, and $9(c/2)^2 - 1 = 84680$.

Also, to bring in ideals we have the following result, which is taken from [92].

Theorem 3.11 *Let* $\Delta_F = 9c^2 - 4$ *and suppose that* c *is a Markov number that is* not *unique. Then there exist relatively prime integers* p *and* q *such that* $1 < p < q < c$ *with* $c = pq$ *such that the following conditions hold.*

(a) There exists $\mathfrak{D}_F$-ideals I and J of norm p^2 and q^2 respectively, such that

$$J \approx J' \approx I\sqrt{\Delta_F}.$$

(b) There exists a form f with $f \sim f^{-1}$ such that f represents both $-p^2$ and q^2.

Proof. If c is not unique, then by Theorem 3.10 on the preceding page, there exist two distinct pairs of principal ideals

$$(\beta_1) = \left(c^2, \frac{-b_1 + \sqrt{\Delta_F}}{2}\right), \text{ and } (\beta_2) = \left(c^2, \frac{-b_2 + \sqrt{\Delta_F}}{2}\right).$$

Since $4c^2 \mid (\Delta_F - b_j^2)$ for $j = 1, 2$, then $b_1^2 \equiv b_2^2 \pmod{2c^2}$. Also, since the ideals are distinct, then $b_1 \not\equiv \pm b_2 \pmod{2c^2}$. Thus, there exist relatively prime integers p, q with $1 < p < q < c$ with $c = pq$ such that

$$b_1 + b_2 \equiv 0 \pmod{2p^2}, \text{ and } b_1 - b_2 \equiv 0 \pmod{2q^2}. \tag{3.27}$$

Set

$$I = \left(p^2, \frac{-b_1 + \sqrt{\Delta_F}}{2}\right), \text{ and } J = \left(q^2, \frac{-b_2 + \sqrt{\Delta_F}}{2}\right),$$

which are $\mathfrak{D}_F$-ideals. Since p and q are relatively prime, then

$$\mathfrak{D}_F \approx (\beta_1)\sqrt{\Delta_F} \approx IJ\sqrt{\Delta_F} \approx (\beta_2)\sqrt{\Delta_F} \approx I'J\sqrt{\Delta_F},$$

and multiplying through by I we get

$$J \approx J' \approx I\sqrt{\Delta_F},$$

which yields part (a).

For part (b), we associate the $\mathfrak{D}_F$-ideal $I(\sqrt{\Delta_F})$ with the form $(-p^2, b_1, c_1) = f$ and the $\mathfrak{D}_F$-ideal J with the form $(q^2, b_2, c_2) = g$, so by the equivalences in part (a), $f \sim f^{-1} \sim g$, which gives us part (b). □

These applications conclude the chapter with forays into other dominions and show the beautiful architecture underlying this mathematics–see the quote by Dyson on page 155.

Biography 3.3 Andrei Andreyevich Markov (1856–1922) *was born in Ryazan, Russia on June 14, 1856. He showed mathematical ability at an early age when he wrote a paper on integration of linear differential equations before he entered university. In 1884, he was awarded his doctorate from St. Petersburg University with a thesis on applications of continued fractions. Markov was a professor at St. Petersburg University from 1886 until his retirement in 1905. He worked in number theory, analysis, continued fractions applied to probability theory, approximation theory, and convergence of series. In particular, his work on what we now call Markov chains began the study of stochastic processes. However, it was not until 1923 when Norbert Weiner first gave a rigorous treatment of Markov processes that the true value of the theory came to light. The general theory can be said to have been established by Andrei Kolmogorov in the 1930s. Markov's son also became a mathematician in his own right (under the same last name). Among the honours in his life was the election to the Russian Academy of Sciences in 1902. Markov died on July 20, 1922 in Petrograd (now St. Petersburg), Russia.*

Biography 3.4 Charles Hermite (1822–1901) *was born on December 24, 1822 in Dieuze, Lorraine, France. In 1840-41, he studied at the same institution where Galois had studied a decade and a half earlier, namely the Collège Louis-le-Grand, and had the same instructor as Galois, Louis Richard. He was also tutored by Catalan in those years. He then went to the École Polytechnique, where he was eventually awarded his degree in 1847. He was appointed there as répétiteur and admissions examiner. His most far-reaching mathematical results were accomplished in that appointment over the next decade. One of these was his proof that doubly periodic functions can be represented as quotients of periodic entire functions. He also worked on quadratic forms, including his result on a reciprocity law relating to binary quadratic forms. In 1855, he established a* theory of transformations, *which found an interface among number theory, theta functions, and transformations of abelian functions, the latter of which he had established. In 1858, he proved that although it was known to Ruffin and Abel that an algebraic equation of the fifth degree cannot be solved by radicals, an algebraic equation of the fifth degree could be solved using elliptic functions. In 1862, he was appointed maître de conference at École Polytechnique, becoming examiner in 1863, and then professor in 1869. He left for the Sorbonne in 1876, where he stayed until he retired in 1897.*

Among his other accomplishments for which he is well known is the proof of the transcendence of e–see Biography 3.6 on page 128 and Theorem 4.6 on page 172. He also is known for a variety of topics that bear his name among which are: Hermite differential equations, Hermite matrices, Hermite polynomials, and his formula for interpolation. He died in Paris, France on January 14, 1901.

Biography 3.5 David Hilbert (1862–1943) *was born in Köningsberg, Prussia, which is now Kaliningrad, Russia. He studied at the University of Köningsberg where he received his doctorate under the supervision of Lindemann. He was employed at Köninsberg from 1886 to 1895. In 1895, he was appointed to fill the chair of mathematics at the University of Göttingen, where he remained for the rest of his life. Hilbert was very eminent in the mathematical world after 1900 and it may be argued that his work was a major influence throughout the twentieth century. In 1900, at the Paris meeting of the Second International Congress of Mathematicians, he delivered his now-famous lecture* The Problems of Mathematics, *which outlined twenty-three problems that continue to challenge mathematicians today. Among these were Golbach's conjecture and the Riemann hypothesis. Some of the Hilbert problems have been resolved and some have not such as the latter two. Hilbert made contributions to many branches of mathematics including algebraic number theory, the calculus of variations, functional analysis, integral equations, invariant theory, and mathematical physics. He also had Hermann Weyl as one of his students–see Biography 1.1 on page 31. Hilbert retired in 1930 at which time the city of Köninsberg made him an honourary citizen. He died on February 14, 1943 in Göttingen.*

Exercises

In Exercises 3.19–3.21, Δ_F denotes the discriminant of quadratic field F with ring of integers $\mathfrak{O}_F$. Also,

$$I = (a, (-b + \sqrt{\Delta_F})/2)$$

is an $\mathfrak{O}_F$-ideal, with

$$I' = (a, (b + \sqrt{\Delta_F})/2)$$

its conjugate ideal.

3.19. Prove that $\mathbf{I'} = \mathbf{I}^{-1}$ in $\mathbf{C}_{\mathfrak{O}_F}$.

(*Hint: Use* The Multiplication formulas *on page 59.*)

3.20. Prove that I has order at most 2 in $\mathbf{C}_{\mathfrak{O}_F}$ if and only if $I \sim I'$.

(*Hint: Use Exercise 3.19.*)

3.21. Let u be a unit in $\mathfrak{O}_F$ such that $N_F(u) = 1$. Prove that there exists an $\alpha \in \mathfrak{O}_F$ such that $\alpha = u\alpha'$, where α' is the algebraic conjugate of α.

(*This is the quadratic analogue of Hilbert's Theorem 90, but is actually due to Kummer–see Biographies 3.2 on page 124 and 3.5.*)

3.22. Prove that if $f = \mathbb{Q}(\sqrt{221})$, then $\mathfrak{O}_F$ has no unit u with $N_F(u) = -1$.

Biography 3.6 Carl Louis Ferdinand von Lindemann (1852–1939) *was born in Hannover, Hanover, which is now Germany. He studied at the University of Göttingen which he entered in 1870. However, as was a practice at the time, he moved from one university to another studying at Munich and at Erlangen. He was awarded his doctorate in 1873 under the direction of Klein at Erlangen. In 1877, he was awarded his habilitation by the University of Würzburg. Also, in 1877, he was appointed as extraordinary professor to the University of Freiburg, and promoted there to ordinary professor in 1879. In 1883, he was appointed professor at the University of Königsberg. In 1893, he accepted a chair at the University of Munich where he remained for the rest of his life.*

Lindemann is probably best known for his proof that π is transcendental–see Theorem 4.7 on page 175. He proved this in 1882, using methods of Hermite who had shown, in 1873, that e is transcendental–see Biography 3.4 on page 126. Lindemann was also interested in physics as well as in the history of mathematics, including the translation and expansion of some of Poincaré's work. Among the honours bestowed upon him were being elected to the Bavarian Academy of Sciences and being given an honourary degree from the University of St. Andrews. As noted above, Hilbert was one of his students, as was Oskar Perron. He died in Munich on March 6, 1939.

3.4 Genus

> *My mind rebels at stagnation. Give me a problem, give me work, give me the most abstrusive cryptogram, or the most intricate analysis, and I am in my own proper atmosphere.*
>
> spoken by Sherlock Holmes in *The Sign of Four* (1890)
> **Sir Arthur Conan Doyle–see page 55.**

In §3.1, we looked at representation of integers by binary quadratic forms. Thus, if the discriminant of forms f and g is D and f and g are not in the same equivalence class, then there is the problem of distinguishing those numbers represented by f from those represented by g. This is, in particular, of value when the forms are positive definite in view of Theorem 3.2 on page 102, when we know $h_D = h_{-4n} > 1$. The notion on the header of this section was created by Gauss to express this type of distinction. In order to be able to precisely define it, we need the following result. As usual, we use the term *form* herein to mean *binary quadratic form*.

Lemma 3.5 | **Jacobi Symbols and Representation**

Let F be a quadratic field with discriminant Δ_F with

$$|\Delta_F| = p_1 p_2 \cdots p_r \text{ if } \Delta_F \equiv 1 \pmod 4$$

and

$$|\Delta_F| = 2^\alpha p_2 p_3 \cdots p_r \text{ if } \Delta_F \equiv 0 \pmod 4,$$

where $\alpha \in \{2,3\}$ and p_j, for $j = 1, 2, \ldots, r \in \mathbb{N}$, are distinct odd primes.

If $n_1, n_2 \in \mathbb{Z}$ are properly represented by a form of discriminant Δ_F with $\gcd(2\Delta_F, n_1) = \gcd(2\Delta_F, n_2) = 1$, then,

$$(\Delta_F/|n_1|) = (\Delta_F/|n_2|) = 1,$$

where $(*/*)$ denotes the Jacobi symbol. Also,

$$(n_1/p_j) = (n_2/p_j) \text{ for } j = 2, \ldots, r,$$

and

$$(\varepsilon_1/\alpha_1) = (\varepsilon_2/\alpha_2)$$

where (ε_j/α_j) are defined by the following with

$$\text{sign}(n_j) = 1 \text{ if } n_j > 0 \text{ and } \text{sign}(n_j) = -1 \text{ if } n_j < 0 \text{ for } j = 1, 2:$$

$$
\left(\frac{\varepsilon_j}{\alpha_j}\right) = \left[
\begin{array}{ll}
\left(\dfrac{n_j}{p_1}\right) & \text{if } \Delta_F \equiv 1 \,(\text{mod } 4), \\[2mm]
\left(\dfrac{-1}{|n_j|}\right) \cdot \text{sign}(n_j) & \text{if } \Delta_F \equiv 12 \,(\text{mod } 16), \\[2mm]
\left(\dfrac{2}{|n_j|}\right) & \text{if } \Delta_F \equiv 8 \,(\text{mod } 32), \\[2mm]
\left(\dfrac{-2}{|n_j|}\right) \cdot \text{sign}(n_j) & \text{if } \Delta_F \equiv 24 \,(\text{mod } 32).
\end{array}
\right]
$$

Proof. Suppose that the integers n_1 and n_2 are properly represented by the form $f = (a, b, c)$. Since there are relatively prime integers x_j, y_j for $j = 1, 2$ such that

$$f(x_j, y_j) = n_j = ax_j^2 + bx_jy_j + cy_j^2,$$

where $\gcd(2\Delta_F, n_1) = \gcd(2\Delta_F, n_2) = 1$, then

$$4af(x_j, y_j) = (2ax_j + by_j)^2 - \Delta_F y_j^2.$$

Therefore, for each odd $p_i \mid \Delta_F$,

$$4an_j \equiv (2ax_j + by_j)^2 \pmod{p_i}.$$

Hence,

$$\left(\frac{an_j}{p_i}\right) = 1,$$

from which it follows that

$$\left(\frac{n_1}{p_i}\right) = \left(\frac{n_2}{p_i}\right) = \left(\frac{a}{p_i}\right).$$

It remains to deal with the case when $\Delta \equiv 0 \pmod 4$ and show that $(\varepsilon_1/\alpha_1) = (\varepsilon_2/\alpha_2)$ and to show that $(\Delta_F/|n_j|) = 1$ for $j = 1, 2$. The latter follows from Exercise 3.9 on page 104. The balance of the result will now follow from the product formula that we establish as follows.

Claim 3.4 *For $j = 1, 2$,*

$$\left(\frac{\varepsilon_j}{\alpha_j}\right) \prod_{i=2}^{r} \left(\frac{n_j}{p_i}\right) = \left(\frac{\Delta_F}{|n_j|}\right).$$

First, we know from the quadratic reciprocity law–see [68, Theorem 4.11, p. 196]–that

$$\prod_{i=2}^{r} \left(\frac{n_j}{p_i}\right) = \left(\frac{(-1)^{((n_j-1)/2)(\Delta_F/2^\gamma-1)/2)}\Delta_F/2^\gamma}{|n_j|}\right) \text{ where } \gamma \in \{2, 3\}. \quad (3.28)$$

If $\Delta_F \equiv 12 \pmod{16}$, then $\gamma = 2$ and $\Delta_F/4 \equiv 3 \pmod 4$, so from (3.28),

$$\left(\frac{\varepsilon_j}{\alpha_j}\right) \prod_{i=2}^{r} \left(\frac{n_j}{p_i}\right) = \left(\frac{\Delta_F}{|n_j|}\right)\left(\frac{(-1)^{(n_j+1)/2}}{|n_j|}\right) \cdot \text{sign}(n_j) = \left(\frac{\Delta_F}{|n_j|}\right).$$

If $\Delta_F \equiv 8 \pmod{32}$, then we get $\gamma = 3$, $\Delta_F/8 \equiv 1 \pmod 4$, and so from (3.28),

$$\left(\frac{\varepsilon_j}{\alpha_j}\right) \prod_{i=2}^{r} \left(\frac{n_j}{p_i}\right) = \left(\frac{\Delta_F}{|n_j|}\right).$$

Lastly, if $\Delta \equiv 24 \,(\mathrm{mod}\ 32)$, then $\gamma = 3$ and $\Delta/8 \equiv 3 \,(\mathrm{mod}\ 4)$, so from (3.28),

$$\left(\frac{\varepsilon_j}{\alpha_j}\right) \prod_{i=2}^{r} \left(\frac{n_j}{p_i}\right) = \left(\frac{\Delta_F}{|n_j|}\right) \left(\frac{(-1)^{(n_j+1)/2}}{|n_j|}\right) \cdot \mathrm{sign}(n_j) = \left(\frac{\Delta_F}{|n_j|}\right).$$

This is Claim 3.4 and so the entire result. $\qquad\square$

We are now in a position to define the salient feature that will provide the mechanism for the primary definition we are seeking. In what follows $\mathrm{sign}(n)$ is as defined in Lemma 3.5.

Definition 3.13 | **Assigned Values of Generic Characters**

Let F be a quadratic field with discriminant Δ_F,

$$|\Delta_F| = p_1 p_2 \cdots p_r \quad \text{if } \Delta_F \equiv 1 \ (\mathrm{mod}\ 4),$$

and

$$|\Delta_F| = 2^{\alpha} p_2 p_3 \cdots p_r \quad \text{if } \Delta_F \equiv 0 \ (\mathrm{mod}\ 4),$$

where $\alpha \in \{2,3\}$ and p_j, for $j = 1, 2, \ldots, r \in \mathbb{N}$, are distinct odd primes. Suppose that n is a nonzero integer with $\gcd(2\Delta_F, n) = 1$. Let χ_1 be defined as the following, where $\left(\frac{*}{*}\right)$ denoted the Jacobi symbol:

$$\chi_1 = \left[\begin{array}{ll} \left(\dfrac{n}{p_1}\right) & \text{if } \Delta_F \equiv 1 \,(\mathrm{mod}\ 4), \\[2mm] \left(\dfrac{-1}{|n|}\right) \cdot \mathrm{sign}(n) & \text{if } \Delta_F \equiv 12 \,(\mathrm{mod}\ 16), \\[2mm] \left(\dfrac{2}{|n|}\right) & \text{if } \Delta_F \equiv 8 \,(\mathrm{mod}\ 32), \\[2mm] \left(\dfrac{-2}{|n|}\right) \cdot \mathrm{sign}(n) & \text{if } \Delta_F \equiv 24 \,(\mathrm{mod}\ 32), \end{array} \right]$$

for $j = 2, 3, \ldots, r$, and let χ_j be the Jacobi symbol (n/p_j). Then the values χ_j are called the *generic characters* of n and their assigned values are given by the r-tuple

$$(\chi_1, \chi_2, \ldots, \chi_r). \tag{3.29}$$

If n is represented by a form f of discriminant Δ_F, then *(3.29) are the generic characters of the form f, denoted by $\chi_j(f)$ for $j = 1, 2, \ldots, r$.*

Remark 3.9 Note that the assigned characters in Definition 3.13 satisfy the multiplicative property

$$\chi_j(fg) = \chi_j(f)\chi_j(g)$$

by the properties of Jacobi symbols. Also, we may view the multiplicative characters as functions mapping from $\mathbb{Z}$ to $\{\pm 1\}$, so $(\chi_1, \ldots, \chi_r)$ may be considered as a vector-valued function from r-tuples of integers to r-tuples with entries ± 1. With this in mind, Lemma 3.5 tells us that the vector of assigned values remains invariant over all integers represented by a form from a class in $\mathbb{C}_{\Delta_F}$. Hence, the following holds.

Corollary 3.6 *All integers n relatively prime to $2\Delta_F$, which are representable by forms in a given equivalence class of C_{Δ_F} have the same assigned values of generic characters, and $(\Delta_F/|n|) = 1$.*

By Corollary 3.6, the characters of f are the same for all integers represented by f so the notion of the characters of f in Definition 3.13 is indeed a well-defined concept. Now we have the tools to define the main topic.

Definition 3.14 $\boxed{\text{Genus}}$

A class of forms in C_{Δ_F} having the same assigned characters is called a *genus* of forms. The genus of forms having all assigned characters $+1$ must contain the principal form called the *principal genus*.

The following is an important consequence from Corollary 3.6.

Corollary 3.7 *The product of the assigned values for the characters for any given genus is $+1$.*

Proof. This is immediate from Claim 3.4 on page 130 and Exercise 3.9 on page 104. $\square$

Remark 3.10 It follows that equivalent forms necessarily represent the same integers, so they are in the same genus. Also, we will see later that each genus consists of a finite number of classes of forms, the same for each genus, and there are only finitely many genera–see Theorem 3.14 on page 142. Also, by Exercise 3.26 on page 145, each genus is a coset of the principal genus. It is also known that the principal genus is actually the subgroup of squares $C_{\Delta_F}^2$ of C_{Δ_F}–see Remark 3.13 on page 143.

The following is a general aspect of genus theory applied to the principal genus. Note that we will be using Dirichlet's Theorem on primes in arithmetic progression–see Chapter 7, where we provide a proof. This result guarantees that every class in $\left(\frac{\mathbb{Z}}{|D|\mathbb{Z}}\right)^*$ includes an odd prime. Moreover, in the proof of the following, we will be using properties of the Jacobi symbol–see [68, pp. 192–200].

Theorem 3.12 $\boxed{\textbf{Principal Forms and Genus}}$

Let Δ_F be the discriminant of a quadratic field F and let f be a primitive form of discriminant Δ_F. Set

$$U_{\Delta_F} = \left\{ \overline{m} \in \left(\frac{\mathbb{Z}}{|\Delta_F|\mathbb{Z}}\right)^* : \text{ there is an odd prime } p \in \overline{m} \text{ and } (\Delta_F/p) = 1 \right\}.$$

Then each of the following hold.

(a) *If $m \in \mathbb{Z}$ with $\gcd(\Delta_F, m) = 1$ is represented (not necessarily properly) by a form of discriminant Δ_F, then $\overline{m} \in U_{\Delta_F}$.*

(b) The elements $\overline{m} \in (\mathbb{Z}/|\Delta_F|\mathbb{Z})^*$ such that $\overline{m}$ is represented by the principal genus of discriminant Δ_F form a subgroup H_{Δ_F} of U_{Δ_F}.

(c) The cosets of H_{Δ_F} in U_{Δ_F} are precisely the elements of

$$L_f = \left\{ \overline{\ell} \in \left(\frac{\mathbb{Z}}{|\Delta_F|\mathbb{Z}} \right)^* : f(x,y) \equiv \ell \pmod{|\Delta_F|} \text{ for some } x, y \in \mathbb{Z} \right\}$$

where f ranges over the primitive forms of discriminant Δ_F which represent distinct values.

(d) Forms f, g of discriminant Δ_F are in the same genus if and only if $L_f = L_g$.[3.1]

Proof. First of all, we show that U_{Δ_F} is a group. If p_1, p_2 are odd primes with $\overline{p_1}, \overline{p_2} \in U_{\Delta_F}$, then it suffices to show that $\overline{p_1} \cdot \overline{p_2}^{-1} \in U_{\Delta_F}$. Let $\overline{p_3} = \overline{p_2}^{-1}$ where p_3 is an odd prime. Then by the quadratic reciprocity law for the Jacobi symbol in the case where $\Delta_F \equiv 1 \pmod 4$–see, for instance, [68, Exercise 4.25, p. 200],

$$\left(\frac{\Delta_F}{p_1 p_3} \right) = \left(\frac{p_1 p_3}{|\Delta_F|} \right) = \left(\frac{p_1 p_3}{|\Delta_F|} \right) = \left(\frac{1}{|\Delta_F|} \right) = 1,$$

since

$$1 \equiv p_1 \cdot p_2^{-1} \equiv p_1 \cdot p_3 \pmod{|\Delta_F|},$$

so

$$\overline{p_1 \cdot p_3} = \overline{p_1 \cdot p_2^{-1}} \in U_{\Delta_F}.$$

Note that $\overline{1} = \overline{p_1 \cdot p_2^{-1}} = \overline{p_4}$ for some odd prime p_4. This comment holds for the remaining cases.

If $\Delta_F \equiv 0 \pmod 8$, then

$$\left(\frac{\Delta_F}{p_1 p_3} \right) = \left(\frac{\Delta_F/4}{p_1 p_3} \right) = \left(\frac{2}{p_1 p_3} \right) \left(\frac{\Delta_F/8}{p_1 p_3} \right)$$

$$= (-1)^{\frac{(p_1 p_3)^2 - 1}{8}} \cdot (-1)^{\frac{p_1 p_3 - 1}{2} \cdot \frac{\Delta_F/8 - 1}{2}} \left(\frac{p_1 p_3}{\Delta_F/8} \right) = \left(\frac{p_1 p_3}{\Delta_F/8} \right) = \left(\frac{1}{\Delta_F/8} \right) = 1,$$

since $p_1 p_3 \equiv 1 \pmod{\Delta_F}$ so $(p_1 p_3 - 1)/2$ is even as is $((p_1 p_2)^2 - 1)/8$.

If $\Delta_F \equiv 4 \pmod 8$, then

$$\left(\frac{\Delta_F}{p_1 p_3} \right) = \left(\frac{\Delta_F/4}{p_1 p_3} \right) = (-1)^{\frac{p_1 p_3 - 1}{2} \cdot \frac{\Delta_F/4 - 1}{2}} \left(\frac{p_1 p_3}{\Delta_F/4} \right) = \left(\frac{p_1 p_3}{\Delta_F/4} \right) = 1,$$

[3.1] An important fact must be highlighted here. Part (d) says that two forms, f, g, are in the same genus if and only if they represent the *same values modulo* $|\Delta_F|$. Therefore, although it is possible that f and g are in the same genus, yet there may exist an integer n such that $g(x, y) = n$ a but f does *not represent* n (meaning that there are no integers X, Y such that $f(X, Y) = n$) it *must hold* that there exist integers u, v such that $\overline{f(u, v)} = \overline{n} \in (\mathbb{Z}/|\Delta_F|\mathbb{Z})^*$. This means that f and g are in the same genus if and only if they represent the same values in U_{Δ_F}, namely if and only if $L_f = L_g$. See Example 3.6 on page 136 for an explicit depiction of these facts.

since $p_1 p_3 \equiv 1 \,(\mathrm{mod}\ \Delta_F)$, forcing $(p_1 p_3 - 1)/2$ to be even. We have shown that U_{Δ_F} is indeed a group.

Now if $m \in \mathbb{Z}$ with $\gcd(\Delta_F, m) = 1$ is represented by a form of discriminant Δ_F, then by Exercise 3.6 on page 103, we may let $m = m_1^2 m_2$ where m_2 is properly represented by a form of discriminant Δ_F. Suppose that $p > 2$ is prime with $\overline{p} = \overline{m_2} \in \left(\frac{\mathbb{Z}}{|\Delta_F|\mathbb{Z}} \right)^*$. By Exercise 3.9 on page 104, $(\Delta_F/p) = 1$, so $\overline{p} = \overline{m_2} \in U_{\Delta_F}$. Also, since

$$\left(\left(\frac{\mathbb{Z}}{|\Delta_F|\mathbb{Z}} \right)^* \right)^2 \subseteq U_{\Delta_F},$$

and U_{Δ_F} has been shown to be a group, then $\overline{m_1^2} \in U_{\Delta_F}$ and $\overline{m_2} \in U_{\Delta_F}$ so

$$\overline{m} = \overline{m_1^2 m_2} = \overline{m_1^2} \cdot \overline{m_2} \in U_{\Delta_F}.$$

Hence, $\overline{m} \in U_{\Delta_F}$ and we have completed the proof of part (a).

For part (b), we have that $H_{\Delta_F} \subseteq U_{\Delta_F}$ by part (a). Also, products of classes in U_{Δ_F} all of whose assigned characters are $+1$ must also be all $+1$. It follows that if $\overline{m}, \overline{n} \in H_{\Delta_F}$, then $\overline{m} \cdot \overline{n}^{-1} \in H_{\Delta_F}$, so H_{Δ_F} is a subgroup of U_{Δ_F}. This is part (b).

For part (c), let $\overline{\ell} \in L_f$. Since $L_f \subseteq U_{\Delta_F}$ by part (a), then there is an odd prime p such that $\overline{p} = \overline{\ell}$ and f properly represents p. By Exercise 3.2, there exist $x, y, b, c \in \mathbb{Z}$ such that

$$f(x, y) = px^2 + bxy + cy^2.$$

Therefore, by setting $\sigma = 2$ if $\Delta_F \equiv 1 \,(\mathrm{mod}\ 4)$ and $\sigma = 1$ if $\Delta_F \equiv 0 \,(\mathrm{mod}\ 4)$, we have

$$\sigma^2 p f(x, y) = (\sigma p x + \sigma b y / 2)^2 - y^2 \Delta_F \sigma^2 / 4. \tag{3.30}$$

Hence, $\overline{\sigma^2 p f(x, y)} \in H_{\Delta_F}$, namely $\overline{f(x, y)} \in (\overline{\sigma^2 p})^{-1} H_{\Delta_F}$. We have shown that

$$L_f \subseteq (\overline{\sigma^2 p})^{-1} H_{\Delta_F}.$$

Conversely, if $\overline{m} \in (\overline{\sigma^2 p})^{-1} H_{\Delta_F}$, then by the discussion in Footnote 3.1 on the preceding page, there are $X, Y \in \mathbb{Z}$ such that

$$\sigma^2 p m \equiv X^2 + (\sigma - 1) X Y + \frac{(\sigma - 1 - \Delta_F)}{4} Y^2 \pmod{|\Delta_F|}. \tag{3.31}$$

Hence, from (3.30)–(3.31), we can find $u, v \in \mathbb{Z}$ such that

$$f(u, v) \equiv m \pmod{|\Delta_F|}.$$

In other words, $\overline{m} \in L_f$. This shows that $(\overline{\sigma^2 p})^{-1} H_{\Delta_F} \subseteq L_f$, so

$$L_f = (\overline{\sigma^2 p})^{-1} H_{\Delta_F},$$

securing part (c).

For part (d), if f and g are in the same genus, then f and g have the same assigned characters $\chi_j(f) = \chi_j(g)$ for each j as in Definition 3.13 on page 131. Therefore, by Remark 3.9 on page 131,

$$\chi_j(f \circ g^{-1}) = \chi_j(f)\chi_j(g^{-1}) = \chi_j(g)\chi_j(g^{-1}) = \chi_j(g \circ g^{-1}) = \chi_j(1_{\Delta_F}) = +1$$

for all such j where 1_{Δ_F} is the principal form in C_{Δ_F} as given in Definition 3.3 on page 100. Thus, $f \circ g^{-1}$ is in the principal genus, so $L_{f \circ g^{-1}} = H_{\Delta_F}$ by part (b). It follows from part (c) that $L_f = L_g$. Conversely, if $L_f = L_g$, then $L_{f \circ g^{-1}} = H_{\Delta_F}$ by part (c), so $f \circ g^{-1}$ is in the principal genus by part (b). This means that $\chi_j(f \circ g^{-1}) = +1$ for all j so

$$1 = \chi_j(f \circ g^{-1}) = \chi_j(f)\chi_j(g^{-1}),$$

which means that

$$\chi_j(g) = \chi_j(f)\chi_j(g^{-1})\chi_j(g) = \chi_j(f)\chi_j(g^{-1} \circ g) = \chi_j(f)\chi_j(i_{\Delta_F}) = \chi_j(f),$$

so f and g have the same assigned values, namely f and g are in the same genus. This completes part (d) and so the total result.   □

Remark 3.11 With L_f and U_{Δ_F} as defined in Theorem 3.12 on page 132,

$$U_{\Delta_F} = \bigcup L_f,$$

where the *disjoint* union is over forms of discriminant Δ_F which represent distinct values. In other words, L_f is a coset of H_{Δ_F} in U_{Δ_F}. This allows the following notion.

Definition 3.15 $\boxed{\text{Genus and Cosets}}$

The *genus of the coset L_f*, as given in Theorem 3.12 on page 132, consists of all the forms of discriminant Δ_F that represent the values of L_f modulo $|\Delta_F|$.

Notice, as well, the nice manner in which the coset approach yields the generic interpretation of forms given in Definition 3.13 on page 131. If $\ell, m \in L_f$ for a form f, then $\ell \cdot m^{-1}$ is in the principal genus, so the assigned characters for $\ell \cdot m^{-1}$ are all $+1$. Hence, the generic characters of ℓ and m are the same.

Historically, it was Lagrange who first introduced the notion of looking at congruence classes in $(\mathbb{Z}/|\Delta_F|)^*$ represented by a *single* form. To do this he gathered together these forms that represent the same equivalence classes in $(\mathbb{Z}/|\Delta_F|)^*$. Thus, Lagrange was prescient in this regard since this was the fundamental idea behind genus theory.

Following the notation of the proof of Theorem 3.12 on page 132, $\overline{x}$ will continue to denote an element in $U_{\Delta_F}/H_{\Delta_F}$ in the ensuing developments.

Example 3.4 Let $\Delta_F = -20$ where $h_{\Delta_F} = 2$, $f = (1, 0, 5)$, and $g = (2, 2, 3)$, with

$$U_{\Delta_F} = \{\overline{1}, \overline{3}, \overline{7}, \overline{9}\}$$

with

$$H_{\Delta_F} = L_f = \{\overline{1}, \overline{9}\},$$

and $L_g = \{\overline{3}, \overline{7}\}$.

Example 3.5 Let $\Delta_F = -35$, which has $h_{\Delta_F} = 2$ with $f = (1,1,9)$ being the principal form and $g = (3,1,3)$ being in a different genus. Here

$$U_{\Delta_F} = \{\overline{1}, \overline{3}, \overline{4}, \overline{9}, \overline{11}, \overline{12}, \overline{13}, \overline{16}, \overline{17}, \overline{27}, \overline{29}, \overline{33}\},$$

and

$$H_{\Delta_F} = \{\overline{1}, \overline{4}, \overline{9}, \overline{11}, \overline{16}, \overline{29}\}.$$

The above illustrate parts (a)–(b) of Theorem 3.12, and what follows illustrates part (c). Also, with reference to Remark 3.11, notice that

$$U_{\Delta_F} = \cup_f L_f = L_{(1,1,9)} \cup L_{(3,1,3)} = \{\overline{1}, \overline{4}, \overline{9}, \overline{11}, \overline{16}, \overline{29}\} \cup \{\overline{3}, \overline{12}, \overline{13}, \overline{17}, \overline{27}, \overline{33}\}.$$

Since f represents $\{1,4,9,11,16,29\}$, then $\chi_1(f) = (1/5) = 1$ and $\chi_2(f) = (1/7) = 1$, so the assigned values for f are $(1,1)$, as stated in Definition 3.14 on page 132. Since g represents $\{3,12,13,17,27,33\}$, $\chi_1(g) = (3/5) = -1$ and $\chi_2(g) = (3/7) = -1$. Thus, the assigned values for g are $(-1,-1)$. Indeed, it follows from Corollary 3.7 on page 132, that if Δ_F has $r = 2$ generic characters, then the assigned values *must be* $(+1,+1)$ and $(-1,-1)$. Furthermore to depict the mechanism of the proof of part (c) in Theorem 3.12, we have the following. Since

$$L_g = L_{(3,1,3)} = \{\overline{3}, \overline{12}, \overline{13}, \overline{17}, \overline{27}, \overline{33}\},$$

then if we let $\ell = 3$, then $\overline{\ell} \in L_g$, and we have, $(\overline{4\ell})^{-1} = \overline{3}$, so

$$(\overline{4\ell})^{-1} H_{\Delta_F} = \{\overline{3 \cdot 1}, \overline{3 \cdot 4}, \overline{3 \cdot 9}, \overline{3 \cdot 11}, \overline{3 \cdot 16}, \overline{3 \cdot 29}\}$$

$$= \{\overline{3}, \overline{12}, \overline{27}, \overline{33}, \overline{13}, \overline{17}\} = L_g.$$

Also, if $\ell = 4$, then $(\overline{4\ell})^{-1} = \overline{11}$, then

$$(\overline{4\ell})^{-1} H_{\Delta_F} = \{\overline{11 \cdot 1}, \overline{11 \cdot 4}, \overline{11 \cdot 9}, \overline{11 \cdot 11}, \overline{11 \cdot 16}, \overline{11 \cdot 29}\}$$

$$= \{\overline{11}, \overline{9}, \overline{29}, \overline{16}, \overline{1}, \overline{4}\} = L_f.$$

We have shown that the cosets of H_{Δ_F} in G are precisely the elements of L_f and L_g as asserted by part (c) of Theorem 3.12.

Example 3.6 Now we illustrate Theorem 3.12 on page 132 part (c) when the principal genus has more than one class of forms in light of Footnote 3.1 on page 133. For instance, if $\Delta_F = -23$, then $h_{\Delta_F} = 3$ and there is a single genus of forms, the principal genus, having the three distinct forms, the principal form $f = (1,1,6)$, as well as $g = (2,1,3)$ and $h = (2,-1,3)$. Also,

$$U_{\Delta_F} = H_{\Delta_F} = \{\overline{1}, \overline{2}, \overline{3}, \overline{4}, \overline{6}, \overline{8}, \overline{9}, \overline{12}, \overline{13}, \overline{16}, \overline{18}\},$$

so the only cosets of H_{Δ_F} in U_{Δ_F} is $U_{\Delta_F} = H_{\Delta_F}$ itself. Moreover, $L_f = L_g = L_h$. It is a direct computation to show, for instance, that f directly represents $\{1,4,6,8,9,12,16,18\}$ in the sense that we can find x, y values for $f(x,y)$ to equal any member of this set. Yet, it is not clear about $\overline{2}, \overline{3}, \overline{13}$ since f does not represent 2, 3, or 13. However, the definition of L_f requires only that we find *any element* in one of these classes (not necessarily the same element for each value) that f *does represent*. Since $\ell = 117 = 2 + 23 \cdot 5 \in \overline{2}$ and

$$f(3,4) = \ell = 3^2 + 3 \cdot 4 + 6 \cdot 4^2,$$

then $\overline{2} \in L_f$; since $f(7,0) = \ell = 49 \in \overline{3}$, then $\overline{3} \in L_f$, where we note that proper representation is not a requirement in Theorem 3.12 on page 132. Also, since $f(5,2) = \ell = 59 \in \overline{13}$, then $\overline{13} \in L_f$. The reader may verify that $L_g = L_h = U_{\Delta_F}$ so, as the genus of a coset given in Definition 3.15 on page 135 tells us, the genus of $L_f = U_{\Delta_F}$ consists of all the forms of discriminant -23 that represent the values of L_f modulo 23, namely all of U_{Δ_F}.

Example 3.7 This example illustrates the $\Delta_F > 0$ case when each genus has a single class as a real analogue of Example 3.5 on the facing page. Let

$$\Delta_F = 105 = 3 \cdot 5 \cdot 7$$

for which

$$U_{\Delta_F} = \{\overline{1}, \overline{2}, \overline{4}, \overline{8}, \overline{13}, \overline{16}, \overline{23}, \overline{26}, \overline{32}, \overline{41}, \overline{46}, \overline{52}, \overline{53}, \overline{59},$$
$$\overline{64}, \overline{73}, \overline{79}, \overline{82}, \overline{89}, \overline{92}, \overline{97}, \overline{101}, \overline{103}, \overline{104}\}.$$

Also, $h_{\Delta_F} = 4$ and there is a single genus in each class, where the inequivalent reduced forms are given by

$$f = (1,1,-26), g = (2,9,-3), h = (7,7,-2) \text{ and } k = (5,5,-4).$$

We have

$$H_{\Delta_F} = L_f = \{\overline{1}, \overline{4}, \overline{16}, \overline{46}, \overline{64}, \overline{79}\}, \quad L_g = \{\overline{2}, \overline{8}, \overline{23}, \overline{32}, \overline{53}, \overline{92}\},$$

$$L_h = \{\overline{13}, \overline{52}, \overline{73}, \overline{82}, \overline{97}, \overline{103}\}, \text{ and } L_k = \{\overline{26}, \overline{41}, \overline{59}, \overline{89}, \overline{101}, \overline{104}\}.$$

Thus,

$$U_{\Delta_F} = L_f \cup L_g \cup L_h \cup L_k.$$

To illustrate a comment made in Remark 3.11 on page 135, we have the following. Since $\overline{2}, \overline{8} \in L_g$ then $\overline{2} \cdot \overline{8}^{-1} = \overline{2} \cdot \overline{92} = \overline{72} \in H_{\Delta_F} = L_f$. In other words, for any of the forms, if $\overline{m}, \overline{n}$ is in one of the cosets, then $\overline{m} \cdot \overline{n}^{-1} \in H_{\Delta_F} = L_f$, the elements of U_{Δ_F} represented by the principal genus H_{Δ_F}, and so by the principal form f, as described in Footnote 3.1 on page 133.

For an example of a discriminant $\Delta_F > 0$ which is a real analogue of Example 3.6 on the preceding page, see Exercise 3.36 on page 146.

The above allows us to state a fundamental result in genus theory.

Theorem 3.13 | Cosets and Genus |

Let Δ_F be the discriminant of a quadratic field F, and let H_{Δ_F} be as in Theorem 3.12. If J is a coset of H_{Δ_F} in U_{Δ_F} and $p > 2$ is a prime not dividing $2\Delta_F$, then $\overline{p} \in J$ if and only if p is represented by a reduced form of discriminant Δ_F in the genus of J.

Proof. By Theorem 3.12 on page 132, $J = L_f$ for some primitive form of discriminant Δ_F. Also,

$$f(x, y) \equiv p \pmod{|\Delta_F|}$$

by the definition of L_f. Therefore, The Legendre symbol

$$(\Delta_F/p) = (\Delta_F/f(x, y)) = 1.$$

Thus, by Exercise 3.9 on page 104, p is properly represented by a primitive form g with $L_g = L_f$, and $p = g(X, Y)$ for some $X, Y \in \mathbb{Z}$. By Corollary 3.2 on page 114, g may be assumed to be reduced. Conversely, if p is represented by a reduced form f of discriminant Δ_F in the genus of J, then $\overline{p} \in L_f = J$ by Theorem 3.12. $\qquad\square$

Corollary 3.8 *Let Δ_F be the discriminant of a quadratic field F, and let p be a prime not dividing $2\Delta_F$. Then p is represented by a form of discriminant Δ_F in the principal genus if and only if there exists an integer z such that*

$$p \equiv z^2 + m \pmod{|\Delta_F|},$$

where $m = 0$ or

$$m = \begin{cases} z + (1 - \Delta_F)/4 & \text{if } \Delta_F \equiv 1 \, (\mathrm{mod} \, 4), \\ -\Delta_F/4 & \text{if } \Delta_F \equiv 0 \, (\mathrm{mod} \, 4). \end{cases}$$

Proof. By Theorem 3.12 on page 132 and Theorem 3.13, p is represented by a form in the principal genus if and only if

$$p \equiv \begin{cases} x^2 + xy + (1 - \Delta_F)y^2/4 & \text{if } \Delta_F \equiv 1 \, (\mathrm{mod} \, 4), \\ x^2 - \Delta_F y^2/4 & \text{if } \Delta_F \equiv 0 \, (\mathrm{mod} \, 4). \end{cases}$$

In the case where y is even, this says

$$p \equiv \begin{cases} x^2 + xy + (y/2)^2 \equiv (x + y/2)^2 \pmod{|\Delta_F|} & \text{if } \Delta_F \equiv 1 \, (\mathrm{mod} \, 4), \\ x^2 \pmod{|\Delta_F|} & \text{if } \Delta_F \equiv 0 \, (\mathrm{mod} \, 4). \end{cases}$$

In the case where y is odd, it says

$$p \equiv \begin{cases} (x + \frac{y-1}{2})^2 + x + \frac{y-1}{2} + (1 - \Delta_F)/4 \pmod{|\Delta_F|} & \text{if } \Delta_F \equiv 1 \, (\mathrm{mod} \, 4), \\ x^2 - \Delta_F/4 \pmod{|\Delta_F|} & \text{if } \Delta_F \equiv 0 \, (\mathrm{mod} \, 4). \end{cases}$$

Setting $z = (x + y/2)$ with $m = 0$, in the first case, and $z = x + (y - 1)/2$, with $m = z + (1 - \Delta_F)/2$, in the second case yields the result for $\Delta_F \equiv 1 \pmod 4$ and setting $z = x$ in both cases yields the result for $\Delta_F \equiv 0 \pmod 4$. □

Example 3.8 Considering Example 3.5 on page 136 again, for $\Delta_F = -35$, we see that the cosets of H_{Δ_F} in U_{Δ_F} are L_f and L_g where, for instance, $\overline{12} = \overline{47} \in L_g$ with

$$47 = p = 3x^2 + xy + 3y^2 = 3 \cdot 1^2 + 1 \cdot (-4) + 3(-4)^2.$$

Also, $\overline{4} = \overline{109} \in L_f$ where

$$109 = p = x^2 + xy + 9y^2 = 4^2 + 4 \cdot 3 + 9 \cdot 3^2, \tag{3.32}$$

and $\overline{16} = \overline{191} \in L_f$ with

$$191 = p = x^2 + xy + 9y^2 = 13^2 + 13 \cdot 1 + 9 \cdot 1^2.$$

To illustrate Corollary 3.8, using the notation therein, we note that in the principal genus, the prime

$$p = 109 \equiv z^2 + z + (1 - \Delta_F)/4 \equiv 5^2 + 5 + 9 \equiv 39 \equiv 4 \pmod{|\Delta_F|},$$

where $z = x + (y - 1)/2$ and $m = z + (1 - \Delta_F)/2$, from the x, y given in *(3.32)*. This illustrates the $\Delta_F \equiv 1 \pmod 4$ with $m \neq 0$ case in the proof of Corollary 3.8. Also, since $\overline{1} = \overline{71}$, we have

$$p = 71 = 5^2 + 5 \cdot 2 + 9 \cdot 2^2 \equiv z^2 \equiv (x + y/2)^2 \pmod{|\Delta_F|},$$

where $z = x + (y - 1)/2 = 5 + 2/2 = 6$, and $m = 0$ which illustrates the $\Delta_F \equiv 1 \pmod 4$ with y even case in the proof of Corollary 3.8.

Example 3.9 To illustrate the case where $\Delta_F \equiv 0 \pmod 4$ in Corollary 3.8, as well as to motivate the next illustration, we let $\Delta_F = -8$, where $U_{\Delta_F} = \{\overline{1}, \overline{3}\}$. Here $\overline{1} = \overline{19}$, and

$$19 = 1^2 + 2 \cdot 3^2 \equiv x^2 - \Delta_F/4 \equiv 3 \pmod{|\Delta_F|},$$

illustrating the case where the y value is odd. On the other hand, if $\Delta_F = -4$, then $U_{\Delta_F} = \{\overline{1}\}$ and $\overline{1} = \overline{5}$ where

$$5 = 1^2 + 2^2 = x^2 - y^2 \Delta_F/4 \equiv x^2 \equiv 1 \pmod{|\Delta_F|},$$

illustrating the remaining case where y is even.

The two discriminants $\Delta_F = -4, -8$ are special from another perspective that we explore in the following depiction of representation of primes and *class numbers*, that we will study in greater detail in §3.5.

For an illustration of Corollary 3.8 on page 138 in the case where $\Delta_F > 0$ see Exercise 3.37 on page 146.

For the following illustration, the reader should solve Exercise 3.23 on page 144 in preparation.

Example 3.10 Corollary 3.8 allows us to categorize the principal genus via congruence conditions, especially when there is exactly one class in the principal genus. For instance, when $\Delta_F = -4$, there is only one class for the principal genus given by the unique reduced form $f(x, y) = x^2 + y^2$ of discriminant -4, which is our problem of representation as a sum of two squares. In this case, an odd prime $p = x^2 + y^2$ for some integers x, y if and only if, by Corollary 3.8, there exists an integer z such that

$$p \equiv z^2 \pmod{8} \text{ or } p \equiv z^2 + 1 \pmod{4}, \text{ i.e., if and only if } p \equiv 1 \pmod{4},$$

a result we have already seen in Theorem 1.13 on page 26. Similarly, if $\Delta_F = -8$, then there is only one class for the principal genus since the unique reduced form of discriminant -8 is $x^2 + 2y^2$. By Corollary 1.13, $p = x^2 + 2y^2$ if and only if there is an integer z such that

$$p \equiv z^2 \pmod{8} \text{ or } p \equiv z^2 + 2 \pmod{8}, \text{ i.e., if and only if } p \equiv 1, 3 \pmod{8}.$$

This is tantamount to saying that the Legendre symbol $(-2/p) = 1$ if and only if $p \equiv 1, 3 \pmod{8}$, a result we know from elementary number theory–see [68, Exercise 4.3, p. 187] for instance. Also, see Exercise 3.10 on page 104.

When $\Delta_F \equiv 1 \pmod{4}$ we have as an illustration the unique reduced form $x^2 + xy + 2y^2$ in the principal genus of discriminant -7. Here, by Corollary 3.10, an odd prime

$$p = x^2 + xy + 2y^2$$

if and only if for some integer z,

$$p \equiv z^2, \text{ or } z^2 + z + 2 \pmod{7}, \text{ i.e., if and only if } p \equiv 1, 2, 4 \pmod{7}.$$

The latter is tantamount to saying that p is a quadratic residue modulo 7, and this holds if and only if -7 is a quadratic residue modulo p. For instance, $p = 29 = (-1)^2 + (-1)(4) + 2 \cdot 4^2$.

Lastly, consider $\Delta_F = -43$ for which there is the unique reduced form $x^2 + xy + 11y^2$. Then, by Corollary 3.10, an odd prime $p = x^2 + xy + 11y^2$ if and only if there is a $z \in \mathbb{Z}$ with $p \equiv z^2 + z + 11 \pmod{43}$ or $p \equiv z^2 \pmod{43}$. However, the former congruence implies $4p \equiv (2z + 1)^2 \pmod{43}$ so this representation occurs if and only if p is a quadratic residue modulo 43, and this holds if and only if -43 is a quadratic residue modulo p.

At this juncture, it is worth pointing out a rather beautiful result by Jacobi–see [68, Biography 4.4, p. 192]. He discovered that if $p \equiv 3 \pmod{4}$ is a prime and $p > 3$, then if R is the sum of all the quadratic residues modulo p, and NR is the sum of the quadratic nonresidues, then

$$h_{-p} = \frac{NR - R}{p}.$$

For instance, for $p = 43$,

$$R = 1 + 4 + 6 + 9 + 10 + 11 + 13 + 14 + 15 + 16 + 17 + 21$$

$$+23 + 24 + 25 + 31 + 35 + 36 + 38 + 40 + 41 = 430,$$

and

$$NR = 2 + 3 + 5 + 7 + 8 + 12 + 18 + 19 + 20 + 22 + 26 + 27$$

$$+28 + 29 + 30 + 32 + 33 + 34 + 37 + 39 + 42 = 473,$$

so

$$h_{-43} = (473 - 430)/43 = 1,$$

which we know from Exercise 3.23 on page 144. The first proof of this remarkable result was provided by Dirichlet in 1838, a result known today as Dirichlet's class number formula–see [68, Biography 1.8, p. 35]. The actual number of discriminants $\Delta_F < 0$ with $h_D = 1$ has been solved for some time and the values for which we have class number one are

$$\Delta_F \in \{-3, -4, -7, -8, -11, -19, -43, -67, -163\}.$$

In 1934, Heilbronn and Linfoot proved that the above list could contain at most one more value—see Biography 1.3 on page 50. In 1966 this was proved by Stark. However, in 1952, a proof was given by Heegner, in [40], but this proof was fragmentary and not well-understood, so it was generally discredited. It turns out that it is a valid proof as was later acknowledged after Deuring cleared it up—see Biography 3.7 on page 146.

Remark 3.12 The conditions in Example 3.10 for representations of primes do not always occur. In other words, given a form $f(x, y) = ax^2 + bxy + cy^2$ of discriminant Δ_F it is not always the case that there exist natural numbers $s, a_1, \ldots, a_s, m$, depending on a, b, and c, such that for an odd prime p not dividing Δ_F we have

$$p = ax^2 + bxy + cy^2 \text{ if and only if } p \equiv a_1, \ldots, a_s \pmod{m}. \tag{3.33}$$

In Example 3.10, we saw several instances where (3.33) *is* satisfied, but these relied on $h_{\Delta_F} = 1$. When the class number is greater than one, we may *not* have (3.33). For instance, if $\Delta_F = -56$, then by Exercises 3.25–3.29, $h_{-56} = 4$, and there are two genera with $x^2 + 14y^2$ and $2x^2 + 7y^2$ being in the one genus and $3x^2 + 2xy + 5y^2$ and $3x^2 - 2xy + 5y^2$ being in the other genus. Moreover, as shown in the very readable [91, Theorem, p. 424], (3.33) fails for $p = x^2 + 14y^2$. The authors, Spearman and Williams, do this by proving that every arithmetic progression $\{a + km : k = 0, 1, \ldots\}$ where m is assumed even without loss of generality, and $\gcd(a, m) = 1$, either contains no primes of the form $x^2 + 14y^2$ or it contains primes of both forms $x^2 + 14y^2$ and $2x^2 + 7y^2$. Note, as well, that $x^2 + 14y^2$ represents 23 but not 79 and $2x^2 + 7y^2$ represents 79 but not 23. However, it is worth observing that this is not to be confused with the fact, noted in Footnote 3.1 on page 133, that since $f = (1, 0, 14)$ and $g = (2, 0, 7)$ are both in the principal genus, they represent $\{\overline{1}, \overline{9}, \overline{15}, \overline{23}, \overline{25}, \overline{39}\}$ *modulo* 56. In other words, even though f does not represent 79, we do have that $f(-1, 3) = 23 \in \overline{79} \in (\mathbb{Z}/56\mathbb{Z})^*$, and similarly, $g(-3, 8) = 191 \in \overline{23} \in (\mathbb{Z}/56\mathbb{Z})^*$, even though g does not represent 23. This latter interpretation via Theorem 3.12 on page 132 allowed us to view the cosets and genera with an ease that the above more rigid interpretation did not allow. Yet to consider the solvability of (3.33), we cannot allow the coset interpretation since it does not apply to this most interesting question.

Now we are ready for the exact number of genera. The following was proved, in greater generality, by Gauss in 1801.

Theorem 3.14 | **The Genus Group**

Suppose that F is a quadratic field of discriminant Δ_F divisible by r distinct primes. Then each of the following holds.

(a) The h_{Δ_F} proper equivalence classes of forms can be subdivided into 2^{r-1} genera consisting of $h_{\Delta_F}/2^{r-1}$ classes of forms each, which comprise a subgroup G_{Δ_F} of C_{Δ_F} under Dirichlet composition.

(b) With U_{Δ_F} and H_{Δ_F} as given in Theorem 3.12 on page 132,

$$G_{\Delta_F} \cong \frac{U_{\Delta_F}}{H_{\Delta_F}},$$

and $|G_{\Delta_F}| = 2^{r-1}$.

Proof. By Exercise 3.9 on page 104 there exists at least one class of forms in each genus. Also, by Exercise 3.26 on page 145, there are an equal number of classes in each genus. Lastly, by Lemma 3.5 on page 129 there are 2^{r-1} possible genera, with the product, given in Claim 3.4 on page 130, being $+1$, since there are that many possible r-tuples of $+1$'s and -1's corresponding to the Jacobi symbols. Under Dirichlet composition given in Definition 3.6 on page 107, with the identity element being the principal genus, **P**, this is a subgroup G_{Δ_F} of C_{Δ_F}, which establishes part (a).

For part (b), let

$$f \sim_{gen} g$$

denote that f and g are in the same genus, namely the same equivalence class in G_{Δ_F}. Also, let $\overline{f}^{gen}$ denote this class and define the map

$$\psi : G_{\Delta_F} \mapsto \frac{U_{\Delta_F}}{H_{\Delta_F}}$$

via

$$\overline{f}^{gen} \mapsto \overline{L_f},$$

where

$$\overline{L_f} = \frac{L_f}{H_{\Delta_F}},$$

observing, from Remark 3.11 on page 135, that

$$U_{\Delta_F} = \cup_f L_f$$

so

$$\frac{U_{\Delta_F}}{H_{\Delta_F}} \cong \cup_f \overline{L_f}.$$

In addition, note that by parts (c)–(d) of Theorem 3.12 on page 132,

$$\cup_f \overline{L_f} \cong \{L_{\overline{f}^{gen}}\}_{\overline{f}^{gen} \in G_{\Delta_F}},$$

and

$$L_{\overline{f}^{gen}} = L_{\overline{g}^{gen}} \text{ if and only if } f \sim_{gen} g,$$

so ψ is not only well defined but is indeed a bijection. By part (a), $|G_{\Delta_F}| = 2^{r-1}$, which is the entire result. $\square$

Example 3.11 If, as in Example 3.6 on page 136, $\Delta_F = -23$, then $r = 1$ so by Theorem 3.14, there is $2^{r-1} = 1$ genus. Since $h_{\Delta_F} = 3$, then the classes of forms $f = (1,1,6), g = (2,1,3)$, and $h = (2,-1,3)$ are all in the principal genus, with

$$U_{\Delta_F} = H_{\Delta_F} = ((\mathbb{Z}/|\Delta_F|\mathbb{Z})^*)^2 = \{\overline{1},\overline{2},\overline{3},\overline{4},\overline{6},\overline{8},\overline{9},\overline{12},\overline{13},\overline{16},\overline{18}\},$$

so there are $h_{\Delta_F}/2^{r-1} = 3$ proper equivalence classes of forms in the principal genus.

On the other hand, if $\Delta_F = -35$ as in Example 3.5 on page 136, then $r = 2 = h_{-35}$, so there are $2^{r-1} = 2$ genera, each having $h_{-35}/2^{r-1} = 1$ proper equivalence class.

Example 3.12 If $\Delta_F = -420$, then it can be shown that $h_{-420} = 8$ and in this case, $r = 4$, so

$$h_{-420} = 2^{r-1} \tag{3.34}$$

and each genus therefore has exactly one class of forms by Theorem 3.14, and $|G_{\Delta_F}| = 8$. Indeed, by Exercise 3.27 on page 145, the criterion for the property that every genus of forms of discriminant $\Delta_F = -4n$ consists of a single class is that (3.34) holds. Also, see Exercise 3.33 for other criteria.

Remark 3.13 Gauss not only proved Theorem 3.14 on the preceding page, but also he showed that the principal genus contains exactly those forms that are squares of some form under Dirichlet composition, sometimes called the *duplication* or *squaring* theorem. In other words, if F is a quadratic field of discriminant Δ_F and if **P** denotes the principal genus of discriminant Δ_F, then

$\mathbf{P} \cong C^2_{\Delta_F}$. It is also the case, related to the above, that the set of ambiguous forms A_{Δ_F} is a subgroup of C_{Δ_F} and has cardinality

$$|A_{\Delta_F}| = 2^{r-1},$$

where r is the number of distinct prime divisors of the discriminant Δ_F. It follows that the genus group G_{Δ_F} and group of ambiguous forms A_{Δ_F} are related by

$$A_{\Delta_F} \cong G_{\Delta_F}.$$

Remark 3.14 Some concluding remarks for this section to summarize the above developments are in order. Roughly speaking, when we look at forms in C_{Δ_F}, we are essentially considering sets of integers represented by forms. In this case, it is sufficient to consider whether or not Δ_F is a quadratic residue modulo a given prime to determine whether or not such a prime is represented by a form of discriminant Δ_F. Essentially this is what Exercise 3.9 on page 104 tells us. When looking at forms in G_{Δ_F}, we are considering sets of congruence classes modulo $|\Delta_F|$ to which the represented integers belong. When there is a single class of forms (from C_{Δ_F}) in each genus, then the question of which primes are represented by a given form of discriminant Δ_F is completely answered by congruence conditions. Many such illustrations were considered in Example 3.10 on page 140. However, if there exist more than one class (from C_{Δ_F}) in a given genus, then it is possible that no such congruence conditions exist to determine which of the forms from the distinct classes, in the same genus, represent a given prime. For instance, the case $\Delta_F = -56$, considered in Remark 3.12 on page 141, is one such case.

Essentially, two forms of discriminant Δ_F are in the same genus if they represent the same values in $(\mathbb{Z}/|\Delta_F|)^*$, and this is what Theorem 3.12 on page 132 tells us. Theorem 3.14 on page 142 groups these forms into equivalence classes related to the results in Theorem 3.12 in a very natural fashion. The cosets L_f of H_{Δ_F} in U_{Δ_F} determine in which genus the form f lies. This is tied to the fact that forms f and g are in the same genus if and only if they have the same assigned character, and this is tantamount to $L_f = L_g$ as cosets, a beautiful interconnection. Furthermore, the duplication theorem mentioned in Remark 3.13 on the previous page tells us that the principal genus consists of just the squares of forms under composition. Also, Remark 3.13 tells us that the genus group is essentially the group of ambiguous forms, the central topic of §3.3.

Exercises

3.23. Prove that for $D \in \{-4, -7, -8, -11, -12, -16, -19, -28, -43\}$ there is exactly one class in the principal genus; indeed that $h_D = 1$.

(Hint: See the solution of Exercise 3.13 on page 413.)

3.24. By Exercise 3.23, the first negative discriminant $D \equiv 1 \pmod 4$ with $h_D > 1$ is $D = -15$. Show that $h_D = 2$, and determine the congruence classes for representation by the principal form $(1, 1, 4)$.

3.25. Prove that $x^2 + 14y^2$ and $2x^2 + 7y^2$ are in the same genus of discriminant -56 by showing that if $p \neq 2, 7$ is a prime then

$$p = x^2 + 14y^2 \text{ or } p = 2x^2 + 7y^2 \text{ for some } x, y \in \mathbb{Z}$$

if and only if

$$p \equiv 1, 9, 15, 23, 25, 39 \pmod{56}.$$

3.26. Prove that the classes belonging to the principal genus P form a subgroup of C_{Δ_F}. Then show that every genus forms a coset of P in C_{Δ_F}. Conclude that there are an equal number of classes in each genus.

3.27. Prove that the following are equivalent:

(a) Every genus of forms with discriminant $\Delta_F = -4n$ for $n \in \mathbb{N}$ consists of a single class.

(b) $h_{-4n} = 2^{r-1}$ where r is the number of distinct prime divisors of Δ_F.

(*Euler found 65 values of n for which this holds and called such n convenient numbers. No others are known.*)

3.28. Prove that $3x^2 + 2xy + 5y^2$ and $3x^2 - 2xy + 5y^2$ are in the same genus of discriminant -56 by showing that if $p \neq 2, 7$ is a prime then

$$p = 3x^2 \pm 2xy + 5y \text{ for some } x, y \in \mathbb{Z}$$

if and only if

$$p \equiv 3, 5, 13, 19, 27, 45 \pmod{56}.$$

3.29. Prove that $h_{-56} = 4$.

3.30. Prove that the product of any two forms of discriminant $-4n$ for $n \in \mathbb{N}$ is of the form $X^2 + nY^2$ for some $X, Y \in \mathbb{Z}$.

3.31. Prove the assertion made in Remark 3.6 on page 108 that $(ac, b, 1) \sim (1, 0, \Delta_F/4)$ when $\Delta_F \equiv 0 \pmod 4$ and $(ac, b, 1) \sim (1, 1, (1-\Delta_F)/4)$ when $\Delta \equiv 1 \pmod 4$.

(*Hint: When $\Delta_F \equiv 0 \pmod 4$, in Definition 3.1 on page 98, select $p = b/2$, $q = 1$, $r = -1$, and $s = 0$, and when $\Delta_F \equiv 1 \pmod 4$ select $p = -(1+b)/2$, $q = -1$, $r = 1$ and $s = 0$.*)

3.32. Let $\Delta_F < 0$. Prove that a reduced positive definite form $f = (a, b, c)$ of discriminant Δ_F has order 1 or 2 if and only if $b = 0$, $a = b$, or $a = c$.

3.33. Prove that the following are equivalent.

(a) Every genus of forms with discriminant $\Delta_F = -4n$ for $n \in \mathbb{N}$ consists of a single class.

(b) Every reduced positive definite form $f = (a, b, c)$ of discriminant $-4n$ satisfies that either $b = 0$, $a = b$, or $a = c$.

(Hint: Use Exercise 3.32 in conjunction with Remark 3.13 on page 143 where it is noted that the principal genus is the group of squares.)

To solve Exercises 3.34–3.35, use the techniques employed in the solution of Exercise 3.13 on page 413.

3.34. Find all primitive reduced forms of discriminant $\Delta_F = -71$.

3.35. Find all primitive reduced forms of discriminant $\Delta_F = -80$.

3.36. Given $\Delta_F = 229$, we have that $h_{\Delta_F} = 3$ where Theorem 3.14 on page 142 tells us there is a single genus of forms. Find three inequivalent reduced forms of discriminant Δ_F and a distinct prime p represented by each form such that $p \equiv 1 \pmod{229}$.

3.37. With reference to Exercise 3.36 find an integer $z \neq 0$ for each of the three primes p found therein such that $p \equiv z^2 + z - 57 \pmod{229}$. This illustrates Corollary 3.8 on page 138 for the case $\Delta_F > 0$.

Biography 3.7 Max Deuring (1907–1984) *was born in Göttingen, Germany on December 9, 1907. He entered the University of Göttingen in 1926, where he studied mathematics and physics. In 1931, under the supervision of Emmy Noether, he received his doctorate entitled* Arithmetische Theorie der algebraischen Funktionen–*see Biography 2.1 on page 73. One of his strengths was the ability to simplify and generalize existing results, one of these being the aforementioned work of Heegner on page 141. In his first paper, published in 1931, he generalized Hilbert's theory of prime divisors in Galois fields to more general fields–see Biography 3.5 on page 127. In 1931, at the University of Leipzig, Deuring was appointed as van der Waerden's assistant. In 1937, Deuring went to the University of Jena where he stayed for six years. In 1937 and 1940 he published two papers which were his greatest contributions. These publications generalized Hasse's results on the Riemann hypothesis for the zeta function associated with elliptic curve over a finite field. To do this he extended Hasse's idea from curves of genus 1 to elliptic curves of higher genus using his algebraic theory of correspondences, which André Weil later used to generalize the Reimann hypothesis to function fields of arbitrary genus. In 1950, Deuring was appointed to the fill the chair vacated by Herglotz at Göttingen, which Deuring held until his retirement in 1976. He died in Göttingen on December 20, 1984.*

Among his honours, were election to the Academy of Science and Literature in Mainz, and the Göttingen Academy of Sciences.

Biography 3.8 Henri Jules Poincaré (1854–1912) *was born in Nancy, France on April 29, 1854. He entered the École Polytechnique in 1873, and graduated in 1875. After receiving his doctorate, he was appointed to teach at the University of Caen, but remained there for only two years. In 1881, he was appointed to a chair in the Faculty of Science in Paris. Also, in 1886, with the support of Hermite–see Biography 3.4 on page 126–he was nominated for a chair at the Sorbonne. He held these two chairs until his untimely death at the age of 58 on July 17, 1912 in Paris.*

Poincaré created the theory of automorphic forms, non-Euclidean geometry, and complex functions. His contributions to algebraic topology are also seminal and the Poincaré conjecture in that area remains a major challenge. In his paper, published in 1890, on the three-body problem, he created new avenues in celestial mechanics, and gave the first mathematical description of chaotic motion, which essentially began the modern study of dynamical systems. Indeed, in three volumes published between 1892 and 1899, he aimed to completely characterize all motions of mechanical systems. He also wrote on the philosophy of mathematics and science in general. In that vein, a quote from an article published in 1904 is germane: "It is by logic we prove, it is by intuition that we invent." Another quote, made at an address from his funeral is a fitting bottom line: "He was a mathematician, geometer, philosopher, and man of letters, who was a kind of poet of the infinite, a kind of bard of science."

3.5   Representation

We have looked at some representation problems already in Example 3.10 on page 140 for positive definite forms where we looked at representation of primes. We now look at the problem more extensively. Indeed, as mentioned in Example 3.10, the problem of representation as a sum of two integer squares is solved via the consideration of $\Delta_F = -4$, the discriminant of $F = \mathbb{Q}(\sqrt{-1})$. We also looked, in that example, at representations of the form $x^2 + 2y^2$, emanating from $\Delta_F = -8$, the discriminant of $\mathbb{Q}(\sqrt{-2})$. Some other special forms were considered as well. We now look at more general results based upon the class numbers of quadratic fields that we linked to the form class group in §3.2. Recall that by Corollary 3.4 on page 117, we know that $h_{\mathfrak{O}_{\mathbf{F}}} < \infty$.

Theorem 3.15 | Prime Representation and $h_{\mathfrak{O}_{\mathbf{F}}}$

Let F be a quadratic field with discriminant Δ_F and (wide) class number $h_{\mathfrak{O}_{\mathbf{F}}}$. Suppose that $p > 2$ is a prime such that $\gcd(\Delta_F, p) = 1$ and Δ_F is a quadratic residue modulo p. Then the following hold.

(a) If either $\Delta_F < 0$ or $\Delta_F > 0$ and there exists a $u \in \mathfrak{U}_F$ with $N_F(u) = -1$, then there exist relatively prime integers a, b such that

$$p^{h_{\mathfrak{O}_{\mathbf{F}}}} = \begin{cases} a^2 - \Delta_F b^2 & \text{if } \Delta_F \equiv 1 \,(\mathrm{mod}\, 8), \\ a^2 - \frac{\Delta_F}{4} b^2 & \text{if } \Delta_F \equiv 0 \,(\mathrm{mod}\, 4), \\ a^2 + ab + \frac{1}{4}(1 - \Delta_F)b^2 & \text{if } \Delta_F \equiv 5 \,(\mathrm{mod}\, 8). \end{cases}$$

(b) If $\Delta_F > 0$ and there does *not* exist a $u \in \mathfrak{U}_F$ with $N_F(u) = -1$, then there exist relatively prime integers a, b such that

$$p^{h_{\mathfrak{O}_{\mathbf{F}}}} = \begin{cases} \pm(a^2 - \Delta_F b^2) & \text{if } \Delta_F \equiv 1 \,(\mathrm{mod}\, 8), \\ \pm(a^2 - \frac{\Delta_F}{4} b^2) & \text{if } \Delta_F \equiv 0 \,(\mathrm{mod}\, 4), \\ \pm(a^2 + ab + \frac{1}{4}(1 - \Delta_F)b^2) & \text{if } \Delta_F \equiv 5 \,(\mathrm{mod}\, 8). \end{cases}$$

Proof. By Theorem 2.4 on page 60, since $p > 2$, then if $(\Delta_F/p) = 1$, we have $(p) = \mathcal{P}_1 \mathcal{P}_2$ where $\mathcal{P}_j$ are distinct prime $\mathfrak{O}_F$-ideals for $j = 1, 2$. Thus,

$$(p^{h_{\mathfrak{O}_{\mathbf{F}}}}) = (p)^{h_{\mathfrak{O}_{\mathbf{F}}}} = \mathcal{P}_1^{h_{\mathfrak{O}_{\mathbf{F}}}} \mathcal{P}_2^{h_{\mathfrak{O}_{\mathbf{F}}}} \sim (1),$$

since $\mathcal{P}_j^{h_{\mathfrak{O}_F}} \sim (1)$ for $j = 1, 2$ by Exercise 3.18 on page 117. Hence, $\mathcal{P}_j^{h_{\mathfrak{O}_F}}$ is a principal ideal for $j = 1, 2$. Let

$$\mathcal{P}_1^{h_{\mathfrak{O}_F}} = \left(\frac{u + v\sqrt{\Delta_F}}{2}\right)$$

where $u \equiv v \pmod 2$, if $\Delta_F \equiv 1 \pmod 4$, and u is even if $\Delta_F \equiv 0 \pmod 4$. Then via the proof of Theorem 2.4 on page 60 we know that $\mathcal{P}_2$ must be the conjugate of $\mathcal{P}_1$, namely

$$\mathcal{P}_2^{h_{\mathfrak{O}_F}} = \left(\frac{u - v\sqrt{\Delta_F}}{2}\right).$$

Hence,

$$(p^{h_{\mathfrak{O}_F}}) = \left(\frac{u^2 - \Delta_F v^2}{4}\right),$$

so there exists an $\alpha \in \mathfrak{U}_F$ such that

$$p^{h_{\mathfrak{O}_F}} = \alpha \left(\frac{u^2 - \Delta_F v^2}{4}\right).$$

However,

$$\alpha = \frac{4p^{h_{\mathfrak{O}_F}}}{u^2 - \Delta_F v^2} \in \mathbb{Q}.$$

But, by Corollary 1.2 on page 4, $\mathfrak{O}_F \cap \mathbb{Q} = \mathbb{Z}$, so $\alpha \in \mathfrak{U}_{\mathbb{Z}} = \{\pm 1\}$. Thus,

$$4p^{h_{\mathfrak{O}_F}} = \pm(u^2 - \Delta_F v^2). \tag{3.35}$$

Claim 3.5 *If $\Delta_F \equiv 0 \pmod 4$, then $\gcd(u/2, v) = 1$, and if $\Delta_F \equiv 1 \pmod 4$, $\gcd(u, v) = 1$ or 2.*

If $\Delta_F \equiv 1 \pmod 4$, let $q > 2$ be a prime such that $q \mid \gcd(u, v)$. Then there exist integers x, y such that $u = qx$ and $v = qy$, where $x \equiv y \pmod 2$. Therefore, by (3.35), $q^2 \mid 4p^{h_{\mathfrak{O}_F}}$, but $q > 2$ so $q = p$. Hence,

$$\mathcal{P}_1^{h_{\mathfrak{O}_F}} = (p)\left(\frac{x + y\sqrt{\Delta_F}}{2}\right) = \mathcal{P}_1 \mathcal{P}_2 \left(\frac{x + y\sqrt{\Delta_F}}{2}\right),$$

which forces $\mathcal{P}_2 \mid \mathcal{P}_1^{h_{\mathfrak{O}_F}}$, contradicting that $\mathcal{P}_1$ and $\mathcal{P}_2$ are distinct $\mathfrak{O}_F$-ideals. We have shown that $\gcd(u, v) = 2^c$ for some integer $c \geq 0$. It follows from (3.35) that $4^c \mid 4$ so $c = 0$ or $c = 1$.

If $\Delta_F \equiv 0 \pmod 4$, and q is a prime such that $q \mid \gcd(u/2, v)$, then there exist integers x, y such that $u = 2qx$ and $v = qy$, so

$$p^{h_{\mathfrak{O}_F}} = \pm((qx)^2 - (\Delta_F/4)(qy)^2)$$

which forces $p = q$ and this leads to a contradiction as above. This is Claim 3.5.

If $\Delta_F < 0$ then the plus sign holds in (3.35), since $u^2 - \Delta_F v^2 > 0$. When $\Delta_F > 0$ and there exists a $\alpha \in \mathfrak{U}_F$ with $N_F(\alpha) = -1$, we may multiply by

$$N_F(\alpha) = N(r + s\sqrt{\Delta_F}) = r^2 - \Delta_F s^2 = -1$$

to get

$$-(u^2 - \Delta_F v^2) = (r^2 - \Delta_F s^2)(u^2 - \Delta_F v^2) = (ru + \Delta_F sv)^2 - \Delta_F(rv + su)^2.$$

To complete the proof, we need only show how the a, b may be selected to satisfy parts (a)–(b) of our theorem.

When $\Delta_F \equiv 1 \,(\mathrm{mod}\ 4)$, then by (3.35), if u and v are odd, so $4p^{h_{\mathfrak{O}F}} \equiv 0$ (mod 8), contradicting that $p > 2$. Thus, by Claim 3.5, $\gcd(u, v) = 2$ so we select $a = u/2$ and $b = v/2$. If $\Delta_F \equiv 0 \,(\mathrm{mod}\ 4)$, then by Claim 3.5, we may select $a = u/2$ and $b = v$. Lastly, when $\Delta_F \equiv 5 \,(\mathrm{mod}\ 8)$, since $u \equiv v \,(\mathrm{mod}\ 2)$, set $u = b + 2a$ and $b = v$ where $a, b \in \mathbb{Z}$. Then (3.35) becomes,

$$\pm 4p^{h_{\mathfrak{O}F}} = u^2 - \Delta_F v^2 = (b + 2a)^2 - \Delta_F b^2 = 4a^2 + 4ab + (1 - \Delta_F)b^2,$$

so

$$p^{h_{\mathfrak{O}F}} = \pm\left(a^2 + ab + \frac{1}{4}(1 - \Delta_F)b^2\right),$$

which secures our result.   □

Remark 3.15 As a counterfoil to Theorem 3.15 on page 148, we note that, by Exercise 3.9 on page 104, if Δ_F is not a quadratic residue modulo a prime $p > 2$, then there is no binary quadratic form that represents p^k for any positive integer k. Hence, there cannot exist integers (a, b, c) such that $p^k = ax^2 + bxy + cy^2$ for any integers x, y.

Theorem 3.15 has certain value when $h_{\mathfrak{O}_F} = 1$. In particular, we have the following results, the first two of which are a recapitulation of what we discussed in Example 3.10 on page 140–and the first of which also appears in Theorem 1.13 on page 26–via Theorem 3.15 this time.

Corollary 3.9 *Let p be a prime. Then there exist relatively prime integers a, b such that*

$$p = a^2 + b^2 \text{ if and only if } p = 2 \text{ or } p \equiv 1 \pmod{4}.$$

Proof. By Theorem 3.2 on page 102 and Theorem 3.6 on page 113, for $\Delta_F = -4$,

$$h_{\mathfrak{O}_F} = h_{\mathbb{Z}[\sqrt{-1}]} = 1.$$

Thus, by Theorem 3.15, if $(\Delta_F/p) = 1$, namely $p \equiv 1 \,(\mathrm{mod}\ 4)$, then $p = a^2 + b^2$ for $a, b \in \mathbb{N}$. Since $2 = 1^2 + 1^2$, then we have one direction. Conversely, if $p = a^2 + b^2$, and $p > 2$, then by Exercise 3.9 on page 104, $(-4/p) = (-1/p) = 1$, which implies that $p \equiv 1 \,(\mathrm{mod}\ 4)$.   □

Corollary 3.10 *Let p be a prime. Then there exist relatively prime integers a, b such that*

$$p = a^2 + 2b^2 \text{ if and only if } p = 2 \text{ or } p \equiv 1, 3 \pmod 8.$$

Proof. First, we know that $(-8/p) = (-2/p) = 1$ if and only if $p \equiv 1, 3$ (mod 8)–see Example 3.10. By Theorem 3.2 and Theorem 3.6, for $\Delta_F = -8$,

$$h_{\mathcal{D}_F} = h_{\mathbb{Z}[\sqrt{-1}]} = 1.$$

Therefore, by Theorem 3.15, if $(-8/p) = 1$, $p = a^2 + 2b^2$ for $a, b \in \mathbb{N}$. Also, $2 = 0^2 + 2 \cdot 1^2$. Conversely, if

$$p = a^2 + 2b^2, \text{ and } p > 2,$$

then by Exercise 3.9 on page 104, $(-8/p) = (-2/p) = 1$. □

Corollary 3.11 *Let p be a prime. Then there exist relatively prime integers a, b such that*

$$p = a^2 + ab + b^2 \text{ if and only if } p = 3 \text{ or } p \equiv 1 \pmod 3.$$

Proof. From Exercise 3.38 on the next page, $(-3/p) = 1$ if and only if $p \equiv 1 \pmod 3$. By Example 3.10, Theorem 1.3 on page 6, and Theorem 3.6 on page 113, we have that

$$h_{-3} = h_{\mathbb{Z}[(1+\sqrt{-3})/2]} = 1.$$

Thus, by Theorem 3.15, if $(\Delta_F/p) = (-3/p) = 1$, then

$$p = a^2 + ab + b^2 \text{ for some integers } a, b.$$

Also $3 = 1^2 + 1 \cdot 1 + 1^2$. Conversely, by Exercise 3.9 on page 104, if $p > 3$ and $p = a^2 + ab + b^2$, then $(-3/p) = 1$. □

Corollary 3.12 *Let p be a prime. Then there exist relatively prime integers a, b such that $p = a^2 + 7b^2$ if and only if $p = 7$ or*

$$p \equiv 1, 9, 11, 15, 23, 25 \pmod{28}.$$

Proof. By Exercise 3.39 on the next page, $(-7/p) = 1$ if and only if

$$p \equiv 1, 9, 11, 15, 23, 25 \pmod{28}.$$

Also, by Theorem 1.3, Theorem 3.6, and Example 3.10, for $\Delta_F = -7$,

$$h_{\mathcal{D}_F} = h_{\mathbb{Z}[(1+\sqrt{-7})/2]} = h_{-7} = 1.$$

Therefore, by Theorem 3.15, if $(-7/p) = 1$, $p = a^2 + 7b^2$ for $a, b \in \mathbb{N}$. Also, $7 = 0^2 + 7 \cdot 1^2$. Conversely, if

$$p = a^2 + 7b^2, \text{ and } p \neq 7,$$

then by Exercise 3.9 on page 104, $(-7/p) = 1$. □

Exercises

3.38. Prove that $(-3/p) = 1$ for a prime $p > 3$ if and only if $p \equiv 1 \,(\mathrm{mod}\ 3)$.

(*Hint: You may use the fact from* [68, Example 4.11, p. 191], *that* $(3/p) = 1$ *if and only if* $p \equiv \pm 1 \,(\mathrm{mod}\ 12)$ *and* $(3/p) = -1$ *if and only if* $p \equiv \pm 5 \,(\mathrm{mod}\ 12)$.)

3.39. Prove that $(-7/p) = 1$ for an odd prime p if and only if $p \equiv 1, 9, 11, 15, 23, 25 \,(\mathrm{mod}\ 28)$.

In Exercises 3.40–3.43, use the techniques of Corollary 3.11 on the previous page to solve the representation problems.

3.40. Prove that a prime p is representable in the form

$$p = a^2 + ab + 3b^2 \text{ for relatively prime } a, b \in \mathbb{Z}$$

if and only if

$$p = 11 \text{ or } p \equiv 1, 3, 5, 9, 15, 21, 23, 25, 27, 31 \pmod{44}.$$

3.41. Prove that a prime p is representable in the form

$$p = a^2 + ab + 5b^2 \text{ for relatively prime } a, b \in \mathbb{N}$$

if and only if $p = 19$ or

$$p \equiv 1, 5, 7, 9, 11, 17, 23, 25, 35, 39, 43, 45, 47, 49, 55, 61, 63, 73 \pmod{76}.$$

3.42. Prove that a prime p is representable in the form

$$p = a^2 + ab + 11b^2 \text{ for relatively prime } a, b \in \mathbb{Z}$$

if and only if $p = 43$ or

$$p \equiv 1, 9, 11, 13, 15, 17, 21, 23, 25, 31, 35, 41, 47, 49, 53, 57, 59, 67, 79, 81,$$

$$83, 87, 95, 97, 99, 101, 103, 107, 109, 111, 117, 121, 127, 133,$$

$$135, 139, 143, 145, 153, 165, 167, 169 \pmod{172}.$$

3.43. Prove that a prime p is representable in the form

$$p = a^2 + ab + 17b^2 \text{ for relatively prime } a, b \in \mathbb{Z}$$

if and only if $p = 67$ or

$p \equiv 1, 9, 15, 17, 19, 21, 23, 25, 29, 33, 35, 37, 39, 47, 49, 55, 59, 65, 71, 73, 77, 81,$

$83, 89, 91, 93, 103, 107, 121, 123, 127, 129, 131, 135, 143, 149, 151, 153, 155,$

$157, 159, 163, 167, 169, 171, 173, 181, 183, 189, 193, 199, 205, 207, 211, 215,$

$217, 223, 225, 227, 237, 241, 255, 257, 261, 263, 265 \pmod{268}.$

3.44. From Theorem 1.3 on page 6, Example 3.10 on page 140, and Theorem 3.6 on page 113, we know that $h_{\mathcal{O}_F} = h_{\mathbb{Z}[(1+\sqrt{-163})/2]} = 1$. Thus, Theorem 3.15 on page 148 informs us that odd prime p with $(\Delta_F/p) = (-163/p) = 1$ satisfy that $p = a^2 + ab + 41b^2$ for some relatively prime integers a, b. Show that for $b = 1$, $a^2 + a + 41$ is indeed prime for $a = 0, 1, \ldots, 39$.

(*This is related to a result of Rabinowitsch [79], which states that for negative Δ_F, with $\Delta_F \equiv 1 \pmod 4$, we have that $h_{\mathcal{O}_F} = 1$ if and only if $x^2 + x + (1 - \Delta_F)/4$ is prime for $x = 0, 1, \ldots, \lfloor |\Delta_F|/4 - 1 \rfloor$. The reader may now go to Exercises 3.40–3.43, and indeed for all values in Example 3.10, and verify this fact for those values as well.*)

(*See Biography 3.9 on the next page.*)

3.45. Related to the Rabinowitsch result in Exercise 3.44 is the following, known as the *Rabiniowitsch–Mollin–Williams* criterion for real quadratic fields–see [63]. If F is a real quadratic field with discriminant $\Delta_F \equiv 1 \pmod 4$, then $|x^2 + x + (1 - \Delta_F)/4|$ is 1 or prime for all $x = 1, 2, \ldots, \lfloor (\sqrt{\Delta_F} - 1)/2 \rfloor$ if and only if $h_{\mathcal{O}_F} = 1$ and either $\Delta_F = 17$ or $\Delta_F = n^2 + r \equiv 5 \pmod 8$ where $r \in \{\pm 4, 1\}$–see [65, Theorem 6.5.13, p. 352]. Verify this primality for the values

$$\Delta_F \in \{17, 21, 29, 37, 53, 77, 101, 173, 197, 293, 437, 677\}.$$

(*See Biography 3.10 on the following page.*)

3.46. It is known that for $\Delta_F = -20$, $h_{\mathcal{O}_F} = 2$ and $P = (2, 1 + \sqrt{-5})$ is an ideal representing the nonprincipal class. Use the identification given in the proof of Theorem 3.5 on page 110 to prove the following, where $p \neq 5$ is an odd prime.

(a) $p = a^2 + 5b^2$ if and only if $p \equiv 1, 9 \pmod{20}$.

(b) $p = 2a^2 + 2ab + 3b^2$ if and only if $p \equiv 3, 7 \pmod{20}$.

(c) Conclude that for $\Delta_F = -20$, there are two genera each consisting of a single class.

Biography 3.9 The following was taken from a most interesting article about G. Rabinowitsch by Mordell [72]. Mordell writes: "In 1923, I attended a meeting of the American Mathematical Society held at Vassar College in New York State. Someone called Rainich from the University of Michigan at Ann Arbor, gave a talk upon the class number of quadratic fields, a subject in which I was very much interested. I noticed that he made no reference to a rather pretty paper written by Rabinowitz from Odessa and published in Crelle's journal. I commented upon this. He blushed and stammered and said, "I am Rabinowitz." He had moved to the U.S.A. and changed his name.... The spelling of Rabinowitsch in this book coincides with that which appears in Crelle [79].

Biography 3.10 Hugh Cowie Williams *was born in London, Ontario, Canada on July 23, 1943. He graduated with a doctorate in computer science from the University of Waterloo in 1969. Since that time, his research interests have been in using computational techniques to solve problems in number theory, and in particular, those with applications to cryptography. Currently he holds a Chair under Alberta Informatics Circle of Research Excellence (iCORE) at the University of Calgary (U of C). He oversees the Centre for Information Security and Cryptography (CISaC), a multi-disciplinary research centre at the U of C devoted to research and development towards providing security and privacy in information communication systems. There are also more than two dozen graduate students and post doctoral fellows being trained at the centre. The iCORE Chair is in algorithmic number theory and cryptography (ICANTC), which is the main funder of CISaC. The initial funding from Icore was $3 million dollars for the first five years and this has been renewed for another five years. In conjunction with this iCORE Chair, Professor Williams has set up a research team in pure and applied cryptography to investigate the high-end theoretical foundations of communications security. Professor Williams comes from the University of Manitoba, where he was Associate Dean of Science for Research and Development, and Adjunct Professor for the Department of Combinatorics and Optimization at the University of Waterloo. He has an extensive research and leadership background and a strong international reputation for his work in cryptography and number theory. CISaC and ICANTC were acronyms coined by this author, who initiated the application for the Chair, and is currently a member of the academic staff of CISaC, as well as professor at the U of C's mathematics department. This author and Professor Williams have coauthored more than two dozen papers in number theory, and computational mathematics, over the past quarter century.*

3.6 Equivalence Modulo p

> *The bottom line for mathematicians is that the architecture has to be right. In all the mathematics that I did, the essential point was to find the right architecture. It's like building a bridge. Once the main lines of the structure are right, then the details miraculously fit. The problem is the overall design.*
>
> —From the interview in [1]
> **Freeman Dyson (1923–)**
> **American physicist, mathematician, and author**

Now we turn to equivalence of forms modulo a prime, a topic that has some rather palatable results.

Definition 3.16 | **Forms Equivalent Modulo a Prime**

Let p be a prime and for $j = 1, 2$, let Δ_{F_j} be the discriminants of a quadratic fields F_j. Also, let

$$f_j = (a_j, b_j, c_j)$$

be primitive forms of discriminant Δ_{F_j} for $j = 1, 2$. If there is a transformation

$$x = rX + sY, y = tX + uY,$$

where

$$f_1(x, y) \equiv f_2(X, Y) \pmod{p}$$

with $\gcd(ru - st, p) = 1$, we say that f_1 and f_2 are *equivalent modulo p*, and we denote this by

$$f_1 \sim f_2 \pmod{p}.$$

Remark 3.16 If the forms f_1 and f_2 are equivalent modulo a prime p, as given in Definition 3.16, then if $p \nmid \Delta_{F_j}$ for $j = 1, 2$,

$$\Delta_{F_2} \equiv (ru - st)^2(b_1^2 - 4a_1c_1) \equiv (ru - st)^2\Delta_{F_1} \pmod{p}. \qquad (3.36)$$

Thus, from (3.36), the following Legendre symbol equality holds,

$$\left(\frac{\Delta_{F_1}}{p}\right) = \left(\frac{\Delta_{F_2}}{p}\right).$$

Lemma 3.6 | **Vanishing Middle Term Modulo p**

If $f = (a, b, c)$ is a primitive form of discriminant Δ_F for a quadratic field F, and p is an odd prime not dividing Δ_F, then for some $a_1, c_1 \in \mathbb{Z}$,

$$(a, b, c) \sim (a_1, 0, c_1) \pmod{p}.$$

Proof. Since f is primitive, then $\gcd(a, b, c) = 1$, so if $p \nmid a$, then by setting

$$X \equiv \left(x + \frac{b}{2a}y\right) \pmod{p}, \text{ and } Y \equiv y \pmod{p},$$

we get

$$ax^2 + bxy + cy^2 \equiv a\left(x + \frac{b}{2a}y\right)^2 - \frac{\Delta}{4a}y^2 \equiv aX^2 - \frac{\Delta}{4a}Y^2 \pmod{p}.$$

Similarly, we get such a result when we assume that $p \nmid c$. On the other hand, if $p \mid \gcd(a, c)$, then by setting

$$x = X + Y \text{ and } y = X - Y,$$

we achieve

$$ax^2 + bxy + cy^2 \equiv bX^2 - bY^2 \pmod{p}.$$

We have shown that we always have f equivalent modulo p to a form of type $(a_1, 0, c_1)$.   □

Remark 3.17 Lemma 3.6 shows we may always assume that if we consider a form (a, b, c) modulo p, we may assume that $p \mid b$ and $p \nmid (ac)$. Now we have sufficient tools to establish the first main result.

Theorem 3.16 | **Canonical Equivalence Modulo p** |

Suppose that F is a quadratic field of discriminant Δ_F and p is an odd prime not dividing Δ_F. If (a, b, c) is a primitive form of discriminant Δ_F, then each of the following holds.

(a) *If Δ_F is a quadratic residue modulo p, then $(a, b, c) \sim (1, 0, -1) \pmod{p}$.*

(b) *If Δ_F is a quadratic nonresidue modulo p, then*

$$(a, b, c) \sim (1, 0, -\Delta_F) \not\sim (1, 0, -1) \pmod{p}.$$

Proof. We begin with a claim.

Claim 3.6 *If $p \nmid (ac)$, then there exist $x, y \in \mathbb{Z}$ such that*

$$ax^2 + cy^2 \equiv 1 \pmod{p}.$$

For $x = 0, 1, \ldots, p - 1$, ax^2 takes on $(p + 1)/2$ distinct values and as y ranges over $0, 1, \ldots, p - 1$, $1 - cy^2$ takes on $(p + 1)/2$ distinct values. Hence, by the Pigeonhole Principal–see [68, p.35]– there must exist $x, y \in \mathbb{Z}$ such that

$$ax^2 \equiv 1 - cy^2 \pmod{p},$$

securing the claim.

By Claim 3.6, we may let r, t be integers such that

$$ar^2 + ct^2 \equiv 1 \pmod{p},$$

and select fixed integers s, u with $p \nmid (ru - st)$. Now set

$$b_1 \equiv 2ars + 2ctu \pmod{p} \text{ and } c_1 \equiv as^2 + cu^2 \pmod{p}.$$

Therefore, via the transformation

$$x = rX + sY, y = tX + uY,$$

we get

$$(a, 0, c) \sim (1, b_1, c_1) \pmod{p}.$$

If we set

$$\Delta_{F_1} = b_1^2 - 4c_1,$$

then since, $p \mid b_1$ and $p \nmid c_1$, via Remark 3.17 on the facing page, we get

$$(1, b_1, c_1) \sim (1, 0, -\Delta_{F_1}/4) \sim (1, 0, -\Delta_{F_1}) \pmod{p}.$$

Thus, if

$$\Delta_{F_1} \equiv z^2 \pmod{p},$$

then via Remark 3.16 on page 155, we know that

$$\Delta_F \equiv z^2 w^2 \pmod{p},$$

so via the transformation $x = X$ and $y = wzY$,

$$(1, 0, -\Delta_{F_1}) \sim (1, 0, -1) \pmod{p}.$$

Since $p \mid b$ and $p \nmid (ac)$ may be assumed via Remark 3.17, then we have shown that when Δ_F is a quadratic residue modulo p,

$$(a, b, c) \sim (a, 0, c) \sim (1, 0, -\Delta_{F_1}) \sim (1, 0, -1) \pmod{p}.$$

This is part (a).

If Δ_F is not a quadratic residue modulo p, then we have shown that

$$(a, b, c) \sim (1, 0, -\Delta_F) \pmod{p}.$$

That

$$(1, 0, -\Delta_F) \not\sim (1, 0, -1)$$

is Exercise 3.47 on the following page. This is (b) and we have secured the result. □

Corollary 3.13 *If p is an odd prime not dividing Δ_F, then any two forms with discriminant Δ_F must be equivalent modulo p.*

The reader may go to Exercises 3.48–3.50 for further results on equivalence modulo 2.

Exercises
In Exercises 3.47–3.49, Δ_F denotes the discriminant of quadratic field F.

3.47. Prove the fact stated in part (b) of Theorem 3.16 on page 156, that

$$(1, 0, -\Delta_F) \not\sim (1, 0, -1).$$

3.48. Prove that any form $f = (a, b, c)$ of odd discriminant Δ_F must satisfy

$$(a, b, c) \sim (0, 1, 0) \pmod{2} \text{ if } 2 \mid (ac),$$

and

$$(a, b, c) \sim (1, 1, 1) \pmod{2} \text{ if } 2 \nmid (ac),$$

in the sense of Defintion 3.16 on page 155.

3.49. With reference to Exercise 3.48, show that

$$(0, 1, 0) \not\sim (1, 1, 1) \pmod{2}.$$

3.50. Prove that any two forms with the same odd discriminant must be equivalent modulo 2.

3.51. Let p be an odd prime and let $f_j = (a_j, b_j, c_j)$ be forms of discriminant Δ_{F_j} where $p \mid \Delta_j$ for $j = 1, 2$. Prove that

$$(a_1, b_1, c_1) \sim (a_2, b_2, c_2) \pmod{p}$$

if and only if the Legendre symbol equality

$$\left(\frac{n_1}{p} \right) = \left(\frac{n_2}{p} \right)$$

holds where n_j is an integer represented by f_j with $\gcd(n_j, \Delta_{F_j}) = 1$ for $j = 1, 2$.

(*Hint: Use Lemma 3.1 on page 105.*)

Chapter 4

Diophantine Approximation

> *We could use up two Eternities in learning all that is to be learned about our own world and the thousands of nations that have arisen and flourished and vanished from it. Mathematics alone would occupy me eight million years.*
>
> from **Notebook No. 22, Spring 1883–September 1884**.
> **Mark Twain (1835–1910), born Samuel Langhorne Clemens,**
> **American writer**

In this chapter,we assume the background on continued fractions, rational approximations, quadratic irrationals, and related topics covered, for instance, in [68, Chapter 5].

4.1 Algebraic and Transcendental Numbers

We have already looked at some Diophantine equations in §1.1. In particular, in Definition 1.10 on page 13, and Theorem 1.8 on page 14, we considered the Ramanujan–Nagell equation, the generalization of which we will study later in the text. The relationship between the solution of Diophantine equations and approximation of algebraic numbers by rational numbers is the focus of this section. In particular, we know from [68, Corollary 5.3, p. 215, Exercise 5.10, p.220], for instance, that there are infinitely many rational number p, q such that

$$\left| \alpha - \frac{p}{q} \right| < \frac{1}{q^2}. \tag{4.1}$$

A natural query is: Can the exponent 2 be increased to get a general result that improves upon (4.1)? In a drive to answer this question, the Fields medal was achieved by Roth in 1958 for his 1955 result: If α is an algebraic number, then for a given $\varepsilon > 0$, there exist at most finitely many rational numbers p, q, with

$q > 0$ such that

$$\left| \alpha - \frac{p}{q} \right| < \frac{1}{q^{2+\varepsilon}} \tag{4.2}$$

–see [21]. Roth's work was preceded by results of Thue in 1909 and Siegel in 1921–see [68, Biography 1.12, p. 45] and Biography 4.4 on page 170. Both of the latter two improved upon the following result of Liouville–see Biography 4.3 on page 168.

Biography 4.1 Klaus Friedrich Roth (1925–) *was born on October 29, 1925 in Breslau, Germany (now Wroclaw, Poland). He achieved his BA in 1945 from Peterhouse, Cambridge. In 1946, he entered University College, London where he was awarded his master's degree in 1948. In that year he was appointed lecturer there and was awarded his doctorate in 1950, under the direction of Davenport. In 1955, when he was a lecturer at University College in London, he proved what is now known as the* Thue–Siegel–Roth Theorem, *or just Roth's Theorem, (4.2), for which he won the Fields medal. Indeed, the medal, was awarded by Davenport at the International Congress of Mathematicians in 1958–see Biography 1.6 on page 54. To date, he is the oldest Fields medalist.*

Roth became a professor at University College, London in 1961. Then he moved to a chair at Imperial College, London, a position he held until his retirement in 1988. He came back as a visiting professor there and remained at Imperial College until 1996 when he returned to Scotland. He is also known for his 1952 proof that subsets of the integers of positive density must contain infinitely many arithmetic progressions of length three, which established the first non-trivial case of what we now call Szemerédi's theorem.

Among Roth's many honours were also fellowship in the Royal Society of London in 1990, and in the Royal Society of Edinburgh in 1993. Moreover, other medals he won were the De Morgan Medal of the London Mathematical Society in 1983, and the Sylvestor Medal of the Royal Society in 1991.

In the aforementioned presentation of the medal by Davenport, he said of Roth's work: "It will stand as a landmark in mathematics for as long as mathematics is cultivated."

Remark 4.1 Before stating the result, we will need an elementary fact from introductory calculus, the *Mean-Value Theorem*, which says that, given a function continuous on the interval $[a, b]$, $a \neq b$, in $\mathbb{R}^2$ and differentiable on the open interval (a, b), then there exists a $\beta \in (a, b)$ such that

$$f'(\beta) = \frac{f(b) - f(a)}{b - a}.$$

Theorem 4.1 Liouville's Theorem

If α is a real algebraic number of degree $n > 1$, then there is a constant

$c_\alpha > 0$ *such that for any rational number* p/q, $q > 0$,

$$\left| \alpha - \frac{p}{q} \right| > \frac{c_\alpha}{q^n}.$$

Proof. Let

$$f(x) = \sum_{j=0}^{n} a_j x^j$$

be the minimal polynomial of α over $\mathbb{Q}$, where $a_j \in \mathbb{Z}$ for $0 \leq j \leq n$ by Lemma 1.1 on page 9. We may assume that

$$|\alpha - p/q| < 1 \tag{4.3}$$

since otherwise we choose $c_\alpha \leq 1/2$, then if

$$|\alpha - p/q| > 1,$$

we must have

$$|\alpha - p/q| > c_\alpha/q^n$$

because $1 > c_\alpha/q^n$. By the Mean-Value Theorem cited in Remark 4.1, there exists a β between p/q and α such that

$$f'(\beta)\left(\alpha - \frac{p}{q} \right) = f(\alpha) - f\left(\frac{p}{q} \right) = -f\left(\frac{p}{q} \right). \tag{4.4}$$

We require the following which is of interest in its own right.

Claim 4.1 *If we set*

$$c_\alpha = \frac{1}{n^2 \max_{0 \leq j \leq n}\{|a_j|\}(1 + |\alpha|)^{n-1}}, \tag{4.5}$$

a positive constant, depending only on α, *then*

$$|f'(\beta)| < \frac{1}{c_\alpha}.$$

By (4.3), $|\beta| < 1 + |\alpha|$, so

$$|f'(\beta)| = \left| \sum_{j=1}^{n-1} j a_j x^{j-1} \right| < \left| \sum_{j=1}^{n-1} n \max_{0 \leq j \leq n}\{|a_j|\}(1 + |\alpha|)^{j-1} \right|$$

$$< n^2 \max_{0 \leq j \leq n}\{|a_j|\}(1 + |\alpha|)^{n-1} = \frac{1}{c_\alpha},$$

since $n > 1$. This secures Claim 4.1.

Since we have that

$$\left| f\left(\frac{p}{q}\right) \right| = \frac{|a_n p^n + \sum_{j=0}^{n-1} a_j p^j q^{n-j}|}{q^n} \geq \frac{1}{q^n} \tag{4.6}$$

then by (4.4), Claim 4.1, and (4.6),

$$\left| \alpha - \frac{p}{q} \right| = \frac{|f(\frac{p}{q})|}{|f'(\beta)|} > \frac{c_\alpha}{q^n},$$

as required.   □

Remark 4.2 In the definition of c_α given in (4.5), there is

$$H(\alpha) = \max_{0 \leq j \leq n} \{|a_j|\},$$

which is called the *height of* α, also known as the *height of the minimal poly-nomial* $f(x)$. Louiville's Theorem actually states that algebraic numbers are *not* too well approximated by rational numbers. Moreover, the statement of the theorem seems to suggest that the degree of approximation depends on the given algebraic number α. However, Roth's Theorem shows that this is not the case – see (4.2) on page 160. Indeed, transcendental numbers can be *better approximated* by rational numbers. (Recall that a transcendental number is a complex number that is not algebraic.) To see this, note that if α is an algebraic number with continued fraction expansion

$$\alpha = \langle q_0; q_1, q_2, \ldots \rangle,$$

having convergents $C_j = A_j/B_j$ for $j = 0, 1, 2, \ldots$, then by Exercise 4.1 on page 167,

$$\left| \alpha - \frac{A_j}{B_j} \right| \leq \frac{1}{q_{j+1} B_j^2},$$

and by Liouville's Theorem,

$$\left| \alpha - \frac{A_j}{B_j} \right| > \frac{c_\alpha}{B_j^n},$$

so by combining the two, we get

$$c_\alpha q_{j+1} < B_j^{n-2}. \tag{4.7}$$

In particular, when $n = 2$, the sequence of partial quotients are bounded, since

$$q_{j+1} < \frac{1}{c_\alpha}.$$

In reference to the Liouville numbers cited in Biography 4.3 on page 168, con-sider the continued fraction expansion of $\gamma \in \mathbb{R}$ given by

$$\gamma = \langle 1, 10^{1!}, 10^{2!}, 10^{3!}, \ldots, \rangle,$$

from which it follows that $q_j = 10^{j!}$ and

$$B_j = 10^{j!(1+o(1))}.$$

(Recall that the *"little oh"* symbol is defined for functions f and g, denoted by $f = o(g)$, to mean that $\lim_{x \to \infty} f(x)/g(x) = 0$.) Hence,

$$\frac{q_{j+1}}{B_j^k} \to \infty \text{ as } j \to \infty \text{ for any } k,$$

so by (4.7), γ must be transcendental. In fact, this motivates a major result later in this section, namely that almost all real numbers are transcendental. Here "almost all" means all but an "enumerable" set, a concept we now define.

Definition 4.1 | Cardinal Numbers and Enumerable Sets

If there exists a bijection between two sets A and B, namely there exists a one-to-one correspondence between them, then the sets are said to have the same *cardinal number*. Equivalently they are said to be *equipotent* to one another. Any set that is equipotent to $\mathbb{N}$, the natural numbers, is called *enumerable*. Any set that is either finite or enumerable is called a *countable* set. If a set is not countable, it is called *uncountable*.

If α is an algebraic number, then there exist a polynomial of degree $d \in \mathbb{N}$,

$$f(x) = a_0 + a_1 x + \cdots + a_d x^d, \tag{4.8}$$

with $a_j \in \mathbb{Z}$ for $j = 0, 1, 2, \ldots, d$ not all zero such that $f(\alpha) = 0$. We define the *rank* of (4.8) to be

$$\ell = d + |a_0| + |a_1| + \cdots + |a_d|, \tag{4.9}$$

where we see that $\ell \geq 2$.

Now we show that the set of algebraic numbers $\overline{\mathbb{Q}}$ is countable–see Definition 1.4 on page 2.

Theorem 4.2 | The Set of Algebraic Numbers is Enumerable

$\overline{\mathbb{Q}}$ *is enumerable.*

Proof. For a given value of $\ell \in \mathbb{N}$ with $\ell \geq 2$, there are only finitely many equations (4.8) for which (4.9) holds. Thus for a given $\ell \in \{2, 3, \ldots\}$ let those finitely many equations be given by

$$E_{\ell,1}, E_{\ell,2}, \ldots, E_{\ell,n_\ell}.$$

For each $\ell = 2, 3, \ldots$, we may arrange the equations in a sequence

$$E_{2,1}, E_{2,2}, \ldots, E_{2,n_2}, E_{3,1} E_{3,2}, \ldots, E_{3,n_3}, E_{4,1}, \ldots$$

and let the set of all of these equations be denoted by $\mathcal{S}$. We may now put $\mathcal{S}$ in a one-to-one correspondence with $\mathbb{N}$ via the mapping, for $\ell = 2, 3, \ldots$, with $j = 1, 2, \ldots, n_\ell$ where $n_1 = 1$, given by

$$\tau : E_{\ell,j} \mapsto \sum_{i=1}^{\ell-1} n_i + j - 1.$$

Clearly $\tau(\mathcal{S}) \subseteq \mathbb{N}$, and now we show that τ is surjective. Let $k \in \mathbb{N}$ be arbitrary, and let $m_k \geq 2$ be the largest value such that $k \geq \sum_{i=1}^{m_k-1} n_i$. Thus, there exists an integer $s_k \geq 0$ such that

$$k = \sum_{i=1}^{m_k-1} n_i + s_k.$$

If $s_k \geq n_{m_k}$, then $k \geq \sum_{i=1}^{m_k} n_i$, contradicting the definition of m_k, so $0 \leq s_k \leq n_{m_k} - 1$. Hence,

$$\tau\left(E_{m_k, s_k+1}\right) = \sum_{i=1}^{m_k-1} n_i + s_k = k,$$

and this shows that $\tau(\mathcal{S}) = \mathbb{N}$. Now we show that τ is injective. If there exists $k, \ell \in \{2, 3, \ldots\}$ and $j, m \in \mathbb{N}$ such that $1 \leq j \leq n_k$, $1 \leq m \leq n_\ell$, and

$$\tau\left(E_{k,j}\right) = \sum_{i=1}^{k-1} n_i + j - 1 = \sum_{i=1}^{\ell-1} n_i + m - 1 = \tau(E_{\ell,m}),$$

then we need to show that $k = \ell$ from which we get that $j = m$ and τ is then shown to be injective. If $k \neq \ell$, then we may assume without loss of generality that $\ell > k$, so

$$0 = \sum_{i=k}^{\ell-1} n_i + m - j \geq \sum_{i=k}^{\ell-1} n_i + 1 - n_k \geq 1,$$

a contradiction. This secures the entire result. □

Corollary 4.1 *The set of all rational numbers is countable.*

Proof. Since $\mathbb{Q} \subseteq \overline{\mathbb{Q}}$, the result follows from Theorem 4.2 and Exercise 4.2 on page 167. □

The following was proved by Cantor.

Theorem 4.3 *The set of real numbers is uncountable.*

Proof. If $\mathbb{R}$ is countable, then by Exercise 4.2, the interval $(0, 1) \subseteq \mathbb{R}$ is countable. Thus, we may let $\alpha_j \in (0, 1)$ for $j = 1, 2, \ldots$ be an enumeration of these unit interval numbers. Each α_j has a decimal expansion which we will denote by

$$\alpha_j = 0.d_{j,1} d_{j,2} \cdots d_{j,n} \cdots, \text{ with } 0 \leq d_{j,n} \leq 9.$$

Now define
$$\alpha = 0.c_1 c_2 \cdots c_n \cdots,$$
where
$$c_j = \begin{cases} d_{j,j} + 1 & \text{if } 0 \leq d_{j,j} \leq 5, \\ d_{j,j} - 1 & \text{if } 6 \leq d_{j,j} \leq 9. \end{cases}$$

Since the j-th decimal place of α differs from that of α_j for any j and $\alpha \in (0,1)$. Also, since $c_j \neq 0, 9$ for any j, then α can have only one decimal representation. Thus, since α is not on the list of α_j, this contradicts the enumerability of $(0,1)$. $\square$

Hence, we have the following result promised in Remark 4.2 on page 162.

Corollary 4.2 *Almost all real numbers are transcendental.*

Proof. By Theorem 4.3, $\mathbb{R}$ is uncountable and by Theorem 4.2, $\overline{\mathbb{Q}}$ is countable. Hence, almost all real numbers are transcendental. $\square$

Biography 4.2 Georg Cantor (1845–1918) *was born in St. Petersburg, Russia. He attended university at Zurich, then later at the University of Berlin, where he studied under Kummer, Weierstrass, and Kronecker–see Biography 4.6 on page 179. In 1867, he obtained his doctorate for his work in number theory. In 1869 he took a position at the University of Halle which he kept until he retired in 1913. Unfortunately, he suffered from mental illness in the later years of his life and died of a heart attack in a psychiatric clinic in 1918.*

Cantor is known to be the founder of set theory, as well as for his contributions to mathematical analysis. Cantor even wrote on the connections between set theory and metaphysics, displaying his interest in philosophy as well. There were some, such as Kronecker, who did not agree with Cantor's views on set theory. Indeed, Kronecker blocked an application by Cantor for a position at Berlin when he applied for a better-paying position there.

In Exercise 4.3 on page 167, we have irreducibility criteria for polynomials that allows us to establish a result on the degree of roots of natural numbers.

Theorem 4.4 | **Rational Roots of Natural Numbers**

If $n \in \mathbb{N}$ and $m > 1$ is an integer such that $m \neq r^d$ for any $r, d \in \mathbb{N}$ such that $d \mid n$ and $d > 1$, then $m^{1/n}$ is an algebraic integer of degree n.

Proof. Let $f(x) = x^n - m$ for a given integer $m > 1$ which is not an dth power of a natural number for any divisor $d > 1$ of n. If $\alpha = m^{1/n}$, then $f(\alpha) = 0$. By Definition 1.4 on page 2, it suffices to show that f is irreducible over $\mathbb{Q}$. Suppose that $f(x) = g_1(x)g_2(x)$, where we may assume via Gauss' Lemma elucidated in Exercise 4.3 on page 167 that $g_j(x) \in \mathbb{Z}[x]$ for $j = 1, 2$. If ζ_n denotes a primitive nth root of unity–see Definition 1.2 on page 2–then we may write

$$f(x) = \prod_{j=0}^{n-1} (x - \zeta_n^j \alpha).$$

Let $\mathcal{S}_j$ for $j = 1, 2$ be sets such that $\mathcal{S}_1 \cup \mathcal{S}_2 = \{0, 1, 2, \ldots, n-1\}$, defined via the following,

$$g_1(x) = \prod_{j \in \mathcal{S}_1} (x - \zeta_n^j \alpha) \text{ and } g_2(x) = \prod_{j \in \mathcal{S}_2} (x - \zeta_n^j \alpha).$$

If $|\mathcal{S}_j| = s_j$, for $j = 1, 2$, then

$$g_1(0) = (-1)^{s_1} \alpha^{s_1} \zeta_n^{\sum_{j \in \mathcal{S}_1} j} \text{ and } g_2(0) = (-1)^{s_2} \alpha^{s_2} \zeta_n^{\sum_{j \in \mathcal{S}_2} j}. \qquad (4.10)$$

Now since $g_1(0)g_2(0) = f(0) = -m$, and $(-1)^{s_1 + s_2} = (-1)^n$ while

$$\zeta_n^{\sum_{j \in \mathcal{S}_1} j + \sum_{j \in \mathcal{S}_2} j} = \zeta_n^{\sum_{j=0}^{n-1} j} = \zeta_n^{n(n-1)/2}$$

(see [68, Theorem 1.1, p. 2] for the last equality), then $(-1)^n \zeta_n^{n(n-1)/2} = -1$, observing that $\zeta_n^{n(n-1)/2} = -1$ when n is even. Therefore, $\alpha^{s_j} \in \mathbb{N}$ for $j = 1, 2$. Let $t \in \mathbb{N}$ be the least value such that $\alpha^t \in \mathbb{Q}$.

Claim 4.2 *If $j \in \mathbb{N}$ with $\alpha^j \in \mathbb{Q}$, then $t \mid j$.*

Since $t \in \mathbb{N}$ is the least such value, then there exist $q, r \in \mathbb{Z}$ such that $j = tq + r$ where $0 \le r < t$. However, $\alpha^r = \alpha^j \alpha^{-tq} \in \mathbb{Q}$, so by the minimality of t, if $r > 0$, then $r \ge t$, a contradiction. Hence, $r = 0$ and $t \mid j$ and we have the claim.

By Claim 4.2, $s_j \mid t$ for $j = 1, 2$ and $t \mid n$. Since $(\alpha^t)^{n/t} = m$, and $\alpha^t = q_1/q_2$ for $q_i = 1, 2$ with $q_i \in \mathbb{N}$ and $\gcd(q_1, q_2) = 1$, then $mq_2^{n/t} = q_1^{n/t}$ so each prime factor of q_2 divides q_1. But since $\gcd(q_1, q_2) = 1$, then this means that $q_2 = 1$, so $m = \alpha^t \in \mathbb{N}$. Now, if we set $d = n/t$, then $m = (\alpha^t)^d$, contradicting that m is not the dth power of any natural number if $n \ne t$. Hence, $n = t$, and either $\mathcal{S}_1 = \varnothing$ or $\mathcal{S}_2 = \varnothing$. In other words, f is irreducible over $\mathbb{Q}$. $\square$

Theorem 4.4 speaks about rational powers of algebraic numbers. A natural question to pose is: what happens when we raise algebraic numbers to irrational powers? In 1934, Gel'fond and Schneider proved, independently, that if $\alpha \ne 0, 1$ is an algebraic integer and β is an irrational algebraic integer, then α^β is transcendental, a result known as the *Gel'fond–Schneider Theorem*. This result was generalized substantively by Baker [3] in 1966, when he showed that if $\{\alpha_j\}_{1 \le j \le n}$ and $\{\gamma_j\}_{1 \le j \le n}$ are algebraic integers where $\{1, \gamma_1, \ldots, \gamma_n\}$ and $\{2\pi i, \log_e(\alpha_1), \ldots, \log_e(\alpha_n)\}$ are linearly independent over $\mathbb{Q}$, then

$$\prod_{j=1}^n \alpha_j^{\gamma_j} \text{ is transcendental}$$

and $\{\log_e(\alpha_j)\}_{1 \le j \ln n}$ are linearly independent over $\overline{\mathbb{Q}}$, where $\overline{\mathbb{Q}}$ is the field of all algebraic numbers. (Recall that a set $\{\alpha_j\}_{j=1}^n$ is *linearly independent* over $\mathbb{Q}$ if $\sum_{j=1}^n q_j \alpha_j = 0$ for $q_j \in \mathbb{Q}$ implies $q_j = 0$ for $j = 1, 2, \ldots, n$.)

Baker's result yields methods that are applicable to Diophantine equations. For instance, one such quantitative result is the following.

Suppose that we have the Diophantine equation with $n \geq 3$,

$$f(x,y) = \sum_{j=0}^{n} a_j x^j y^{n-j} \in \mathbb{Z}[x,y]. \tag{4.11}$$

Then if $m \in \mathbb{N}$, a solution $(X,Y) \in \mathbb{Z}^2$ to (4.11) satisfies

$$\log_e\{\max\{|X|,|Y|\}\} \leq C$$

for some constant C depending on m, n and

$$H = \max_{0 \leq j \leq n} \{|a_j|\},$$

the *height* of f–see Remark 4.2 on page 162. Indeed, it can be shown that we may select

$$C = (nH)^{(10n)^5} + (\log_e m)^{2n+2}.$$

In §4.2, we examine the role of transcendental numbers, including the contributions of Liouville and others discussed above.

Exercises

4.1. Given a simple continued fraction expansion

$$\alpha = \langle q_0; q_1, \ldots \rangle,$$

of an algebraic number α, with convergents A_j/B_j for $j = 0, 1, \ldots$, prove that

$$\left| \alpha - \frac{A_j}{B_j} \right| \leq \frac{1}{q_{j+1} B_j^2}.$$

(*Hint: you may use the fact that*

$$\alpha - \frac{A_j}{B_j} = \frac{(-1)^j}{B_j(\alpha_{j+1} B_j + B_{j-1})},$$

where $\alpha_{j+1} = q_{j+1} + 1/\alpha_{j+2}$, *which is a fact that follows from* [68, Theorem 1.12, p. 25].)

4.2. Prove that every subset of a countable set is countable.

4.3. Let $f_1(x), f_2(x) \in \mathbb{Z}[x]$, set $f_3(x) = f_1(x)f_2(x)$, and define $\gcd(f_j)$ to be the gcd of the coefficients of $f_j(x)$ for $j = 1, 2, 3$. Prove that

$$\gcd(f_3) = \gcd(f_1)\gcd(f_2).$$

Conclude that if $f(x) \in \mathbb{Z}[x]$ and $f(x) = h(x)g(x)$, for $h(x), g(x) \in \mathbb{Q}[x]$, then $f(x) = G(x)H(x)$ for some $G(x), H(x) \in \mathbb{Z}[x]$.

(This is often called Gauss' Lemma *on integral polynomial factorization. Essentially, it says that any polynomial that is irreducible in $\mathbb{Z}[x]$ is also irreducible in $\mathbb{Q}[x]$, or speaking in the contrapositive, if $f(x)$ is reducible in $\mathbb{Q}[x]$, then it is already reducible in $\mathbb{Z}[x]$.)*

Biography 4.3 Joseph Liouville (1809–1882) *was born in Saint-Omer, France on March 24, 1809. He entered the École Polytechnique in 1825 and graduated in 1827 with Poisson being one of his examiners—see* [68, Biography 1.22, p.68]. *After graduation, he suffered some health problems, but in 1831 found his first academic post with an appointment as assistant to Claude Mathieu, who held a chair at École Polytechnique after succeeding Ampère. This and other positions he held were largely teaching positions with up to 40 hours a week of instruction. Yet in 1836, he founded the* Journal de Mathmatiques Pures et Appliques, *now commonly called* Journal de Liouville, *which was influential in France in the nineteenth century. In 1837, he was appointed Professor of Analysis and Mechanics at the École Polytechnique, and in 1838 he was elected to the astronomy section of the Académie des Sciences. Then Poisson died and Liouville was appointed to the Bureau des Longitudes to fill the vacancy in 1840. During much of the next decade, he was involved in politics. In 1851, he won the bid for the chair vacated by Libri at the Collège de France, beating Cauchy, and began lecturing there in 1851. In that year he published results on transcendental numbers that eliminated their dependence on continued fractions. In particular, he presented the first proof of the existence of a transcendental number, now called the* Liouvillian number, 0.110001 . . . *where there are zeros except in the n! place, for each $n \in \mathbb{N}$, where there is a 1–see Remark 4.2 on page 162.*

Liouville's mathematical interests ranged widely from mathematical physics to astronomy and pure mathematics. For instance, his work on differential equations resulted in the Sturm–Liouville theory, *used in solving integral equations, which have applications to mathematical physics. As well, he made inroads in differential geometry when he studied conformal transformations. There he proved a major result involving the measure-preserving property of Hamiltonian dynamics, which is fundamental in statistical mechanics. He published more than four hundred papers of which half were in number theory. He died in Paris, France on September 8, 1882.*

☆   4.4. Use Roth's result (4.2) to prove the following. Let $n \geq 3$ and assume that

$$f(x,y) = \sum_{j=0}^{n} a_{n-j} x^j y^{n-j} \in \mathbb{Z}[x,y]$$

is an irreducible (homogeneous) polynomial. Suppose furthermore that

$$g(x,y) = \sum_{0 \leq k+\ell \leq n-3} b_{k\ell} x^k y^\ell \in \mathbb{Q}[x].$$

Prove that $f(x,y) = g(x,y)$ has only finitely many solution $(x,y) \in \mathbb{Z}^2$.

(*Hint: Suppose that α_j for $j = 1, 2, \ldots, n$ are all solutions of $f(x, 1) = 0$. Show that there is a constant K such that*

$$\left| a_0 \prod_{j=0}^{n} (x - \alpha_j y) \right| \leq K y^{n-3}.$$

Proceed to conclude that there must exist some natural number $m \leq n$ such that

$$\left| \alpha_m - \frac{x}{y} \right| < \frac{C}{y^3},$$

for some constant C.)

4.5. Show that

$$\sum_{j=1}^{\infty} a^{-j^2}$$

is irrational where $a > 1$ is an integer.

4.6. Prove the following result due to Thue. Let $n \geq 3$ and let

$$f(x, y) = \sum_{j=0}^{n} a_j x^{n-j} y^j \in \mathbb{Q}[x]$$

be an irreducible homogeneous polynomial. If $m \in \mathbb{Q}$, then $f(x, y) = m$ has only finitely many solutions $(x, y) \in \mathbb{Z}^2$.

Biography 4.4 Carl Ludwig Siegel (1896–1981) *was born in Berlin, Germany. He entered the University of Berlin in 1915, attending lectures by Frobenius and Planck. In 1917, his studies ended when he was called to military duties. After being discharged, he returned to his studies in Göttingen in 1919 under the supervision of Landau—see Biography 3.1 on page 104—achieving his doctorate in 1920. Siegel improved upon Thue's result that in turn extended Liouville's Theorem 4.1 on page 160. Thue proved that, given an algebraic number α of degree $n \geq 2$, there exists a positive constant c_α such that for all rational numbers p/q and any $\varepsilon > 0$,*

$$\left| \alpha - \frac{p}{q} \right| > \frac{c_\alpha}{q^{n/2+1+\varepsilon}}.$$

Siegel improved this by showing that the above exponent on q could be replaced by $2\sqrt{n} + \varepsilon$. In 1947, Dyson improved this to show that the exponent could be replaced by $\sqrt{2n} + \varepsilon$—see page 155.

In 1922, Siegel was appointed professor at Johann-Wolfgang-Goethe University of Frankfurt to succeed Schönflies. For over more than a decade Siegel collaborated with his colleagues Hellinger, Epstein, and Dehn at Frankfurt. This included a history of mathematics seminar they held for thirteen years.

On January 30, 1933, Hitler came to power enacting the Civil Service Law on April 7, 1933. This was used as a mechanism for firing Jewish teachers from positions at universities. Although Siegel was not Jewish, he vehemently disagreed with the Nazi policies so much that he left for the U.S.A. in 1935 where he spent a year at the Institute for Advanced Study at Princeton. However, in 1937, he accepted a professorship at the University of Göttingen. But when Germany went to war in 1939, he felt he could not stay in his homeland. In 1940, he spent a brief time in Norway, then went back to the Institute at Princeton where he remained from 1940 to 1951. In that year, he returned to Göttingen where he remained until his death on April 4, 1981.

Siegel contributed to many areas of mathematics including: as noted above, approximation of algebraic numbers by rational numbers, but also transcendence theory, zeta functions, the geometry of numbers, quadratic forms, and celestial mechanics. Siegel never married and had very few doctoral students. He had devoted his life to research. Perhaps his most prestigious honour was the Wolf Prize bestowed on him in 1978.

4.2 Transcendence

> *The meaning doesn't matter if it's only idle chatter of a transcendental kind.*
> From act I of **Patience (1881)**
> **William Schwenck Gilbert (1836–1911)**
> –**English writer of comic and satirical verse**

In Corollary 4.2 on page 165, we proved that almost all real numbers are transcendental. We now look more closely at such numbers. In Remark 4.2 on page 162 and Biography 4.3 on page 168, we defined a *Liouvillian number*, which is transcendental. We now generalize this notion.

Definition 4.2 | **Liouville Number**

A real number α is called a *Liouville number* if for all $m \in \mathbb{N}$ there exist $a_m, b_m \in \mathbb{Z}$ such that $b_m > 1$ and

$$0 < \left| \alpha - \frac{a_m}{b_m} \right| < \frac{1}{b_m^m}. \tag{4.12}$$

The Liouvillian number cited above from §4.1 is a special case of Definition 4.2. Given that

$$\lim_{m \to \infty} \frac{1}{b_m^m} = 0,$$

then α is approximated by a_m/b_m better as m grows large, which shows that the set of these b_m is unbounded. We establish this via Liouville's Theorem 4.1 on page 160 in the following result.

Theorem 4.5 | **Liouville Numbers Are Transcendental**

Every Liouville number is transcendental.

Proof. Assume that α is a Liouville number that is not transcendental. Then α is an algebraic number of degree $n > 1$. By (4.12) and Liouville's Theorem 4.1,

$$\frac{c_\alpha}{b_m^n} < \left| \alpha - \frac{a_m}{b_m} \right| < \frac{1}{b_m^m},$$

so $0 < c_\alpha < b_m^{n-m}$. However, as noted above, $\lim_{m \to \infty} 1/b_m^m = 0$, a contradiction. $\square$

Now that we have established the existence of transcendental numbers and provided sets thereof, we turn to the problem of the transcendence of specific numbers such as e and π—see Biographies 3.4 on page 126 and 3.6 on page 128 for background on the solution of these two problems. Compared to the methodology for establishing existence above and in §4.1, establishing the transcendence of individual numbers is a more intricate problem. One open question is the transcendence of

$$\gamma = \lim_{n \to \infty} \left(1 + \frac{1}{2} + \frac{1}{3} + \cdots + \frac{1}{n} - \log_e(n)\right), \tag{4.13}$$

called *Euler's constant*. Indeed it is unknown if γ is irrational. Other well known numbers that have resisted attempts to prove transcendence are the values of the Riemann Zeta function $\zeta(2n + 1)$ for $n = 1, 2, \ldots$, although $\zeta(3)$ has been proved irrational by Apéry, and thus is known as *Apéry's constant*—see §5.3.

The following proof is essentially due to Hermite, and we follow the approach given in [93, Theorem 9.5, p. 145]. We assume knowledge of elementary calculus, including the following generalization of the product formula known as the *Leibniz formula*–see Biography 4.5 on page 175.

$$(fg)^{(i)}(x) = \sum_{k=0}^{i} \binom{i}{k} f(x)^{(k)} g(x)^{(i-k)}, \tag{4.14}$$

where $f^{(j)}$ denotes the j-th derivative of f. Also, recall the *integration by parts formula*

$$\int d(uv) = uv = \int u\,dv + \int v\,du. \tag{4.15}$$

Theorem 4.6 ⏐ **The Transcendence of e** ⏐

The real number e is transcendental.

Proof. We use properties of the following integral defined for $t \geq 0$,

$$I(t) = \int_0^t e^{t-x} f(x)dx,$$

where $f(x) \in \mathbb{R}[x]$. Employing integration by parts we get

$$I(t) = -e^{t-x} f(x)|_{x=0}^t + \int_0^t e^{t-x} f'(x)dx$$

$$= e^t f(0) - f(t) - e^{t-x} f'(x)|_{x=0}^t + \int_0^t e^{t-x} f''(x)|_{x=0}^t$$

$$\vdots$$

$$= e^t \sum_{i=0}^{d} f^{(i)}(0) - \sum_{i=0}^{d} f^{(i)}(t).$$

Therefore,

$$I(t) = e^t \sum_{i=0}^{d} f^{(i)}(0) - \sum_{i=0}^{d} f^{(i)}(t). \tag{4.16}$$

Now if we let $f^{ab}(x)$ be $f(x)$ with absolute values around the coefficients of $f(x)$, then

$$|I(t)| \leq \int_0^t |e^{t-x} f(x)| dx \leq t e^t f^{ab}(t). \tag{4.17}$$

We will employ the above for a specific function f that we will define below.

We proceed to prove that e is transcendental by contradiction. Assume, to the contrary, that e is algebraic. Then there is a minimal polynomial

$$P(x) = \sum_{j=0}^{d} b_j x^j \in \mathbb{Z}[x],$$

with

$$P(e) = \sum_{j=0}^{d} b_j e^j = 0. \tag{4.18}$$

Note that by Exercise 4.7 on page 179, e is irrational so we may assume that $d > 1$, allowing for the following. We arbitrarily select a large prime p, which we may specify later, and set

$$f(x) = x^{p-1} \prod_{j=1}^{d} (x - j)^p. \tag{4.19}$$

Observe that the degree of f is

$$d_f = (d+1)p - 1. \tag{4.20}$$

Now consider the sum

$$J = \sum_{j=0}^{d} b_j I(j). \tag{4.21}$$

Thus, by (4.16) and (4.18),

$$J = \sum_{j=0}^{d} b_j \left(e^j \sum_{i=0}^{d_f} f^{(i)}(0) - \sum_{i=0}^{d_f} f^{(i)}(j) \right)$$

$$= \sum_{i=0}^{d_f} f^{(i)}(0) \sum_{J=0}^{d} b_j e^j - \sum_{i=0}^{d_f} \sum_{j=0}^{d} b_j f^{(i)}(j) = -\sum_{i=0}^{d_f} \sum_{j=0}^{d} b_j f^{(i)}(j).$$

Thus,

$$J = -\sum_{i=0}^{d_f} \sum_{j=0}^{d} b_j f^{(i)}(j). \tag{4.22}$$

Now suppose that $0 \leq k \leq d$ and define

$$h_k(x) = \frac{f(x)}{(x-k)^p} = x^{p-1} \prod_{\substack{m=1 \\ m \neq k}}^{d} (x - m)^p \in \mathbb{Z}[x].$$

By (4.14),

$$f^{(i)}(x) = ((x-k)^p \cdot h_k(x))^{(i)} = \sum_{\ell=0}^{i} \binom{i}{\ell} ((x-k)^p)^{(\ell)} (h_k(x))^{(i-\ell)}. \qquad (4.23)$$

If $0 \le i < p$, then $f^{(i)}(k) = 0$ since the sum in (4.23) has $(k-k)^p = 0$ in each term. On the other hand, if $i \ge p$, then

$$f^{(i)}(k) = \binom{i}{p} p! \cdot h_k^{(i-p)}(k).$$

Hence, for any $i \ge 0$, $p! \mid f^{(i)}(k)$. By a similar analysis, $f^{(i)}(0) = 0$ for any $i < p-1$. Also, if $i \ge p-1$ and we set

$$m(x) = \frac{P(x)}{x^{p-1}} \in \mathbb{Z}[x],$$

then

$$f^{(i)}(0) = \binom{i}{p-1} (p-1)! \cdot m^{(i-p-1)}(0).$$

Hence, $p \mid m^{(i)}(0) \in \mathbb{Z}$ for $i > 0$ and

$$m(0) = (-1)^{dp}(d!)^p.$$

It follows that for $i \ne p-1$, $p! \mid f^{(i)}(0) \in \mathbb{Z}$ and that $(p-1)! \mid f^{(p-1)}(0) \in \mathbb{Z}$, but for $p > d$, $p \nmid f^{(p-1)}(0)$. Since we may select $p > d$ as large as we like, it follows from (4.22) that $(p-1)! \mid J$, so $|J| \ge (p-1)!$. However, we also have from (4.19)–(4.20), that

$$f^{ab}(k) \le k^{p-1} \prod_{j=1}^{d} (k+j)^p < (2d)^{2p-1} \le (2d)^{d_f} < (2d)^{2dp}. \qquad (4.24)$$

Thus, (4.24) tells us via (4.17) and (4.21) that

$$|J| \le \sum_{k=0}^{d} |b_k| k e^k f^{ab}(k) \le d(d+1) K e^d (2d)^{2dp} < C^p,$$

where

$$K = \max_{0 \le k \le d} \{|b_k|\},$$

and

$$C = d(d+1) K e (2d)^{2d},$$

which is a constant not depending on p. We have shown that

$$(p-1)! \le |J| \le C^p,$$

which bounds p, a contradiction to the fact that we have arbitrarily chosen p large. This contradiction establishes the result. $\square$

Remark 4.3 Theorem 4.6 on page 172 illustrates a few of the techniques involved in the theory of transcendental numbers. Although the proof of the transcendence of π does not really use any deeper results, more is needed in the proof in terms of algebraic conjugates of $\alpha \in \overline{\mathbb{Q}}$ and the use of symmetric polynomials—see Exercise 4.8 on page 180. The following is due to Lindemann —see Biography 3.6 on page 128

Biography 4.5 Gottfried Wilhelm von Leibniz (1646–1716) *was born on July 1, 1646 in Leipzig, Saxony (now Germany). He studied law at Leipzig from 1661 to 1666 and ultimately received a doctorate in law from the University of Altdorf in February 1667. Then he pursued a career in law at the courts of Mainz from 1667 to 1672. From 1672 to 1676, he spent his time in Paris where he studied mathematics and physics under Christian Huygens (1629–1695). In 1676, he left for Hannover, Hanover (now Germany), where he remained for the balance of his life. Leibniz began looking for a uniform and useful notation for the calculus in 1673, and by the autumn of 1676, he discovered the differential notation $d(x^n) = nx^{n-1}dx$ for $n \in \mathbb{Q}$. In 1684, he published the details of the differential calculus, the year before Newton published his famed Principia. There remained a bitter dispute over priority concerning discovery of the calculus. In 1700, Leibniz created the Brandenburg Society of Sciences, which led to the creation of the Berlin Academy some years later. Then he became increasingly reclusive until his death in Hannover on November 14, 1716.*

Much of the mathematical activity in his last years involved the aforementioned priority dispute over the invention of the calculus. In 1714, he published a pamphlet indicating a mistake made by Newton in understanding higher order derivatives, an error that was discovered by Johann Bernoulli, as evidence of his case.

Theorem 4.7 | **The Transcendence of π** |

    The real value π is transcendental.

Proof. If π is algebraic, then given that $\overline{\mathbb{Q}}$ is a field, then $\alpha = i\pi$ is also algebraic where $i^2 = -1$. Let the algebraic conjugates of α be

$$\alpha = \alpha_1, \alpha_2, \ldots, \alpha_d, \text{ for some } d \in \mathbb{N}.$$

Since

$$e^{\alpha_1} = e^{i\pi} = -1,$$

called *Euler's identity* for e, then

$$(1 + e^{\alpha_1})(1 + e^{\alpha_2}) \cdots (1 + e^{\alpha_d}) = 0.$$

We may write

$$\prod_{j=1}^{d}(1 + e^{\alpha_j}) = \sum_{\substack{\rho = \sum_{i=1}^{d} \delta_i \alpha_i \\ \text{where } \delta_i \in \{0,1\}}} e^\rho. \tag{4.25}$$

If we let

$$\{\rho_1, \rho_2, \ldots, \rho_n\}$$

be the exponents in the sum (4.25) that are nonzero, then

$$2^d - n + \sum_{j=1}^{n} e^{\rho_j} = 0.$$

Now we may invoke the techniques used in the proof of Theorem 4.6 by comparing

$$\sum_{i=1}^{n} I(\rho_i) \text{ (where } I(t) \text{ is given in (4.16) on page 172)}$$

with

$$f(x) = a^n x^{p-1} \prod_{i=1}^{n} (x - \rho_i)^p,$$

where a is the leading coefficient of the minimal polynomial of α and p is an arbitrarily chosen large prime to be specified later. Since $a\rho_i \in \mathbb{A} \cap \mathbb{Q} = \mathbb{Z}$ by Corollary 1.1 on page 4, and since

$$\prod_{\rho} (x - \rho) = x^{2^d - n} \prod_{i=1}^{n} (x - \rho)$$

is symmetric with respect to $\alpha_1, \ldots, \alpha_d$, then by Exercise 4.8 on page 180, $f(x) \in \mathbb{Z}[x]$. Now we let

$$n_f = (n+1)p - 1,$$

and

$$g = -(2^d - n) \sum_{j=0}^{n_f} f^{(j)}(0) - \sum_{j=0}^{n_f} \sum_{i=0}^{n} f^{(j)}(\rho_i) \in \mathbb{Z}[x], \qquad (4.26)$$

where

$$\sum_{i=0}^{n} f^{(j)}(\rho_i)$$

is symmetric in the $a\rho_i$. Hence, by Exercise 4.8 again,

$$\sum_{i=0}^{n} f^{(j)}(\rho_i) \in \mathbb{Z}[x].$$

However, for $j < p$, $f^{(j)}(\rho_i) = 0$, so

$$\sum_{i=0}^{n} f^{(j)}(\rho_i) \equiv 0 \pmod{p!}.$$

Also, if $j \neq p-1$, $f^{(j)}(0) \equiv 0 \pmod{p!}$. As well, for p sufficiently large,

$$f^{(p-1)}(0) = (p-1)!(-a)^{np}(\rho_1 \cdots \rho_n)^p \equiv 0 \pmod{(p-1)!},$$

but

$$f^{(p-1)}(0) = (p-1)!(-a)^{np}(\rho_1 \cdots \rho_n)^p \not\equiv 0 \pmod{p!}.$$

Hence,

$$|g| \le \sum_{i=1}^{n} |\rho_i| f^{ab}(|\rho_i|) \le c^p$$

where $c \in \mathbb{R}$ is independent of p. Then we proceed as in the proof of Theorem 4.6, this time using Exercise 4.9 on page 180, to get a contradiction to π being algebraic. □

Lindemann proved a stronger result than Theorem 4.7, namely that if $\alpha \in \mathbb{C}$, $\alpha \ne 0$, then at least one of α or e^α is transcendental. Then this result was generalized considerably by Weierstrass to linear combinations stated in our next result.

Theorem 4.8 | **The Lindemann–Weierstrass Result**

Given $\alpha_i, \beta_j \in \overline{\mathbb{Q}}$, where α_i, for $i = 1, 2, \ldots, n$, are distinct and $\beta_j \ne 0$ for $j = 1, 2, \ldots, n$,

$$\sum_{j=1}^{n} \beta_j e^{\alpha_j} \ne 0.$$

Proof. See [54]. □

Theorem 4.6 on page 172 is immediate from Theorem 4.8, and Theorem 4.7 follows from Theorem 4.8 via Euler's identity $e^{i\pi} + 1 = 0$, cited on page 175. The very notion of transcendence itself can be generalized as follows.

Definition 4.3 | **Algebraic Independence**

If $\alpha_j \in \mathbb{R}$ for $j = 1, 2, \ldots, n \in \mathbb{N}$, then $\{\alpha_j\}_{j=1}^{n}$ is said to be *algebraically independent* over $\mathbb{Q}$ if there does *not* exist a polynomial

$$f(x_1, x_2, \ldots, x_n) \in \mathbb{Q}[x_1, x_2, \ldots, x_n]$$

with $f(\alpha_1, \alpha_2, \ldots, \alpha_n) = 0$.

Since the concept of a single α being transcendental is included in Definition 4.3, then we have our generalization.

To take the theory of transcendental numbers to its pinnacle, we state a result that is more general still than Theorem 4.8, namely the following open conjecture, the verification of which would fell numerous open questions in the theory of transcendental numbers.

Conjecture 4.1 | Schanuel's Conjecture |

If $\alpha_j \in \mathbb{C}$ are linearly independent over $\mathbb{Q}$ for $j = 1, 2, \ldots, n \in \mathbb{N}$, then there exists a subset $\mathcal{S}$ of $\{\alpha_1, \alpha_2, \ldots, \alpha_n, e^{\alpha_1}, e^{\alpha_2}, \ldots, e^{\alpha_n}\}$ such that $|\mathcal{S}| \geq n$ where $\mathcal{S}$ is algebraically independent over $\mathbb{Q}$.

We conclude with some numbers known to be transcendental, and some that are not. From Theorem 4.8 on the preceding page, we know that e^α is transcendental if $\alpha \in \overline{\mathbb{Q}}$ is nonzero. It also follows from Theorem 4.8 that $\sin(\alpha), \cos(\alpha), \tan(\alpha)$ are transcendental for any nonzero $\alpha \in \overline{\mathbb{Q}}$, as well as $\log_e(\alpha)$ for any $\alpha \in \overline{\mathbb{Q}}$ with $\alpha \neq 0, 1$. Gel'fond constant e^π and the Gel'fond–Schneider constant $\sqrt{2}^{\sqrt{2}}$ are known to be transcendental by the Gel'fond–Schneider Theorem, a result that follows from Conjecture 4.1 – see page 166. Also, Gel'fond's constant and the Gel'fond–Schneider constant were noted in Hilbert's seventh problem as examples of numbers whose transcendence was unknown at the turn of the twentieth century – see Biography 3.5 on page 127.

The number whose binary expansion is given by

$$\mathfrak{p} = 0.0110100110010011010010011001101001\ldots$$

is known as the *Proulet–Thue–Morse constant*. To see how this number is defined, let the first term be $t_0 = 0$ and for $n \in \mathbb{N}$, define $t_n = 1$ if the number of ones in the binary expansion of n is odd, and $t_n = 0$ if the number of ones is even. Thus, the *Thue–Morse sequence* t_n is given by $t_0 = 0, t_{2n} = t_n$, and $t_{2n+1} = 1 - t_n$ for all $n \in \mathbb{N}$. The generating function for the t_n is given by

$$\tau(x) = \sum_{n=0}^{\infty} (-1)^{t_n} x^n = \prod_{n=0}^{\infty} (1 - x^{2^n}),$$

– see [68, §1.7]. The sequence was independently discovered by P. Proulet, Axel Thue, and Marston Morse. This constant $\mathfrak{p}$ was shown to be transcendental by Mahler in 1929 – see Biography 4.7 on page 181.

Some numbers *unknown* to be transcendental are the Euler constant, discussed on page 172, as well as Apéry's constant mentioned there. There is also *Catalan's constant* defined by

$$K = \sum_{j=0}^{\infty} \frac{(-1)^j}{(2j+1)^2}$$

which, like Euler's constant, is not known to be irrational. Also, sums, products, and powers of π and e, except Gel'fond's constant, such as π^π and $\pi + e$ or e^e are not known to be transcendental. It is of interest to note that since π is known to be transcendental, then it is not possible to get the square root of π from rational numbers, so it is impossible to find the length of the side of a square having the same area as a given circle using ruler and compass. This means that the classical problem of squaring the circle cannot be accomplished.

For a nice discussion of many open problems in diophantine analysis, see [100].

Biography 4.6 Karl Theodor Wilhelm Weierstrass (1815–1897) *was born on October 31, 1815 in Ostenfelde, Westphalia (now Germany). His early education was spotty in terms of his commitment. He entered the Catholic Gymnasium in Pederborn in 1829, and graduated in 1834. Then he entered the University of Bonn, where he was enrolled in the study of law, finance, and economics largely to satisfy the wishes of his father, which conflicted with his love of mathematics. This led to a conflict within him that resulted in his not studying any subjects, rather spending four years of exhaustive drinking and fencing. He left the Bonn in 1838 without taking the examinations. In 1839, he entered the Academy at Münster to study to become a secondary school teacher, and began his career as such in 1842 at the Pro-Gymnasium in West Prussia (now Poland). In 1848, he moved to the Collegium Hoseanum in Brandenburg. During much of this time he studied mathematics on his own, including his reading of* Crelle's Journal, *for instance. Given his lack of formal training, his publication on abelian functions in the Brandenburg school prospectus was largely ignored. However, he published a paper in* Crelle's Journal *in 1854 on his (partial) theory of inversion of hyperelliptic integrals, which was more than noticed. On the basis of this paper alone, the University of Königsberg presented him with an honorary doctorate on March 31, 1854. This made Weierstrass decide to ultimately leave secondary school teaching never to return to it. When he published his full theory of inversion of hyperelliptic integrals in* Crelle's Journal *in 1856, he began receiving many offers for chairs at various universities. He accepted an offer of a professorship at the University of Berlin in October of 1856. His lectures on applications of Fourier series and integrals to mathematical physics, the theory of analytic functions, and of elliptic functions, as well as applications to problems in geometry and mechanics were received with enthusiasm from the many students from around the globe who came to attend. Among those who benefited from his teaching were Cantor, Frobenius, Hensel, Hurwitz, Klein, Lie, Mertens, Minkowski, Mittag-Leffler, and Stolz. Indeed, together with his colleagues, Kummer and Kronecker at Berlin, the university was provided with a reputation as a leader for excellence in mathematics. He died on February 19, 1897 in Berlin*

Weierstrass is known as the father of modern analysis. He created tests for convergence of series, established fundamental work in the theory of periodic functions, functions of a real variable, elliptic functions, abelian functions, converging infinite products, and the calculus of variations, not to mention the theory of quadratic forms. He set a standard of rigour, for instance, establishing irrational numbers as limits of convergent series, that is with us today.

Exercises

4.7. Prove the result first established by Euler that e is irrational.

(*Hint: Prove that* e^{-1} *is irrational by using the formula* $e^{-1} = \sum_{i=0}^{\infty} \frac{(-1)^i}{i!}$,
and breaking it into two parts, $\alpha_n = \sum_{i=0}^{n} \frac{(-1)^i}{i!}$ *and* $\beta_n = \sum_{i=n+1}^{\infty} \frac{(-1)^i}{i!}$,
demonstrating that $n!\alpha_n + n!\beta_n(-1)^{n+1}$ *cannot be an integer.*)

4.8. This exercise deals with *symmetric polynomials*. These are defined to be
those $f(x_1, x_2, \ldots, x_n) \in R[x]$, for a given commutative ring with identity
R, such that for any permutation σ of $\{1, 2, \ldots, n\}$,

$$f(x_{\sigma(1)}, x_{\sigma(2)}, \ldots, x_{\sigma(n)}) = f(x_1, x_2, \ldots, x_n),$$

denoted succinctly by $f^{\sigma} = f$. The *elementary symmetric polynomials*
s_j in the variables x_j for $j = 1, 2, \ldots, n$, are the coefficients of the monic
polynomial: $(X-x_1)(X-x_2)\cdots(X-x_n) = X^n - s_1 X^{n-1} \pm \cdots + (-1)^n s_n$,
which are homogeneous, symmetric, and

$$s_1 = \sum_{j=1}^{n} x,$$

$$\vdots$$

$$s_k = \sum_{1 \le i_1 < i_2 < \cdots < i_k \le n} x_{i_1} x_{i_2} \cdots x_{i_k}$$

$$\vdots$$

$$s_n = \prod_{j=1}^{n} x_j.$$

The *Fundamental Theorem of Symmetric Polynomials* is the following.

Let $f(x_1, x_2, \ldots, x_n) \in \mathbb{Q}[x_1, x_2, \ldots, x_n]$ *be symmetric. Then there
exists a polynomial* $g(x_1, x_2, \ldots, x_n) \in \mathbb{Q}[x_1, x_2, \ldots, x_n]$ *such that*
$f(x_1, x_2, \ldots, x_n) = g(s_1, s_2, \ldots, s_n)$.

Prove the fundamental theorem.

(*Hint: Since* f *is a sum of monomials* $ax_1^{a_1} x_2^{a_2} \cdots x_n^{a_n}$ *where* $a \in \mathbb{Q}$ *and
$a_j \ge 0$ for all $j = 1, 2, \ldots, n$, order them according to the exponents a_n,
called a* dictionary ordering. *Select a largest one* $ax_1^{a_1} x_2^{a_2} \cdots x_n^{a_n}$. *Then
consider* $as_1^{a_1-a_2} s_2^{a_2-a_3} \cdots s_n^{a_n} = g_1$ *which is symmetric in $x_1, x_2, \ldots, x_n$
and is a sum of monomials in $x_1, x_2, \ldots, x_n$. Then the largest one appear-
ing in f is* $ax_1^{a_1-a_2}(x_1 x_2)^{a_2-a_3} \cdots (x_1 x_2 \cdots x_n)^{a_n}$. *Consider* $f_1 = f - g_1$
and repeat the process which must terminate.)

4.9. Prove that $\pi \notin \mathbb{Q}$.

(*Hint: Assume to the contrary that* $\pi = a/b$ *and let* $f(x) = x^n(a-bx)^n/n!$.
Consider the sum $\sum_{j=0}^{n}(-1)^j f^{(2j)}(x)$ *and show that the sum at $x = 0, \pi$
are integers so that you may demonstrate that* $\int_0^{\pi} f(x)dx$ *is an integer.
Reach a contradiction by showing that for large enough n the integral lies
between 0 and 1.*)

Biography 4.7 Kurt Mahler (1903–1988) *was born in Krefeld, Prussian Rhineland on July 26, 1903. From an early age he taught himself mathematics by reading the masters such as Landau, Klein, and Hilbert as well as many others. In 1925, he moved to Göttingen where he attended lectures by many including Emmy Noether, Landau, Heisenberg, Hilbert, and Ostroski. In particular, Noether was influential in that she taught him about p-adic numbers. By 1927 he had enough to submit a thesis to Frankfurt on zeros of the gamma function. This was sufficient for his doctoral requirements. His first appointment was to the University of Königsberg in 1933. However, with Hitler's rise to power he had to leave Germany. Mordell invited him to Manchester where he stayed from 1933 to 1934. Then he went to Groningen in the Netherlands for 1934-1936, and retuned to Manchester in 1937, where he remained until 1962 when he went to Canberra, Australia for the last six years of his career. He died there in his eighty-fifth year on February 25, 1988.*

Among his works were the proof of the transcendence of $\sqrt{2}^{\sqrt{2}}$. Also, he classified real and complex numbers into classes which are algebraically independent. As well, he worked on p-adic numbers, p-adic Diophantine approximation, the geometry of numbers, and measure of polynomials. Among the honours in his life was the De Morgan medal awarded in 1971. Moreover, he was elected a Fellow of the Australian Academy of Science in 1965 and received its Lyle Medal in 1977. In November 1977, he received a diploma at a special ceremony in Frankfurt to mark the golden jubilee of his doctorate. The Dutch Mathematical Society made him an honorary member in 1957, as did the Australian Mathematical Society in 1986. Among his nonmathematical activities was photography. Indeed, many of his pictures are displayed at the University House of Australian National University where he lived for more than two decades.

4.3   Minkowski's Convex Body Theorem

> *Poetry is a subject as precise as geometry.*
> From a letter to Louise Colet, August 14, 1853 in **Correspondence**
> **1853–1856**, M. Nadeau (ed.) (1964)
>
> **Gustave Flaubert (1821–1880)**
> –French novelist

Minkowski coined the term *geometry of numbers* to mean the use of geometric methods, especially in Euclidean n-space, to solve deep problems in number theory–see Biography 4.8 on page 190. Perhaps the most celebrated of these is the *convex body theorem* which he proved in 1896. Before presenting this result, we need to develop some basic ideas in the theory of the geometry of numbers, the first of which is given as follows. Some of the material in this section is adapted from [64]. The reader should be familiar with the basics of vector spaces such as that to be found in [68, Appendix A].

Definition 4.4 | **Lattices and Parallelotopes**

Let $\ell_1, \ell_2, \ldots, \ell_m \in \mathbb{R}^n$ $(m, n \in \mathbb{N}, m \leq n)$ be $\mathbb{R}$-linearly independent vectors. If

$$L = \{\ell \in \mathbb{R}^n : \ell = \sum_{j=1}^m z_j \ell_j \text{ for some } z_j \in \mathbb{Z}\} = \mathbb{Z}[\ell_1, \ldots, \ell_m],$$

then L is called a lattice of dimension m in $\mathbb{R}^n$. When $m = n$, L is called a full lattice. In other words, a full lattice L is a free abelian group of rank n having a $\mathbb{Z}$-basis that is also an $\mathbb{R}$-basis for $\mathbb{R}^n$. Furthermore, the set

$$\mathcal{P} = \left\{ \sum_{j=1}^n r_j \ell_j : r_j \in \mathbb{R}, 0 \leq r_j < 1 \text{ for } j = 1, 2, \ldots, n \right\}$$

is called the fundamental parallelotope, or fundamental parallelepiped, or fundamental domain of L. An invariant of $\mathcal{P}$ is

$$V(\mathcal{P}) = |\det(\ell_j)|,$$

called the volume of $\mathcal{P}$, and also called the discriminant of L, denoted by $D(L)$.

Remark 4.4 Recall that a *free abelian group with a basis of n elements* is an additive abelian group with a linearly independent subset $\mathcal{S}$ of order n that *generates* it, meaning that G equals the intersection of all subgroups containing $\mathcal{S}$. See Exercise 2.5 on page 66 for a reminder of the definition of $GL(n, \mathbb{Z})$, if needed.

As well, note that the term "invariant" in Definition 4.4 means that, irrespective of which basis we choose for L, the volume of $\mathcal{P}$ remains the same. It is an easy exercise for the reader to verify that the determinant remains the same under change of basis using Exercise 4.10 on page 189. For the reader with a knowledge of measure theory, or *Lebesgue measure* in $\mathbb{R}^n$,

the volume of a so-called *measurable set* $S \subseteq \mathbb{R}^n$ is called the *measure of S*.

This measure can be shown to be the absolute value of the determinant of the matrix with rows ℓ_j for $j = 1, 2, \ldots, n$ for *any* basis $\{\ell_j\}$ of S. Thus, the Lebesgue measure of S is called the volume of S.

Example 4.1 $\mathbb{Z}^n$ is a full lattice in $\mathbb{R}^n$ for any $n \in \mathbb{N}$. In other words, a free abelian group of rank n in $\mathbb{R}^n$ is a full lattice. Hence, $\mathfrak{O}_F$ is a full lattice in $\mathbb{R}^n$, where $|F : \mathbb{Q}| = n$. Also, note that any lattice of dimension $m \in \mathbb{N}$ is full in $\mathbb{R}^m$.

We will now show that lattices as subsets of $\mathbb{R}^n$ are characterized by the following property, where the notation for a cardinality of a set $|\mathfrak{G}| < \infty$ means $\mathfrak{G}$ has finitely many elements.

Definition 4.5 | **Discrete Sets**

Suppose that $S \subseteq \mathbb{R}^n$, $n \in \mathbb{N}$, $r \in \mathbb{R}^+$, and

$$\mathcal{S}_r = \{s \in \mathbb{R}^n : |s| \leq r\}$$

is the sphere or ball in $\mathbb{R}^n$, with radius r, centered at the origin. Then S is called discrete if

$$|S \cap \mathcal{S}_r| < \infty,$$

for all $r \in \mathbb{R}^+$.

Remark 4.5 For what follows, the reader is asked to recall that if

$$s = (s_1, s_2, \ldots, s_n) \in \mathbb{R}^n,$$

then $|s| \leq r$ means that

$$\sum_{j=1}^{n} s_j^2 \leq r^2,$$

since

$$|s| = \left(\sum_{j=1}^{n} s_j^2 \right)^{1/2},$$

so $|s_j| \leq r$ for each such j. Also, the symbol

$$G \oplus H$$

denotes the additive free abelian group structure on free abelian groups G, H, called a *direct sum* of G and H.

Theorem 4.9 | Lattices are Discrete |

Let $L \subseteq \mathbb{R}^n$, $L \neq \varnothing$. Then L is a lattice if and only if L is a discrete, additive subgroup of $\mathbb{R}^n$.

Proof. Let L be a lattice of dimension n, namely a full lattice in $\mathbb{R}^n$. If

$$L = \ell_1 \mathbb{Z} \oplus \cdots \oplus \ell_n \mathbb{Z},$$

then

$$\{\ell_1, \ldots, \ell_n\}$$

is an $\mathbb{R}$-basis for $\mathbb{R}^n$. Thus, any $\alpha \in \mathbb{R}^n$ can be written in the form

$$\alpha = \sum_{j=1}^n r_j \ell_j \quad (r_j \in \mathbb{R}).$$

If $\alpha \in L \cap \mathcal{S}_r$ for any $r \in \mathbb{R}^+$, then each $r_j \in \mathbb{Z}$ and $|r_j| \leq r$ for each $j = 1, 2, \ldots, n$. Hence, there exist only finitely many points in $L \cap \mathcal{S}_r$. In other words, L is discrete.

Conversely, assume that L is a discrete, additive subgroup of $\mathbb{R}^n$. We use induction on n. For $n = 1$, let $\{\ell\}$ be a basis for $\mathbb{R}$, namely

$$\mathbb{R}^1 = \mathbb{R}\ell.$$

Since $\mathcal{S}_r \cap L$ is finite for all $r \in \mathbb{R}^+$, there exists a smallest positive value r_1 such that $r_1 \ell \in L$. Therefore,

$$\mathbb{Z}r_1 \ell \subseteq L.$$

Since any $s \in \mathbb{R}$ may be written as

$$s = \left\lfloor \frac{s}{r_1} \right\rfloor r_1 + s_1 r_1,$$

for some real number s_1 with $0 \leq s_1 < 1$, then any $s\ell \in L$ may be written in the form

$$s\ell = nr_1\ell + s_1 r_1 \ell,$$

with

$$n = \left\lfloor \frac{s}{r_1} \right\rfloor \in \mathbb{Z},$$

and $0 \leq s_1 < 1$. Therefore, by the minimality of r_1, we must have that $s_1 = 0$, so

$$L = \mathbb{Z}[r_1 \ell].$$

This establishes the induction step. Assume the induction hypothesis, namely that any discrete subgroup of $\mathbb{R}^k$ for $k < n$ is a lattice. Hence, we may assume that

$$L \subseteq \mathbb{R}^n \text{ is discrete and } L \not\subseteq \mathbb{R}^k \text{ for any } k < n.$$

Hence, we may choose a basis

$$\{\ell_1, \ldots, \ell_n\}$$

of $\mathbb{R}^n$ with $\ell_j \in L$ for each $j = 1, 2, \ldots, n$. Set

$$V = \mathbb{R}[\ell_1, \ldots, \ell_{n-1}].$$

By the induction hypothesis,

$$L_V = L \cap V$$

is a lattice of dimension $n - 1$. Let

$$\{\beta_1, \ldots, \beta_{n-1}\}$$

be a basis for L_V. Therefore, any element $\gamma \in L$ may be written

$$\gamma = \sum_{j=1}^{n-1} r_j \beta_j + r_n \ell_n \quad (r_j \in \mathbb{R}).$$

By the discreteness of L, there exist only finitely many such γ with all r_j bounded. Thus, we may choose one with $r_n > 0$, and minimal with respect to $|r_j| < 1$ for all $j \neq n$. Let β_n denote this choice. Thus,

$$\mathbb{R}^n = \mathbb{R}[\beta_1, \ldots, \beta_n].$$

Then for any $\delta \in L$,

$$\delta = \sum_{j=1}^{n} t_j \beta_j \quad (t_j \in \mathbb{R}).$$

Let

$$\sigma = \delta - \sum_{j=1}^{n} \lfloor t_j \rfloor \beta_j = \sum_{j=1}^{n} s_j \beta_j.$$

Therefore, $0 \le s_j < 1$ for all $j = 1, \ldots, n$. By the minimality of r_n, we must have that $s_n = 0$. Hence,

$$\sigma \in L_V,$$

so

$$\delta \in L_V \oplus \mathbb{Z}\beta_n.$$

This gives us, in total, that

$$L \subseteq L_V \oplus \mathbb{Z}\beta_n \subseteq L.$$

Therefore,

$$L = L_V \oplus \mathbb{Z}\beta_n$$

is a lattice. $\qquad\square$

We also need other fundamental notions from geometry.

Definition 4.6 | **Bounded, Convex, and Symmetric Sets** |

A set S in $\mathbb{R}^n$ is said to be convex if, whenever $s, t \in S$, the point

$$\lambda s + (1 - \lambda)t \in S$$

for all $\lambda \in \mathbb{R}$ such that $0 \leq \lambda \leq 1$. In other words, S is convex if it satisfies the property that, for all $s, t \in S$, the line segment joining s and t is also in S. The volume of a convex set S is given by the multiple integral

$$V(S) = \int_S \cdots \int dx_1 dx_2 \cdots dx_n$$

carried out over the set S. A set S in $\mathbb{R}^n$ is said to be bounded if there exists a sufficiently large $r \in \mathbb{R}$ such that $|s| \leq r$ for all $s \in S$. Another way of looking at this geometrically is that S is bounded if it can fit into a sphere with center at the origin of $\mathbb{R}^n$ and radius r.

A set S in $\mathbb{R}^n$ is symmetric provided that, for each $s \in S$, we have $-s \in S$.

Remark 4.6 A theorem of W. Blanschke says that the volume of every bounded, convex set exists. Hence, the integral in Definition 4.6 always exists for convex sets.

Example 4.2 Clearly, ellipses and squares are convex in $\mathbb{R}^2$, but a crescent shape, for instance, is not. Also, an n-dimensional cube

$$S = \left\{ s = (s_1, \ldots, s_n) \in \mathbb{R}^n : -1 \leq s_j \leq 1 \text{ for } j = 1, 2, \ldots, n \right\}$$

is a bounded, symmetric convex set, as is an n-dimensional unit sphere

$$\{ s \in \mathbb{R}^n : |s| \leq 1 \}.$$

Before proceeding to the main result, we need a technical lemma.

Lemma 4.1 | **Translates and Volume** |

Let $S \subseteq \mathbb{R}^n$ be a bounded set and let L be an n-dimensional lattice. If the translates of S by L, given by

$$S_z = \{ s + z : s \in S \},$$

for a given $z \in L$, are pairwise disjoint, namely

$$S_z \cap S_y = \varnothing,$$

for each $y, z \in L$ with $y \neq z$, then

$$V(S) \leq V(\mathcal{P})$$

where $\mathcal{P}$ is a fundamental parallelotope of L.

Proof. Since $\mathcal{P}$ is a fundamental parallelotope of L, we have the following description of S as a disjoint union:

$$S = \cup_{z \in L}(S \cap \mathcal{P}_{-z}),$$

where

$$\mathcal{P}_{-z} = \{x - z : x \in \mathcal{P}\},$$

so it follows that

$$V(S) = \sum_{z \in L} V(S \cap \mathcal{P}_{-z}).$$

Since the translate of the set

$$S \cap \mathcal{P}_{-z}$$

by the vector z is

$$S_z \cap \mathcal{P},$$

then

$$V(S \cap \mathcal{P}_{-z}) = V(S_z \cap \mathcal{P}). \tag{4.27}$$

Therefore,

$$V(S) = \sum_{z \in L} V(S_z \cap \mathcal{P}).$$

If the translates S_z are pairwise disjoint, then so are $S_z \cap \mathcal{P}$. Since

$$S_z \cap \mathcal{P} \subseteq \mathcal{P},$$

then Equation (4.27) tells us that

$$\sum_{z \in L} V(S_z \cap \mathcal{P}) \leq V(\mathcal{P}),$$

so the result is proved. $\qquad\square$

Remark 4.7 The interested reader will note that the term convex *body*, used in what follows, refers to a nonempty, convex bounded and *closed* subset S of $\mathbb{R}^n$. The topological term "closed" means that every *accumulation point* of a sequence of elements in S must also be in S. This is equivalent to saying that S is closed in the topological space $\mathbb{R}^n$, with its natural topology. However, we do not need to concern ourselves here with this, since it is possible to state and prove the result without such topological considerations. It can also be shown that if S is "compact," namely every "cover" (a union of sets containing S) contains a *finite* cover, then it suffices to assume that

$$V(S) \geq 2^n V(\mathcal{P}).$$

Now we are in a position to state the central result of this section.

Theorem 4.10 | Minkowski's Convex Body Theorem

Suppose that L is a lattice of dimension n, and let $V(\mathcal{P})$ be the volume of a fundamental parallelotope $\mathcal{P}$ of L. If S is a symmetric, convex set in $\mathbb{R}^n$ with volume $V(S)$ such that

$$V(S) > 2^n V(\mathcal{P}),$$

there exists an $x \in S \cap L$ such that $x \neq 0$.

Proof. It suffices to prove the result for a bounded set S. To see this, we observe that when S is unbounded, we may restrict attention to the intersection of S with an n-dimensional sphere, centered at the origin, having a sufficiently large radius. Let

$$T = \tfrac{1}{2}S = \{s/2 : s \in S\}.$$

Then

$$V(T) = \frac{V(S)}{2^n} > V(\mathcal{P}).$$

If the translates

$$T_z = \frac{1}{2}S + z$$

were pairwise disjoint, then by Lemma 4.1,

$$V(\mathcal{P}) \geq V(T),$$

a contradiction. Therefore, there must exist two distinct elements $s, t \in L$ such that

$$(\tfrac{1}{2}S - s) \cap (\tfrac{1}{2}S - t) \neq \varnothing.$$

Let $x, y \in S$ such that

$$\frac{1}{2}x - s = \frac{1}{2}y - t.$$

Then

$$t - s = \frac{1}{2}y - \frac{1}{2}x.$$

Since S is symmetric, then $-x \in S$, and since S is convex, then

$$\frac{1}{2}y + \frac{1}{2}(-x) \in S.$$

Hence,

$$t - s \in S \cap L,$$

and $t - s \neq 0$, as required. $\square$

We summarize the contents of this section as a closing feature of this chapter. Minkowski's convex body result given in Theorem 4.10 is an exceptionally simple test to guarantee a convex symmetric set contains a nonzero lattice point. It has a broad range of applications some of which are beyond the scope of this book–see [64], for instance. However, we may conclude with the application of Minkowski's result to verify (4.1) on page 159. Let α be a real number such that $0 < \alpha < 1$ and let $n \in \mathbb{N}$. Define

$$\mathcal{S} = \left\{ (x, y) \in \mathbb{R}^2 : -n - \frac{1}{2} \le x \le n + \frac{1}{2}, \text{ and } |x\alpha - y| < \frac{1}{n} \right\}.$$

This is a convex, symmetric set with area

$$(2n + 1)\frac{2}{n} = 4 + \frac{2}{n} > 4.$$

Therefore, Minkowski tells us that there is a nonzero lattice point (p, q), say, and by symmetry we may assume without loss of generality that $q > 0$. Hence, by the definition of $\mathcal{S}$, $q \le n$ and

$$\left| \alpha - \frac{p}{q} \right| < \frac{1}{qn} < \frac{1}{q^2},$$

which is (4.1).

Exercises

4.10. Let G be a free abelian group with basis

$$\mathcal{S} = \{g_1, g_2, \ldots, g_n\}.$$

Suppose that $A = (a_{i,j})$ is an $n \times n$ matrix with entries from $\mathbb{Z}$. Prove that the elements

$$h_i = \sum_{j=1}^{n} a_{i,j} g_j \text{ for } i = 1, 2, \ldots, n$$

form a basis for G if and only if $A \in GL(n, \mathbb{Z})$.

4.11. Let G be free abelian group of rank n, and let H be a subgroup of G. Prove that G/H is finite if and only if the rank of H is n. Conclude that a subgroup H of a lattice L that has finite index in L must also be a lattice. (See Exercise 4.10.)

Biography 4.8 Hermann Minkowski (1864–1909) *was born on June 22, 1864 in Alexotas of what was then the Russian empire, but is now Kaunas, Lithuania. He studied at the Universities of Berlin, then Königsberg where he received his doctorate in 1885. He taught at both Bonn and Zürich, until Hilbert created a chair for him at Göttingen, which he accepted in 1902 and remained there for the rest of his life. He pioneered the area we now call the geometry of numbers. This led to work on convex bodies and to packing problems—see Remark 4.7 on page 187. He is also known for having laid the groundwork for relativity theory by thinking of space and time as linked together in a four-dimensional space-time continuum. Indeed by 1907, he came to the conclusion that the work of Einstein and others could be best formulated in a non-euclidean space. Later Einstein used these ideas to formulate the general theory of relativity (see also Biography 2.1 on page 73 for Noether's influence on Einstein's theory). Furthermore, his geometric insights paved the way for modern functional analysis. He died from a ruptured appendix on January 12, 1909 in Göttingen.*

Minkowski is best known for his ideas applied as cited above, especially his creation of the geometry of numbers in 1890. However, he had an early interest in pure mathematics such as his study of binary quadratic forms and continued fractions. In 1907, he published Diophantische Approximationen: Eine Einführung in die Zahlenthorie, *which provided an elementary discussion of his work on the geometry of numbers, and the applications to the theories of Diophantine approximation and algebraic numbers.*

Chapter 5

Arithmetic Functions

> To still be searching what we know not, by what we know, still closing up truth to truth as we find it (for all her body is homogeneal and proportional), this is the golden rule in theology as well as in arithmetic, and makes up the best harmony in a church.
>
> from **Areopagitica (1644)**.
> **John Milton (1608–1674)**
> **British poet**

Arithmetic functions, studied in a first course in number theory, are those functions whose domain is $\mathbb{N}$ and whose range is a subset of $\mathbb{C}$— for instance, see [68, §2.3 –§2.5]. In this chapter we look at a more in-depth analysis of these functions, especially from the perspective of their behaviour for large values of n. Actually plotting an arithmetic function seems to show chaotic behaviour, but most such functions do behave well on "average," a term we will define precisely in §5.2. First, we need a strong result from the number-theoretic toolkit provided in the following.

5.1   The Euler–Maclaurin Summation Formula

We seek to establish the formula in the title, and explore some of the applications such as Fourier series of Bernoulli polynomials–see Definitions 5.2 on the next page and 5.3 on page 194 as well as Biographies 5.1 on page 197 and 5.4 on page 207. First, we need to introduce the following, which first appeared in the posthumous work *Ars Conjectandi* by Jacob (Jacques) Bernoulli in 1713. Also, the reader should be familiar with the background on the basics concerning series–for instance, see [68, Appendix A, pp. 307–310].

Definition 5.1 $\boxed{\textbf{Bernoulli Numbers}}$

In the Taylor series, for a complex variable x,

$$F(x) = \frac{x}{e^x - 1} = \sum_{j=0}^{\infty} \frac{B_j x^j}{j!},$$

the coefficients B_j are called the *Bernoulli numbers*.

Example 5.1 Using the recursion formula given in Exercise 5.2 on page 206, we calculate the first few Bernoulli numbers:

n	0	1	2	3	4	5	6	7	8	9	10
B_n	1	$-\frac{1}{2}$	$\frac{1}{6}$	0	$-\frac{1}{30}$	0	$\frac{1}{42}$	0	$-\frac{1}{30}$	0	$\frac{5}{66}$

n	11	12	13	14	15	16	17	18	19
B_n	0	$-\frac{691}{2730}$	0	$\frac{7}{6}$	0	$-\frac{3617}{510}$	0	$\frac{43867}{798}$	0

Example 5.1 suggests that $B_{2n+1} = 0$ for all $n \in \mathbb{N}$ and this is indeed the case—see Exercise 5.1 on page 205.

Suppose that x, s are complex variables and set

$$F(s,x) = \frac{s e^{xs}}{e^s - 1} = \sum_{n=0}^{\infty} B_n(x) \frac{s^n}{n!}, \text{ for } |s| < 2\pi. \tag{5.1}$$

Then by comparing coefficients of x^n in

$$\sum_{n=0}^{\infty} B_n(x) \frac{s^n}{n!} = F(s,x) = F(s) e^{xs} = \sum_{n=0}^{\infty} B_n \frac{s^n}{n!} \sum_{j=0}^{\infty} x^j \frac{s^j}{j!},$$

we get the following.

Definition 5.2 $\boxed{\textbf{Bernoulli Polynomials}}$

For $x \in \mathbb{C}$,

$$B_n(x) = \sum_{j=0}^{n} \binom{n}{j} B_j x^{n-j},$$

called the *n*-th *Bernoulli polynomial*.

Example 5.2 Using the recursion formula in Exercise 5.2 on page 206 again, we calculate the first few Bernoulli polynomials:

$$B_0(x) = 1, B_1(x) = x - \frac{1}{2}, B_2(x) = x^2 - x + \frac{1}{6},$$

$$B_3(x) = x^3 - \frac{3}{2}x = x(x-1)\left(x - \frac{1}{2}\right),$$

$$B_4(x) = x^4 - 2x^3 + x^2 - \frac{1}{30},$$

$$B_5(x) = x^5 - \frac{5}{2}x^4 + \frac{5}{3}x^3 - \frac{1}{6}x.$$

$$B_6(x) = x^6 - 3x^5 + \frac{5}{2}x^4 - \frac{1}{2}x^2 + \frac{1}{42}.$$

Now we are in a position to prove the result in the section's header. We will be invoking the integration by parts formula several times in what follows–see (4.15) on page 172. The following formula has the dual attribution since it was discovered independently and almost simultaneously by the two authors in the first half of the eighteenth century, but neither of them obtained the remainder term displayed in the second line of the theorem, and that is an essential ingredient.

Theorem 5.1 | **The Euler–Maclaurin Summation Formula**

Let $a < b$ be integers and let $n \in \mathbb{N}$. If $f(x)$ has n continuous derivatives on the interval $[a, b]$, then

$$\sum_{j=a+1}^{b} f(j) = \int_a^b f(x)dx + \sum_{i=1}^{n} (-1)^i \frac{B_i}{i!} \left(f^{(i-1)}(b) - f^{(i-1)}(a)\right)$$

$$+ \frac{(-1)^{n-1}}{n!} \int_a^b B_n(x - \lfloor x \rfloor) f^{(n)}(x)dx.$$

Proof. If we set

$$\int_0^1 f(x)dx = \int_0^1 B_0(x)f(x)dx,$$

then we may integrate by parts n times,

$$\int_0^1 f(x)dx = \sum_{i=1}^{n} (-1)^{i-1} \frac{B_i(x)}{i!} f^{(i-1)}(x) \Big|_0^1 + (-1)^n \int_0^1 \frac{B_n(x)}{n!} f^{(n)}(x)dx$$

$$= \sum_{i=1}^{n} (-1)^{i-1} \frac{B_i}{i!} \left(f^{(i-1)}(1) - f^{(i-1)}(0)\right) + f(1) + (-1)^n \int_0^1 \frac{B_n(x)}{n!} f^{(n)}(x)dx,$$

where the $f(1)$ comes from the fact that we must add it back on given that $B_1 = -1/2$, but $B_1(1) = 1/2$ by Exercise 5.4, whereas $B_i(1) = B_i$ for $i > 1$, and $B_i(0) = B_i$ by Definition 5.2 on the facing page. Now by replacing $f(x)$ by $f(j-1+x)$, we obtain that $f(1)$ becomes $f(j)$ so by the above,

$$f(j) = \int_0^1 f(j-1+x)dx + \sum_{i=1}^{n} (-1)^i \frac{B_i}{i!} \left(f^{(i-1)}(j) - f^{(i-1)}(j-1)\right)$$

$$+(-1)^{n-1} \int_0^1 \frac{B_n(x)}{n!} f^{(n)}(j-1+x)dx.$$

Since we have

$$\sum_{j=a+1}^b \int_0^1 f(j-1+x)dx = \int_a^b f(x)dx,$$

$$\sum_{j=a+1}^b \left(f^{(i-1)}(j) - f^{(i-1)}(j-1) \right) = f^{(i-1)}(b) - f^{(i-1)}(a),$$

and

$$\sum_{j=a+1}^b \int_0^1 B_n(x)f^{(n)}(j-1+x)dx = \int_a^b B_n(x - \lfloor x \rfloor)f^{(n)}(x)dx,$$

then we have secured the result.   □

In order to be able to apply Theorem 5.1 to Fourier series, we need to know more about the functions $f_n(x) = B_n(x - \lfloor x \rfloor)$ in the remainder term of the Euler-Maclaurin summation formula. Thus, we need the formal definition in order to introduce such expansions for $f_n(x)$.

Definition 5.3 | Fourier Series |

A *Fourier series* is a periodic function f, defined for $x \in [-\pi, \pi]$, given by the convergent series

$$f(x) = \frac{a_0}{2} + \sum_{j=1}^{\infty} (a_j \cos(\pi j x) + b_j \sin(\pi j x)).$$

The study of Fourier series is known as *harmonic analysis*.

It is known that one may compute the Fourier series of a 2π-periodic function f via the following:

$$f(x) = \frac{a_0}{2} + \sum_{j=1}^{\infty} (a_j \cos(2\pi j x) + b_j \sin(2\pi j x)),$$

where

$$a_0 = 2 \int_0^1 f(x)dx,$$

$$a_j = 2 \int_0^1 f(x) \cos(2\pi j x)dx,$$

and

$$b_j = 2 \int_0^1 f(x) \sin(2\pi j x)dx.$$

Since $f_n(x) = B_n(x - \lfloor x \rfloor)$ is periodic with period length 1, we have

$$f_n(x) = \frac{a_0^{(n)}}{2} + \sum_{j=1}^{\infty} \left(a_j^{(n)} \cos(2\pi j x) + b_j^{(n)} \sin(2\pi j x) \right),$$

with

$$a_0^{(n)} = 2 \int_0^1 B_n(x) dx,$$

$$a_j^{(n)} = 2 \int_0^1 B_n(x) \cos(2\pi j x) dx,$$

and

$$b_j^{(n)} = 2 \int_0^1 B_n(x) \sin(2\pi j x) dx.$$

However, by Exercises 5.4 and 5.6 on page 206 in conjunction with Definition 5.2 on page 192, it holds for any $n \in \mathbb{N}$ that

$$a_0^{(n)} = 2 \int_0^1 B_n(x) = 2 \int_0^1 \frac{B_{n+1}'(x)}{n+1} dx = \frac{2}{n+1} (B_{n+1}(x)) \bigg|_0^1 =$$

$$\frac{2}{n+1}(B_{n+1}(1) - B_{n+1}(0)) = \frac{2}{n+1}(B_{n+1} - B_{n+1}) = 0.$$

Also, using integration by parts

$$a_j^{(n)} = 2 \int_0^1 B_n(x) \cos(2\pi j x) dx = 2 \int_0^1 B_n(x) d \left(\frac{\sin(2\pi j x)}{2\pi j} \right)$$

$$= 2 B_n(x) \frac{\sin(2\pi j x)}{2\pi j} \bigg|_0^1 - \frac{1}{\pi j} \int_0^1 B_n'(x) \sin(2\pi j x) dx$$

$$= -\frac{n}{\pi j} \int_0^1 B_{n-1}(x) \sin(2\pi j x) dx = -\frac{n}{2\pi j} b_j^{(n-1)},$$

for any $n \geq 2$ and $a_j^{(1)} = 0$ for any $j \in \mathbb{N}$. Furthermore, again employing integration by parts,

$$b_j^{(1)} = -2 B_1(x) \frac{\cos(2\pi j x)}{2\pi j} \bigg|_0^1 + \frac{1}{\pi j} \int_0^1 \cos(2\pi j x) dx = -\frac{1}{\pi j},$$

and for any $n \geq 2$,

$$b_j^{(n)} = -2 B_n(x) \frac{\cos(2\pi j x)}{2\pi j} \bigg|_0^1 + \frac{1}{\pi j} \int_0^1 B_n'(x) \cos(2\pi j x) dx$$

$$= \frac{n}{\pi j} \int_0^1 B_{n-1}(x) \cos(2\pi j x) dx = \frac{n}{2\pi j} a_j^{(n-1)}.$$

Thus far, we have demonstrated that for any $j \in \mathbb{N}$,

$$a_0^{(n)} = 0, \quad a_j^{(1)} = 0, \quad b_j^{(1)} = -\frac{1}{\pi j},$$

$$a_j^{(n)} = -\frac{n}{2\pi j} b_j^{(n-1)} = -\frac{n(n-1)}{(2\pi j)^2} a_j^{(n-2)} \text{ for any } n \geq 2,$$

and

$$b_j^{(n)} = \frac{n}{2\pi j} a_j^{(n-1)} = -\frac{n(n-1)}{(2\pi j)^2} b_j^{(n-2)} \text{ for any } n \geq 2.$$

Continuing in this fashion, an inductive process gives us that for any $j, k \in \mathbb{N}$,

$$a_j^{(2k-1)} = 0, \quad a_j^{(2k)} = (-1)^{k-1} \frac{2(2k)!}{(2\pi j)^{2k}},$$

$$b_j^{(2k)} = 0, \text{ and } b_j^{(2k-1)} = (-1)^k \frac{2(2k-1)!}{(2\pi j)^{2k-1}}.$$

We have therefore proved the following.

Theorem 5.2 $\boxed{\textbf{Fourier Series for Bernoulli Polynomials}}$

For all $x \in \mathbb{R}$ and $k \in \mathbb{N}$,

$$B_{2k-1}(x - \lfloor x \rfloor) = (-1)^k 2(2k-1)! \sum_{j=1}^{\infty} \frac{\sin(2\pi j x)}{(2\pi j)^{2k-1}}, \text{ for } k \geq 2, \qquad (5.2)$$

$$B_{2k}(x - \lfloor x \rfloor) = (-1)^{k-1} 2(2k)! \sum_{j=1}^{\infty} \frac{\cos(2\pi j x)}{(2\pi j)^{2k}}, \text{ for } k \geq 1. \qquad (5.3)$$

Remark 5.1 We have deliberately left out from Theorem 5.2 the case of

$$B_1(x - \lfloor x \rfloor) = \frac{1}{2}(x - \lfloor x \rfloor)$$

since the Fourier series vanishes, while $B_1(x - \lfloor x \rfloor)$ jumps between $+1/2$ and $-1/2$ for $x \notin \mathbb{Z}$, and is 0 at integer values of x. This is the only case where $B_n(x - \lfloor x \rfloor)$ is *not* continuous of period 1. Note, as well, that by setting $x = 0$ in *(5.2)*, we get that $B_{2k-1} = 0$ for any $k \geq 2$, which is Exercise 5.1 on page 205. Similarly, we have the next result.

Corollary 5.1 *If $k \in \mathbb{N}$, then $(-1)^{k-1} B_{2k} > 0$.*

Proof. Set $x = 0$ in (5.3) to get

$$B_{2k} = 2(-1)^{k-1} \frac{(2k)!}{(2\pi)^{2k}} \sum_{n=1}^{\infty} \frac{1}{n^{2k}} \qquad (5.4)$$

from which the result follows.   □

Biography 5.1 Jean Baptiste Joseph Fourier (1768–1830) *was born on March 21, 1768 in Auxerre, Bourgogne, France. His early teenage schooling began at the École Militaire of Auxerre, and he later became a teacher at the Benedictine college there. Unfortunately, he got enmeshed in the politics of the French revolution. By July of 1794, he was arrested and imprisoned, then freed later that year but was arrested again and imprisoned in 1795. However, by September 1, 1795, he was teaching at the École Polytechnique where he had been during his brief stint of freedom earlier. He stayed out of trouble, remained free, and in 1797 succeeded Lagrange to the chair of analysis and mechanics. However, in 1797, he joined Napleon's army in its invasion of Egypt, acting as a scientific advisor. While Fourier was in Cairo, he assisted in the founding of the Cairo Institute, and was one of the members of the division of mathematics, later being elected secretary to the Institute. He held this position during the entirety of France's occupation of Egypt. In 1801, Fourier returned to his position as Professor of Analysis at the École Polytechnique. However, Napoleon requested that Fourier go to Grenoble as Prefect. Although he did not want to leave the world of academe, he could not refuse the request and so he went, where he spent an inordinate amount of time on the historical document* Description of Egypt, *which was completed in 1810, largely a rewriting of Napoleon's influence there. Yet it was in Grenoble that Fourier accomplished his best work on the theory of heat. By 1807 he had completed his memoir* On the Propagation of Heat in Solid Bodies, *which contained expansions of functions, which we now call* Fourier series. *In 1811, he was awarded a prize by the Paris Institute for this work. When Napoleon was defeated on July 1, 1815, Fourier returned to Paris, where he was elected to the Académie des Sciences in 1817. In 1822, Fourier filled the post as Secretary to the mathematical section of the Académie des Sciences, a vacancy created by the death of Delambre. In 1822, Fourier published* Théorie analytique de las chaleur, *which was a prize winning essay. Fourier continued his mathematical output during his eight years in Paris. He died there on May 16, 1830. Fourier's work paved the way for subsequent work on trigonometric series and the theory of functions of a real variable, which are vital areas in today's modern world.*

Remark 5.2 Bernoulli numbers are among the most distinguished and important numbers in all of mathematics. Indeed, they play a vital role in number theory, especially in connection with Fermat's last theorem, see Remark 1.17 on page 41, as well as Biography 5.6 on page 228.

The Bernoulli numbers may also be calculated from the integral

$$B_n = \frac{n!}{2\pi i} \int \frac{z}{e^z - 1} \frac{dz}{z^{n+1}},$$

as well as from the derivative

$$B_n = \left[\frac{d^n}{dx^n} \left(\frac{x}{e^x - 1} \right) \right]_{x=0},$$

and they have connections to the Riemann ζ-function

$$\zeta(s) = \sum_{j=1}^{\infty} j^{-s} = \prod_{p=prime} (1 - p^{-s})^{-1},$$

via the identity given in (5.4), namely the following formula first proved by Euler — see Exercise 10.14 on page 346,

$$\zeta(2k) = \frac{(2\pi)^{2k}}{2(2k)!} |B_{2k}| \tag{5.5}$$

– see [68, §1.9, pp. 65–72]. We will look, in detail, at the Riemann ζ-function in §5.3.

Now we proceed to demonstrate yet more applications of the Maclaurin sum formula by deriving, from it, a well-known and very accurate approximation for $n!$. First of all, we need the following basic formula from elementary calculus.

Lemma 5.1 | Integral of Powers of Sine |

For any $n \in \mathbb{N}$,

$$\int_0^{\pi/2} \sin^n(x)dx = \begin{cases} \dfrac{(n-1)(n-3)\cdots 3 \cdot 1}{n(n-2)\cdots 4 \cdot 2} \cdot \dfrac{\pi}{2} & \text{if } 2 \mid n, \\[2mm] \dfrac{(n-1)(n-3)\cdots 4 \cdot 2}{n(n-2)\cdots 5 \cdot 3} & \text{if } 2 \nmid n. \end{cases} \tag{5.6}$$

Proof. If we set

$$I_n = \int_0^{\pi/2} \sin^n x \, dx,$$

then using integration by parts we get

$$I_n = \int_0^{\pi/2} (\sin^{n-1} x)(\sin x)dx = -(\sin^{n-1} x)(\cos x)\Big|_0^{\pi/2}$$

$$+ (n-1) \int_0^{\pi/2} (\sin^{n-2} x)(\cos^2 x)dx = (n-1)(I_{n-2} - I_n).$$

Therefore,

$$I_n = \frac{n-1}{n} \cdot I_{n-2} \tag{5.7}$$

for any integer $n \geq 2$. By including $I_0 = \pi/2$, and $I_1 = 1$, we get (5.6) from the recursion in (5.7).   ☐

From the above we are able to obtain the following renowned formula–see Biography 5.3 on page 205.

Theorem 5.3 | **The Wallis Formula**

Given $n \in \mathbb{N}$,

$$\lim_{n\to\infty} \frac{2^{2n}(n!)^2}{(2n)!\sqrt{n}} = \sqrt{\pi}. \tag{5.8}$$

Proof. Since for any $n \in \mathbb{N}$ we have,

$$0 < I_{2n+1} < I_{2n} < I_{2n-1},$$

by Lemma 5.1 we have,

$$0 < \frac{(2n)(2n-2)\cdots 4\cdot 2}{(2n+1)(2n-1)\cdots 5\cdot 3} < \frac{(2n-1)(2n-3)\cdots 3\cdot 1}{2n(2n-2)\cdots 4\cdot 2}\cdot\frac{\pi}{2}$$

$$< \frac{(2n-2)(2n-4)\cdots 4\cdot 2}{(2n-1)(2n-3)\cdots 5\cdot 3}$$

By inverting this inequality and multiplying through by

$$\frac{(2n-2)(2n-4)\cdots 4\cdot 2}{(2n-1)(2n-3)\cdots 5\cdot 3\cdot 1}\cdot\pi$$

we get

$$\frac{2n+1}{2n}\cdot\pi > \frac{1}{n}\left(\frac{(2n)(2n-2)(2n-4)\cdots 4\cdot 2}{(2n-1)(2n-3)\cdots 5\cdot 3\cdot 1}\right)^2 > \pi.$$

By letting $n \to \infty$ and observing the outside values go to π, then the center is squeezed to π as well. Therefore,

$$\lim_{n\to\infty}\frac{1}{n}\left(\frac{(2n)(2n-2)(2n-4)\cdots 4\cdot 2}{(2n-1)(2n-3)\cdots 5\cdot 3\cdot 1}\right)^2 = \pi,$$

namely

$$\lim_{n\to\infty}\frac{1}{n}\left(\frac{2^{2n}(n!)^2}{(2n)!}\right)^2 = \pi,$$

and by taking square roots we get (5.8).   ☐

Now we have one more result before we present the approximation for $n!$.

Definition 5.4 | **Asymptotically Equal**

In what follows the notation

$$f(n) \sim g(n)$$

will signify that

$$\lim_{n \to \infty} \frac{f(n)}{g(n)} = 1,$$

which is sometimes referenced as f and g being *asymptotically equal*.

The following is a renowned constant–see Biography 5.2 on page 204.

Theorem 5.4 | **Stirling's Constant**

For $N \in \mathbb{N}$,

$$\lim_{N \to \infty} \left(\log_e(N!) - \left(N + \frac{1}{2}\right) \log_e(N) + N \right) = \log_e \sqrt{2\pi}.$$

Proof. Let

$$C = \lim_{N \to \infty} \left(\log_e(N!) - \left(N + \frac{1}{2}\right) \log_e(N) + N \right).$$

Then

$$e^C = \lim_{N \to \infty} \frac{N! e^N}{N^{N+1/2}}.$$

In other words,

$$N! \sim e^{C-N} N^{N+1/2}. \tag{5.9}$$

Also, by inverting Wallis' formula (5.8) on page 199, we get

$$\lim_{n \to \infty} \frac{(2n)!}{(2^n n!)^2} \sqrt{n} = \frac{1}{\sqrt{\pi}}.$$

Now by using $N = 2n$ and $N = n$ in the latter employing (5.9), we get

$$\lim_{n \to \infty} \frac{e^{C-2n} (2n)^{2n+1/2}}{(2^n e^{C-n} n^{n+1/2})^2} \sqrt{n} = \frac{1}{\sqrt{\pi}},$$

which simplifies to

$$\frac{\sqrt{2}}{e^C} = \frac{1}{\sqrt{\pi}},$$

from which we get

$$e^C = \sqrt{2\pi},$$

yielding that $C = \log_e \sqrt{2\pi}$. $\square$

Now we are ready for the approximation for the factorial.

Theorem 5.5 $\boxed{\textbf{Stirling's Formula}}$

For any $N \in \mathbb{N}$

$$N! \sim \sqrt{2\pi} \cdot e^{-N} \cdot N^{N+1/2} = \sqrt{2\pi N}\left(\frac{N}{e}\right)^N. \qquad (5.10)$$

Proof. For

$$f(x) = \log_e(x)$$

and $n \in \mathbb{N}$, , the n-th derivative is given by

$$f^{(n)}(x) = (-1)^{n-1}x^{n-1}(n-1)!.$$

We now apply Theorem 5.1 on page 193 to $f(x)$, with $a = 1$, $b = N \geq 2$, and $n = 2k$, to get

$$\log_e(N!) = \sum_{j=2}^{N}\log_e(j) = \int_1^N \log_e(x)dx + \sum_{j=1}^{2k}(-1)^j\frac{B_j}{j!}\left(f^{(j-1)}(N) - f^{(j-1)}(1)\right)$$

$$+\frac{1}{2k}\int_N^\infty B_{2k}(x - \lfloor x\rfloor)x^{-2k}dx = \int_1^N \log_e(x)dx - B_1\left(f(N) - f(1)\right)$$

$$+\sum_{j=2}^{2k}(-1)^j\frac{B_j}{j!}\left((-1)^{j-2}N^{1-j}(j-2)! - (-1)^{1-j}(j-2)!\right)$$

$$= \int_1^N \log_e(x)dx + \frac{\log_e(N)}{2} + \sum_{i=1}^{k}\frac{B_{2i}(2i-2)!(N^{1-2i}-1)}{(2i)!}$$

$$+\int_1^N \frac{B_{2k}(x - \lfloor x\rfloor)x^{-2k}}{2k}dx,$$

and using integration by parts on the first integral while rewriting the remainder yields that the above equals

$$\left(N + \frac{1}{2}\right)\log_e(N) - N + 1 + \sum_{i=1}^{k}\frac{B_{2i}}{(2i-1)2i}N^{1-2i}$$

$$+\frac{1}{2k}\int_1^N B_{2k}(x - \lfloor x\rfloor)x^{-2k}dx - \sum_{i=1}^{k}\frac{B_{2i}}{(2i-1)2i}. \qquad (5.11)$$

Claim 5.1 *For* $k \in \mathbb{N}$,

$$\frac{1}{2k}\int_1^\infty B_{2k}(x - \lfloor x\rfloor)x^{-2k}dx = \log_e(\sqrt{2\pi}) + \sum_{i=1}^{k}\frac{B_{2i}}{(2i-1)2i} - 1.$$

From (5.11),

$$\frac{1}{2k}\int_1^\infty B_{2k}(x-\lfloor x\rfloor)x^{-2k}dx = \lim_{N\to\infty}\frac{1}{2k}\int_1^\infty B_{2k}(x-\lfloor x\rfloor)x^{-2k}dx =$$

$$\lim_{N\to\infty}\left[\log_e(N!)-\left(N+\frac{1}{2}\right)\log_e(N)+N\right]-1$$

$$-\lim_{N\to\infty}\left(\sum_{i=1}^{k}\frac{B_{2i}}{(2i-1)2i}N^{1-2i}\right)+\sum_{i=1}^{k}\frac{B_{2i}}{(2i-1)2i} = \log_e(\sqrt{2\pi})-1+\sum_{i=1}^{k}\frac{B_{2i}}{(2i-1)2i},$$

by Theorem 5.4 on page 200, which is the claim.

Plugging the result of Claim 5.1 into (5.11), we get,

$$\lim_{N\to\infty}\log_e(N!) = \lim_{N\to\infty}\left[\left(N+\frac{1}{2}\right)\log_e(N)-N\right]+\log_e(\sqrt{2\pi}),$$

and by rewriting using the laws for logs,

$$\lim_{N\to\infty}\left[\log_e\left(\frac{N!}{N^{N+1/2}}\right)+N\right] = \log_e(\sqrt{2\pi}),$$

and raising to the power of e,

$$\lim_{N\to\infty}\left(\frac{N!}{N^{N+1/2}}\cdot e^N\right) = \sqrt{2\pi}.$$

namely,

$$\frac{N!}{N^{N+1/2}}\cdot e^N \sim \sqrt{2\pi}.$$

In other words,

$$N! \sim \sqrt{2\pi}e^{-N}N^{N+1/2} = \sqrt{2\pi N}\left(\frac{N}{e}\right)^N,$$

as required.   □

One of the really slick applications of the Euler–Maclaurin summation formula is Euler's constant (4.13) which we introduced in the discussion of transcendence on page 172. (We do not know if this constant is irrational, let alone transcendental.) The definition given in (4.13) is an exceptionally bad method for computing γ given by

$$\lim_{N\to\infty}\left(\sum_{n=1}^{N}1/n-\log_e N\right)$$

since we are taking the limit of a quantity that is within a constant times N^{-1} of γ. This means that we require approximately 10^{10} summation terms to compute γ to ten decimal places. Even using a computer to do this will lead to astronomical round-off errors and so the loss of significant figures is devastating. Euler–Maclaurin comes to the rescue. In Theorem 5.1 on page 193, take $f(x) = 1/x$, $n = m$, $a = 1$, and $b = N$. Using the techniques of this section such as used in the derivation of Stirling's approximation, it follows that

$$\sum_{n=1}^{N} \frac{1}{n} = \log_e N + \gamma + \frac{1}{2N} - \sum_{j=1}^{m} \frac{B_{2j}}{2j} N^{-2j} + R_{2m}(N), \tag{5.12}$$

where

$$|R_{2m}(N)| \le \frac{|B_{2m}|}{2m} N^{-2m}. \tag{5.13}$$

We now demonstrate how the estimates given by (5.12) are far more precise than that given in (4.13) can be for γ. We choose small values for pedagogical reasons, but larger values for m and N yield more precision.

Example 5.3 *Let* $m = 5$ *and* $N = 8$. *Then* $2.717857142857 =$

$$\sum_{n=1}^{8} \frac{1}{n} = \log_e 8 + \gamma + \frac{1}{16} - \frac{B_2}{2} 8^{-2} - \frac{B_4}{4} 8^{-4} - \frac{B_6}{6} 8^{-6} - \frac{B_8}{8} 8^{-8} - \frac{B_{10}}{10} 8^{-10} + R_{10}(8).$$

By Example 5.1 on page 192, we know the values of B_{2j} for $j = 1, 2, 3, 4, 5$, so we know from (5.13) that with error no greater than $|R_{10}(8)| \le \frac{|B_{10}|}{10} 8^{-10} = 0.000000000007055$, we have

$$\gamma = 2.717857142857143 - \log_e 8 - \frac{1}{16} + \frac{6^{-1}}{2} 8^{-2} - \frac{30^{-1}}{4} 8^{-4} + \frac{42^{-1}}{6} 8^{-6}$$

$$-30^{-1} 8^{-9} + \frac{66^{-1} \cdot 5}{10} 8^{-10} \sim 0.577215664901822.$$

Since higher values for N and m will yield more accurate estimates, we note that the above is accurate within the error expected since

$$\gamma = 0.5772156649015328606065120900824024310 42 \ldots.$$

This value is sometimes called the Euler–Mascheroni *constant since it was calculated to sixteen digits of decimal accuracy by Euler in 1781, but later by Mascheroni to double that length in 1790. However, Mascheroni's calculations were correct only to the first nineteen digits. In 1809, Soldner correctly computed it to forty decimal digits, which Gauss verified in 1812. The latest calculation was by Alexander Yee and Raymond Chan done March 13, 2009, accurate to $29,844,489,545$ decimal digits, the world record at the time of this writing. To check for future updates see:*

http://en.wikipedia.org/wiki/Euler-Mascheroni_constant#Known_digits.

Biography 5.2 James Stirling (1692–1770) *was born in Garden, near Stirling, Scotland. Little is known of his early education, or even his exact birth date. It is known that he matriculated at Balliol College in Oxford on January 18, 1711 with two scholarships, one of which was the Bishop Warner Exhibition and the other was the Snell Exhibition. However, he lost both of them when he refused to swear an oath of allegiance to the king since it went against his Jacobite sympathies. The Jacobite cause was that of King James II of England, also known as James the VII of Scotland (Jacobus in Latin) and his descendants. This king was one of the Stuarts, who were Scottish but not Roman Catholics, and who offered an alternative to the British crown. Stirling's father was a strong Jacobite supporter and was even imprisoned for his sympathies and accused of high treason when Stirling was only seventeen, but was later acquitted. Stirling himself was charged with blaspheming the British King George, but was acquitted as well. In 1717, Stirling published* Lineae Tertii Neutonianae, *a generalization of Newton's theory of plane curves of degree three, as well as results on curves of quickest descent, and on orthogonal trajectories. The latter problem was coined by Leibniz, and was advanced not only by Stirling, but also by Johann Bernoulli, Nicolaus (I) Bernoulli, Nicolaus (II) Bernoulli, and Euler. Stirling solved the problem in 1716. He held the chair at the University of Padua from 1716 to 1722, when he returned to Glasgow. What he did between 1722 and 1724 is not clearly known. Yet he went to London in 1724 where he stayed for the next decade. There he was friends with Newton and was very active mathematically. Indeed, Newton supported Stirling in a bid for fellowship of the Royal Society of London, and on November 3, 1726, Stirling was elected. In 1730, he published* Methodus Differentialis, *a book on infinite series, summation, interpolation and quadrature, including results on the Gamma function and the Hypergeometirc function. Theorem 5.5 on page 201 appears in this book as Example 2 of Proposition 28. Thus, this was Stirling's most important work. In 1735, Stirling returned to Scotland where he was appointed manager of the Scottish mining company, Leadhills, in Lanarkshire. In 1745, he published a paper on the ventilation of mine shafts. In that year arose the greatest of the Jacobite rebellions. On September 17, 1745, Charles Edward, the Young Pretender, entered Edinburgh with his army. Maclaurin played a very active part in the defence of the city against the Jacobites. In fact, he died in 1746 from consequences of the battles in the previous year. Stirling was subsequently considered for his chair at Edinburgh. However, his Jacobite sympathies prevented that from happening. In 1746, Stirling was elected to membership of the Royal Academy of Berlin. In 1752 was his last work in the realm of science when he conducted the first survey of the River Clyde for the Corporation of Glasgow. He fell ill later in his life and died on December 5, 1770 in Edinburgh where he was buried at Greyfriars Churchyard. There his contributions to the theory of infinite series are honoured by a small plaque in the cemetery wall.*

In §5.2, we will use results in this section to get asymptotic facts for certain arithmetic functions.

Biography 5.3 John Wallis (1616–1703) *was born in Ashford, Kent, England, the son of a minister, who died when John was only six years old. His mother left Ashford when there was an outbreak of the plague in the area. When he was only thirteen, he felt that he was ready for university. However, it was not until 1632 that he entered Emmanuel College Cambridge. In 1637 he was awarded his bachelor's degree and received his master's degree in 1640. In that year he was also ordained by the bishop of Winchester and appointed chaplain to Sir Richard Darley at Butterworth in Yorkshire. During the next few years he excelled as a cryptanalyst by deciphering messages sent by the Royalists who were engaged in a civil war with the Parliamentarians.* (For background on this and related historical and crytological issues, see [67].) *By 1649, his support for the Parliamentarians paid off when he was appointed to the Savilian Chair of Geometry at Oxford by Cromwell, who had dismissed the previous chair holder for his Royalist views. (Oliver Cromwell (1599–1658) was a soldier and statesman who was instrumental in the execution of King Charles I on January 30, 1649. Then the monarchy was abolished and Cromwell made himself chairman of the Council of State of the new Commonwealth. By 1653, he had reorganized the Church of England, established Puritanism, brought prosperity to Scotland, and granted Irish representation in Parliament.) Indeed, Wallis held this chair for fifty years until his death. Yet, in 1657, he was appointed as keeper of the University archives there. Wallis is known for his contributions to the foundations of the calculus and was, arguably, the most prominent English mathematician before Newton. His most renowned work was* Arithmetica Infinitorum, *published in 1656, which built upon Cavalieri's methods of indivisibles. He contributed further to the history of mathematics by restoring some Greek texts from antiquity such as Ptolemy's* Harmonics, *as well as Archimedes'* Sand-reckoner, *among others. In the mathematics that he did, Wallis may be said to have helped to build a calculus established upon arithmetical, rather than geometrical conceptions. This work won the respect and support of his contemporaries such as James Gregory. Those who saw the solution of problems through geometric means opposed this point of view including Thomas Hobbes, with whom Wallis had an ongoing public dispute that lasted over twenty years. Hobbes' views of mathematics were rooted in the Greek thought that accepted mathematics as derived from the senses by abstraction from real objects, rather than an abstract branch of formal logic. Yet the analytic symbolism of Descartes, Fermat, and Wallis may be seen today in the calculus as embodying the rules of differentiation and integration, even the fundamental theorem of calculus. For this and many other contributions, Wallis will be remembered. He died on October 28, 1703 in Oxford, England.*

Exercises

5.1. Without using Theorem 5.2 on page 196, prove that the odd-indexed Bernoulli numbers bigger than one are equal to zero, namely $B_{2n+1} = 0$ for all $n \in \mathbb{N}$.

5.2. Prove the following *recursion formula for Bernoulli numbers* for $n \in \mathbb{N}$,

$$\sum_{i=0}^{n-1} \binom{n}{i} B_i = \begin{cases} 1 & \text{if } n = 1, \\ 0 & \text{if } n > 1, \end{cases}$$

where $\binom{n}{i}$ is the binomial coefficient.

(*Hint: Use the fact that* $e^x = \sum_{i=0}^{\infty} \frac{x^i}{i!}$.)

5.3. Prove that $\sum_{j=1}^{\infty}(1/j)$ diverges.

(*Hint: Assume* $\sum_{j=1}^{\infty}(1/j) = d \in \mathbb{R}$ *and reach a contradiction.*)

5.4. Prove that, from Definition 5.2 on page 192,

$$B_n(1) = \begin{cases} 1/2 & \text{if } n = 1, \\ B_n & \text{if } n > 1. \end{cases}$$

(*Hint: Use Exercise 5.2.*)

5.5. Prove the following result by Jacob Bernoulli on the sums of n-th powers, namely that, for every nonnegative $n \in \mathbb{Z}$ and $k \in \mathbb{N}$,

$$S_n(k) = \sum_{j=1}^{k-1} j^n = \frac{B_{n+1}(k) - B_{n+1}}{n+1} = \frac{1}{n+1} \sum_{j=0}^{n} \binom{n+1}{j} B_j k^{n+1-j}.$$

(*Hint: Compare the coefficients of* s^n *on both sides of* $F(s, x) - F(s, x - 1)$—*see* (5.1) *on page 192.*)

5.6. Prove the following derivative formula for Bernoulli polynomials.

$$B'_{n+1}(x) = (n + 1)B_n(x).$$

(*Hint: Replace* x *by* $x+1$ *in Equation* (S21) *on page 422 and differentiate with respect to* x.)

5.7. Prove that for any real $a \le b$, and integers $n \ge 0$,

$$\int_a^b B_n(t)dt = \frac{1}{n+1}(B_{n+1}(b) - B_{n+1}(a)).$$

(*Hint: Use Exercise 5.6.*)

Biography 5.4 Jacob Bernoulli (1654–1705) *was born on December 27, 1654 in Basel, Switzerland. He was one of ten children of Nicolaus and Margaretha Bernoulli. His brother Johann (1667–1748) was the tenth child of the union, and the two brothers had an influence on each other's mathematical development. Jacob was the first to explore the realms of mathematics, and being the pioneer in the family in this regard, he had no tradition to follow as did his brothers after him. In fact, his parents forced him to study philosophy and theology, which he silently resented. However, he obtained a licentiate in theology in 1676, after which he moved to Geneva where he was employed as a tutor. Then he travelled to France where he studied with Nicholas Malebranche, a leader among René Descartes' followers. (Malebranche represented the synthesis of the philosophies of St. Augustine and Descartes. This resulted in the Malebranche doctrine, which says that we see bodies through ideas in God and that God is the only real cause.) In 1681, Bernoulli travelled to the Netherlands where he met the mathematician Hudde, then to England where he met with Boyle and Hooke. This began a correspondence with numerous mathematicians that continued over several years. In 1683, he returned to Switzerland to teach at the University in Basel. He studied the work of leading mathematicians there and cultivated an increasing love of mathematics. In 1687, his brother Johann was appointed professor of mathematics at Basel. The two brothers embarked upon a study of mathematical publications, including the calculus proposed by Leibniz—see Biography 4.5 on page 175. However, their collaboration turned to rivalry with numerous public and private recriminations. Yet they both made significant contributions. Jacob's first such important work was in his 1685 publications on logic, algebra, and probability. In 1689, he published significant work on infinite series and on his law of large numbers. The latter is a mathematical interpretation of probability as relative frequency. This means that if an experiment is carried out for a large number of trials, then the relative frequency with which an event occurs equals the probability of the event. By 1704, Jacob had published five works on infinite series containing such fundamental results such as that $\sum_{j=1}^{\infty} 1/j$ diverges—see Exercise 5.3 on the preceding page. Although Jacob thought he had discovered the latter, it had been already discovered by Mengoli some four decades earlier. In 1690, Jacob published an important result in the history of mathematics by solving a differential equation using, in modern terms,* separation of variables. *This was the first time that the term* integral *was employed with its proper meaning for integration. In 1692, he investigated curves, including the logarithmic spiral, and in 1694, conceived of what we now call the* lemniscate of Bernoulli.*By 1696, he had solved what we now call the* Bernoulli equation: $y' = p(x)y + q(x)y^n$. *Eight years after his death, the* Ars Conjectandi *was published in 1713, a book in which the Bernoulli numbers first appear—see Definition 5.1 on page 192. In the book, they appear in his discussion of exponential series. Jacob held his chair at Basel until his death on August 16, 1705, when it was filled by his brother Johann. Jacob was always enthralled with the logarithmic spiral mentioned above. Indeed, he requested that it be carved on his tombstone with the (Latin) inscription* I shall arise the same though changed.

5.2   Average Orders

> *If all the arts aspire to the condition of music, all the sciences aspire to the*
> *condition of mathematics.*
> from **Some Turns of Thought in Modern Philosophy (1933)**
> **George Santayana (1863–1952)**
> **Spanish-born American skeptical philosopher**

In this section, we look at methods for getting accurate estimates for the behaviour of arithmetic functions for large n. More precisely, we look at the following notion.

Definition 5.5 | **Average Order of Arithmetic Functions**

If $f(n)$ is an arithmetic function and $g(n)$ is an elementary function, then we say that $f(n)$ is of the *average order* of $g(n)$ if

$$\sum_{j=1}^{n} f(j) \sim \sum_{j=1}^{n} g(j),$$

where $\sim$ is given by Definition 5.4 on page 200.

One of the arithmetic functions, studied in a first course in number theory, is the *number of divisors function* $\tau(n)$, which is the number of the positive divisors of $n \in \mathbb{N}$. This is the first arithmetic function we explore from the perspective of Definition 5.5. If we were to simply look at $\tau(n)$ as n gets large, we see that $\tau(n)$ is equal to 2 infinitely often since there are infinitely many primes. Furthermore, since it holds that for any prime p and $a \in \mathbb{N}$, $\tau(p^a) = a + 1$, then $\tau(n)$ can be made to be as large as desired infinitely often. However, looking at the average order of $\tau(n)$ tames down the process considerably. In order to determine this, we first need the following result—see Biography 3.4 on page 126.

Lemma 5.2 | **Hermite's Formula**

For $n \in \mathbb{N}$

$$\sum_{j=1}^{n} \tau(j) = 2 \sum_{j=1}^{\lfloor \sqrt{n} \rfloor} \left\lfloor \frac{n}{j} \right\rfloor - \lfloor \sqrt{n} \rfloor^2. \tag{5.14}$$

Proof. It is easy to see that the number of solutions to $rs = j$ for $r, s \in \mathbb{N}$ is the same as $\tau(j)$. Hence, $\sum_{j=1}^{n} \tau(j)$ is the number of solutions of the inequality

$$rs \leq n, \tag{5.15}$$

for $r, s \in \mathbb{N}$. We partition the number of solutions of the inequality (5.15) into sets for each given $s \leq n$, as follows. Define

$$\mathfrak{T}_s = \{r \in \mathbb{N} : rs \leq n\},$$

and let t_s be the cardinality of $\mathfrak{T}_s$. We now calculate t_s explicitly.

If $s \in \mathbb{N}$, $s \le n$ is fixed, then the number of solutions of $r \le \frac{n}{s}$ is clearly

$$t_s = \left\lfloor \frac{n}{s} \right\rfloor.$$

Hence,

$$\sum_{j=1}^{n} \tau(j) = \sum_{s=1}^{n} \left\lfloor \frac{n}{s} \right\rfloor. \tag{5.16}$$

Also, we can split this sum as follows.

$$\sum_{j=1}^{n} \tau(j) = \sum_{s=1}^{\lfloor \sqrt{n} \rfloor} \left\lfloor \frac{n}{s} \right\rfloor + \sum_{s=\lfloor \sqrt{n} \rfloor + 1}^{n} \left\lfloor \frac{n}{s} \right\rfloor.$$

In the second summand, we have for each $r \in \mathfrak{T}_s$, with $r \le \sqrt{n}$, that

$$\sqrt{n} < s \le n/r.$$

There are

$$\left\lfloor \frac{n}{s} \right\rfloor - \lfloor \sqrt{n} \rfloor$$

such pairs r, s, since the cardinality of the set of those $s \le \sqrt{n}$ is $\lfloor \sqrt{n} \rfloor$. Thus,

$$\sum_{s=\lfloor \sqrt{n} \rfloor + 1}^{n} \left\lfloor \frac{n}{s} \right\rfloor = \sum_{s=1}^{\lfloor \sqrt{n} \rfloor} \left(\left\lfloor \frac{n}{s} \right\rfloor - \lfloor \sqrt{n} \rfloor \right) =$$

$$\sum_{s=1}^{\lfloor \sqrt{n} \rfloor} \left\lfloor \frac{n}{s} \right\rfloor - \sum_{s=1}^{\lfloor \sqrt{n} \rfloor} \lfloor \sqrt{n} \rfloor = \sum_{s=1}^{\lfloor \sqrt{n} \rfloor} \left\lfloor \frac{n}{s} \right\rfloor - \lfloor \sqrt{n} \rfloor^2.$$

Hence,

$$\sum_{s=1}^{n} \left\lfloor \frac{n}{s} \right\rfloor = \sum_{j=1}^{n} \tau(j) = 2 \sum_{s=1}^{\lfloor \sqrt{n} \rfloor} \left\lfloor \frac{n}{s} \right\rfloor - \lfloor \sqrt{n} \rfloor^2,$$

which is Hermite's formula. □

In what follows, we remind the reader that the *big O* symbol for positive real-valued functions f and g, denoted by $f = O(g)$, means that there is a constant $c \in \mathbb{R}$ such that $f(x) < cg(x)$ for all sufficiently large x. — see [68, Appendix B].

Remark 5.3 Comparing the symbols $f \sim g$ with $f = O(g)$, we see that the former is generally weaker than the latter. For instance, from (5.12) on page 203, we may deduce that

$$\sum_{j=1}^{n} \frac{1}{j} = \log_e n + \gamma + O\left(\frac{1}{n}\right). \tag{5.17}$$

However, since it may also be deduced from that (4.13) on page 172 that

$$\sum_{j=1}^{n} \frac{1}{j} - \log_e n \sim \gamma, \tag{5.18}$$

(5.17) is a stronger statement than (5.18), the reason being that the former cannot be deduced from the latter. In fact, (5.18) is tantamount to merely saying

$$\sum_{j=1}^{n} \frac{1}{j} \sim \log_e n. \tag{5.19}$$

In other words, terms may not be transposed in a relation between asymptotically equal functions.

Now we are in a position to derive the average order for the number of divisors function.

Theorem 5.6 | **Average Order of the Number of Divisors Function**

If $n \in \mathbb{N}$ and $\tau(n)$ is the number of divisors function, then

$$\sum_{j=1}^{n} \tau(j) \sim n \log_e n, \tag{5.20}$$

and the average order of $\tau(n)$ is $\log_e n$.

Proof. From (5.16), we know that

$$\sum_{j=1}^{n} \tau(j) = \sum_{j=1}^{n} \left\lfloor \frac{n}{j} \right\rfloor,$$

and the latter equals

$$n \sum_{j=1}^{n} \frac{1}{j} + O(n) = n \log_e n + O(n),$$

since removal of the floor function introduces an error of less than 1 for each j. Note that the last equality may be deduced from (5.19). Hence,

$$\sum_{j=1}^{n} \tau(j) \sim n \log_e n$$

which is (5.20).
    Since

$$\sum_{j=1}^{n} \log_e(j) = \log_e(n!),$$

then by Stirling's formula (5.10) on page 201,

$$\sum_{j=1}^{n} \log_e(j) \sim \left(n + \frac{1}{2}\right) \log_e n - n \sim n \log_e n.$$

Hence,

$$\sum_{j=1}^{n} \tau(j) \sim \sum_{j=1}^{n} \log_e(j),$$

so by Definition (5.5), $\log_e n$ is the average order of τ. □

Remark 5.4 Although Theorem 5.6 tells us that the average order of $\tau(n)$ is $\log_e n$, this should not be interpreted as saying that almost all $n \in \mathbb{N}$ have approximately $\log_e n$ divisors. Here the term "almost all," when used in reference to $n \in \mathbb{N}$ satisfying a certain property P, means that the proportion of natural numbers *not* possessing property P for $n \le x$ is $o(x)$ — see Remark 4.2 on page 162. In other words, if $P(x)$ denotes the number of $n \le x$ satisfying property P and

$$P(x) \sim x$$

then almost all $n \in \mathbb{N}$ have property P. Indeed, it can be shown that almost all $n \in \mathbb{N}$ have approximately $(\log_e n)^{\log_e 2}$ divisors, since it holds that for any $\varepsilon > 0$ that

$$(\log_e n)^{-\varepsilon} < \frac{\tau(n)}{(\log_e n)^{\log_e 2}} < (\log_e n)^{\varepsilon}.$$

The reason that the average order of $\tau(n)$ is $\log_e n$ arises from the contributions of a small proportion of $n \in \mathbb{N}$ where $\tau(n)$ is unusually big. What this means is that for a very small minority of $n \in \mathbb{N}$, $\tau(n)$ is closer to a power of n than of $\log_e n$.

We use Lemma 5.2 and results of the last section to derive the following more accurate estimate for $\tau(n)$, which was proved by Dirichlet in 1838.

Theorem 5.7 | **A Precise Estimate for $\tau(\mathbf{n})$** |

If $n \in \mathbb{N}$ and γ is the Euler constant given by (4.13) on page 172, then

$$\sum_{j=1}^{n} \tau(j) = n \log_e n + (2\gamma - 1)n + O(\sqrt{n}).$$

Proof. From Hermite's formula (5.14),

$$\sum_{j=1}^{n} \tau(j) = 2 \sum_{j=1}^{\lfloor \sqrt{n} \rfloor} \left\lfloor \frac{n}{j} \right\rfloor - \lfloor \sqrt{n} \rfloor^2,$$

and this in turn equals

$$2n \sum_{j=1}^{\lfloor \sqrt{n} \rfloor} \frac{n}{j} - n + O(\sqrt{n}),$$

which, by (5.12) on page 203, equals

$$2n \log_e(\sqrt{n}) + 2\gamma n + O(n/\sqrt{n}) - n + O(\sqrt{n}) = n \log_e n + (2\gamma - 1)n + O(\sqrt{n}),$$

as required.   □

Remark 5.5 The value

$$\Delta(x) = \sum_{n \leq x} \tau(n) - x \log_e x - (2\gamma - 1)x$$

is called the *error* term in Theorem 5.7 on the preceding page, which says that $\Delta(x) = O(\sqrt{x})$. The problem of estimating $\Delta(x)$ is known as the *Dirichlet divisor problem*, a celebrated area of research that is largely open. The difficulty of solving this problem has led to much more complex problems involving what are called *exponential sums*, which have intimate connections with other problems such as the Riemann hypothesis. Hence, any progress on the Dirichlet divisor problem will probably have implications for a variety of other unsolved problems. Typically, estimates are of the type

$$\Delta(x) = O(x^\varepsilon).$$

Theorem 5.7 shows us that $\varepsilon = 1/2$ may be chosen. The consensus is that $\varepsilon = 1/4$ works, but this is still open. However, G.H. Hardy showed that $\varepsilon \geq 1/4$. Also, G.F. Voronoii proved, in 1903, that $\varepsilon = 1/3$ may be selected, but since that time about a century ago, there has not been much advancement. To date the best known value is

$$\varepsilon \leq \frac{131}{416} = 0.314903846\ldots$$

obtained by M.N. Huxley in 2003.

Now we turn our attention to the *sum of divisors function* $\sigma(n)$, which is the sum of all the positive divisors of n, where the irregularities are far less pronounced than those for $\tau(n)$ discussed in Remark 5.4 on the previous page.

Theorem 5.8 | **Average Order of $\sigma(n)$** |

For $n \in \mathbb{N}$,

$$\sum_{j=1}^{n} \sigma(j) = \frac{(\pi n)^2}{12} + O(n \log_e n), \tag{5.21}$$

and the average order of $\sigma(n)$ is $\pi^2 n/6$.

Proof. We know from a first course in number theory that

$$\sum_{j=1}^{n} \sigma(j) = \sum_{k=1}^{n} k \left\lfloor \frac{n}{k} \right\rfloor,$$

— see [68, Corollary 2.4, p. 110], for instance. Also, since

$$\sum_{k=1}^{n} k \left\lfloor \frac{n}{k} \right\rfloor = (1+2+3+\cdots+n) + \left(1+2+3+\cdots+\left\lfloor \frac{n}{2} \right\rfloor\right) + (1+2+3+\cdots+\left\lfloor \frac{n}{3} \right\rfloor)$$

$$+ \cdots + \left(1+2+3+\cdots+\left\lfloor \frac{n}{k} \right\rfloor\right) + \cdots + (1),$$

then

$$\sum_{j=1}^{n} \sigma(j) = \sum_{j=1}^{n} \sum_{k=1}^{\lfloor n/j \rfloor} k.$$

However, since we know that

$$\sum_{\ell=1}^{m} \ell = m(m+1)/2$$

– see [68, Theorem 1.1, p.2], for instance, then

$$\sum_{j=1}^{n} \sigma(j) = \sum_{j=1}^{n} \frac{\lfloor n/j \rfloor (\lfloor n/j \rfloor + 1)}{2},$$

and by the same reasoning as in the proof of Theorem 5.6 on page 210, the latter equals

$$\frac{1}{2} \sum_{j=1}^{n} \left(\frac{n}{j} + O(1)\right) \left(\frac{n}{j} + O(1)\right) = \frac{n^2}{2} \sum_{j=1}^{n} \frac{1}{j^2} + O\left(n \sum_{j=1}^{n} \frac{1}{j}\right) + O(n).$$

Given that (5.5) on page 198 tells us

$$\sum_{j=1}^{n} \frac{1}{j^2} = \sum_{j=1}^{\infty} \frac{1}{j^2} + O\left(\frac{1}{n}\right) = \zeta(2) + O\left(\frac{1}{n}\right) = \frac{\pi^2}{6} + O\left(\frac{1}{n}\right),$$

and since

$$O(n \sum_{j=1}^{n} 1/j) + O(n) = O(n)O(\sum_{j=1}^{n} 1/j) = O(n)O(\log_e n) = O(n \log_e n),$$

we have,

$$\sum_{j=1}^{n} \sigma(j) = \frac{\pi^2 n^2}{12} + O\left(n \log_e n\right),$$

which is (5.21). Also, since

$$\sum_{j=1}^{n} j \sim n^2/2,$$

then

$$\sum_{j=1}^{n} \sigma(j) \sim \frac{\pi^2 n^2}{12} \sim \frac{\pi^2}{6} \sum_{j=1}^{n} j = \sum_{j=1}^{n} \frac{\pi^2 j}{6},$$

and then the average order of $\sigma(n)$ is $\pi^2 n/6$. $\qquad\qquad\qquad\qquad\qquad\qquad$ $\square$

Lastly, in this section, we look at Euler's totient $\phi(n)$ from the perspective of average order. Recall that the totient is equal to the number of positive integers less than n and relatively prime to it. In what follows $\mu(n)$ denotes the Möbius function defined by

$$\mu(n) = \begin{cases} 1 & \text{if } n = 1, \\ 0 & \text{if } n \text{ is not squarefree,} \\ (-1)^k & \text{if } n = \prod_{j=1}^{k} p_j \text{ where the } p_j \text{ are distinct primes.} \end{cases} \qquad (5.22)$$

We remind the reader of a fundamental relationship between the totient and the Möbius function given by the following that we will use in the closing result,

$$\phi(n) = n \sum_{d \mid n} \frac{\mu(d)}{d} \qquad (5.23)$$

— see [68, Theorem 2.17, p. 99]. There is also a relationship between the Möbius function and the zeta function (studied in detail in §5.3) that is an important component of what follows. It is given by

$$\sum_{d=1}^{\infty} \frac{\mu(d)}{d^s} = \frac{1}{\zeta(s)}, \text{ for } s \in \mathbb{R},\ s > 1 \qquad (5.24)$$

— see [68, top formula, page 112].

Theorem 5.9 $\boxed{\text{The Average Order of the Totient}}$

For $n \in \mathbb{N}$,

$$\sum_{j=1}^{n} \phi(j) = \frac{3n^2}{\pi^2} + O(n \log_e n), \qquad (5.25)$$

and the average order of $\phi(n)$ is $6n/\pi^2$.

Proof. From (5.23), we get

$$\sum_{j=1}^{n} \phi(j) = \sum_{j=1}^{n} j \sum_{d \mid j} \frac{\mu(d)}{d} = \sum_{1 \leq dd' \leq n} d'\mu(d) = \sum_{d=1}^{n} \mu(d) \sum_{d'=1}^{\lfloor n/d \rfloor} d'$$

and the latter is equal to the following by the same reasoning as in the proof of Theorem 5.8

$$\frac{1}{2}\sum_{d=1}^{n}\mu(d)\left(\left\lfloor\frac{n}{d}\right\rfloor^2+\left\lfloor\frac{n}{d}\right\rfloor\right)=\frac{1}{2}\sum_{d=1}^{n}\mu(d)\left(\frac{n^2}{d^2}+O\left(\frac{n}{d}\right)\right)$$

$$=\frac{n^2}{2}\sum_{d=1}^{n}\frac{\mu(d)}{d^2}+O\left(n\sum_{d=1}^{n}\frac{1}{d}\right)=\frac{n^2}{2}\sum_{d=1}^{\infty}\frac{\mu(d)}{d^2}+O\left(n^2\sum_{d=n+1}^{\infty}\frac{1}{d^2}\right)+O(n\log_e n).$$

However, by (5.24) on page 214, and (5.5) on page 198,

$$\sum_{d=1}^{\infty}\frac{\mu(d)}{d^2}=\frac{1}{\zeta(2)}=\frac{6}{\pi^2}$$

and since

$$O\left(n^2\sum_{d=n+1}^{\infty}\frac{1}{d^2}\right)+O(n\log_e n)=O(n^2)O\left(\sum_{d=n+1}^{\infty}\frac{1}{d^2}\right)+O(n)O(\log_e n)$$

$$=O(n^2)O\left(\frac{1}{n}\right)+O(n)O(\log_e n)=O(n)+O(n)O(\log_e n)=O(n\log_e n),$$

then

$$\sum_{j=1}^{n}\phi(j)=\frac{3n^2}{\pi^2}+O(n\log_e n),$$

which is (5.25). Also, the same reasoning as in the proof of Theorem 5.8, we have that

$$\sum_{j=1}^{n}\phi(j)\sim\frac{3n^2}{\pi^2}\sim\frac{6}{\pi^2}\sum_{j=1}^{n}j=\sum_{j=1}^{n}\frac{6j}{\pi^2},$$

so the average order of $\phi(n)$ is $6n/\pi^2$. $\qquad\square$

Remark 5.6 An application of Theorem 5.9 is given in Exercise 5.9 on the next page where it is shown that two integers, less than $n\in\mathbb{N}$, have a probability of being relatively prime equal to $6/\pi^2$. Here the *probability* means the following. If $A(n)$ is the total number of pairs of integers less than n and $B(n)$ is the number of them that are relatively prime (in lowest terms) then the probability that any two are coprime is

$$\lim_{n\to\infty}\frac{A(n)}{B(n)}.$$

Another application is to Farey sequences— see [68, Page 239]. The number of terms in a Farey sequence of order n is $\sum_{j=1}^{n}\phi(j)+1$, so by Theorem 5.9, the number of terms in a Farey sequence of order n is approximately $3n^2/\pi^2$.

Exercises

5.8. Prove that for $x \in \mathbb{R}^+$, if $SF(x)$ denotes the number of squarefree $n \in \mathbb{N}$ with $n \le x$, then

$$\lim_{x \to \infty} SF(x) = \frac{6x}{\pi^2} + O(\sqrt{x}).$$

(*Hint: Use* (5.24) *on page 214 and the Möbius inversion formula which says: If f and g are arithmetic functions, then*

$$f(n) = \sum_{d \mid n} g(d) \ \text{for every } n \in \mathbb{N},$$

if and only if

$$g(n) = \sum_{d \mid n} \mu(d) f\left(\frac{n}{d}\right) \ \text{for every } n \in \mathbb{N}$$

— *see* [68, *Theorem 2.16, p. 98*].)

5.9. Given $n \in \mathbb{N}$ and integers x and y satisfying $1 \le x \le y \le n$, prove that the probability they are relatively prime is $6/\pi^2$.

(*Hint: See Remark 5.6 on the preceding page and note that the total number of pairs $1 \le x \le y \le n$ is equal to $n(n+1)/2$, and the number of them that are relatively prime is $\sum_{j=1}^{n} \phi(j)$.*)

5.10. Given arithmetic functions f and g related by

$$f(n) = \sum_{d \mid n} d \cdot g\left(\frac{n}{d}\right) = \sum_{d_1 d_2 = n} d_1 g(d_2),$$

and given $\sum_{n=1}^{\infty} |g(n)|n < \infty$, prove that

$$\lim_{x \to \infty} \frac{1}{x} \sum_{n \le x} f(n) = \sum_{n=1}^{\infty} \frac{g(n)}{n}.$$

Remark 5.7 *The result in this exercise is known as* Wintner's mean value theorem— *see* [104]. *The* mean value *of an arithmetic function f is defined to be*

$$\lim_{x \to \infty} \frac{1}{x} \sum_{n \le x} f(n),$$

provided the limit exists. For instance, by Exercise 5.8, the mean value of SF is $6/\pi^2$.

(*Hint: You may use the fact, following from the hint to Exercise 5.8, that*

$$\sum_{n \le x} f(n) = \sum_{d \le x} g(d) \lfloor x/d \rfloor,$$

and that

$$\lim_{x \to \infty} \frac{1}{x} \sum_{d \le x} |g(d)| = 0,$$

the latter of which follows from a result known as Kronecker's Lemma, *which states that if f is an arithmetic function and*

$$\sum_{s=1}^{\infty} \frac{f(n)}{n^s}$$

converges for a complex number s with $Re(s) > 0$, *then*

$$\lim_{x \to \infty} \frac{1}{x^s} \sum_{n \le x} f(n) = 0.$$

In particular, if

$$\sum_{n=1}^{\infty} \frac{f(n)}{n}$$

converges, then f has mean value zero since

$$\lim_{x \to \infty} (1/x) \sum_{n \le x} f(n) = 0.)$$

5.11. Find $\sum_{n \le x} |\mu(n)|$, and show that the mean value of μ^2 is $6/\pi^2$.
(*Hint: Use Theorem 5.9 on page 214.*)

5.3   The Riemann ζ-function

To see a world in a grain of sand
And a heaven in a wild flower
Hold infinity in the palm of your hand
An eternity in an hour.

from **Auguries of Innocence** (1803)
William Blake (1757–1827)
English Poet

In this section, we will be looking at infinite series, especially the renowned zeta function. To this end, we remind the reader of a few salient facts. The term *analytic*, or *holomorphic* function of a complex variable, is one which has derivatives whenever the function is defined. Also, *absolutely convergent* series are those with the property that the series formed by the absolute values of the terms converges. *Convergence* of an infinite series means that the sequence formed by the partial sums of the terms of the sequence converges, in which case this limit is the *sum* of the series. In other words, if we have an infinite series given by

$$\sum_{n=1}^{\infty} a_n,$$

then the partial sums are

$$s_m = \sum_{n=1}^{m} a_n,$$

and if

$$\lim_{m \to \infty} s_m = S \in \mathbb{R},$$

then S is the sum of the (convergent) series. Series that do not converge are said to *diverge*. Exercises 5.12–5.14 are designed to test some basic knowledge of series and establish a foundation for the establishment of some facts below.

Herein, we explore the Riemann ζ-function given for $s = a + bi \in \mathbb{C}$ with $\Re(s) = a > 1$ by

$$\zeta(s) = \sum_{n=1}^{\infty} \frac{1}{n^s} = \prod_{p=prime} (1 - p^{-s})^{-1}, \tag{5.26}$$

which we discussed briefly in Remark 5.2 on page 197, as well as in §5.2.

The last equality, which follows from Exercise 5.13 on page 227, is known as the *Euler product*, which provides a fundamental relationship between the primes and the zeta function.

The series on the left is absolutely convergent, which implies that $\zeta(s)$ is analytic on the half plane $\Re(s) > 1$. To see this we may employ Theorem 5.1 on page 193 and in addition this will provide us a formula which is a means of computationally evaluating the Riemann ζ-function as well as extending its domain of definition to the entire complex plane, with one singularity. The following proof follows the line of reasoning given in [74, Section 3.3].

Theorem 5.10 | **A Formula for $\zeta(s)$ from Euler–Maclaurin**

For $s \in \mathbb{C}$ and $\Re(s) > 1 - n$, for $n \in \mathbb{N}$, $\zeta(s)$ is convergent, except at $s = 1$, and

$$\zeta(s) = \frac{1}{s-1} + \frac{1}{2} + \sum_{j=2}^{n} \frac{B_j}{j!} s(s+1)\cdots(s+j-2)$$

$$-\frac{1}{n!}s(s+1)\cdots(s+n-1)\int_1^\infty B_n(t - \lfloor t \rfloor)t^{-s-n}dt. \qquad (5.27)$$

Proof. Let $n \in \mathbb{N}$ and set $f(x) = x^{-s}$, $a = 1$, and $b = N$ in Theorem 5.1. Since

$$f^{(n)}(x) = (-1)^n s(s+1)(s+2)\cdots(s+n-1)x^{-s-n}$$

and

$$\zeta(s) = 1 + \lim_{N\to\infty} \sum_{j=2}^{N} f(j),$$

then

$$\zeta(s) - 1 = \lim_{N\to\infty}\left[\int_1^N x^{-s}dx - \sum_{j=1}^{N}\frac{B_j}{j!}s(s+1)\cdots(s+j-2)(N^{-s+1-j}-1)\right.$$

$$\left.-\frac{1}{n!}s(s+1)\cdots(s+n-1)\int_1^N B_n(x - \lfloor x \rfloor)x^{-s-n}dx\right]$$

$$= \lim_{N\to\infty}\left[\frac{N^{1-s}-1}{1-s} + \frac{N^{-s}-1}{2} - \sum_{j=2}^{n}\frac{B_j}{j!}s(s+1)\cdots(s+j-2)(N^{-s-j+1}-1)\right.$$

$$\left.-\frac{1}{n!}s(s+1)\cdots(s+n-1)\int_1^N B_n(x - \lfloor x \rfloor)x^{-s-n}dx\right].$$

For $\Re(s) > 1$, we may pull the limit through. Thus, since $\lim_{N\to\infty} N^{1-s} = 0 = \lim_{N\to\infty} N^{-s}$, we get (5.27), the right-hand side of which converges for $\Re(s) > 1 - n$, except at $s = 1$. $\qquad \square$

Remark 5.8 To delve into some deeper complex analysis, Theorem 5.10 says that $\zeta(s)$ can be *analytically continued* to a *meromorphic function* in the whole complex plane with its only singularity a simple *pole* at $s = 1$. The principle of analytic continuation says that two analytic functions that agree on a *sufficiently dense* set are identical. A set S is said to be "dense" in a set T if the smallest "closed" set in T containing S is equal to T. Think of a *closed* set as one that contains all of its limit points. A function is "meromorphic" on a region if it is analytic there except for some "poles" which are singularities that behave like the singularity of $f(x) = 1/x^n$ at $x = 0$.

Theorem 5.10 may be employed as a useful tool to calculate the zeta function for values of s—see Exercise 5.15 on page 227 for instance. Note that we may estimate the error term via Theorem 5.2 on page 196 as follows. For $n > 1 - \Re(s)$, we have from Theorem 5.10:

$$|B_n(x - \lfloor x \rfloor)| \leq \frac{2n!}{(2\pi)^n} \sum_{j=1}^{\infty} \frac{1}{j^n} = \frac{2n!}{(2\pi)^n} \zeta(n) \leq \frac{2n!}{(2\pi)^n} \zeta(2) = \frac{n!}{12(2\pi)^{n-2}},$$

and for even n we have from (5.5) on page 198 that $|B_{2m}(x - \lfloor x \rfloor)| \leq |B_{2m}|$.

Remark 5.9 Another application of the Riemann ζ-function is to probability as discussed in Remark 5.6 on page 215. Via Exercise 5.9 on page 216, we showed that the probability of two randomly selected integers being relatively prime is approximately equal to

$$\frac{1}{\zeta(2)} = \frac{6}{\pi^2} = 0.608\ldots.$$

This is also the probability that a randomly selected integer is squarefree—see Exercise 5.11 on page 217. The reason the latter is true, in terms of the Riemann ζ-function given by the Euler product in (5.26), is that for a number to be squarefree it must not be divisible by the same prime more than once. In other words, either it is not divisible by p or it is divisible by p but not divisible by it again. Thus, the probability that an integer is not divisible by the square of a prime p equals

$$\left(1 - \frac{1}{p}\right) + \frac{1}{p}\left(1 - \frac{1}{p}\right) = 1 - \frac{1}{p^2},$$

and taking the product over all primes (assuming the independence of the divisibility by different primes) the probability then that an integer is squarefree tends to

$$\prod_{p=prime} \left(1 - p^{-2}\right) = \zeta(2)^{-1}.$$

This has a generalization, which can be shown by the same reasoning as for $n = 2$, namely that the probability that n randomly selected integers are coprime is

$$P_n \sim \zeta(n)^{-1}. \tag{5.28}$$

Thus, we may calculate $P_3 \sim 0.832$, $P_4 \sim 90/\pi^2 \sim 0.9239$, and so forth. Again using similar reasoning to the above, the probability that a randomly selected integer is cube-free, or fourth-power free, etc., is also given by (5.28). Thus, the probability that a randomly selected integer is cube-free equals roughly 83%, and that an integer is fourth-power free is roughly 92%.

A more general question still is the following. What is the probability that n randomly selected integers have greatest common divisor equal to g? Let this probability be denoted by $P_n(g)$ — see [84, page 48]. To resolve this question, let $I_N = [1, N]$ where $N \geq 1$ is real, and let $P_n^N(g)$ be the probability of selecting n random integers from I_N with gcd equal to g. Then

$$P_n^N(g) = \frac{1}{g^n} P_n^{\lfloor N/g \rfloor}(1) + o(N),$$

observing that $o(N) = 0$ if $N \in \mathbb{N}$ with $N \equiv 0 \,(\mathrm{mod}\ g)$, so,

$$P_n(g) = \lim_{N \to \infty} P_n^N(g) = \frac{1}{g^n} P_n(1) = \frac{1}{g^n \zeta(n)},$$

with thanks to Thomas Hagedorn of the College of New Jersey, USA for the idea behind the proof of the above generalization.

Now we turn to the relationship between the Riemann ζ-function and the distribution of primes, extending what is covered in [68, §1.9, pp. 65–72], to which we refer the reader for background, especially pertaining to the *Riemann hypothesis* that we will discuss with the covered material from [68] in mind. We begin by reminding the reader that $\pi(x)$ denotes the number of primes $\leq x$. The first celebrated result, the history of the proof of which is given in detail in [68], is our starting point.

Theorem 5.11 **The Prime Number Theorem (PNT)**

For $x \in \mathbb{R}^+$,

$$\pi(x) \sim \frac{x}{\log_e x}.$$

The close relationship between the Riemann ζ-function and $\pi(x)$ is given by

$$\log_e \zeta(s) = s \int_2^{\infty} \frac{\pi(x)}{x(x^s - 1)} dx, \tag{5.29}$$

for $\Re(s) > 1$—see Exercise 5.20 on page 227.

It is noteworthy that the Euler product (5.26) for the Riemann ζ-function tells us that since $\zeta(s) \to \infty$ as $s \to 1$ then there are infinitely many primes. To see this consider $\zeta(s)$ for $s \in \mathbb{R}^+$. By the series expansion in (5.26), $\zeta(s)$ diverges as $s \to 1^+$ since the harmonic series $\sum_{j=1}^{\infty} 1/j$ diverges. Actually, more can be said, namely

$$\log_e \left(\prod_{p=prime} (1 - p^{-s})^{-1} \right) = - \sum_{p=prime} \log_e(1 - p^{-s})$$

$$= \sum_{p=prime} p^{-s} + O(1) < \sum_{p=prime} p^{-1} + O(1) \text{ when } s > 1. \tag{5.30}$$

Understanding the sums $\sum_{p \leq x} p^{-1}$ is implicit in the development of Theorem 5.11. Indeed, the following predates the PNT, and follows from it.

Theorem 5.12 | **Merten's Theorem** |

$$\sum_{p \leq x} \frac{1}{p} = \log_e \log_e x + M + o(1),$$

and

$$M = \gamma + \sum_{p=prime} \left(\log_e \left(1 - \frac{1}{p} \right) + \frac{1}{p} \right),$$

where γ is Euler's constant and M is called Merten's constant.

Note that Theorem 5.12 is equivalent to the asymptotic relationship

$$\prod_{p \leq x} \left(1 - \frac{1}{p} \right) \sim \frac{e^{-\gamma}}{\log_e x}.$$

There is an equivalent formulation of Theorem 5.11 via the following function called *Merten's function*–see [68, Biography 2.4, p. 100].

$$M(x) = \sum_{n \leq x} \mu(x),$$

for any $x \in \mathbb{R}$, where μ is the Möbius function defined in (5.22) on page 214. It can be shown that Theorem 5.11 is equivalent to the following.

Theorem 5.13 | **Merten's Equivalence to the PNT** |

$$M(x) = o(x).$$

Even more, Theorem 5.11 is also equivalent to the following.

Theorem 5.14 | **Möbius' Equivalence to the PNT** |

$$\sum_{n=1}^{\infty} \frac{\mu(n)}{n} = 0.$$

The relationship between the Riemann ζ-function and Merten's function is evoked from (5.24) on page 214, namely

$$\frac{1}{\zeta(s)} = \sum_{d=1}^{\infty} \frac{\mu(d)}{d^s} = s \int_1^{\infty} \frac{M(x)}{x^{s+1}} dx,$$

for $\Re(s) > 1$. This brings us to one of the most important and celebrated outstanding problems.

Conjecture 5.1 | **The Riemann Hypothesis (RH)** |

All of the zeros of $\zeta(s)$ in the critical strip $0 < \Re(s) < 1$ lie on the *critical* line $\Re(s) = 1/2$.

In terms of the Merten's function we may reformulate Conjecture 5.1 as being equivalent to the following.

Conjecture 5.2 | **Merten's Equivalence to the RH** |

$$M(x) = O_\varepsilon\left(x^{\frac{1}{2}+\varepsilon}\right),$$

for any fixed $\varepsilon > 0$, where O_ε means that, in the big O notation, the constant depends on ε only.

Also, Riemann postulated the following in 1859, which is also equivalent to Conjecture 5.1.

Conjecture 5.3 | **Integral Equivalence to the RH** |‘

$$\pi(x) = li(x) + O\left(\sqrt{x}\log_e x\right),$$

where

$$li(x) = \int_2^\infty \frac{dt}{\log_e t},$$

called the *logarithmic integral.*

Conjecture 5.3 will hold if and only if the Riemann ζ-function does not vanish on the half plane $\Re(s) > 1/2$. In other words, Conjecture 5.1 is equivalent to the statement that the error which occurs, when $\pi(x)$ is estimated by $li(x)$, is $O(\sqrt{x}\log_e x)$.

Now we are in a position to establish a fundamental equation, which puts the above more in focus. Indeed, with Remark 5.9 on page 220 in mind, the following shows the central role that the Riemann ζ-function plays in analytic number theory via the *functional equation*, $\zeta(s) = f(s)\zeta(1-s)$, where we define $f(s)$ below.

Let $n = 3$ in (5.27) to get

$$\zeta(s) = \frac{1}{s-1} + \frac{1}{2} + \frac{B_2 s}{2} - \frac{s(s+1)(s+2)}{6}\int_1^\infty B_3(t - \lfloor t\rfloor)t^{-s-3}dt,$$

for $\Re(s) > -2$. By Exercise 5.21 on page 227,

$$\frac{s(s+1)(s+2)}{6}\int_0^1 B_3(t - \lfloor t\rfloor)t^{-s-3}dt = -\frac{B_2 s}{2} - \frac{1}{2} - \frac{1}{s-1}, \qquad (5.31)$$

so

$$\zeta(s) = -\frac{s(s+1)(s+2)}{6}\int_0^\infty B_3(t - \lfloor t\rfloor)t^{-s-3}dt.$$

Replacing $B_3(t - \lfloor t \rfloor)$ by the Fourier series in Theorem 5.2 on page 196, we get

$$\zeta(s) = -\frac{s(s+1)(s+2)}{6} \sum_{j=1}^{\infty} 12 \int_0^{\infty} \frac{\sin(2\pi jt)}{(2\pi j)^3} t^{-s-3} dt$$

and by setting $x = 2\pi jt$, this equals

$$-2s(s+1)(s+2) \sum_{j=1}^{\infty} \frac{1}{(2\pi j)^{1-s}} \int_0^{\infty} x^{-s-3} \sin x \, dx.$$

Since $\sin x = \sum_{i=0}^{\infty} (-1)^i x^{2i+1}/(2i+1)!$ is the analytic continuation of the usual trigonometric function, and converges for all $x \in \mathbb{C}$, then we may interchange the sum and integral above so the latter equals

$$-2^s \pi^{s-1} s(s+1)(s+2) \left(\int_0^{\infty} x^{-s-3} \sin x \, dx \right) \sum_{j=1}^{\infty} \frac{1}{j^{1-s}}.$$

Hence,

$$\zeta(s) = -2^s \pi^{s-1} s(s+1)(s+2) \left(\int_0^{\infty} x^{-s-3} \sin x \, dx \right) \cdot \zeta(1-s). \qquad (5.32)$$

In order to complete the derivation of the functional equation, we need the following concept due to Euler.

Definition 5.6 | The Gamma Function |

For $s \in \mathbb{C}$ and $\Re(s) > 0$, $\Gamma(s) = \int_0^{\infty} e^{-t} t^{s-1} dt$ is called the *gamma function*.

We will employ two well-known formulas for the gamma function given as follows.

For $0 < \Re(z) < 1$, the *Wolfskehl equation*—see Biography 5.6 on page 228 —is given by

$$\left(\sin \frac{\pi z}{2} \right) \cdot \Gamma(1-z) = z \int_0^{\infty} y^{-z-1} \sin y \, dy, \qquad (5.33)$$

a formula known since 1886—see [105], and

$$(-z)\Gamma(-z) = \Gamma(1-z) \qquad (5.34)$$

—see Exercise 5.23 on page 227.
Now we are ready for the functional equation.

Biography 5.5 Andrew John Wiles (1953–) *was born on April 11, 1953 in Cambridge, England. When he was merely ten years old he had an interest in FLT. In 1971, he entered Merton College, Oxford and achieved his B.A. in 1974, after which he entered Clare College, Cambridge and studied under John Coates, obtaining his doctorate in 1980. However, he did not work on FLT at that time. In 1981, he took a position at the Institute for Advanced Study at Princeton and was appointed professor in 1981 there. Wiles learned, in the mid 1980s, that the works of G. Frey and K. Ribet established that FLT would follow from the Shimura–Taniyama conjecture, namely that every elliptic curve defined over the rational numbers is modular. Eventually, Wiles proved that all semistable elliptic curves defined over the rational numbers are modular, from which FLT follows. On June 23, 1993, he announced he had a proof of FLT and wrote up the results for publication. However, a subtle error was discovered. Over the next year, with help from R. Taylor, he eventually filled the gap and the proof was published in the* Annals of Mathematics *in 1995. Wiles commented: "There's no other problem that will mean the same to me. I had this very rare privilege of being able to pursue in my adult life what had been my childhood dream. I know it's a rare privilege but I know if one can do this it's more rewarding than anything one can imagine." See §10.4 for details.*

Theorem 5.15 | **Riemann's Functional Equation for $\zeta(\mathrm{s})$**

For $s \in \mathbb{C}$, $\zeta(s) = 2^s \pi^{s-1} \Gamma(1-s)\zeta(1-s) \cdot \left(\sin \frac{\pi s}{2}\right)$.

Proof. From (5.32), we have to show only that

$$-s(s+1)(s+2)\int_0^\infty x^{-s-3}\sin x \, dx = \left(\sin \frac{\pi s}{2}\right)\Gamma(1-s).$$

To this end, we employ (5.33)–(5.34) as follows,

$$-s(s+1)(s+2)\int_0^\infty x^{-s-3}\sin x \, dx = s(s+1)\left(\sin \frac{\pi s}{2}\right)\Gamma(1-(s+2))$$

$$= s(s+1)\left(\sin \frac{\pi s}{2}\right)\Gamma(-1-s) = -s\left(\sin \frac{\pi s}{2}\right)(-(s+1))\Gamma(-(s+1)))$$

$$= -s\left(\sin \frac{\pi s}{2}\right)\Gamma(-s) = \left(\sin \frac{\pi s}{2}\right)(-s\Gamma(-s)) = \left(\sin \frac{\pi s}{2}\right)\Gamma(1-s),$$

and we have our functional equation. □

Note that the standard form for the functional equation is given by

$$\pi^{-s/2}\Gamma\left(\frac{s}{2}\right)\zeta(s) = \pi^{-(1-s)/2}\Gamma\left(\frac{1-s}{2}\right)\zeta(1-s),$$

which can be derived from the form in Theorem 5.15 via *Legendre's duplication formula* given by

$$\Gamma(2z) = (2\pi)^{-1/2}2^{2z-1/2}\Gamma(z)\Gamma\left(z + \frac{1}{2}\right). \tag{5.35}$$

Remark 5.10 The functional equation is valid for all complex numbers s where both sides are defined. We know that $\zeta(s)$ has no zeros for $\Re(s) \geq 1$ and has only trivial zeros for $\Re(s) \leq 0$, which correspond to poles of $\Gamma(s/2)$, and has infinitely many zeros on the critical strip $0 < \Re(s) < 1$. We may define a related function, which shows symmetry properties more readily than does Theorem 5.15. If we define

$$\xi(s) = \pi^{-s/2}\Gamma(s/2)\zeta(s),$$

then by Exercise 5.24 on the facing page,

$$\xi(s) = \pi^{-(1-s)/2}\Gamma\left(\frac{1-s}{2}\right)\zeta(1-s) = \xi(1-s) \tag{5.36}$$

showing that $\xi(s)$ is symmetric about the critical line $\Re(s) = 1/2$. Note that $\xi(s)$ is analytic on the whole plane (such functions are called *entire*), since the factor of $s - 1$ eliminates the pole of $\zeta(s)$ at $s = 1$. (Often $\xi(s)$ is called the *completed zeta function*.) The functional equation given in Theorem 5.15 shows that if s is a zero in the critical strip, then so is $1 - s$, since by Theorem 5.10 on page 219, zeros occur in complex conjugate pairs. So if it were to be (incredibly) that the Reimann hypothesis is false, then zeros in the critical strip that are not on the critical line would occur in four-tuples corresponding to vertices of rectangles in the complex plane.

The zeros of the ζ-function are intimately connected with the distribution of primes. If $\mathfrak{U}$ denotes the upper bound of the real parts of the zeros of $\zeta(s)$, with $1/2 \leq \mathfrak{U} \leq 1$, then $|\pi(x) - \mathrm{li}(x)| \leq cx^{\mathfrak{U}}\log_e x$ for a constant $c \in \mathbb{R}^+$. The Riemann hypothesis is tantamount to $\mathfrak{U} = 1/2$.

Furthermore, as discussed on page 178 in reference to transcendence, the values of $\zeta(2n+1)$ are largely a mystery, with the exception of a notable—Apéry's constant

$$\zeta(3) = \frac{5}{2}\sum_{j=1}^{\infty}\frac{(-1)^{j-1}}{j^3\binom{2j}{j}} \notin \mathbb{Q}.$$

Exercises

5.12. Suppose that $f(n)$ is a multiplicative arithmetic function and the series $S = \sum_{n=1}^{\infty} f(n)$ is absolutely convergent. Prove that the product

$$P = \prod_{p=prime}\left(\sum_{j=0}^{\infty} f(p^j)\right)$$

is also absolutely convergent and $S = P$. (Recall that an *arithmetic multiplicative function* $f : \mathbb{N} \mapsto \mathbb{C}$ where $f(ab) = f(a)f(b)$ when $\gcd(a, b) = 1$.)

(*Hint: Set $\mathcal{S}_N = \{n \in \mathbb{N} : n \prod_{j=1}^{n} p^{a_j}$ where $a_j \geq 0\}$, and reformulate S in terms of $\mathcal{S}_N$. Then look at the limit as n goes to infinity of $|S - P|$.*)

5.13. Suppose that $f(n)$ is a completely multiplicative arithmetic function and the series S in Exercise 5.12 is absolutely convergent. Prove that

$$S = \prod_{p=prime} \frac{1}{1 - f(p)}.$$

(Recall that a *completely multiplicative* function f is one for which $f(ab) = f(a)f(b)$ for any $a, b \in \mathbb{N}$.) (*Hint: Prove that $|f(p)| < 1$ for all primes p.*)

5.14. If $f(n)$ is a multiplicative function, $s \in \mathbb{C}$ and the following series is absolutely convergent,

$$\sum_{n=1}^{\infty} f(n)n^{-s}, \tag{5.37}$$

then

$$\sum_{n=1}^{\infty} f(n)n^{-s} = \prod_{p=prime} \sum_{j=0}^{\infty} f(p^j)p^{-js}.$$

(*Hint: Use Exercise 5.12.*) (*The series in* (5.37) *is called a* Dirichlet series, *a special case of which is our Riemann ζ-function given in* (5.26) *on page 218.*)

5.15. Use Theorem 5.10 on page 219 to evaluate $\zeta(-k)$ for any $k \in \mathbb{N}$. (*Hint: Use Exercise 5.5 on page 206 once a formulation from Theorem 5.10 is obtained.*)

5.16. Prove that for any $N \in \mathbb{N}$, $\zeta(1 - 2N) = -B_{2N}/(2N)$. (*Hint: Use Exercise 5.15.*)

5.17. Prove that $\zeta(-2N) = 0$ for any $N \in \mathbb{N}$, called the *trivial zeros* or *real zeros* of the Riemann ζ-function. (*Hint: Use Exercise 5.15.*)

5.18. Prove that $\lim_{s \to 1}(s - 1)\zeta(s) = 1$.

5.19. Prove that $\zeta(0) = -1/2$.

5.20. Prove that (5.29) on page 221 holds.

5.21. Prove that (5.31) on page 223 holds.

5.22. Prove that $\Gamma(s) = (s - 1)\Gamma(s - 1)$. (*Hint: Use integration by parts on Definition 5.6 on page 224 for a real argument.*)

5.23. Establish formula (5.34) on page 224. (*Hint: Use Exercise 5.22.*)

5.24. Prove the formula for $\xi(s)$ given in (5.36), displayed on page 226.

(*Hint: You may use the formula $\Gamma(z)\Gamma(1 - z) = \pi/\sin \pi z$—see* [101, *Formula* (25), *page 697*]), *as well as* (5.35) *on page 226.*

5.25. Prove that for any $n \in \mathbb{N}$, $\Gamma(n) = (n-1)!$. (*Hint: Use Exercise 5.22.*)

Biography 5.6 Paul Wolfskehl (1856–1906) *was born on June 30, 1856 in Darmstadt, Germany. He acheived a doctorate in medicine around 1880. (It is difficult to be accurate since documentation of some parts of his life do not exist.) However, he suffered from multiple sclerosis (MS) and decided to leave medicine for the more solitary study of mathematics. In 1881, in Berlin, he began his mathematical journey. He was deeply influenced by the lectures of Kummer in 1883–84, and largely due to that connection, he decided to study number theory. Indeed, Wolfskehl himself gave lectures in number theory at the Institute of Technology in Darmstadt starting in 1887. However, his MS worsened and he was completely paralyzed by 1890, giving up his lectures there. In January of 1905, he added to his will "whosoever first succeeds in proving the great Theorem of Fermat" would receive 100,000 marks. He entrusted the Royal Society of Science in Göttingen with the money and with the task of judging and awarding the prize. This speaks to the influence that Kummer's work must have had on him in Kummer's failed attempts to prove Fermat's last theorem (FLT). Wolfskehl died on September 13, 1906. On June 27, 1908, the Göttingen Royal Society of Science published their conditions for awarding the prize. Ironically, exactly eighty-nine years—to the day—later, the prize was awarded to Andrew Wiles for his solution of FLT on June 27, 1997, a total of DM 75,000. The value had decreased due to the hyperinflation of the Weimar Republic in the early 1920s. See* [68, Biography 1.10, p. 38] *for background on FLT and the life of Fermat.*

Chapter 6

Introduction to p-Adic Analysis

> *The Analytical Engine weaves algebraic patterns just as the Jacquard loom[a]*
> *weaves flowers and leaves.*
>
> from Luigi Manabrea's **Sketch of the Analytical Engine invented**
> **by Charles Babbage (1843)** translated and annotated by **Ada Lovelace**
> **(1815–1852)**.
>
> **English mathematician, and daughter of Lord Byron**
>
> ---
>
> [a]Joseph Marie Jacquard was a silk-weaver, who invented an improved textile loom in
> 1801. Jacquard's loom used interchangeable punched cards that controlled the weaving of
> the cloth so that any desired pattern could be obtained automatically to produce beautiful
> patterns in a style previously accomplished only with very hard manual labour. These
> punched cards were adopted by the pioneer English inventor Charles Babbage as an input-
> output medium for his proposed "analytical engine." They were eventually used as a means
> of inputting data into digital computers but were later replaced by electronic devices.

6.1 Solving Modulo p^n

The topic of this chapter is due to Hensel. The theory of p-adic numbers is
rich with numerous applications, not only to number theory, but also to algebra
in general, as well as to algebraic functions and algebraic geometry. This section
is devoted to motivating the definitions of the theory by starting with elementary
congruential arithmetic—see [68, Chapter 2]. In particular, we look at integral
polynomial congruences

$$f(x) \equiv 0 \pmod{p^k}, \text{ for } k \in \mathbb{N} \tag{6.1}$$

for a prime p. The goal is to begin with $k = 1$ and build upon solutions of
(6.1) for successively higher powers of p, then show how this translates into a

power series in p that will be the foundation for the theory. In order to do this, we call upon the pioneering work of Hensel, and remind the reader of his fundamental result presented in an introductory course in number theory, such as [68, Theorem 2.24, p. 115].

Lemma 6.1 │ Hensel's Lemma │

Let $f(x)$ be an integral polynomial, p a prime, and $k \in \mathbb{N}$. Suppose that $r_1, r_2, \ldots, r_m$ for some $m \in \mathbb{N}$ are all of the incongruent solutions of $f(x)$ modulo p^k, where $0 \le r_i < p^k$ for each $i = 1, 2, \ldots, m$. If $a \in \mathbb{Z}$ is such that

$$f(a) \equiv 0 \pmod{p^{k+1}} \text{ with } 0 \le a < p^{k+1}, \tag{6.2}$$

then there exists $q \in \mathbb{Z}$ such that

(a) for some $i \in \{1, 2, \ldots, m\}$, $a = qp^k + r_i$ with $0 \le q < p$, and

(b) $f(r_i) + qf'(r_i)p^k \equiv 0 \pmod{p^{k+1}}$.

Additionally, if $f'(r_i) \not\equiv 0 \pmod{p}$, then

$$f(qp^k + r_i) \equiv 0 \pmod{p^{k+1}} \tag{6.3}$$

has a unique solution for the value of q given by

$$q \equiv -\frac{f(r_i)}{p^k}(f'(r_i))^{-1} \pmod{p}, \tag{6.4}$$

with $(f'(r_i))^{-1}$ being a multiplicative inverse of $f'(r_i)$ modulo p.

If $f'(r_i) \equiv 0 \pmod{p}$ and $f(r_i) \equiv 0 \pmod{p^{k+1}}$, then all values of $q = 0, 1, 2, \ldots, p-1$ yield incongruent solutions to (6.3).

If $f'(r_i) \equiv 0 \pmod{p}$ and $f(r_i) \not\equiv 0 \pmod{p^{k+1}}$, then $f(x) \equiv 0 \pmod{p^{k+1}}$ has no solutions.

Remark 6.1 Note that in Hensel's Lemma for $k = m$,

$$q \equiv -\frac{f(r_{m-1})}{p}(f'(r_{m-1}))^{-1} \pmod{p}$$

uniquely determines q and

$$x = r_1 + r_2 p + r_3 p^2 + \cdots + r_m p^{m-1} \tag{6.5}$$

is a solution of (6.1) for $k = m$.

Example 6.1 Consider $f(x) = x^3 + 5x^2 + 1$ and solve for $f(x) \equiv 0 \pmod{7^3}$. By inspection, we see that $x = 1 = r_1$ is a solution of $f(x) \equiv 0 \pmod 7$. Also, we observe that $f'(1) \equiv -1 \pmod 7$, so we set $r_2 = 1 + 7q$, where $k = 1$ and q is uniquely determined by

$$q \equiv -\frac{f(r_1)}{p^k}(f'(r_1))^{-1} \equiv -1 \cdot (13)^{-1} \equiv -6 \equiv 1 \pmod 7.$$

Thus, $r_2 = 8$ we set $r_3 = 8 + 7^2 q$, where $k = 2$ and q is uniquely determined by

$$q \equiv -\frac{f(r_2)}{p^k}(f'(r_2))^{-1} \equiv -3 \cdot 6 \equiv 3 \pmod 7,$$

so $r_3 \equiv 155 \pmod{7^3}$, which is the solution we sought to find. Moreover, since $m = 1$ this is the only such solution.

We need not stop the process illustrated in Example 6.1, since it may be continued indefinitely to obtain a power series in p,

$$x = \sum_{j=0}^{\infty} r_{j+1}p^j = r_1 + r_2 p + \cdots + r_j p^{j-1} + \cdots, \text{ where } 0 \le r_j < p, \quad (6.6)$$

which we call a *p-adic solution* to $f(x) \equiv 0 \pmod p$. The power series solutions may be approximated to higher degrees of accuracy as we solve modulo p^k for successively higher powers k. The values in (6.6) are known, formally, as *p-adic numbers*. However, in the most general sense of the term, we allow for a finite number of negative powers k. Therefore, a *p-adic number* is formally an expression of the form

$$r_{-m}p^{-m-1} + \cdots + r_1 + r_2 p + \cdots + r_n p^{n-1} + \cdots \text{ for } m, n \in \mathbb{N}. \quad (6.7)$$

The reader will note that this is akin to binary expansions of $\alpha \in \mathbb{R}$ given by

$$a_{-m}2^{-m-1} + \cdots + a_0 2^{-1} + a_1 + \cdots + a_n 2^{n-1} + \cdots \text{ where for } 0 \le a_j < 2,$$

or decimal expansions or expansions to any base $b > 1$, given the base representation theorem—see [68, Theorem 1.5, p. 8]. Indeed, if $q \in \mathbb{Q}$, then the *p-adic representation* of q is the p-adic solution of $x = q$, and if $q = z \in \mathbb{N}$ this is just the representation of z in base p. Thus, the methodology for finding a p-adic representation of z is to divide it by p and set $q_0 = \lfloor z/p \rfloor$, and write $z = q_0 p + z_0$ where $0 \le z_0 < p$. Then, divide q_0 by p and write $q_1 = \lfloor q_0/p \rfloor$, so $q_0 = q_1 p + z_1$ with $0 \le z_1 < p$, and $z = z_0 + z_1 p + q_1 p^2$. Continuing in this fashion, we get the unique p-adic representation of z, namely

$$z = z_0 + z_1 p + z_2 p^2 + \cdots + z_\ell p^\ell.$$

Note that the addition and subtraction of p-adic numbers is just obtained in the usual way by increasing the next coefficient by 1 if a given coefficient is greater than p such as $(5+3)5^2 = 3 \cdot 5^2 + 1 \cdot 5^3$. Similarly, when subtracting, if a given coefficient becomes negative, we "borrow" from the next term such as $(2-4)3^2 + (2-1)3^3 = (2-4+3)3^2 + (2-2)3^3 = 3^2$. Also, multiplication of p-adic numbers is just formal multiplication of power series, allowing for shifting terms to ensure all coefficients are nonnegative and less than p.

Example 6.2 A 7-adic solution of $3x = 5$ is given by

$$4 + 2 \cdot 7 + 2 \cdot 7^2 + 2 \cdot 7^3 + 2 \cdot 7^3 + 2 \cdot 7^4 + \cdots .$$

Looking at a power series expansion of $3x$, we have

$$3(4 + 2 \cdot 7 + 2 \cdot 7^2 + 2 \cdot 7^3 + 2 \cdot 7^3 + 2 \cdot 7^4 + \cdots) = 12 + 6 \cdot 7 + 6 \cdot 7^2 + 6 \cdot 7^3 + 6 \cdot 7^3 + 6 \cdot 7^4 + \cdots$$

$$= 5 + 7 + 6 \cdot 7 + 6 \cdot 7^2 + 6 \cdot 7^3 + 6 \cdot 7^3 + 6 \cdot 7^4 + \cdots = 5 + 0 \cdot 7 + 7 \cdot 7^2 + 6 \cdot 7^3 + 6 \cdot 7^3 + 6 \cdot 7^4 + \cdots$$

$$= 5 + 0 \cdot 7 + 0 \cdot 7^2 + 7 \cdot 7^3 + 6 \cdot 7^3 + 6 \cdot 7^4 + \cdots = 5 + 0 \cdot 7 + 0 \cdot 7^2 + 0 \cdot 7^3 + 7 \cdot 7^3 + 6 \cdot 7^4 + \cdots$$

$$= 5 + 0 \cdot 7 + 0 \cdot 7^2 + 0 \cdot 7^3 + 7 \cdot 7^4 + \cdots = 5 + 0 \cdot 7 + 0 \cdot 7^2 + 0 \cdot 7^3 + 0 \cdot 7^4 + \cdots = 5.$$

Example 6.3 A 5-adic solution to $x^2 = 11$ is

$$1 + 5 + 2 \cdot 5^2 + 2 \cdot 5^5 + 3 \cdot 5^7 + 3 \cdot 5^8 + 2 \cdot 5^9 + 5^{11} + 4 \cdot 5^{12} + 3 \cdot 5^{13} + 2 \cdot 5^{14} + 3 \cdot 5^{15} + 3 \cdot 5^{16} + \cdots ,$$

which corresponds to the positive square root. Another 5-adic solution corresponding to the negative square root is given by

$$4 + 3 \cdot 5 + 2 \cdot 5^2 + 4 \cdot 5^3 + 4 \cdot 5^4 + 2 \cdot 5^5 + 4 \cdot 5^6 + 5^7 + 5^8 + 2 \cdot 5^9 + 4 \cdot 5^{10} + 3 \cdot 5^{11} + 5^{13} + 2 \cdot 5^{14} + \cdots .$$

Exercises

6.1. Find all solutions of $x^3 + 3x^2 + 12 \equiv 0 \pmod{7^2}$.

6.2. Find all solutions of $x^3 + 4x + 1 \equiv 0 \pmod{5^3}$.

6.3. Find all solutions of $x^3 + x^2 + x + 1 \equiv 0 \pmod{13^3}$.

6.4. Find all solutions of $x^3 + x^2 - 11 \equiv 0 \pmod{17^3}$.

6.5. Find the 7-adic solution to $2x = 3$.

6.6. Find the 11-adic solution to $2x = 3$.

6.7. Find the 5-adic solution to $x^3 + 4x + 1 = 0$.

6.8. Find the 13-adic solution to $x^3 + x^2 + x + 1 = 0$.

6.2 Introduction to Valuations

> *Eureka!* [I've got it!]
> from Preface, Section 10 of **Vitruvius Pollio** *De Architectura* **book 9**
> **Archimedes (c. 287–212 B.C.)**
> **Greek mathematician and philosopher**

In this section, we address the problem of convergence of the power series we considered in §6.1. Indeed if we look at Example 6.2 on the preceding page, then we see that we are getting higher and higher powers of 7 as we zero out the previous terms. Thus, we need a formal definition of the notion.

Definition 6.1 | Valuations Over $\mathbb{Q}$ |

If v is a function mapping $\mathbb{Q}$ to $\mathbb{Q}$, satisfying the following conditions,

(a) $v(x) \geq 0$ with equality if and only if $x = 0$,

(b) $v(xy) = v(x)v(y)$ for any $x, y \in \mathbb{Q}$,

(c) $v(x + y) \leq v(x) + v(y)$ for any $x, y \in \mathbb{Q}$,

then v is called a *valuation* on $\mathbb{Q}$.

Two important types of valuations are isolated as follows.

Definition 6.2 | Absolute Value on a Field |

An *absolute value on a field* F is a function $|\cdot| : F \mapsto \mathbb{R}$ satisfying each of the following.

(a) $|x| \geq 0$ for all $x \in F$ and $|x| = 0$ if and only if $x = 0$.

(b) $|x \cdot y| = |x| \cdot |y|$ for all $x, y \in F$.

(c) $|x + y| \leq |x| + |y|$ for all $x, y \in F$. (Triangle inequality)

If the triangle inequality can be replaced by the condition

$$|x + y| \leq \max\{|x|, |y|\} \text{ for all } x, y \in F, \tag{6.8}$$

then the absolute value is said to be *non-Archimedean*, and otherwise it is called *Archimidean*.

Definition 6.3 | p-Adic Absolute Value and Valuations |

Let $x \in \mathbb{Q}$, and set

$$x = \pm \prod_{p=prime} p^{\nu_p(x)}, \text{ where } \nu_p(x) \in \mathbb{Z},$$

then for a fixed prime p, observing that there are only finitely many of the $v_p(x)$ that are not zero, there exist nonzero integers a, b such that

$$x = \frac{a}{b} \cdot p^{\nu_p(x)} \text{ with } ab \not\equiv 0 \pmod{p}. \tag{6.9}$$

Then the *p-adic absolute value on* $\mathbb{Q}$ is given by

$$|x|_p = \begin{cases} p^{-\nu_p(x)} & \text{if } x \neq 0, \\ 0 & \text{if } x = 0. \end{cases}$$

The function that maps $x \mapsto \nu_p(x)$ is called a *p-adic valuation.*

Example 6.4 If we define $v(x) = 1$ if $x \neq 0$ and $v(0) = 0$, this is known as the *unitary, identical,* or *trivial* absolute value, which is non-Archimedean. See Exercise 6.10 on page 238 for some more elementary properties of valuations.

Example 6.5 The ordinary absolute value is given by

$$|x|_\infty = \begin{cases} x & \text{if } x \geq 0, \\ -x & \text{if } x < 0, \end{cases}$$

where the symbol $|\cdot|_\infty$ is used in the context of "*p*-adic numbers" which we define below, where we typically allow $p = \infty$ to denote the ordinary absolute value in what follows.

Remark 6.2 Given a fixed prime p any rational number x may be uniquely written in the form (6.9). By Exercise 6.9, the p-adic absolute value $|\cdot|_p$ is indeed an absolute value in the sense of Definition 6.2. Hensel's idea was to ensure that the number x has *small* p-adic absolute value precisely when x is divisible by a *large* power of p, so the magnitude of x has no effect in this context. The p-adic absolute value gives us an arithmetical notion of "distance." Two rationals are close together under the p-adic absolute value if the numerator of their difference has a power of p as a factor. Indeed, if we look only at integers, then the following holds. If p is a prime and $z, w \in \mathbb{Z}$, then

$$|z - w|_p \leq 1/p^n \text{ if and only if } z \equiv w \pmod{p^n}$$

for some nonnegative integer n—see Exercise 6.18.

Definition 6.4 | **Cauchy Sequences**

Let $|\cdot|_p$ be a p-adic absolute value on $\mathbb{Q}$ for $p \leq \infty$. Then a sequence of rational numbers $\{q_j\}_{j=1}^\infty$ is called a *Cauchy sequence* (relative to $|\cdot|_p$, also called a *p-adic Cauchy sequence*) if for every rational $\varepsilon > 0$, there exists an integer $n = n(\varepsilon)$ such that

$$|q_j - q_k|_p < \varepsilon \text{ for all } j, k > n.$$

A Cauchy sequence is called a *null sequence* if $\lim_{j\to\infty} |q_j|_p = 0$. Two Cauchy sequences $\{q_j\}_{j=1}^{\infty}$ and $\{q_j'\}_{j=1}^{\infty}$ are said to be *equivalent* if they differ by a null sequence, namely if

$$\lim_{j\to\infty} |q_j - q_j'|_p = 0,$$

and we denote equivalence of sequences by

$$\{q_j\} \sim \{q_j'\},$$

using the notation $\overline{\{q_j\}}$ for the class containing $\{q_j\}$.

Definition 6.4 tells us that a sequence is Cauchy if the terms get arbitrarily close to each other with respect to the p-adic absolute value.

By Exercises 6.12– 6.13 on page 238, Cauchy sequences are partitioned into equivalence classes since $\{q_j\} \sim \{q_j'\}$ is an equivalence relation. Let

$$\mathbb{Q}_p = \{\overline{\{q_j\}} : q_j \in \mathbb{Q} \text{ and } \{q_j\} \text{ is a Cauchy sequence}\}.$$

If $p = \infty$, then an equivalence class $\overline{\{q_j\}}$ is called a *real number*, and if $p < \infty$, then it is called a *p-adic number*. (In fact Cantor employed Cauchy sequences to provide a constructive definition of $\mathbb{R}$ without using Dedekind cuts, which are more difficult to manipulate than Cauchy sequences.) Also, by the aforementioned exercises,

$$\overline{\{q_j\}} \cdot \overline{\{q_j'\}} = \overline{\{q_j \cdot q_j'\}}$$

and

$$\overline{\{q_j\}} \pm \overline{\{q_j'\}} = \overline{\{q_j \pm q_j'\}}$$

are well defined. This makes the classes of Cauchy sequences into a commutative ring with identity. Here, the class of the null sequence is the zero element, and the constant sequence $q_j = 1$ for all $j \in \mathbb{N}$ provides the unity element. It follows that when $\overline{\{q_j'\}} \neq \overline{\{0\}}$, then

$$\overline{\{q_j\}} \cdot \overline{\{q_j'\}}^{-1} = \overline{\{q_j \cdot (q_j')^{-1}\}},$$

so the classes, excluding the null sequence class, form a multiplicative commutative group. Hence, $\mathbb{Q}_p$ is a field, called the *field of p-adic numbers*. When $p = \infty$, then $\mathbb{Q}_p = \mathbb{R}$. The p-adic fields $\mathbb{Q}_p$ are known as *completions* of $\mathbb{Q}$ with respect to the p-adic valuation. This larger field contains $\mathbb{Q}$.

We define

$$\overset{(p)}{\underset{j\to\infty}{\lim}} \{q_j\} = \overline{\{q_j\}}, \text{ and } \overset{(p)}{\underset{j\to\infty}{\lim}} |q_j|_p = |\overline{\{q_j\}}|_p \tag{6.10}$$

and say the *sequence* $\{q_j\}_{j=1}^{\infty}$ *converges p-adically*. The p-adic field is *complete* in the sense that all Cauchy sequences converge to a p-adic number. Observe that $\mathbb{Q}$ is not complete with respect to the p-adic valuation because Cauchy sequences may have irrational limit points. For instance, the well-known sequence $q_j = F_j/F_{j-1}$ where F_j is the j-th Fibonacci number converges to the golden ratio

$$\mathfrak{g} = (1 + \sqrt{5})/2,$$

which is clearly irrational—see [68, p. 4, ff.]. Note as well that exponential and trigonometric functions such as $f(x) = e^x$ and $g(x) = \sin x$ are known to be irrational for any rational value of x, but may be defined as the limit of a Cauchy sequence via Maclaurin series—see §5.1. In other words, there are "holes" in $\mathbb{Q}$, missing some points to which Cauchy sequences converge in $\mathbb{R}$. We filled those holes by *completing* $\mathbb{Q}$ to the fields $\mathbb{Q}_p$ for each $p \leq \infty$, a much larger field. In the case of $p = \infty$, we get $\mathbb{Q}_p = \mathbb{R}$, so we can build the real numbers by using the rationals and the notion of distance in the reals provided by the usual absolute value function. The notion of distance provided by p-adic valuations is also an absolute value, as noted in Remark 6.2 on page 234. When $\nu_p(x) \geq 0$, then x is called a *p-adic integer*, and the set $\mathcal{O}_p$ of all p-adic integers is easily checked to be an integral domain whose units are the integers with $\nu_p(x) = 1$, and $\mathbb{Q}_p$ is the quotient field of $\mathcal{O}_p$. For $p < \infty$, $\mathbb{Q}_p$ is the non-Archimedean analogue of $\mathbb{R}$.

This is summarized in the following.

Theorem 6.1 | **The** p-**Adic Fields and Domains**

For any prime $p \leq \infty$, $\mathbb{Q}_p$, *the field of* p *-adic numbers, forms a field where* $\mathbb{Q}_p = \mathbb{R}$ *when* $p = \infty$ *and each of the p-adic fields, for* $p < \infty$, *has an isomorphic copy of* $\mathbb{Q}$ *via the embedding*

$$q \in \mathbb{Q} \mapsto (q, q, q, \ldots) \in \mathbb{Q}_p,$$

(where $(q, q, q, \ldots)$ is a Cauchy sequence). *Furthermore, if*

$$\mathcal{O}_p = \{x \in \mathbb{Q}_p : \nu_p(x) \geq 0\} = \{x \in \mathbb{Q}_p : |x|_p \leq 1\},$$

then $\mathcal{O}_p$ *is an integral domain and the units in* $\mathcal{O}_p$ *are those for which* $|x|_p = 1$, *and* $\mathbb{Q}_p$ *is the quotient field of* $\mathcal{O}_p$.

In order to classify valuations, we need the following concept.

Definition 6.5 | **Equivalent Valuations**

If υ and υ' are valuations, then we say that υ and υ' are *equivalent* provided that for any $x, y \in \mathbb{Q}$,

$$\upsilon(x) < \upsilon(y) \text{ if and only if } \upsilon'(x) < \upsilon'(y).$$

Theorem 6.2 | **Equivalent Valuations are Powers**

A nontrivial valuation v is equivalent to a valuation v' on $\mathbb{Q}$, if and only if there exists a positive real number r such that $v' = v^r$.

Proof. Since v is nontrivial, then there exists a $q_0 \in \mathbb{Q}$ such that $v(q_0) \neq 0, 1$. If $v(q_0) > 1$, then by property (b) of Definition 6.1 on page 233, $v(1/q_0) < 1$. Hence, we may assume without loss of generality that $0 < v(q_0) < 1$. Let q be an arbitrarily chosen nonzero rational number and set

$$\mathcal{S} = \{(m, n) \in \mathbb{N}^2 : v(q_0^m) = v(q_0)^m < v(q)^n = v(q^n)\},$$

where the equalities in the definition of $\mathcal{S}$ above also come from property (b). Thus, if $(m, n) \in \mathcal{S}$, then

$$\frac{m}{n} > \frac{\log_e v(q)}{\log_e v(q_0)}.$$

If v' is equivalent to v, then for any nonzero $q \in \mathbb{Q}$,

$$\frac{\log_e v(q)}{\log_e v(q_0)} = \frac{\log_e v'(q)}{\log_e v'(q_0)},$$

so there exists a constant $r \in \mathbb{R}^+$, depending solely upon v and v', such that

$$\frac{\log_e v'(q)}{\log_e v(q)} = \frac{\log_e v'(q_0)}{\log_e v(q_0)} = r > 0.$$

Hence, since we know from elementary calculus that

$$\frac{\log_e v'(q)}{\log_e v(q)} = \log_{v(q)}(v'(q)),$$

then $v'(q) = v(q)^r$.

Conversely, if $v' = v^r$ for some $r \in \mathbb{R}+$, then $v'(x) < v'(y)$ if and only if $v^r(x) < v^r(y)$ if and only if $v(x) < v(y)$, whch secures the result. $\square$

Remark 6.3 Exercise 6.19 on the following page tells us that for p a prime, all triangles are p-adically isosceles. This shows the difference between Archimedean and non-Archimedean geometry. We explore this difference in more depth in §6.3.

Exercises

6.9. Prove that $|\cdot|_p$ given in Definition 6.3 on page 233 is an absolute value in the sense of Definition 6.2, and that the absolute value is non-Archimedean.

6.10. Prove that if v is a valuation on $\mathbb{Q}$, then $v(1) = v(-1) = 1$, $v(-x) = v(x)$ for any $x \in \mathbb{Q}$, and if $n \in \mathbb{N}$ then $v(n) \leq n$.

6.11. Prove that all Cauchy sequences are bounded. In other words, if $\{q_j\}$ is a Cauchy sequence, then there exists an $M \in \mathbb{R}^+$ such that $|q_j|_p < M$ for all $j \in \mathbb{N}$.

6.12. Show the sum $\{q_j\} + \{q'_j\} = \{q_j + q'_j\}$ and the product $\{q_j\} \cdot \{q'_j\} = \{q_j \cdot q'_j\}$ of Cauchy sequences is again a Cauchy sequence.

   (*Hint: Use Exercise 6.11.*)

6.13. Prove that equivalence of Cauchy sequences, given in Definition 6.4 on page 234, is an equivalence relation, namely that it satisfies the three properties of being reflexive, symmetric, and transitive.

6.14. Prove that every Cauchy sequence is convergent in $\mathbb{R}$.

   (*Recall that a sequence $\{q_j\}$ is convergent in $\mathbb{R}$ if there exists an $L \in \mathbb{R}$ satisfying the property that for any $\varepsilon > 0$, there exists an $N \in \mathbb{N}$ such that $|q_j - L| < \varepsilon$ for all $j \geq N$.*)

   (*Hint: Use Exercise 6.11 and the fact that every bounded sequence has a convergent subsequence, which is the interpretation of the well-known Bolzano–Weierstrass theorem for $\mathbb{R}$.*)

6.15. Prove that every sequence that converges in $\mathbb{R}$ is a Cauchy sequence.

6.16. Prove that if p is prime, then when $|x|_p \neq |y|_p$, we have

$$|x + y|_p = \max\{|x|_p, |y|_p\}.$$

6.17. In Exercise 6.16, provide an example where $|x|_p = |y|_p$ and $|x + y|_p < \max\{|x|_p, |y|_p\}$, called the *strong triangle inequality*.

6.18. Prove that if $z, w \in \mathbb{Z}$, then

$$|z - w|_p \leq \frac{1}{p^n} \text{ if and only if } z \equiv w \pmod{p^n} \text{ for some integer } n \geq 0.$$

6.19. Prove that all p-adic triangles are isosceles, i.e., all sets of vertices x, y, z with $x, y, z \in \mathbb{Q}_p$ are isosceles. In other words, demonstrate that, with respect to a p-adic valuation as a measure of distance, the length of two of the sides must always be the same.

   (*Hint: Use Exercise 6.16.*)

6.20. Prove that Exercise 6.16 holds if $|\cdot|_p$ is replaced by any non-Archimedean absolute value on a field F.

Biography 6.1 Augustine-Louis Cauchy (1789–1857) *was born on August 21, 1789 in Paris, France. When still a teenager, Laplace and Lagrange were visitors to the Cauchy home. Indeed, it was on the recommendation of Lagrange that Cauchy's father took his advice to have the young Cauchy well educated in languages before studying mathematics in earnest. Thus, in 1802, he entered the École Centrale du Panthéon where he devoted two years to the study of classical languages. Then he went on to study mathematics graduating from École Polytechnique in 1807, after which he entered the École des Ponts Chassées. In 1810, he assumed his first job in Cherbourgh to work on port facilities for Napoleon's English invasion fleet. Despite what was a heavy workload in this position, he engaged in mathematical research. One well-known result that he proved in 1811 was that the angles of a complex polyhedron are determined by its faces. In 1812, Cauchy returned to Paris when his health took a turn for the worse. By 1814, he had published his now-famous memoir on definite integrals that became the foundation for our modern theory of complex functions. In 1815, he was appointed assistant professor of analysis at the École Polytechnique, and there, in 1816, he was awarded the Grand Prix of the French Academy of Sciences for his work on waves. In 1817, he took a post at the Collège de France. There he lectured on his integration methodology that involved the first rigorous scheme for convergence of infinite series and a formal definition of the integral. By 1829, he defined the meaning of a complex function of a complex variable, which he published in* Leçons sur le Calcul Différential, *which was a culmination, among other works, of the study of the calculus of residues begun in 1824. Politics intervened in 1830 when he left for Switzerland and after refusing to swear an oath of allegiance to the new regime and failing to return to Paris, he lost all his positions there. In 1831, he went to Turin and taught there in 1832–33, after which he left for Prague on an order from Charles X to tutor his grandson. In 1838, he returned to Paris, and reclaimed his position at the Academy, but was not allowed to teach since he continued to refuse to take the aforementioned oath. Between 1840 and 1847, he published his renowned four-volume* Exercises d'analyse et de physique mathématique. *In 1848, when Louise Phillpe was overthrown, Cauchy reclaimed his university positions. In 1850, he lost an election to Liouville for the chair at the Collège de France, which led to bad temperament between the two of them from that time on. Also, during the last years of his life, he had a dispute with Duhamel over a priority claim on a result in inelastic shocks, a claim, it turns out, about which Cauchy was wrong. He died in Sceaux outside of Paris on May 23, 1857. He managed to publish 789 papers in mathematics. Indeed, Cauchy's name is present on various terms in modern-day mathematics including, the Cauchy integral theorem, the Cauchy-Kovalevskaya existence theorem, the Cauchy-Riemann equations, and the Cauchy sequences that we are studying in this section. Also, his contributions to the foundation of mathematical physics and theoretical mechanics via his work on the theory of light and his theory of elasticity necessitated that he develop not only his calculus of residues, but also new techniques such as Fourier transforms and diagonalization of matrices.*

6.3 Non-Archimedean vs. Archimedean Valuations

> *Philosophy is written in that great book which ever lies before our eyes—I mean the universe...This book is written in mathematical language and its characters are triangles, circles, and other geometrical figures, without whose help...one wanders in vain through a dark labyrinth.*
>
> from **The Asayer (1623)**
> **Galileo Galilei (1564–1642)**
> **Italian astronomer and physicist**

In §6.2 we got a taste of the difference between Archimedean and non-Archimedean valuations. In particular, the counterintuitive result in Exercise 6.19 on page 238, which says that all p-adic triangles are isosceles is seemingly incredible. Let us explore the differences at greater length. The non-Archimedean case $\mathbb{Q}_p$ for $p < \infty$ has no analogue when $p = \infty$ and $\mathbb{Q}_p = \mathbb{R}$ since there is no proper subdomain of $\mathbb{R}$ that has $\mathbb{R}$ as its quotient field, whereas by Theorem 6.1 on page 236, $\mathcal{O}_p$ is a subdomain of $\mathbb{Q}_p$, which is its quotient field. The fields $\mathbb{R}$ and $\mathbb{Q}_p$ for $p < \infty$ are all uncountable and no two of them are isomorphic. Furthermore, and most importantly, we have exhausted all possible valuations on $\mathbb{Q}$ since every such valuation is equivalent to a $|\cdot|_p$ for some $p \leq \infty$. This is the following, proved in 1918—see Biography 6.2 on page 242. Recall the definition of a trivial absolute value given in Example 6.4 on page 234 in what follows.

Theorem 6.3 | Ostrowski's Theorem |

Every nontrivial valuation on $\mathbb{Q}$ is equivalent to one of the absolute values $|\cdot|_p$ for a prime p or $p = \infty$.

Proof. First assume that for every integer $n > 1$ we have that $|n| > 1$.

Claim 6.1 *There exists an $r \in \mathbb{R}^+$ such that for any integer $m > 1$, $|m| = m^r$.*

Let $n > 1$ and $t \geq 1$ be integers. Write n^t to base m,

$$n^t = \sum_{j=0}^{\ell} c_j m^j,$$

where the $c_j \in \mathbb{Z}$ with $0 \leq c_j \leq m - 1$ and $c_\ell \neq 0$. By the triangle inequality, $|c_j| = |1 + 1 + \cdots + 1| \leq c_j|1| = c_j$ for each j. Also, since $n^t \geq m^\ell$, then

$$\ell \leq \frac{\log_e(n^t)}{\log_e m},$$

so by the triangle inequality again,

$$|n^t| \leq \sum_{j=0}^{\ell} |c_j m^j| = \sum_{j=0}^{\ell} |c_j||m^j| < m \sum_{j=0}^{\ell} |m^j| \leq m \sum_{j=0}^{\ell} |m|^\ell$$

$$\leq m(\ell + 1)|m|^{(\log_e(n^t)/\log_e m)}.$$

Hence,

$$|n| \leq \lim_{t\to\infty} \left[m^{1/t}(\ell + 1)^{1/t}|m|^{\log_e(n^t)/(t\log_e m))} \right]$$

$$= \lim_{t\to\infty} \left[m^{1/t}(\ell + 1)^{1/t} \right] \cdot \left[\lim_{t\to\infty} |m|^{(t\log_e(n))/(t\log_e m))} \right] = \lim_{t\to\infty} |m|^{\log_e n/\log_e m}$$

$$= |m|^{\log_e n/\log_e m}.$$

By reversing the roles of m and n in the above argument, we also get

$$|m| \leq |n|^{\log_e m/\log_e n},$$

so

$$|m|^{1/\log_e m} = |n|^{1/\log_e n},$$

for every $m > 1$ and $n > 1$. By setting the constant

$$K = |m|^{1/\log_e m} = |n|^{1/\log_e n},$$

we get

$$|m| = K^{\log_e m} = e^{(\log_e m)\cdot(\log_e K)} = m^{\log_e K} = m^r,$$

where $r = \log_e K \in \mathbb{R}^+$. This establishes the claim.

By Claim 6.1 in the case where $n > 1$ implies $|n| > 1$, we must have that $|m| = |m|_\infty$ by Theorem 6.2 on page 237.

Now assume that $|n| < 1$ for some integer $n > 1$. Since $|\cdot|$ is nontrivial, then there exists a least value $q \in \mathbb{N}$ such that $|q| < 1$. Assume that q is *the least* such value. If $q = q_1 \cdot q_2$ for $q_1, q_2 \in \mathbb{N}$ with $q_j < q$ for $j = 1, 2$, then $|q_1| = 1 = |q_2|$, by the minimality of q, so $|q| = |q_1| \cdot |q_2| = 1$, contradicting that $|q| < 1$. Hence, q is prime. Let $p \neq q$ be a prime with $|p| < 1$, then for sufficiently large $N \in \mathbb{N}$, we have that $|p^N| = |p|^N < 1/2$. Similarly, $|q^M| < \frac{1}{2}$, for sufficiently large $M \in \mathbb{N}$. Hence, since $\gcd(p, q) = 1$, there exist $u, v \in \mathbb{Z}$ such that $up^N + vq^M = 1$, so by the triangle inequality,

$$1 = |1| = |up^N + vq^M| \leq |up^N| + |vp^M| < \frac{1}{2} + \frac{1}{2} = 1,$$

a contradiction. Hence, our assumption that there exists a prime p different from q with $|p| < 1$ is false. This proves that $|p| = 1$ for all primes $p \neq q$. Hence, for any $z \in \mathbb{Z}$ with $q \nmid z$, $|z| = 1 = |z|_q$. Since any $x \in \mathbb{Q}$ may be written uniquely in the form

$$x = \frac{a}{b}q^{\nu_q(x)} \text{ where } |a| = |a|_q = 1 = |b|_q = |b|,$$

then

$$|x| = \frac{|a|}{|b|}|q|^{\nu_q(x)} = |q|^{\nu_q(x)}$$

and since $|q| < 1$, then for some $r \in \mathbb{R}^+$, $|q| = q^{-r}$, where r is independent of x. Therefore, $|x| = q^{-r\nu_q(x)} = (q^{-\nu_q(x)})^r = |q|_q^r = |x|_q^r$, so $|\cdot|$ is equivalent to the q-adic valuation by Theorem 6.2. $\qquad\square$

Biography 6.2 Alexander Markowich Ostrowski (1893–1986) *was born on September 25, 1893 in Kiev, Ukraine. He began his post-secondary studies at Marburg University in Germany in 1912 under Hensel's supervision. However, after the outbreak of World War I, Ostrowski was imprisoned as a hostile foreigner. When the war ended in 1918, he was allowed his freedom, and went to Göttingen, where he worked on his doctorate under Hilbert and Landau—see Biographies 3.5 on page 127 and 3.1 on page 104. In 1920, his doctoral dissertation was published in* Mathematische Zeitschrift, *and this was already his fifteenth publication, having written his first paper before he even entered university. In that year, he went to Hamburg to work for his habilitation as Hecke's assistant, and was awarded it in 1922. In 1923, he accepted a lecturing position at Göttingen. He moved around in the mid 1920s and finally settled on a position offered to him at the University of Basil in Switzerland, where he stayed until he retired in 1958. He published approximately 275 papers in his career, in diverse areas such as determinants, algebraic equations, number theory, topology, differential equations, conformal mappings, among many others. In particular, concerning the topic of this section, he provided a comprehensive description of valuations in 1934. He also worked on the Euler-Maclaurin formula and the Fourier integral formula, among other valued topics. He died on November 20, 1986 in Montagnola, Lugano, Switzerland.*

Exercises

6.21. Prove that a sequence of rational numbers $\{q_j\}_{j=1}^{\infty}$ is a Cauchy sequence with respect to the p-adic absolute value $|\cdot|_p$ for a prime $p < \infty$ if and only if

$$\lim_{j \to \infty}^{(p)} |q_{j+1} - q_j|_p = 0.$$

Conclude that $\{q_j\}_{j=1}^{\infty}$ is p-adically convergent. (*See* (6.10) *on page 235.*)

6.22. Prove that the series

$$\gamma = \sum_{j \geq k \in \mathbb{Z}} c_j p^j \tag{6.11}$$

for a prime $p < \infty$ with $c_j \in \mathbb{Z}$ with $0 \leq c_j \leq p-1$ is p-adically convergent, and that the partial sums $\gamma_n = \sum_{j \geq k}^{n} c_j p^j$ are Cauchy sequences for all $n \in \mathbb{N}$. (*Note that a series $\sum_{j=1}^{\infty} q_j$ with $q_j \in \mathbb{Q}_p$ converges in $\mathbb{Q}_p$ if and only if $\lim_{j \to \infty}^{(p)} |q_j|_p = 0$.*)

6.4 Representation of p-Adic Numbers

> *To get practice in being refused—*
>
> on being asked why he was begging for alms from a statue
> from **Digenese Laertius Lives of the Philosophers**
> **Diogenes (c. 400–c. 325 B.C.)**
> **Greek cynic philosopher**

In this section, we explore the methodology for representation of p-adic numbers as power series. The series representation given by (6.11) in Exercise 6.22 on the facing page tells us, via Exercise 6.21, that such series are limits of Cauchy sequences of elements in $\mathbb{Q}$. We now demonstrate that a number has a representation as a series given in (6.11) if and only if it is a p-adic number. First we need the following.

Lemma 6.2 *Given a prime $p < \infty$, every $\alpha \in \mathbb{Q}$ has a representation as a power series in p.*

Proof. Suppose first that $\alpha = a/b$ where $\gcd(a, b) = 1$, and $p \nmid b$. For a given $j \in \mathbb{N}$, we know from (6.5) on page 230 that a solution to $bx_j \equiv a \,(\mathrm{mod}\ p^j)$ is given by

$$x_j = \sum_{i=0}^{j-1} c_i p^i \text{ with } 0 \leq c_i < p$$

so that

$$\left| \frac{a}{b} - x_j \right| \leq p^{-j}.$$

Thus,

$$\overset{(p)}{\underset{j \to \infty}{\lim}} \left(\frac{a}{b} - x_j \right) = 0.$$

In other words,

$$\frac{a}{b} = \overset{(p)}{\underset{j \to \infty}{\lim}} x_j.$$

Now suppose that for $j < j'$, $x_j, x_{j'}$ are two solutions as above, then

$$|x_j - x_{j'}|_p = \left| \sum_{i=j}^{j'-1} c_i p^i \right|_p \leq \sum_{i=j}^{j'-1} p^{-i} |c_i|_p \leq \sum_{i=j}^{j'-1} p^{-i} = \frac{\frac{1}{p^j} - \frac{1}{p^{j'}}}{1 - \frac{1}{p}} < \varepsilon,$$

where j' may be assumed to be larger than some constant $J(\varepsilon)$ for any given $\varepsilon > 0$. Hence, by Exercise 6.21, the sequence $\{x_j\}_{j=1}^{\infty}$ is p-adically convergent. In other words,

$$\frac{a}{b} = \sum_{i=0}^{\infty} c_i p^i \text{ with } 0 \leq c_i < p$$

is the p-adic representation of the rational number α.

Now we consider the case where $\alpha = a/b$ where $\gcd(a,b) = 1$, and $p^\ell \,\|\, b$ for some $\ell \in \mathbb{N}$. From (6.11) in Exercise 6.22 on page 242 we know that a general p-adic representation of α is given by

$$\alpha = p^{-\ell}\left(\sum_{j=0}^{\infty} c_j p^j\right), \text{ where } 0 \le c_j < p. \tag{6.12}$$

□

If it is the case that in (6.12) in the above proof, there exist fixed $m, n \in \mathbb{N}$ such that for any integer $r \ge 0$,

$$c_{j+m} = c_{j+m+1} = \cdots = c_{j+m+rn} = \cdots \text{ for } j = 1, 2, \ldots, n,$$

then we call this p-adic representation of α *periodic*. In this case we may rewrite (6.12) as

$$\alpha = p^{-\ell}\left(\left(\sum_{j=0}^{m} c_j p^j\right) + p^{m+1}\left(\sum_{j=m+1}^{m+n} c_j p^{j-m-1}\right)\right.$$

$$\left. + p^{m+n+1}\left(\sum_{j=m+1}^{m+n} c_j p^{j-m-1}\right) + \cdots\right)$$

$$= p^{-\ell}\left(\left(\sum_{j=0}^{m} c_j p^j\right) + \sum_{j=0}^{\infty} p^{m+1+jn} C\right),$$

where

$$C = \sum_{j=m+1}^{m+n} c_j p^{j-m-1}.$$

In what follows we prove that every rational number must be so represented.

Theorem 6.4 $\boxed{p\text{-Adic Numbers as Periodic Power Series}}$

For a prime $p < \infty$, $\alpha \in \mathbb{Q}$ if and only if α has a representation as a periodic power series in p.

Proof. First assume that $\alpha \in \mathbb{Q}^+$, say $\alpha = a/b$ where $\gcd(a,b) = 1$, and $p^\ell \,\|\, b$ for some $\ell \ge 0$. By Lemma 6.2 on the preceding page, α has a representation via

$$p^\ell \alpha = \sum_{j=0}^{n} c_j p^j + \frac{u}{w}, \text{ where } 0 \le c_j < p,$$

and either $u/w = 0$ or $\gcd(u,w) = 1$, $w > 0 > u$, $0 > u/w > -1$, and $p \nmid w$. Assuming $u/w \ne 0$, let $i \in \mathbb{N}$ be the least value such that $p^i \equiv 1 \pmod{w}$, and there is a negative integer j with $1 - p^i = jw$, so $u/w = ju/(1 - p^i)$. Since the above conditions imply that $0 < u(1 - p^i)/w < p^i - 1$, then

$$ju = \sum_{j=0}^{i-1} a_j p^j, \text{ with } 0 \le a_j < p.$$

Hence, since $\sum_{j=0}^{\infty} p^{ij}(1 - p^i) = 1$, then

$$\frac{u}{w} = \left(\sum_{j=0}^{i-1} a_j p^j\right) \cdot \left(\sum_{j=0}^{\infty} p^{ij}\right) = \left(\sum_{j=0}^{i-1} a_j p^j\right) + p^i \left(\sum_{j=0}^{i-1} a_j p^j\right) + p^{2i} \left(\sum_{j=0}^{i-1} a_j p^j\right) \cdots,$$

so α has a periodic power series in p.

If $\alpha < 0$, then we perform the above to obtain the power series for $-\alpha$. We obtain that $\alpha = 0 - (-\alpha)$ has a power series in p since we may represent 0 as

$$0 = p + (p - 1) \sum_{j=1}^{\infty} p^j.$$

Conversely, assume that

$$\alpha = p^{-\ell} \left(\left(\sum_{j=0}^{m} c_j p^j\right) + \sum_{j=0}^{\infty} p^{m+1+jn} C \right),$$

where

$$C = \sum_{j=m+1}^{m+n} c_j p^{j-m-1}.$$

Therefore,

$$\alpha p^{\ell} - \left(\sum_{j=0}^{m} c_j p^j\right) = \sum_{j=0}^{\infty} p^{m+1+jn} C = \sum_{j=0}^{\infty} p^{m+1} C \left(\sum_{j=0}^{\infty} p^{jn}\right).$$

However,

$$\sum_{j=0}^{t} p^{jn} = \frac{1 - p^{(t+1)n}}{1 - p^n},$$

and for $t \ge t_0$, namely for t sufficiently large, we have for any $\varepsilon > 0$ that

$$\left| \frac{1}{1 - p^n} - \frac{1 - p^{(t+1)n}}{1 - p^n} \right|_p = p^{-(t+1)n} < \varepsilon.$$

So

$$\sum_{j=0}^{\infty} p^{jn} = \frac{1}{1 - p^n}.$$

Hence,

$$\alpha p^\ell - \left(\sum_{j=0}^{m} c_j p^j \right) = p^{m+1} C \frac{1}{1 - p^n},$$

namely

$$\alpha = p^{-\ell} \left(\sum_{j=0}^{m} c_j p^j \right) + p^{m+1-\ell} C \frac{1}{1 - p^n} \in \mathbb{Q},$$

as required.   □

It is worth isolating a fact proved in the above.

Corollary 6.1 *In $\mathbb{Q}_p$ for a prime $p < \infty$, given any $n \in \mathbb{N}$,*

$$\sum_{j=0}^{\infty} p^{jn} = \frac{1}{1 - p^n}.$$

Exercises

6.23. Prove that for a prime $p < \infty$, $\alpha \in \mathcal{O}_p$ if and only if $\alpha = a/b$ where $p \nmid b$.

6.24. Prove that if $p < \infty$ is prime then

$$\mathcal{P} = \{\alpha \in \mathbb{Q}_p : |\alpha|_p < 1\}$$

is an ideal in $\mathcal{O}_p$.

(*Hint: See Theorem 6.1 on page 236.*)

6.25. Prove that the ideal $\mathcal{P}$ in Exercise 6.24 is maximal in $\mathcal{O}_p$.

(*Hint: Use Hensel's Lemma 6.1 on page 230 with modulus $\mathcal{P}$ to show that $\mathcal{O}_p/\mathcal{P}$ is the set of invertible elements $\mathcal{U}_p$ in $\mathcal{O}_p$, then employ Theorem 2.7 on page 68.*)

6.26. With reference to Exercises 6.24–6.25, prove that every nonzero ideal of $\mathcal{O}_p$ is of the form

$$I = p^n \mathcal{O}_p = \mathcal{P}^n$$

for some integer $n \geq 0$.

(*Hint: Prove that $I = \mathcal{P}_n = \{x : |x| \leq p^{-n}\}$ then use induction on n to establish that $\mathcal{P}_n = \mathcal{P}^n$.*)

6.27. Prove that every nonzero $\alpha \in \mathbb{Q}_p$ may be written uniquely in the form $\alpha = u p^n$ where $n \in \mathbb{Z}$, $u \in \mathcal{U}_p$.

Chapter 7

Dirichlet: Characters, Density, and Primes in Progression

> *Talent develops in quiet places, character in the full current of human life.*
> translation from Act 1, Scene 2 of **Torquato Tasso (1790)**
> **Johann Wolfgang von Goethe (1749–1832)**
> **German poet, novelist, and dramatist**

7.1 Dirichlet Characters

A principal goal of the chapter is to establish the renowned Dirichlet Theorem on primes in arithmetic progression. In order to do so, we need to generalize the Riemann ζ-function, which we studied in detail in §5.3. The generalization requires the introduction of the following notion, the topic of this section.

Definition 7.1 | **Dirichlet Characters**

If $D \in \mathbb{N}$ is fixed and

$$\chi : \mathbb{N} \mapsto \mathbb{C}$$

is a function satisfying the following for each $m, n \in \mathbb{N}$, then χ is called a *Dirichlet character modulo D.*

(a) $\chi(mn) = \chi(m)\chi(n)$.

(b) If $m \equiv n \,(\mathrm{mod}\ D)$, then $\chi(m) = \chi(n)$.

(c) $\chi(n) = 0$ if and only if $\gcd(n, D) > 1$.

Example 7.1 If $D > 1$ is odd then the Jacobi symbol (n/D) is a Dirichlet character for the modulus D.

Remark 7.1 If $\phi(D)$ denotes the Euler totient, and we have that $\gcd(n, D) = 1$, then
$$\chi(n)^{\phi(D)} = \chi(n^{\phi(D)}) = \chi(1)$$
by parts (a)–(b) of Definition 7.1 in conjunction with *Euler's generalization of Fermat's little theorem*, namely
$$n^{\phi(D)} \equiv 1 \pmod{D}.$$

Moreover, since
$$\chi(1) = \chi(1^2) = \chi(1)^2,$$
and $\chi(1) \neq 0$, then $\chi(1) = 1$. In particular, this shows that $\chi(n)$ is a $\phi(D)$-th root of unity for all nonvanishing values of χ. Note, as well, that Dirichlet characters are completely multiplicative—see Exercise 5.13 on page 227.

Example 7.2 The character $\chi_0(n) = 1$ for all $n \in \mathbb{N}$ relatively prime to D and $\chi_0(n) = 0$ otherwise is called the *principal character* for the modulus D. (When referring to a character χ modulo D, the modulus D will be understood in context.) Moreover, if χ_1, χ_2 are Dirichlet characters modulo D, then it is clear that $\chi_1\chi_2$ is also a Dirichlet character modulo D, where
$$\chi_1\chi_2(n) = \chi_1(n)\chi_2(n).$$

In fact, by Remark 7.1, a Dirichlet character is a $\phi(D)$-th root of unity whenever it is nonvanishing. Also, it is completely multiplicative, and is constant on residue classes modulo D. Thus, the Dirichlet characters form a multiplicative group where χ_0 is the identity element, and for any character χ the complex conjugate $\overline{\chi}$ is also a character with
$$\chi\overline{\chi}(n) = \chi(n)\overline{\chi(n)} = |\chi(n)|^2 = 1$$
when $\gcd(n, D) = 1$. In other words, $\chi\overline{\chi} = \chi_0$, so $\overline{\chi}$ is the multiplicative inverse of χ. The group of characters maps homomorphically into the roots of unity in $\mathbb{C}$. We denote this group by G_{char}^D, the *group of Dirichlet characters modulo* D. In what follows we establish the cardinality of G_{char}^D.

Remark 7.2 If χ_1 is a Dirichlet character modulo D_1 and χ_2 is a Dirichlet character modulo D_2, then $\chi_1\chi_2$ is a Dirichlet character modulo $\text{lcm}(D_1, D_2)$, where
$$\chi_1\chi_2(n) = \chi_1(n)\chi_2(n).$$
We will use this fact in the following result.

Theorem 7.1 $\boxed{\textbf{The Number of Dirichlet Characters}}$

For a given integer $D > 1$, there exist exactly $\phi(D)$ distinct Dirichlet characters modulo D. In other words,

$$|G^D_{char}| = \phi(D).$$

Proof. Suppose that $p^a || D$ for a prime p and $a \in \mathbb{N}$. If $q = p^a$ is 2, 4, or an odd prime power, then there exists a primitive root modulo q, a fact from elementary number theory—see [68, Theorem 3.7, p. 151]. Let g be one such primitive root. Then the values g^i for $i = 1, 2, \ldots, \phi(q)$ form a complete set of reduced residues modulo q, namely those residues relatively prime to q—see [68, Theorem 3.1, p. 142]. Therefore, by selecting

$$\chi_i(g) = g^i \text{ for } i = 1, 2, \ldots, \phi(q),$$

we have defined $\phi(q)$ distinct Dirichlet characters modulo q. If $q = 2^a$ where $a > 2$, then q has no primitive root. However, ± 5 have order 2^{a-2} modulo 2^a—see [68, Exercise 3.8, p. 152]. Thus, together $\pm 5^j$ for $j = 1, 2, \ldots, 2^{a-2}$ generate all odd residues modulo 2^a. By selecting a primitive 2^{a-2}-th root of unity $\zeta_{2^{a-2}}$ and defining characters χ_i via

$$\chi_i(5) = \zeta^i_{2^{a-2}}$$

and

$$\chi_i(2^a - 1) = \pm 1, \text{ for } i = 1, 2, \ldots, 2^{a-2},$$

we have constructed

$$\phi(2^a) = 2^{a-1} = 2 \cdot 2^{a-2}$$

distinct characters modulo 2^a. By Remark 7.2, these characters may be put together to form $\phi(D)$ distinct characters modulo D. $\qquad\square$

Immediate from the above is the following—see [68, p. 81 ff.]. Recall that the symbol $\bar{a}$ means the residue class of a.

Corollary 7.1 *For any integer $D > 1$,*

$$G^D_{char} \cong (\mathbb{Z}/D\mathbb{Z})^* = \{\bar{a} \in \mathbb{Z}/D\mathbb{Z} : 0 < a < D \text{ and } \gcd(a, D) = 1\},$$

the group of units of $\mathbb{Z}/D\mathbb{Z}$.

Next are identities involving characters that will allow us to introduce another celebrated function due to Dirichlet.

Theorem 7.2 $\boxed{\textbf{Orthogonality Identities for Dirichlet Characters}}$

If $D > 1$ is an integer, then the following both hold.

(a) $\displaystyle\sum_{n=1}^{D} \chi(n) = \begin{cases} \phi(D) & \text{if } \chi = \chi_0, \\ 0 & \text{if } \chi \neq \chi_0. \end{cases}$

(b) $\displaystyle\sum_{\chi \in G^D_{\text{char}}} \chi(n) = \begin{cases} \phi(D) & \text{if } n \equiv 1 \,(\text{mod } D), \\ 0 & \text{if } n \not\equiv 1 \,(\text{mod } D). \end{cases}$

Proof. (a) If $\chi = \chi_0$, then the sum picks only those nonzero elements prime to D, for which $\chi(n) = 1$, so the result is immediate. If $\chi \neq \chi_0$, then there exists an integer z relatively prime to D such that $\chi(z) \neq 1$. Thus,

$$\chi(z) \sum_{n=1}^{D} \chi(n) = \sum_{n=1}^{D} \chi(zn) = \sum_{n=1}^{D} \chi(n),$$

since $\chi(zn)$ runs over all values of χ as does $\chi(n)$ for $n = 1, 2, \ldots, D$. Therefore,

$$(\chi(z) - 1) \sum_{n=1}^{D} \chi(n) = 0,$$

and since $\chi(z) \neq 1$, we may divide both sides by $(\chi(z) - 1)$ to get the result for part (a).

(b) If $n \equiv 0 \,(\text{mod } D)$, then the result is obvious. Assume that $n \not\equiv 0$ (mod D). By Theorem 7.1 on the previous page, there are $\phi(D)$ distinct characters $\chi \in G^D_{\text{char}}$, so if $n \equiv 1 \,(\text{mod } D)$, then by part (b) of Definition 7.1 on page 247,

$$\sum_{\chi \in G^D_{\text{char}}} \chi(n) = \sum_{\chi \in G^D_{\text{char}}} \chi(1) = \phi(D).$$

On the other hand, if $n \not\equiv 1 \,(\text{mod } D)$, then by Exercise 7.1 on the next page, there exists a character $\chi_n \in G^D_{\text{char}}$ such that $\chi_n(n) \neq 1$. Thus,

$$\chi_n(n) \sum_{\chi \in G^D_{\text{char}}} \chi(n) = \sum_{\chi \in G^D_{\text{char}}} \chi_n(n)\chi(n) = \sum_{\chi \in G^D_{\text{char}}} \chi_n\chi(n) = \sum_{\chi \in G^D_{\text{char}}} \chi(n),$$

by Example 7.2 on page 248, since $\chi_n\chi$ is again a Dirichlet character for each χ. Hence,

$$(\chi_n(n) - 1) \sum_{\chi \in G^D_{\text{char}}} \chi(n) = 0,$$

and since $\chi_n(n) \neq 1$, then we divide both sides by $(\chi_n(n) - 1)$ to secure the result. $\qquad\square$

Corollary 7.2 *If $G = G^D_{\text{char}}$, and $\chi, \psi \in G$, each of the following holds.*

(a) *If $\delta(\chi, \chi) = 1$ and $\delta(\chi, \psi) = 0$ if $\chi \neq \psi$, then*

$$\sum_{n=1}^{D} \chi(n)\overline{\psi(n)} = \phi(D)\delta(\chi, \psi).$$

(b) If $\delta(m, n) = 1$ when $m \equiv n \,(\text{mod } D)$ and $\delta(m, n) = 0$ when $m \not\equiv n \,(\text{mod } D)$, then

$$\sum_{\chi \in G} \chi(m)\overline{\chi(n)} = \phi(D)\delta(m, n).$$

Proof. (a) Since

$$\sum_{n=1}^{D} \chi(n)\overline{\psi(n)} = \sum_{n=1}^{D} \chi(n)\psi(n)^{-1} = \sum_{n=1}^{D} \chi\psi^{-1}(n),$$

then by part (a) of Theorem 7.2, this sum is equal to 0 if $\chi\psi^{-1} \neq \chi_0$ and is $\phi(D)$, otherwise. In other words, the sum is 0 if $\chi \neq \psi$, and is $\phi(D)$ if $\chi = \psi$.

(b) Since

$$\sum_{\chi \in G} \chi(m)\overline{\chi(n)} = \sum_{\chi \in G} \chi(mn^{-1}),$$

then by part (b) of Theorem 7.2, this sum is equal to 0 if

$$mn^{-1} \not\equiv 1 \pmod{D}$$

and if

$$mn^{-1} \equiv 1 \pmod{D}$$

it equals $\phi(D)$. In other words, it is 0 if $m \not\equiv n \,(\text{mod } D)$ and is $\phi(D)$ otherwise. $\square$

Now we have the tools to proceed to §7.2 where we will provide the generalization of Riemann's function promised at the outset of this section.

Exercises

7.1. Let $n \in \mathbb{N}$ and $D > 1$ an integer such that $n \not\equiv 0, 1 \,(\text{mod } D)$. Prove that there exists a $\chi_n \in G_{\text{char}}^D$ such that $\chi_n(n) \neq 1$.

7.2. Prove that if χ is a Dirichlet character modulo D, and $s \in \mathbb{C}$, then the series

$$\sum_{n=1}^{\infty} \chi(n)n^{-s}$$

converges absolutely for $\Re(s) > 1$.

(*Hint: Use Theorem 5.10 on page 219 by bounding $|\chi(n)n^{-s}|$.*)

7.2   Dirichlet's L-Function and Theorem

> *This frightful word* [function] *was born under other skies than those I have loved—those where the sun reigns supreme.*
>   from the introduction of **Le Corbusier (1974) Stephen Gardiner**
>   **Le Corbusier (Charles-Édouard Jeanneret) (1887–1965)**
>   **French architect**

In §7.1 we laid the groundwork for the next notion that will be a generalization of the ζ-function promised therein.

Definition 7.2 $\boxed{\textbf{Dirichlet L-Functions}}$

If χ is a Dirichlet character modulo $D > 1$ and $s \in \mathbb{C}$, then

$$L(s, \chi) = \sum_{n=1}^{\infty} \frac{\chi(n)}{n^s}$$

is called a *Dirichlet L-function.*

Dirichlet defined and studied these L-functions primarily to prove Theorem 7.7 on page 258, which is a principal result of this chapter. We now develop some salient features of these functions. First, we note that by Exercise 7.2 on the previous page, $\sum_{n=1}^{\infty} \chi(n)n^{-s}$ converges absolutely for $\Re(s) > 1$. Note that these L-functions are special cases of the Dirichlet series we encountered in Exercise 5.14 on page 227. Indeed, we have the following.

Theorem 7.3 $\boxed{\textbf{L-Functions and Euler Products}}$

If $s \in \mathbb{C}$ and $\Re(s) > 1$, then for a Dirichlet character χ modulo $D > 1$,

$$L(s, \chi) = \prod_{p=prime} (1 - \chi(p)p^{-s})^{-1}.$$

Proof. This is Exercise 7.3 on page 260.   □

When we restrict to the principal character, then we have a close relationship with the Riemann ζ-function as follows.

Corollary 7.3 *If $\chi = \chi_0$ in Theorem 7.3, then for $\Re(s) > 1$,*

$$L(s, \chi_0) = \prod_{p|D}(1 - p^{-s}) \cdot \zeta(s).$$

Proof. By Theorem 7.3,

$$L(s, \chi_0) = \prod_{p=\text{prime}} (1 - \chi_0(p)p^{-s})^{-1}.$$

However, $\chi_0(p) = 1$ except when $p \mid D$, where we have $\chi_0(p) = 0$. Therefore, since

$$\zeta(s) = \prod_{p=\text{prime}} (1 - p^{-s})^{-1} = \prod_{p \mid D} (1 - p^{-s})^{-1} \cdot \prod_{p \nmid D} (1 - p^{-s})^{-1}$$

$$= \prod_{p \mid D} (1 - p^{-s})^{-1} \cdot \prod_{p \nmid D} (1 - \chi_0(p) p^{-s})^{-1} = \prod_{p \mid D} (1 - p^{-s})^{-1} L(s, \chi_0),$$

we have the result. □

The following provides a functional equation for L-functions based upon Corollary 7.3.

Corollary 7.4 *If χ_0 is the principal character modulo D, then $L(\chi_0, s)$ satisfies the functional equation*

$$L(s, \chi_0) = 2^s \pi^{s-1} \prod_{p \mid D} \frac{1 - p^{-s}}{1 - p^{s-1}} \cdot \Gamma(1-s) \left(\sin \frac{\pi s}{2} \right) L(1 - s, \chi_0),$$

which is tantamount to

$$L(1-s, \chi_0) = 2^{1-s} \pi^{-s} \prod_{p \mid D} \frac{1 - p^{s-1}}{1 - p^{-s}} \cdot \Gamma(s) \left(\cos \frac{\pi s}{2} \right) L(s, \chi_0).$$

Proof. By Theorem 5.15 on page 225, for $s \in \mathbb{C}$,

$$\zeta(s) = 2^s \pi^{s-1} \Gamma(1-s) \zeta(1-s) \cdot \left(\sin \frac{\pi s}{2} \right),$$

and via Corollary 7.3, we may replace the zeta functions by L-functions to get

$$L(s, \chi_0) \prod_{p \mid D} (1 - p^{-s})^{-1} = 2^s \pi^{s-1} L(1 - s, \chi_0) \prod_{p \mid D} (1 - p^{s-1})^{-1} \cdot \Gamma(1-s) \left(\sin \frac{\pi s}{2} \right),$$

from which the first result easily follows. For the second result, we rearrange the above to get

$$L(1-s, \chi_0) = 2^{-s} \pi^{1-s} \prod_{p \mid D} \frac{1 - p^{s-1}}{1 - p^{-s}} \cdot \left(\frac{1}{\Gamma(1-s) \left(\sin \frac{\pi s}{2} \right)} \right) L(s, \chi_0).$$

By Exercise 7.5 on page 260,

$$\left(\frac{1}{\Gamma(1-s) \left(\sin \frac{\pi s}{2} \right)} \right) = \Gamma(s) \pi^{-1} 2 \cos \frac{\pi s}{2},$$

so

$$L(1-s, \chi_0) = 2^{1-s} \pi^{-s} \prod_{p \mid D} \frac{1 - p^{s-1}}{1 - p^{-s}} \cdot \Gamma(s) \left(\cos \frac{\pi s}{2} \right) L(s, \chi_0),$$

which is the entire result. □

Remark 7.3 Corollary 7.3 and Exercise 7.4 on page 260 provide an analytic continuation of $L(s, \chi_0)$ as a meromophic function in the whole plane with a sole singularity at $s = 1$—see Remark 5.8 on page 219. Now we need to look at analytically continuing $L(s, \chi)$ to the region $\Re(s) > 0$ for arbitrary Dirichlet characters χ. This will provide an essential step in the development of material to prove Theorem 7.7 on page 258.

Theorem 7.4 $\boxed{\textbf{Analytic Continuation of L-Functions}}$

If $\chi \neq \chi_0$ is a Dirichlet character modulo $D > 1$, then

$$L(s, \chi) = \sum_{n=1}^{\infty} \chi(n) n^{-s}$$

converges for all $\Re(s) > 0$.

Proof. We begin with a necessary bound.

Claim 7.1 *If $\chi \neq \chi_0$, then for any $N \in \mathbb{N}$,*

$$\left| \sum_{n=1}^{N} \chi(n) \right| \leq \phi(D).$$

Let $N = qD + r$ where q and r are integers with $0 \leq r < D$. Thus,

$$\sum_{n=1}^{N} \chi(n) = q \left(\sum_{n=1}^{D} \chi(n) \right) + \sum_{n=1}^{r} \chi(n)$$

since $\chi(n) = \chi(m)$ for $m \equiv n \,(\text{mod } D)$ by part (a) of Definition 7.1 on page 247. By part (a) of Theorem 7.2 on page 249, if $\chi \neq \chi_0$, then $\sum_{n=1}^{D} \chi(n) = 0$, so by the triangle inequality,

$$\left| \sum_{n=1}^{N} \chi(n) \right| = \left| \sum_{n=1}^{r} \chi(n) \right| \leq \sum_{n=1}^{D} |\chi(n)| \leq \phi(D),$$

which is the claim.

Now define for any real $x \geq 1$ and $m \in \mathbb{N}$, $S(0) = 0$, and

$$S(x) = \sum_{m \leq x} \chi(m).$$

Then

$$\chi(n) = S(n) - S(n - 1).$$

Now suppose that $N \in \mathbb{N}$. Then

$$\sum_{n=1}^{N} \frac{\chi(n)}{n^s} = \sum_{n=1}^{N} \frac{S(n) - S(n-1)}{n^s} = \sum_{n=1}^{N} \frac{S(n)}{n^s} - \sum_{n=1}^{N} \frac{S(n-1)}{n^s}$$

$$= \sum_{n=1}^{N} \frac{S(n)}{n^s} - \sum_{n=0}^{N-1} \frac{S(n)}{(n+1)^s} = \frac{S(N)}{N^s} + \sum_{n=1}^{N-1} \frac{S(n)}{n^s} - \sum_{n=1}^{N-1} \frac{S(n)}{(n+1)^s}$$

$$= \sum_{n=1}^{N-1} S(n) \left(\frac{1}{n^s} - \frac{1}{(n+1)^s} \right) + \frac{S(N)}{N^s}.$$

Hence,

$$\sum_{n=1}^{\infty} \chi(n) n^{-s} = \lim_{N \to \infty} \left(\sum_{n=1}^{N-1} S(n) \left(\frac{1}{n^s} - \frac{1}{(n+1)^s} \right) + \frac{S(N)}{N^s} \right)$$

$$= \sum_{n=1}^{\infty} S(n) \left(\frac{1}{n^s} - \frac{1}{(n+1)^s} \right).$$

Moreover, we have that

$$\sum_{n=1}^{\infty} S(n) \left(\frac{1}{n^s} - \frac{1}{(n+1)^s} \right) = s \sum_{n=1}^{\infty} S(n) \int_{n}^{n+1} x^{-s-1} dx = s \int_{1}^{\infty} S(x) x^{-s-1} dx.$$

Now by Claim 7.1, $|S(x)| \leq \phi(D)$ for all x. Thus, the integral converges and defines an analytic function for all s with $\Re(s) > 0$. $\qquad \square$

Remark 7.4 Just as we commented in Remark 5.10 on page 226 to the effect that the zeros of Riemann's ζ-function are intimately connected with the distribution of primes, so too the zeros of the L-functions $L(s, \chi)$ speak about the distribution of primes in arithmetic progression. In fact, the principal feature of the proof of Dirichlet's theorem on primes in arithmetic progression is the validation that $L(1, \chi) \neq 0$ when $\chi \neq \chi_0$. This is encapsulated in the following generalization of Conjecture 5.1 on page 223. See Remark 7.6 on page 258 and Exercise 7.7 on page 261.

Conjecture 7.1 | **The Generalized Riemann Hypothesis (GRH)**

If χ is a Dirichlet character, then the zeros of $L(s, \chi)$ for $\Re(s) > 0$ lie on the line $\Re(s) = 1/2$.

Remark 7.5 Note that in the literature, Conjecture 7.1 is sometimes called the *Extended Riemann hypothesis* (ERH), and sometimes there is a distinction made between Conjecture 7.1 and a yet more general conjecture involving the

Dedekind-zeta function for an algebraic number field F, which is given by the following sum over all nonzero ideals I of $\mathfrak{O}_F$,

$$\zeta_F(s) = \sum_I \frac{1}{(N(I))^s}$$

for every $s \in \mathbb{C}$ with $\Re(s) > 1$, where $N(I) = |\mathfrak{O}_F/I|$ is the *norm* of the ideal I—see Exercise 8.32 on page 292. The more general assertion is: *If F is a number field and $s \in \mathbb{C}$ with $\zeta_F(s) = 0$ and $0 < \Re(s) < 1$, then $\Re(s) = 1/2$.* Conjecture 5.1 follows from this with $F = \mathbb{Q}$ and $\mathfrak{O}_F = \mathbb{Z}$. Depending on the source in the literature, Conjecture 7.1 is sometimes called the ERH and the last more general one the GRH and sometimes this is reversed. We maintain the GRH label for Conjecture 7.1 since it appears to be the most ubiquitous label. Indeed, for computational relevance and the historical significance of Conjecture 7.1, see [62, §5.4, pp. 172–186].

Now we proceed to verify the contents of the assertions made above in our quest to prove Dirichlet's theorem. In preparation, the reader should solve Exercise 7.6 on page 261.

Theorem 7.5 | **Nonvanishing of L$(1, \chi)$ for Complex χ**

If χ is a nontrivial complex Dirichlet character modulo D, then $L(1, \chi) \neq 0$.

Proof. By Theorem 7.4 on page 254,

$$L(s, \chi) = \sum_{n=1}^{\infty} \chi(n) n^{-s},$$

so for $s \in \mathbb{R}$, $s > 1$,

$$\overline{L(s, \chi)} = \sum_{n=1}^{\infty} \overline{\chi}(n) n^{-s} = L(s, \overline{\chi}).$$

Thus, if $L(1, \chi) = 0$, then $L(1, \overline{\chi}) = 0$.

Assume that $L(1, \chi) = 0$ for a complex character χ. Then $L(s, \chi) \neq L(s, \overline{\chi})$ for $s \in \mathbb{C}$, both have a pole at $s = 1$, and $L(1, \chi) = 0 = L(1, \overline{\chi})$. In the product

$$F(s) = \prod_{\chi \in G_{\text{char}}^D} L(s, \chi),$$

the term $L(s, \chi_0)$ has a pole at $s = 1$ and by Theorem 7.4 on page 254, $L(s, \chi)$ for $\chi \neq \chi_0$ is analytic about $s = 1$. Hence, $F(1) = 0$. However, by Exercise 7.6, $F(s) \geq 1$ for all $s \in \mathbb{R}$ with $s > 1$, so

$$\lim_{s \to 1+} F(s) = F(1) \neq 0,$$

a contradiction, so $L(1, \chi) \neq 0$ for any complex character χ.   □

Now in the final bid to establish the key result in the proof of Dirichlet's theorem, we need to establish the nonvanishing of $L(1, \chi)$ real characters χ. This is the more difficult case.

Theorem 7.6 $\boxed{\textbf{Nonvanishing } \mathbf{L}(1, \chi) \textbf{ for Real } \chi}$

If χ is a nontrivial real Dirichlet character modulo D, then $L(1, \chi) \neq 0$.

Proof. Suppose that χ is a real character and $L(1, \chi) = 0$. Now define the function
$$f(s) = \frac{L(s, \chi)L(s, \chi_0)}{L(2s, \chi_0)}.$$
Since $L(1, \chi) = 0$ and $L(s, \chi_0)$ has a simple pole at $s = 1$ means that the two events cancel out, so $L(s, \chi)L(s, \chi_0)$ is analytic on $\Re(s) > 0$. Also, $L(2s, \chi_0)$ is analytic on $\Re(s) > 1/2$, has a pole at $s = 1/2$, and by Theorem 7.4 on page 254 may be continued to an interval containing $1/2$ with a simple pole at $s = 1/2$. Hence, $\lim_{s \to 1/2^+} f(s) = 0$.

If $s \in \mathbb{R}$ with $s > 1$, then f has an infinite product expansion,
$$f(s) = \prod_{p = \text{prime}} (1 - \chi(p)p^{-s})^{-1}(1 - \chi_0(p)p^{-s})^{-1}(1 - \chi_0(p)p^{-2s})$$

$$= \prod_{p \nmid D} \frac{(1 - p^{-2s})}{(1 - p^{-s})(1 - \chi(p)p^{-s})}. \tag{7.1}$$

By Exercise 7.8 on page 262, $\chi(p) = \pm 1$. If $\chi(p) = -1$, then from (7.1),
$$\frac{(1 - p^{-2s})}{(1 - p^{-s})(1 - \chi(p)p^{-s})} = 1.$$

Hence, from (7.1),
$$f(s) = \prod_{\chi(p)=1} \frac{(1 - p^{-2s})}{(1 - p^{-s})(1 - p^{-s})} = \prod_{\chi(p)=1} \frac{(1 - p^{-s})(1 + p^{-s})}{(1 - p^{-s})(1 - p^{-s})} = \prod_{\chi(p)=1} \frac{1 + p^{-s}}{1 - p^{-s}}.$$

However,
$$\frac{1 + p^{-s}}{1 - p^{-s}} = (1 + p^{-s})\left(\sum_{j=0}^{\infty} p^{-js}\right) = \sum_{j=0}^{\infty} p^{-js} + \sum_{j=0}^{\infty} p^{-(j+1)s}$$

$$= 1 + \sum_{j=1}^{\infty} p^{-js} + \sum_{j=1}^{\infty} p^{-js} = 1 + \sum_{j=1}^{\infty} 2p^{-js}.$$

By Exercise 7.9 on page 262, $f(s) = \sum_{n=1}^{\infty} g_n n^{-s}$ where g_n is nonnegative for all n and converges for $s > 1$. Indeed, since $g_1 = 1$, and $f(s)$ is analytic for $\Re(s) > 1/2$, then $f(s) \geq 1$ for $s > 1/2$, whereas $\lim_{s \to 1/2^+} f(s) = 0$, a contradiction. Hence, $L(1, \chi) \neq 0$.   □

Remark 7.6 Exercise 7.7 speaks to the comments made in Remark 7.4 on page 255. Equation (7.4) tells us that if we can prove that

$$\sum_{\substack{\chi \in G^D_{\text{char}} \\ \chi \neq \chi_0}} \overline{\chi(a)} \log_e L(s, \chi) \mapsto \infty \text{ as } s \to 1^+,$$

then there are infinitely many primes $p \equiv a \,(\text{mod } D)$. We know from Exercise 7.4 that the term

$$L(s, \chi_0) \mapsto \infty \text{ as } s \to 1^+,$$

but the other terms could cancel out this fact, so we get to the comments made in Remark 7.4 to the effect that the core of the proof of Dirichlet's theorem is to show that $L(s, \chi) \neq 0$ when $\chi \neq \chi_0$. This is what we proved in Theorems 7.5 on page 256 and Theorem 7.6 on the preceding page. We are now ready for the main result.

Theorem 7.7 $\boxed{\textbf{Dirichlet: Primes in Arithmetic Progression}}$

If $a, m \in \mathbb{Z}$ with $\gcd(a, m) = 1$, then there are infinitely many primes of the form $p = mn + a$ for $n \in \mathbb{N}$.

Proof. By Exercise 7.7 on page 261,

$$\log_e L(s, \chi_0) + \sum_{\substack{\chi \in G^D_{\text{char}} \\ \chi \neq \chi_0}} \overline{\chi(a)} \log_e L(s, \chi) = \phi(D) \sum_{p \equiv a \ (\text{mod } D)} \frac{1}{p^s} + O\left(\phi(D)\right). \quad (7.2)$$

By Theorems 7.5–7.6,

$$\lim_{s \to 1^+} L(s, \chi) > 0$$

for $\chi \neq \chi_0$. Hence,

$$\lim_{s \to 1^+} \sum_{\chi \neq \chi_0} \overline{\chi(a)} \log_e L(s, \chi) < \infty$$

while by Exercise 7.4 on page 260, $\lim_{s \to 1^+} L(s, \chi_0) = \infty$, so the left hand side of (7.2) increases indefinitely as $s \to 1^+$. If the number of primes in the arithmetic progression $p \equiv a \,(\text{mod } m)$ is finite, then

$$\lim_{s \to 1^+} \sum_{p \equiv a (\text{mod } m)} \frac{1}{p^s} = \sum_{p \equiv a (\text{mod } m)} \frac{1}{p} < \infty,$$

indeed it is rational, but this contradicts (7.2) since the left side is goes to ∞ while the right side is finite. $\square$

Remark 7.7 Although Dirichlet's L-functions are generalizations of the Riemann ζ-function, Dirichlet introduced them before Riemann developed complex function theory. Thus, Dirichlet did not have the complex variable tools at his disposal to establish the nonvanishing of $L(1,\chi)$. He did this by looking at class numbers h_D of binary quadratic forms of discriminant D–see §3.1. He defined, for a quadratic number field F, the character given as follows—see Remark 2.2 on page 63 for a reminder of the terms used below,

$$\chi_F(p) = \begin{cases} 1 & \text{if } (p) \text{ is a split prime in } F, \\ -1 & \text{if } (p) \text{ is an inert prime in } F, \\ 0 & \text{if } (p) \text{ is an ramified prime in } F. \end{cases} \tag{7.3}$$

Then Dirichlet proved that

$$L(1,\chi_F) = Nh_D,$$

where

$$N = \begin{cases} 2\log_e \varepsilon_D/\sqrt{D} & \text{if } D > 0, \\ 2\pi/(w\sqrt{|D|}) & \text{if } D < 0, \end{cases}$$

where $w = 4$ if $D = -4$, $w = 6$ if $D = -3$, and $w = 2$ otherwise. The value ε_D is the smallest unit in $\mathfrak{D}_F$ that exceeds 1 when F is real. Also, when F is real,

$$R_F = \log_e \varepsilon_D$$

is called the *regulator* of F and ε_D is called the *fundamental unit* of F. Clearly, $h_D > 0$ and $R_F > 0$, so $L(1,\chi) > 0$ is immediate.

It also follows that

$$L(s,\chi_F) = \frac{\zeta_F(s)}{\zeta(s)},$$

where $\zeta_F(s)$ is the Dedekind-zeta function given in Remark 7.5 on page 255. In Theorem 7.4 on page 254 we saw that $L(s,\chi)$ may be continued analytically for $\Re(s) > 0$. Riemann showed how to continue it to the entire complex plane. Thus, every zero of $\zeta(s)$ is cancelled by a zero of $\zeta_F(s)$ with at least the same multiplicity. If we look at the more general case where $\mathbb{Q}$ is replaced by any number field $K \subseteq F$, then is it still true that

$$\zeta_F(s)/\zeta_K(s)$$

is analytic on the whole complex plane? The affirmative answer to this is the, still open, *Artin Conjecture*.

The above notion (7.3) of a character defined for a quadratic field may be generalized to any algebraic number field in order to, therefore, associate a given (generalized) Dirichlet character χ_F with any number field F. Once done it can be shown that they form a group G_F and

$$\zeta_F(s) = \prod_{\chi_F \in G_F} L(s,\chi_F),$$

and if N is the order of a given character χ_F in G_F, then it can also be demonstrated that

$$\zeta_F(s) = \zeta(s) \prod_{n=1}^{N-1} L(s, \chi_F^n).$$

Therefore, since $\zeta(s)$ has only a simple pole at $s = 1$, none of the factors $L(s, \chi_F^n)$ can vanish at $s = 1$. In particular, $L(s, \chi_F) \neq 0$, providing a simple proof of the results we achieved in Theorems 7.5–7.6, albeit by employing a more general ζ-function with ostensibly deeper results.

Exercises

7.3. Prove Theorem 7.3 on page 252.

  (*Hint: Use Exercise 7.2 on page 251 in conjuction with Exercises 5.12–5.13 on page 227.*)

7.4. Prove that if χ_0 is the principal Dirichlet character modulo $D > 1$, and $s \in \mathbb{C}$ then

$$\lim_{s \to 1^+} (s - 1)L(s, \chi_0) = \prod_{p|D}(1 - p^{-1}) = \frac{\phi(D)}{D}.$$

  Conclude that $\lim_{s \to 1^+} L(s, \chi_0) = \infty$.

  (*Hint: Use Corollary 7.3 on page 252 in conjunction with Exercise 5.18 on page 227 and the fact that:*

$$\phi(D) = D \prod_{p|D}(1 - 1/p)$$

  —*see* [68, *Corollary 2.1, p. 92*].)

7.5. Prove that for the Gamma function given in Definition 5.6 on page 224,

$$\Gamma(s)\Gamma(1 - s) = \frac{\pi}{\sin \pi s} = \frac{\pi}{2(\sin \frac{\pi s}{2})(\cos \frac{\pi s}{2})}.$$

  (*Hint: You may use the fact that*

$$\int_0^\infty \frac{u^{s-1}}{1 + u}\,du = \frac{\pi}{\sin \pi s},$$

  for $0 < \Re(s) < 1$. This integral is derivable from the relationship between the Beta function

$$B(x, y) = \int_0^1 t^{x-1}(1 - t^{y-1})dt,$$

and the following relationship with the Gamma function

$$B(x, y) = \frac{\Gamma(x)\Gamma(y)}{\Gamma(x+y)}.\Bigg)$$

7.6. If

$$F(s) = \prod_{\chi \in G^D_{\text{char}}} L(s, \chi),$$

namely the product over all Dirichlet characters modulo D, show that $F(s) \geq 1$ for $s \in \mathbb{R}$, $s > 1$.

(Hint: Form the sum, $G(s, \chi) = \sum_p \sum_{n=1}^{\infty} \frac{1}{n} \chi(p^n) p^{-ns}$ and use Corollary 7.2 on page 250, observing that for $z \in \mathbb{C}$ with $|z| < 1$,

$$\exp\left(\sum_{n=1}^{\infty} \frac{1}{n} z^n\right) = \frac{1}{1-z}$$

where $\exp(x) = e^x$. Note that $G(s, \chi)$ converges uniformly for $\Re(s) > 1$ since $\zeta(s)$ converges uniformly for $s \geq 1 + \varepsilon > 1$. Recall that a series converges uniformly if the sequence of partial sums converges uniformly, and a sequence $\{s_n\}_{n=1}^{\infty}$ converges uniformly for a set S of values of x provided that for each $\varepsilon > 0$, there exists an $N \in \mathbb{Z}$ with $|s_n(x) - s(x)| < \varepsilon$ for $n \geq N$ and all $x \in S$. From these considerations, $G(s, \chi)$ is continuous for $\Re(s) > 1$.)

7.7. Prove that if $s \in \mathbb{C}$ with $\Re(s) > 1$ and $a \in \mathbb{Z}$ such that $\gcd(a, D) = 1$, then the following equation holds for Dirichlet characters modulo D,

$$\log_e L(s, \chi_0) + \sum_{\substack{\chi \in G^D_{\text{char}} \\ \chi \neq \chi_0}} \overline{\chi(a)} \log_e L(s, \chi) = \phi(D) \sum_{p \equiv a \pmod{D}} \frac{1}{p^s} + O\left(\phi(D)\right).$$

$$(7.4)$$

(Hint: Use Theorem 7.2 on page 249, Corollary 7.2 on page 250, and Theorem 7.3 on page 252.)

(Note that the left-hand side of (7.4) is a special case of another ζ-function called the *Hurwitz ζ-function* defined for $s, q \in \mathbb{C}$ with $\Re(s) > 1$ and $\Re(q) > 0$ by

$$\zeta(s, q) = \sum_{n=0}^{\infty} \frac{1}{(q+n)^s},$$

which is absolutely convergent for the aforementioned values of s, q and can be extended to a meromorphic function defined for all $s \neq 1$. The Riemann ζ-function is the case where $q = 1$, and (7.4) is given by $q = a/D$ when $D > 2$, namely

$$\zeta(s, a/D) = \sum_{\chi \in G^D_{\text{char}}} \overline{\chi(a)} L(s, \chi).$$

Moreover, we can write the Dirichlet L-functions in terms of the Hurwitz ζ-function as follows,

$$L(s, \chi) = \frac{1}{n^s} \sum_{j=1}^{n} \chi(j) \zeta\left(s, \frac{j}{n}\right),$$

and

$$\zeta(s) = \frac{1}{n^s} \sum_{j=1}^{n} \zeta\left(s, \frac{j}{n}\right),$$

as well.)

7.8. If χ is a Dirichlet character such that $\chi(n)$ is real for all $n \in \mathbb{Z}$, then prove that $\chi(n) = \pm 1$ when $\gcd(n, D) = 1$, and $\chi^2 = \chi_0$.

7.9. Let f be a nonnegative multiplicative arithmetic function and assume there exists a $K \in \mathbb{R}^+$ such that $f(p^j) < K$ for all prime powers p^j. Prove that $\sum_{n=1}^{\infty} f(n) n^{-s}$ converges for all $s \in \mathbb{R}$ with $s > 1$. Also, prove that

$$\sum_{n=1}^{\infty} f(n) n^{-s} = \prod_{p} \left(1 + \sum_{j=1}^{\infty} f(p^j) p^{-js} \right).$$

(*Hint: Use Exercise 5.14 on page 227 for the last assertion.*)

7.3   Dirichlet Density

> *That all things are changed, and that nothing really perishes, and that the*
> *sum of matter remains exactly the same, is sufficiently certain.*
>   translation from **Cogitationes de Natura Rerum Cogitatio**
>   in **The Works of Francis Bacon, Volume 5 (1858)**
>   J. Spedding, editor
>   **Francis Bacon (1561–1626)**
>   **English lawyer, courtier, philosopher, and essayist**

This section deals with a concept that allows us to measure the size of a
set of primes in an accurate fashion and will provide another interpretation of
Dirichlet's Theorem 7.7 on page 258.

Definition 7.3 ⏐Dirichlet Density⏐

If S is a subset of the primes in $\mathbb{Z}$, and if

$$\lim_{s \to 1^+} \frac{\sum_{p \in S} p^{-s}}{\log_e(s-1)^{-1}} = k \in \mathbb{R},$$

then we say that S has *Dirichlet density k*, and we denote this by $\mathcal{D}(S)$. If the
limit does not exist then S has no Dirichlet density. Dirichlet density is often
called *analytic density*.

Remark 7.8 Note that Definition 7.3 may be reformulated in terms of Definition 5.4 on page 200, asymptotic equality, to say that as $s \to 1$,

$$-\mathcal{D}(S) \log_e(s-1) \sim \sum_{p \in S} p^{-s}.$$

One may also define another notion of "density" for two sets relative to one
another in the following fashion. If $S \subseteq W \subseteq \mathbb{N}$ with $|W| = \infty$, then if

$$\lim_{N \to \infty} \frac{|\{n \in S : n \le N\}|}{|\{n \in W : n \le N\}|} = \ell \in \mathbb{R},$$

then we say that S has *natural density* or *asymptotic density* ℓ in W, denoted
by $\mathcal{ND}(S, W)$. In other words, in terms of asymptotic equality, S has natural
density in W if

$$\mathcal{ND}(S, W)|\{n \in W : n \le N\}| \sim |\{n \in S : n \le N\}|.$$

Natural density is a more restrictive notion than Dirichlet density. For instance,
it can be shown that for any integer $b > 2$, the set of primes with first digit 1
when written in base b has Dirichet density but does not have natural density.
Yet any set of primes that has natural density, has Dirichlet density equal to
the same value.

We digress from the main topic to provide an example of natural density and some most interesting consequences with the following, which was proved in 1926—see [5].

Theorem 7.8 | **Beatty's Theorem** |

Suppose that $\alpha, \beta \in R^+$ are irrational and

$$1/\alpha + 1/\beta = 1.$$

If

$$\{s_n\}_{n=1}^{\infty} = \{n\alpha\}_{n=1}^{\infty} \text{ and } \{t_n\}_{n=1}^{\infty} = \{n\beta\}_{n=1}^{\infty},$$

then for any $N \in \mathbb{N}$ there is exactly one element of the sequence

$$\{s_n\}_{n=1}^{\infty} \cup \{t_n\}_{n=1}^{\infty}$$

in the interval $(N, N+1)$.

Proof. Set

$$\mathcal{S}_\alpha = \{\lfloor n\alpha \rfloor : n \in \mathbb{N}\}, \tag{7.5}$$

and

$$\mathcal{S}_\beta = \{\lfloor n\beta \rfloor : n \in \mathbb{N}\}. \tag{7.6}$$

Then for each $N \in \mathbb{N}$, if

$$\mathcal{S}_\alpha^N = \{x \in \mathcal{S}_\alpha : x \leq N\},$$

and

$$\mathcal{S}_\beta^N = \{x \in \mathcal{S}_\beta : x \leq N\},$$

we have cardinalities

$$|\mathcal{S}_\alpha^N| = \left\lfloor \frac{N}{\alpha} \right\rfloor \text{ and } |\mathcal{S}_\beta^N| = \left\lfloor \frac{N}{\beta} \right\rfloor.$$

We have

$$\frac{N}{\alpha} - 1 < \left\lfloor \frac{N}{\alpha} \right\rfloor < \frac{N}{\alpha} \tag{7.7}$$

and

$$\frac{N}{\beta} - 1 < \left\lfloor \frac{N}{\beta} \right\rfloor < \frac{N}{\beta}. \tag{7.8}$$

Adding (7.7) and (7.8) and using the fact that $1/\alpha + 1/\beta = 1$, we get

$$N - 2 < \left\lfloor \frac{N}{\alpha} \right\rfloor + \left\lfloor \frac{N}{\beta} \right\rfloor < N,$$

so

$$|\mathcal{S}_\alpha^N \cup \mathcal{S}_\beta^N| = \left\lfloor \frac{N}{\alpha} \right\rfloor + \left\lfloor \frac{N}{\beta} \right\rfloor = N - 1.$$

Hence,

$$\left|\mathcal{S}_\alpha^{N+1} \cup \mathcal{S}_\beta^{N+1}\right| = \left\lfloor \frac{N+1}{\alpha} \right\rfloor + \left\lfloor \frac{N+1}{\beta} \right\rfloor = N.$$

Thus, the number of elements of

$$\{s_n\}_{n=1}^\infty \cup \{t_n\}_{n=1}^\infty$$

in the interval $(N, N+1)$ is

$$\left|\mathcal{S}_\alpha^{N+1} \cup \mathcal{S}_\beta^{N+1}\right| - \left|\mathcal{S}_\alpha^{N} \cup \mathcal{S}_\beta^{N}\right| = 1,$$

which secures the result. $\qquad\square$

Corollary 7.5 *With $\mathcal{S}_\alpha$ and $\mathcal{S}_\beta$ given by (7.5)–(7.6),*

$$\mathcal{S}_\alpha \cup \mathcal{S}_\beta = \mathbb{N} \text{ and } \mathcal{S}_\alpha \cap \mathcal{S}_\beta = \varnothing.$$

Also,

$$\mathcal{ND}(\mathcal{S}_\alpha) = \frac{1}{\alpha} \text{ and } \mathcal{ND}(\mathcal{S}_\beta) = \frac{1}{\beta}.$$

Proof. Immediate from Theorem 7.8 is the first assertion. Also, from the proof,

$$\mathcal{ND}(\mathcal{S}_\alpha) = \lim_{N\to\infty} \frac{|\mathcal{S}_\alpha^N|}{N} = \lim_{N\to\infty} \frac{\left\lfloor \frac{N}{\alpha} \right\rfloor}{N} = \frac{1}{\alpha},$$

and similarly,

$$\mathcal{ND}(\mathcal{S}_\beta) = \frac{1}{\beta},$$

as required. $\qquad\square$

Remark 7.9 What is remarkable about the Beatty result is that the sequences *complement* each other in $\mathbb{N}$ as explicitly stated in Corollary 7.5. Indeed, two sequences that complement each other in $\mathbb{N}$ are called *complementary*.

Now that we have illustrated the natural density case, we return to Theorem 7.7 on page 258 from the perspective of Dirichlet density.

Theorem 7.9 $\boxed{\textbf{Dirichlet: Primes and Density}}$

If $a, m \in \mathbb{Z}$ with $\gcd(a, m) = 1$, and

$$\mathcal{S}_p^a = \{p \in \mathbb{N} : p \text{ is prime and } p \equiv a \pmod{m}\},$$

then

$$\mathcal{D}(\mathcal{S}_p^a) = \frac{1}{\phi(m)}.$$

Proof. We begin with some claims that will resolve the issue.

Claim 7.2 $\displaystyle\sum_{p \equiv a \,(\mathrm{mod}\, m)} p^{-s} = \frac{1}{\phi(m)} \sum_{\chi \in G_\chi^m} \chi^{-1}(a) \sum_{p \nmid m} \frac{\chi(p)}{p^s}.$

We have

$$\frac{1}{\phi(m)} \sum_{\chi \in G_\chi^m} \chi^{-1}(a) \sum_{p \nmid m} \frac{\chi(p)}{p^s} = \frac{1}{\phi(m)} \sum_{\chi \in G_\chi^m} \sum_{p \nmid m} \frac{\chi(a^{-1}p)}{p^s}.$$

However, by Therorem 7.2 on page 249,

$$\sum_{\chi \in G_\chi^m} \chi(a^{-1}p) = \begin{cases} \phi(m) & \text{if } a^{-1}p \equiv 1 \,(\mathrm{mod}\, m), \\ 0 & \text{otherwise.} \end{cases}$$

Thus,

$$\frac{1}{\phi(m)} \sum_{\chi \in G_{char}^m} \chi^{-1}(a) \sum_{p \nmid m} \frac{\chi(p)}{p^s} = \sum_{p \equiv a \,(\mathrm{mod}\, m)} p^{-s},$$

which is the claim.

Claim 7.3 *For $\chi \neq \chi_0$,*

$$\sum_{p \nmid m} \frac{\chi(p)}{p^s}$$

remains bounded as $s \to 1$.

We have

$$\sum_{p \nmid m} \frac{\chi(p)}{p^s} = \sum_{p \nmid m} \sum_{n=1}^{\infty} \frac{\chi(p)^n}{p^{ns}} - \sum_{p \nmid m} \sum_{n=2}^{\infty} \frac{\chi(p)^n}{p^{sn}}. \qquad (7.9)$$

However,

$$\sum_{p \nmid m} \sum_{n=2}^{\infty} \frac{\chi(p)^n}{p^{sn}} \leq \sum_{p \nmid m} \sum_{n=2}^{\infty} \frac{1}{p^{ns}} = \sum_{p \nmid m} \frac{1}{p^s(p^s - 1)},$$

where the last equality comes from a fact about geometric series—see [68, Theorem 1.2, p. 2]:

$$\sum_{n=2}^{\infty} \frac{1}{p^{ns}} = \sum_{n=0}^{\infty} \frac{1}{p^{ns}} - 1 - p^{-s} = \lim_{N \to \infty} \frac{p^{-(N+1)s} - 1}{p^{-s} - 1} - 1 - p^{-s}$$

$$= \frac{1}{1 - p^{-s}} - 1 - p^{-s} = \frac{1}{p^s(p^s - 1)}.$$

Also, since for $s \geq 1$,

$$\sum_{p \nmid m} \frac{1}{p^s(p^s - 1)} \leq \sum_{p \nmid m} \frac{1}{p(p - 1)} < \sum_{n=2}^{\infty} \frac{1}{n(n - 1)} = \sum_{n=2}^{\infty} \left(\frac{1}{n - 1} - \frac{1}{n} \right)$$

$$= \sum_{n=2}^{\infty} \frac{1}{n-1} - \sum_{n=2}^{\infty} \frac{1}{n} = 1 + \sum_{n=2}^{\infty} \frac{1}{n} - \sum_{n=2}^{\infty} \frac{1}{n} = 1,$$

we have shown that

$$\sum_{p \nmid m} \sum_{n=2}^{\infty} \frac{1}{p^{ns}}$$

is bounded as $s \to 1^+$. To complete Claim 7.3, it remains to show that the left-hand sum in (7.9) is bounded—see Exercise 7.6 on page 261 for some background into the following. Since

$$\exp(\sum_{n=1}^{\infty} z^n/n) = 1/(1-z),$$

then by substituting $z = \chi(p)p^{-s}$, we have

$$\exp\left(\sum_{n=1}^{\infty} \frac{\chi(p)^n}{np^{ns}}\right) = (1 - \chi(p)p^{-s})^{-1},$$

so it follows by taking products over primes then taking logarithms that

$$\sum_{p} \sum_{n=1}^{\infty} \frac{\chi(p)^n}{np^{ns}} = \log_e L(s, \chi),$$

and by Theorems 7.5 on page 256 and 7.6 on page 257, $L(1, \chi)$ remains bounded for $\chi \neq \chi_0$. Hence, the same is true for

$$\sum_{p \nmid m} \sum_{n=1}^{\infty} \frac{\chi(p)^n}{np^{ns}},$$

so we have Claim 7.3.

Now by (5.30) on page 221 and Exercise 5.18 on page 227,

$$\sum_{p \nmid m} p^{-s} \sim \log_e (s-1)^{-1},$$

so by Claim 7.3,

$$\sum_{p \nmid m} \frac{\chi(p)^n}{np^{ns}}$$

remains bounded for all $\chi \in G_{char}^m$ as $s \to 1$. Now we may use Claim 7.2 to conclude that

$$\sum_{p \equiv a \pmod{m}} p^{-s} \sim \frac{1}{\phi(m)} \log_e (s-1)^{-1},$$

namely

$$\mathcal{D}(\mathcal{S}_p^a) = \lim_{s \to 1} \frac{\sum_{p \equiv a \pmod{m}} p^{-s}}{\log_e (s-1)^{-1}} = \frac{1}{\phi(m)},$$

which secures our density result. $\qquad \square$

Corollary 7.6 | Dirichlet's Theorem 7.7 on page 258 |

$|S_p^a| = \infty.$

Proof. If S_p^a were finite, then by Exercise 7.11, $\mathcal{D}(S_p^a) = 0$, contradicting Theorem 7.9. □

Exercises

7.10. Prove that $\mathcal{D}(\mathbb{N}) = 1$. Conclude that any $S \subseteq \mathbb{N}$ where S contains all but finitely many primes must also satisfy $\mathcal{D}(S) = 1$.

7.11. Prove that if S is a set of primes in $\mathbb{Z}$, and $S \subseteq W \subseteq \mathbb{N}$ with $|S| < \infty = |W|$, then
$$\mathcal{N}\mathcal{D}(S, W) = 0 = \mathcal{D}(S).$$

7.12. Given sets S and W of primes in $\mathbb{Z}$ with $S \cap W = \varnothing$, and such that $\mathcal{D}(S), \mathcal{D}(W)$ both exist, prove that
$$\mathcal{D}(S \cup W) = \mathcal{D}(S) + \mathcal{D}(W).$$

7.13. In general a *multiplicative character* is a mapping from $\mathbb{F}_p$ (the finite field of p elements for a prime p) into $\mathbb{C}$ such that
$$\chi(ab) = \chi(a)\chi(b)$$
for all $a, b \in \mathbb{F}_p$. For instance, the Legendre symbol (a/p) for an odd prime p is such a character. The *principal character* χ_0 satisfies $\chi_0(a) = 1$ for all $a \in \mathbb{F}_p$, including $a = 0$, whereas $\chi(0) = 0$ for all $\chi \neq \chi_0$.

Prove that each of the following hold if $a \in \mathbb{F}_p^*$, the multiplicative group of nonzero elements of $\mathbb{F}_p$.

(a) $\chi(1_p) = 1$, where 1_p is the unit in $\mathbb{F}_p^*$.

(b) $\chi(a) = \zeta_{p-1}^j$ for some $j = 1, 2, \ldots, p-1$, where ζ_{p-1} is a primitive $p-1$-st root of unity.

(c) $\chi(a^{-1}) = \chi(a)^{-1} = \overline{\chi(a)}$.

Exercises 7.14–7.18 are all with reference to Exercise 7.13.

7.14. If χ is a multiplicative character prove that
$$\sum_{a \in \mathbb{F}_p^*} \chi(a) = \begin{cases} 0 & \text{if } \chi \neq \chi_0, \\ p & \text{if } \chi = \chi_0. \end{cases}$$

(*Hint: use the same technique as given in the proof of Theorem 7.2 on page 249.*)

7.15. If χ, γ are multiplicative characters, define the map

$$\chi\gamma : \mathbb{F}_p^* \mapsto \mathbb{C}$$

to be defined by $\chi\gamma(a) = \chi(a)\gamma(a)$, for $a \in \mathbb{F}_p^*$. Also, define the map

$$\chi^{-1} : \mathbb{F}_p^* \mapsto \mathbb{C}$$

to be defined by $\chi^{-1}(a) = \chi(a)^{-1}$ for $a \in \mathbb{F}_p^*$. Prove that these are again multiplicative characters and that the set of all multiplicative characters is in fact a cyclic group G of order $p - 1$.

7.16. With reference to Exercise 7.15, prove that if $a \in \mathbb{F}_p^*$ and $a \neq 1_p$, then

$$\sum_{\chi \in G} \chi(a) = 0.$$

(*Hint: Use the same technique as given in the proof of Theorem 7.2 on page 249.*)

7.17. Prove that if $a \in \mathbb{F}_p^*$, $m \mid (p-1)$, and $x^m \neq a$ for any $x \in \mathbb{F}_p^*$, then there is a character χ on $\mathbb{F}_p$ such that $\chi^m = \chi_0$ and $\chi(a) \neq 1$.

7.18. For $a \in \mathbb{F}_p$, let $N(m, a)$ denote the number of solutions of $x^m = a$ in $\mathbb{F}_p$, where $m \mid (p-1)$. Prove that

$$N(m, a) = \sum_{\chi^m = \chi_0} \chi(a).$$

7.19. With reference to Exercise 7.18, prove that if $p > 2$ is prime, then

$$N(2, a) = 1 + (a/p)$$

where $(*/p)$ is the Legendre symbol.

In Exercises 7.20–7.24, we will be referring to the following concept. If χ is a multiplicative character on $\mathbb{F}_p$, $a \in \mathbb{F}_p$, and ζ_p is a primitive p-th root of unity, then

$$\mathcal{G}_a(\chi) = \sum_{j \in \mathbb{F}_p} \chi(j)\zeta_p^{aj}$$

is called a Gauss sum over F_p *belonging to the character* χ.

7.20. Prove that if $a \neq 0$ and $\chi \neq \chi_0$, then $\mathcal{G}_a(\chi) = \chi(a^{-1})\mathcal{G}_1(\chi)$.

7.21. Prove that

$$\sum_{j\in\mathbb{F}_p} \zeta_p^{aj} = \begin{cases} 0 & \text{if } a \neq 0 \\ p & \text{if } a = 0. \end{cases}$$

7.22. Prove that

$$\mathcal{G}_a(\chi) = \begin{cases} 0 & \text{if } \chi = \chi_0, \text{ and } a \neq 0 \\ 0 & \text{if } \chi \neq \chi_0, \text{ and } a = 0 \\ p & \text{if } \chi = \chi_0, \text{ and } a = 0. \end{cases}$$

(*Hint: Use Exercises 7.21 and 7.14 on page 268.*)

7.23. Prove that for a prime p,

$$p^{-1} \sum_{j\in\mathbb{F}_p} \zeta_p^{j(a-b)} = \begin{cases} 0 & \text{if } a \neq b \\ 1 & \text{if } a = b. \end{cases}$$

(*Hint: Use Exercise 7.21.*)

7.24. Prove that if $\chi \neq \chi_0$, then

$$|\mathcal{G}_a(\chi)| = \sqrt{p}.$$

(*Hint: Use Exercises 7.20–7.21 and Exercise 7.23 to evaluate*

$$\sum_{j\in\mathbb{F}_p} \mathcal{G}_j(\chi)\overline{\mathcal{G}_j(\chi)}$$

in two ways.)

Chapter 8

Applications to Diophantine Equations

In a first course in number theory, elementary Diophantine equations are studied and we assume herein familiarity with the fundamentals such as in [68, Chapters 1, 5, & 7], where norm-form equations, including Pell's equation, are completely solved via continued fractions, as are linear equations by congruence conditions. We have already encountered some nonlinear Diophantine equations in our developments in Chapter 1, especially in Theorem 1.8 on page 14, where we looked at the Ramanujan–Nagell equation and its solutions. We revisit this equation in §8.2, where we study solutions of the *generalized* Ramanujan–Nagell equation introduced in Definition 1.10 on page 13. We begin with a theory to solve these latter equations.

8.1 Lucas–Lehmer Theory

Let α and β be the roots of

$$x^2 - \sqrt{R}x + Q = 0, \tag{8.1}$$

where $R, Q \in \mathbb{Z}$, with $\gcd(R, Q) = 1$. By Exercise 8.1 on page 275,

$$\alpha + \beta = \sqrt{R}, \quad \alpha\beta = Q, \text{ and } \alpha - \beta = \sqrt{R - 4Q}. \tag{8.2}$$

Set

$$\sqrt{\Delta} = \sqrt{R - 4Q}.$$

By Exercise 8.2 on page 275,

$$2\alpha = \sqrt{R} + \sqrt{\Delta} = \sqrt{R} + \sqrt{R - 4Q}, \tag{8.3}$$

and

$$2\beta = \sqrt{R} - \sqrt{\Delta} = \sqrt{R} - \sqrt{R - 4Q}. \tag{8.4}$$

Definition 8.1 | Lucas Functions

Let $n \geq 0$ be an integer. Then the following are called *Lucas functions*:

$$U_n = (\alpha^n - \beta^n)/(\alpha - \beta),$$

and

$$V_n = \alpha^n + \beta^n.$$

The above were dubbed functions rather than sequences by Lucas, then later extended by D.H. Lehmer—see [68, Biographies 1.18–1.19, pp. 63–64].

Remark 8.1 Note that when discussing divisibility properties of Lucas functions in what follows, in order to avoid confusion, we assume that a factor of $\sqrt{R}$ may be ignored in U_n or V_n. For instance, if $R = 5$, and $Q = -3$, then $U_3 = 8$, and $U_6 = 112\sqrt{5}$. We say that $\gcd(U_3, U_6) = 8$, and U_6 is called even, ignoring $\sqrt{5}$. Also, m, n are nonnegative integers throughout.

Theorem 8.1 | Properties of Lucas Functions

(a) $U_{n+2} = \sqrt{R}U_{n+1} - QU_n$.

(b) $V_{n+2} = \sqrt{R}V_{n+1} - QV_n$.

(c) $2Q^m V_{n-m} = V_n V_m - \Delta U_n U_m$ $(n > m)$.

(d) $V_n^2 - \Delta U_n^2 = 4Q^n$.

(e) $2U_{m+n} = U_n V_m + V_n U_m$.

(f) $2V_{m+n} = V_m V_n + \Delta U_m U_n$.

(g) For all $m \in \mathbb{N}$, $((V_1 + U_1\sqrt{\Delta})/2)^m = (V_m + U_m\sqrt{\Delta})/2$.

Proof. **(a)**: From (8.1)–(8.4) and Definition 8.1, we have that

$$\sqrt{R}U_{n+1} - QU_n = (\alpha + \beta)\frac{\alpha^{n+1} - \beta^{n+1}}{\alpha - \beta} - \alpha\beta\frac{\alpha^n - \beta^n}{\alpha - \beta}$$

$$= \frac{\alpha^{n+2} - \alpha\beta^{n+1} - \beta^{n+2} + \beta\alpha^{n+1} - \alpha^{n+1}\beta + \alpha\beta^{n+1}}{\alpha - \beta}$$

$$= \frac{\alpha^{n+2} - \beta^{n+2}}{\alpha - \beta} = U_{n+2}.$$

(b): From (8.1)–(8.4) we also have that,

$$\sqrt{R}V_{n+1} - QV_n = (\alpha + \beta)(\alpha^{n+1} + \beta^{n+1}) - \alpha\beta(\alpha^n + \beta^n)$$

$$= \alpha^{n+2} + \alpha\beta^{n+1} + \beta\alpha^{n+1} + \beta^{n+2} - \alpha^{n+1}\beta - \alpha\beta^{n+1}$$

$$= \alpha^{n+2} + \beta^{n+2} = V_{n+2}.$$

(c): We use induction on m. If $m = 0$, then the result is clear. Assume that

$$2Q^{m-j}V_{n-m+j} = V_n V_{m-j} - \Delta U_n U_{m-j},$$

for $1 \le j < m$. Then by parts (a)–(b),

$$V_n V_m - \Delta U_n U_m = V_n(\sqrt{R}V_{m-1} - QV_{m-2}) - \Delta U_n(\sqrt{R}U_{m-1} - QU_{m-2})$$

$$= \sqrt{R}(V_n V_{m-1} - \Delta U_n U_{m-1}) - Q(V_n V_{m-2} - \Delta U_n U_{m-2})$$

$$= \sqrt{R}(2Q^{m-1}V_{n-m+1}) - Q(2Q^{m-2}V_{n-m+2}),$$

where the last equality is from the induction hypothesis, and this equals

$$2Q^{m-1}(\sqrt{R}V_{n-m+1} - V_{n-m+2}) = 2Q^m V_{n-m},$$

where the last equality is from part (b).

(d): Use induction on n. The induction step is

$$V_0^2 - \Delta U_0^2 = 4, \text{ with } U_0 = 0, V_0 = 2.$$

The induction hypothesis is $V_i^2 - \Delta U_i^2 = 4Q^i$ for all $i < n$. By parts (a)–(b)

$$V_n^2 - \Delta U_n^2 = (\sqrt{R}V_{n-1} - QV_{n-2})^2 - \Delta(\sqrt{R}U_{n-1} - QU_{n-2})^2$$

$$= R(V_{n-1}^2 - \Delta U_{n-1}^2) - 2\sqrt{R}Q(V_{n-1}V_{n-2} - \Delta U_{n-1}U_{n-2}) + Q^2(V_{n-2}^2 - \Delta U_{n-2}^2),$$

which, by induction hypothesis and part (c), must equal

$$4RQ^{n-1} - 2\sqrt{R}Q(2Q^{n-2}V_1) + Q^2(4Q^{n-2}),$$

and since $V_1 = \sqrt{R}$ by (8.2), and Definition 8.1, then the latter equals $4Q^n$, which secures part (d).

(e): We have from Definition 8.1,

$$U_n V_m + V_n U_m = \frac{(\alpha^n - \beta^n)(\alpha^m + \beta^m)}{\alpha - \beta} + \frac{(\alpha^n + \beta^n)(\alpha^m - \beta^m)}{\alpha - \beta} =$$

$$\frac{\alpha^{n+m} + \alpha^n\beta^m - \alpha^m\beta^n - \beta^{m+n} + \alpha^{n+m} - \alpha^n\beta^m + \alpha^m\beta^n - \beta^{m+n}}{\alpha - \beta} =$$

$$2\frac{\alpha^{n+m} - \beta^{n+m}}{\alpha - \beta} = 2U_{n+m}.$$

(f): From Definition 8.1 and (8.2),

$$V_m V_n + \Delta U_n U_m =$$

$$(\alpha^m + \beta^m)(\alpha^n + \beta^n) + (\alpha - \beta)^2 \frac{(\alpha^n - \beta^n)(\alpha^m - \beta^m)}{(\alpha - \beta)^2} =$$

$$\alpha^{n+m} + \alpha^m\beta^n + \alpha^n\beta^m + \beta^{m+n} + \alpha^{n+m} - \alpha^n\beta^m - \alpha^m\beta^n + \beta^{m+n} =$$

$$2(\alpha^{n+m} + \beta^{n+m}) = 2V_{n+m}.$$

(g): We use induction on m. For $m = 1$, the result is obvious. Assume that

$$\left(\frac{V_1 + U_1\sqrt{\Delta}}{2}\right)^{m-1} = \frac{V_{m-1} + U_{m-1}\sqrt{\Delta}}{2}.$$

Then

$$\left(\frac{V_1 + U_1\sqrt{\Delta}}{2}\right)^{m} = \left(\frac{V_1 + U_1\sqrt{\Delta}}{2}\right)\left(\frac{V_1 + U_1\sqrt{\Delta}}{2}\right)^{m-1} =$$

$$\left(\frac{V_1 + U_1\sqrt{\Delta}}{2}\right)\left(\frac{V_{m-1} + U_{m-1}\sqrt{\Delta}}{2}\right) =$$

$$\frac{(V_1 V_{m-1} + U_1 U_{m-1}\Delta) + (U_1 V_{m-1} + V_1 U_{m-1}\sqrt{\Delta})}{4}.$$

By parts (e)–(f), this equals

$$\frac{2V_m + 2U_m\sqrt{\Delta}}{4} = \frac{V_m + U_m\sqrt{\Delta}}{2},$$

which secures the entire result.   □

In §8.2, we will use the properties given in Theorem 8.1 to solve the generalized Ramanujan–Nagell equation for certain cases as well as some related equations that we will develop therein. The exercises below are designed to give the reader a grounding in the properties developed above by expanding the theory.

Exercises

8.1. Verify the equations in (8.2) on page 271, where the positive square root in the formula for $\alpha - \beta$ is guaranteed by an appropriate selection of α and β.

8.2. Verify (8.3)–(8.4) on page 272.

In Exercises 8.3–8.12, prove each of the statements involving the Lucas functions given in Definition 8.1 on page 272.

8.3. (a) $U_{2n+1} \in \mathbb{Z}$, $V_{2n} \in \mathbb{Z}$.
   (b) U_{2n} and V_{2n+1} are integer multiples of $\sqrt{R}$.

8.4. $2Q^m U_{n-m} = U_n V_m - V_n U_m$ $(n > m)$.

8.5. For $n \in \mathbb{N}$,

$$2^{n-1} U_n = \sum_{j=1}^{\lfloor (n+1)/2 \rfloor} \binom{n}{2j-1} V_1^{n-2j+1} \Delta^{j-1},$$

and

$$2^{n-1} V_n = \sum_{j=0}^{\lfloor n/2 \rfloor} \binom{n}{2j} V_1^{n-2j} \Delta^j.$$

8.6. $\gcd(U_n, Q) = 1 = \gcd(V_n, Q)$, and $\gcd(U_n, V_n)$ divides 2.

8.7. If U_n is even, then one of the following must hold:

   (a) $R \equiv 0 \pmod 4$, Q is odd and n is even.
   (b) $R \equiv 2 \pmod 4$, Q is odd and $n \equiv 0 \pmod 4$.
   (c) R is odd, Q is odd and $n \equiv 0 \pmod 3$.

8.8. If V_n is even, then one of the following must hold:

   (a) $R \equiv 0 \pmod 4$ and Q is odd.
   (b) $R \equiv 2 \pmod 4$, Q is odd and n is even.
   (c) R and Q are odd and $n \equiv 0 \pmod 3$.

8.9. If $m|n$, $m \geq 1$, then $U_m | U_n$.

8.10. If $m \mid n$ and n/m is odd, then $V_m | V_n$.

8.11. If $\gcd(m, n) = g$, then $\gcd(U_m, U_n) = U_g$.

8.12. (a) Assume that $|Q| > 1$. Prove that $U_n \neq 0$ for any $n \in \mathbb{N}$.
   (b) Give an example for each of the cases $Q = \pm 1$ to show that $U_n = 0$ for some $n \in \mathbb{N}$.
   (c) Assume that $|Q| > 1$, and $m \in \mathbb{N}$. Prove that if $U_m = U_n$, and $V_m = V_n$, then $m = n$.

8.2   Generalized Ramanujan–Nagell Equations

> *All generalizations are dangerous, even this one.*
>   **Alexandre Dumas (*Dumas fils*) (1824–1895)**
>   **French dramatist, novelist, and principal**
> **creator of the 19-th century *comedy of manners*—illegitimate son**
> **of *Dumas Père*, also named Alexandre Dumas (1802–1870), author**
> **of *The Count of Monte Cristo* and *The Three Musketeers*.**

Recall from Definition 1.10 on page 13 that the *generalized Ramanujan–Nagell equation* is given by

$$x^2 - D = p^n. \tag{8.5}$$

In Theorem 1.8 on page 14, we provided all solutions for $p = 2$ and $D = -7$, which were known by Ramanujan and later proved by Nagell to indeed be *all* of them. For the odd prime p case, the history is varied–see [62, p. 70ff] for details.

We may use the result of §8.1 to solve certain of the equations in the title of this section. We begin with a result due to Alter and Kubota from 1973 [2], albeit they use different methods than the Lucas–Lehmer theory coupled with ideal theory that we employ below. Some of the following is adapted from [65].

Remark 8.2 With reference to Exercises 2.2–2.4 on page 66, a primitive R-ideal $I = (a, (\ell + \sqrt{D})/2)$ with $\ell \equiv D \pmod 2$ is called *invertible* if

$$\gcd\left(a, \ell, \frac{\ell^2 - D}{4a}\right) = 1.$$

Then the multiplication formulas on page 59 hold for such ideals, given the discussion therein. It can be shown that the invertible ideals form a group in the same fashion as in Definition 3.7 on page 109, and equivalence of such ideals is similarly denoted by $I \sim J$. Also, the order d of an ideal I in this class group of $\mathcal{O}_D = \mathbb{Z}[\omega_D]$ is defined by the property that $I^d \sim 1$ and $I^n \not\sim 1$ for any $n < d$. Furthermore, if $I^n \sim 1$, then $d \mid n$. Moreover, as with the ideal theory developed in Chapter 2, invertible ideals can be uniquely factored into products of prime ideals–see [62, §1.5] for the general development of these notions. We will use these facts for our special case below to prove our desired result on the equations in the title, and pave the way for the use of Lucas-Lehmer theory.

Theorem 8.2 | Generalized Ramanujan–Nagell Equations: Solutions

Suppose that p is an odd prime, $D \in \mathbb{Z}$ with $D < 0$, and $D \equiv 5 \pmod 8$. If $d \in \mathbb{N}$ is the least value such that

$$a^2 - Db^2 = 4p^d \tag{8.6}$$

for some $a, b \in \mathbb{N}$ with $\gcd(bD, 2p) = 1$, and $(D, p) \neq (-3, 7)$, then the generalized Ramanujan–Nagell equation

$$x^2 - D = p^n \tag{8.7}$$

has a solution $x, n \in \mathbb{N}$ if and only if $b = 1$ and $D = -3a^2 \pm 8$. The unique solution is given by

$$x = \left| \frac{a(a^2 + 3D)}{8} \right| \text{ and } n = 3d.$$

Proof. By Exercise 2.4 on page 66, $I = (p, (a + \sqrt{b^2 D})/2)$ is an ideal in the ring $\mathbb{Z}[(1 + \sqrt{b^2 D})/2]$. Since d is the least natural number such that (8.6) holds, then we have $I^d = (p^d, (a + \sqrt{b^2 D})/2) = (p^d)$ and for no smaller value m do we have I^m equal to a principal ideal. Thus, d is the order of I. If (8.7) holds for some $x \in \mathbb{N}$, then $(p^n) = (x - \sqrt{D})(x - \sqrt{D})$. Therefore, $(p^n) = I^n(I')^n$, where $I' = (p, (a - \sqrt{b^2 D})/2)$. Hence,

$$(x + \sqrt{D})(x - \sqrt{D}) = I^n(I')^n, \tag{8.8}$$

and we claim that $(x + \sqrt{D})$ and $(x - \sqrt{D})$ are relatively prime. If not, then by Remark 8.2, there is a prime ideal $\mathcal{P}$ dividing both of them. Hence, both $(x + \sqrt{D})$ and $(x - \sqrt{D})$ are in $\mathcal{P}$, by the same reasoning as in the proof of Corollary 2.5 on page 76. Thus, both $p \in \mathcal{P}$ and $D \in \mathcal{P}$. However, $\gcd(p, D) = 1$, so there exist $r, s \in \mathbb{Z}$ such that $pr + Ds = 1 \in \mathcal{P}$. Hence, $\mathcal{P} = \mathcal{O}_D$, a contradiction so they are indeed relatively prime. Therefore, by (8.8), we may assume that $I^n = (x + \sqrt{D})$ and $(I')^n = (x - \sqrt{D})$, without loss of generality since p is prime and the only units in $\mathcal{O}_D$ are ± 1—see Exercise 8.18 on page 285. Thus, $I^n \sim (I')^n \sim 1$, so by Remark 8.2, $d \mid n$.

Now we may invoke the Lucas–Lehmer theory. Let $\Delta = b^2 D$, $R = a^2$, $V_1 = a$, and $U_1 = 1$ in the notation of §8.1. Then

$$p^d = N((V_1 + \sqrt{\Delta})/2),$$

so

$$N(x + \sqrt{D}) = p^n = N([(V_1 + \sqrt{\Delta})/2]^{n/d}) = N([(V_{n/d} + U_{n/d}\sqrt{\Delta})/2]),$$

where the last equality follows from part (g) of Theorem 8.1 on page 272. Also, since $x + \sqrt{D}$ and $x - \sqrt{D}$ are relatively prime, then

$$(V_{n/d} + U_{n/d}\sqrt{\Delta})/2 = \pm(x + \sqrt{D}) \text{ or } \pm(x - \sqrt{D}). \tag{8.9}$$

Claim 8.1 $n > d$.

Suppose that $n = d$. Then by (8.9), $U_1 b = b = \pm 2$, but b is odd, so $b = 1$. Therefore, $a^2 - D = 4p^d$ and $x^2 - D = p^d$. By subtracting the two equations, we get $a^2 - x^2 = 3p^d$. If $a - x = 3p^r$ and $a + x = p^s$, then $2a = 3p^r + p^s$. Since $p \nmid 2a$, then $r = 0$, and $s = d$. Therefore, $x = a - 3$, and $p^d = a + x = 2a - 3$, so

$$D = x^2 - p^d = (a - 3)^2 - 2a + 3 = a^2 - 8a + 12 < 0.$$

The latter can hold only when $a < 6$. Since a is odd and $D \equiv 5 \,(\mathrm{mod}\, 8)$, then only $a = 5$ works, namely when $D = -3$, $x = 2$, $p = 7$, and $d = 1$. This is the entire analysis. The reason is that 3 can only divide one of $a - x$ or $a + x$, and since p^d can only divide one of the factors, we would have $a + x = 3p^d$ and $a - x = 1$ otherwise, which one can easily show to be impossible.

By Claim 8.1, $n > d$ so $bU_{n/d} = \pm 2$, and $V_{n/d} = \pm 2x$. Since b is odd, then $b = 1$. By Exercises 8.7–8.8 on page 275,

$$n/d \equiv 0 \pmod 3 \quad \text{and} \quad U_3 | U_{n/d}.$$

However, $U_3 = (3a^2 + D)/4$. Therefore, $(3a^2 + D)/4$ divides ± 2. If $(3a^2 + D)/4 = \pm 1$, then $3a^2 \equiv 7 \,(\mathrm{mod}\, 8)$, so $a^2 \equiv 5 \,(\mathrm{mod}\, 8)$, a contradiction. It follows that

$$D = -3a^2 \pm 8. \tag{8.10}$$

Now we consider

$$\frac{V_{n/d} + U_{n/d}\sqrt{\Delta}}{2} = \left(\frac{V_{n/(3d)} + U_{n/(3d)}\sqrt{\Delta}}{2}\right)^3 =$$

$$\frac{1}{8}[V_{n/(3d)}^3 + 3U_{n/(3d)}^2 V_{n/(3d)} D + (3V_{n/(3d)}^2 U_{n/(3d)} + U_{n/(3d)}^3 D)\sqrt{\Delta}].$$

Hence,

$$(3V_{n/(3d)}^2 U_{n/(3d)} + U_{n/(3d)}^3 D)/4 = U_{n/d},$$

and

$$(V_{n/(3d)}^3 + 3U_{n/(3d)}^2 V_{n/(3d)} D)/4 = V_{n/d}.$$

In other words,

$$3V_{n/(3d)}^2 U_{n/(3d)} + U_{n/(3d)}^3 D = \pm 8, \tag{8.11}$$

and

$$V_{n/(3d)}^3 + 3U_{n/(3d)}^2 V_{n/(3d)} D = \pm 8x. \tag{8.12}$$

From (8.11), $U_{n/(3d)} = \pm 1$ or ± 2. If $U_{n/(3d)} = \pm 1$, then (8.11) becomes

$$3V_{n/(3d)}^2 + D = \pm 8. \tag{8.13}$$

In view of (8.10), we must have $V_{n/(3d)} = a = V_1$. Since $U_{n/(3d)} = \pm 1$, then by part (d) of Theorem 8.1, $4p^{n/3} = a^2 - D$. However, $a^2 - D = 4p^d$, by (8.6). Hence, $n = 3d$. Furthermore, from (8.12), $x = |a(a^2 + 3D)/8|$, as required.

If $U_{n/(3d)} = \pm 2$, then (8.10) forces (8.11) to become

$$(V_{n/(3d)}/2)^2 - a^2 = \pm 3,$$

for which only $a = 1$ is possible. From (8.10), we get $D = -11$, $p^d = 3 = n$, so $\pm 2 = U_{n/(3d)} = U_1 = 1$, a contradiction. (Notice that the case where $D = -11$ is covered by $U_{n/(3d)} = \pm 1$.)

The converse is clear. $\qquad\qquad\qquad\qquad\qquad\qquad\qquad\qquad\qquad\qquad\qquad \square$

Example 8.1 If $D = -19$, $a = 3$, $b = 1$, and $p = 5$, then $d = 1$ and

$$x^2 + 19 = 7^n$$

has the unique positive solution $x = 18$ and $n = 3$.

Example 8.2 If $D = -83$, then the unique positive solution to

$$x^2 + 83 = 3^n$$

is $x = 140$ where $b = 1$, $n = 9$, $d = 3$, and $a = 5$.

Example 8.3 If $D = -41075 = -5^2 \cdot 31 \cdot 53$, then

$$x^2 + 41075 = 13961^n$$

has the unique positive solution $x = 1601964$ with $n = 3$, $b = d = 1$, and $a = 117$.

Remark 8.3 In Theorem 8.2, we only considered the case where $D \equiv 5 \pmod 8$ and $D < 0$. We observe that if $D \equiv 3 \pmod 4$, then *(8.6)* cannot hold since $D \not\equiv a^2 \pmod 4$ in that case. Also, if $D \equiv 1 \pmod 8$, then *(8.6)* implies that $p = 2$. Hence, we need an equation different from *(8.6)* to treat other cases. We have a partial solution to the remaining cases in what follows.

Theorem 8.3 | **More Solutions to Ramanujan–Nagell** |

Let $p > 2$ be prime, $D \in \mathbb{Z}$, $D < 0$, and $p \nmid D$. Suppose that $d \in \mathbb{N}$ is the smallest solution to

$$a^2 - Db^2 = p^d,$$

for $a, b \in \mathbb{N}$. Then the Diophantine Equation

$$x^2 - D = p^n \qquad (8.14)$$

has a solution $x, n \in \mathbb{N}$ with $n > d$, if and only if $b = 1$, and $n = dq$, where $q > 2$ is prime. In particular, if $n = 3m$, then (8.14) has a solution $x, m \in \mathbb{N}$ with $n > d$ if and only if

$$d = m = 1, \quad D = -3a^2 \pm 1, p = 4a^2 \pm 1, \quad and \ x = 8a^3 \pm 3a.$$

Proof. Suppose that q is any prime dividing $n \in \mathbb{N}$. Also, let $m = n/q$ be the least value such that $x^2 - D = p^{qm}$ has a solution $x \in \mathbb{N}$. The first part of this proof employs essentially the same reasoning as that of Theorem 8.2 on page 276, namely there is a primitive $\mathfrak{O}_{4D} = [1, \sqrt{D}]$-ideal $I = [p, c + \sqrt{D}]$, with $I^d \sim 1$ and $d \mid n$. The only difference here is that we are working in the order $\mathfrak{O}_{4D}$ rather than the order $[1, (1 + \sqrt{D})/2]$ used in Theorem 8.2. Now we invoke Lucas–Lehmer theory again.

Set $\Delta = 4b^2 D$, $R = 4a^2$, $Q = p^d$, $U_1 = 1$, and $V_1 = 2a$. Then

$$p^d = a^2 - b^2 D = N[(V_1 + \sqrt{\Delta})/2]$$

and

$$N(x + \sqrt{D}) = p^n = N[(V_{n/d} + U_{n/d}\sqrt{\Delta})/2].$$

Hence, $bU_{n/d} = \pm 1$ and $V_{n/d} = \pm 2x$. Thus, $b = 1$. If $n/d \not\equiv 0 \,(\mathrm{mod}\ q)$, then $q|d$, and so by the minimality of m, we must have $qm = n = d$, contradicting the hypothesis. If $q = 2$, then $U_2 = V_1 = 2a|U_{n/d}$ by Exercise 8.9 on page 275. This contradiction ensures that $q > 2$, and $n/d \equiv 0 \,(\mathrm{mod}\ q)$. By Exercise 8.9, again $U_q|U_{n/d}$, so $U_q = \pm 1$. Thus $(V_q/2)^2 - D = p^{dq}$. By the minimality of m, we must have $n = dq$.

If $q = 3$, then $U_3 = U_{n/d}$. Since $U_3 = 3a^2 + D$, then $D = -3a^2 \pm 1$. Therefore, $p^d = a^2 - D = 4a^2 \pm 1$. If $p^d = 4a^2 - 1$, then $p^d = (2a-1)(2a+1)$, which is possible only for $a = 1 = d$, since p is an odd prime. Hence, $D = -2$. The solutions are

$$a^2 - D = 1^2 + 2 = 3 = p^d \text{ with } x^2 - D = 5^2 + 2 = 3^3 = p^n.$$

This exhausts the case where $p = 4a^2 - 1$, namely $D = -3a^2 + 1$. We assume henceforth that $p^d = 4a^2 + 1$, and $D = -3a^2 - 1$.

Claim 8.2 *Since $p^d = 4a^2 + 1$, then $d = 1$.*

By repeated use of the equation for sums of two squares given in Remark 1.12 on page 27, and a simple induction argument, we see that no prime power p^d can be a sum of two squares with 1^2 as one of the summands unless $d = 1$.

If n is even, then $U_2|U_n$, by Exercise 8.9 again. However, by part (e) of Theorem 8.1, $U_2 = V_1 = 2a$ divides $U_n = \pm 1$, a contradiction. Hence n is odd. By Exercise 8.10, $(2a^3 + 6aD) = V_3|V_n = \pm 2x$. Thus,

$$\pm 2x = 2a^3 + 6aD = 2a^3 + 6a(-3a^2 - 1) = -16a^3 - 6a.$$

Therefore, since $x \in \mathbb{N}$, $x = 8a^3 + 3a$. It remains to show that $n = 3$. If $n \neq 3$, then by Exercise 8.4, $2p^3 U_{n-3} = U_n V_3 - V_n U_3$. However, $U_3 = -1$, $V_3 = -2x$, $U_n = \pm 1$, and $V_n = \pm 2x$, from the above analysis. Hence, $p^3 U_{n-3} = 0$ or $\pm 2x$, a contradiction in any case since $p \nmid 2x$, and by part (a) of Exercise 8.12, $U_{n-3} \neq 0$. Hence, $n = 3$. $\qquad\square$

The following is immediate from Theorem 8.3.

Corollary 8.1 *Suppose that $p > 2$ is a prime not dividing $D \in \mathbb{Z}$, where $D < 0$. If there exist $a, b \in \mathbb{N}$ such that $p = a^2 - b^2 D$, then*

$$x^2 - D = p^{3d}$$

has a solution $x, d \in \mathbb{N}$ if and only if $b = d = 1$, $D = -3a^2 \pm 1$, $p = 4a^2 \pm 1$, and $x = 8a^3 \pm 3a$.

Example 8.4 Let $D = -2$. By Corollary 3.10 on page 151, $p = a^2 + 2b^2$ is solvable for any prime p such that $p \equiv 1, 3 \pmod 8$. Therefore, by Corollary 8.1,

$$x^2 + 2 = p^{3m}$$

is solvable if and only if $b = m = 1$, $x = 5$, and $p = 3$, namely

$$1 + 2 = 3 \text{ and } 5^2 + 2 = 3^3.$$

Here $a = 1$, $D = -3a^2 + 1$, $x = 8a^3 - 3a$, and $p = 4a^2 - 1$.

Example 8.5 If $D = -5$, and p is a prime with $p \equiv 1, 9 \pmod{20}$, then by part (a) of Exercise 3.46 on page 153, $p = a^2 + 5b^2$ for some $a, b \in \mathbb{N}$. Therefore, by Corollary 8.1,

$$x^2 + 5 = p^{3m}$$

has no solutions $x, m \in \mathbb{N}$.

Remark 8.4 Note that in [12], Bugeaud and Shorey look at the generalized Ramanujan–Nagell equations of the form $D_1 x^2 + D_2 = k^n$ in unknowns $x, n \in \mathbb{N}$. They provide necessary and sufficient conditions on D_1, D_2, and k for the equation to have at most $2^{\omega(k)-1}$ solutions where $\omega(k)$ denotes the number of distinct prime divisors of k. It follows that when k is prime the necessary and sufficient conditions determine when the equation has at most one solution. They also completely solve the related equation $x^2 + 7 = 4y^n$, demonstrating that there are no solutions for $y > 2$, $n > 1$, and $x \in \mathbb{N}$. There are a couple of errors however in the paper, corrected by this author in [69], which closes the door on the equation in the title.

Exercises

8.13. If $D = -43$, and $x^2 + 43 = 47^{3d}$ find solutions if they exist.

8.14. If $D = -49$, and $x^2 + 49 = 53^{3d}$ find solutions if they exist.

8.15. If $D = -225$, and $x^2 + 225 = 17^{3d}$ find solutions if they exist.

8.16. If $D = -2209$, and $x^2 + 2209 = 17^n$ find solutions if they exist.

8.17. Find solutions of $x^2 + 161047 = 11^n$ if they exist.

8.3   Bachet's Equation

> *Science is one thing, wisdom is another. Science is an edged tool, with which*
> *men play like children, and cut their own fingers.*
>
> **Arthur Eddington (1882–1944)**
> **British astrophysicist**

We covered instances of Bachet's equation—see [68, Biography 7.2, p. 279],

$$y^2 = x^3 + k \tag{8.15}$$

in §1.4, and [68, §7.3, pp. 277–280]. We extend that investigation by looking
at more advanced use of techniques to solve Bachet's equation. In a beginning
course in number theory Bachet's equation is solved via elementary congruence
conditions. Now that we have algebraic numbers at our disposal, we may proceed to show how those techniques may be applied. This falls in line with §8.2,
where we applied the ideal theory and Lucas–Lehmer theory to solve instances
of the generalized Ramanjuan–Nagell equations. The reader should prepare by
looking at Exercises 8.18–8.20 on page 285 to be reminded of the theory we
developed in Chapters 1–2 and the facts we will use in the following.

Theorem 8.4 | **Solutions of Bachet's Equation**

*Let $F = \mathbb{Q}(\sqrt{k})$ be a complex quadratic field with radicand $k < -1$ such that
$k \not\equiv 1 \,(\mathrm{mod}\ 4)$, and $h_{\mathfrak{O}_F} \not\equiv 0 \,(\mathrm{mod}\ 3)$. Then there are no solutions of (8.15) in
integers x, y except in the following cases: there exists an integer u such that*

$$(k, x, y) = (\pm 1 - 3u^2, 4u^2 \mp 1, \varepsilon u(3 \mp 8u^2)),$$

*where the $\pm$ signs correspond to the $\mp$ signs and $\varepsilon = \pm 1$ is allowed in either
case.*

Proof. Suppose that for k as given in the hypothesis, (8.15) has a solution.

Claim 8.3 $\gcd(x, 2k) = 1$.

Given that $y^2 \equiv 0, 1 \,(\mathrm{mod}\ 4)$, and $k \equiv 2, 3 \,(\mathrm{mod}\ 4)$, then

$$x^3 = y^2 - k \equiv 1, 2, 3 \pmod{4}.$$

However, $x^3 \equiv 2 \,(\mathrm{mod}\ 4)$ is not possible. Hence, x is odd. Now let p be a prime
such that $p \mid \gcd(x, 2k)$, where $p > 2$ since x is odd. Since k is a radicand, it is
squarefree, so

$$p \| k = y^2 - x^3. \tag{8.16}$$

However, $p \mid x$ so $p \mid y$, which implies that $p^2 \mid (y^2 - x^3)$, a contradiction to
(8.16), that establishes the claim.

By Claim 8.3, there exist integers r, s such that

$$rx + 2ks = 1. \tag{8.17}$$

Claim 8.4 *The $\mathfrak{O}_F$-ideals $(y + \sqrt{k})$ and $(y - \sqrt{k})$ are relatively prime.*

If the claim does not hold, then there is a prime $\mathfrak{O}_F$-ideal $\mathcal{P}$ dividing both of the given ideals by Theorem 2.13 on page 80. Therefore, by Corollary 2.5 on page 76, $y \pm \sqrt{D} \in \mathcal{P}$. Therefore, $2\sqrt{k} = y + \sqrt{k} - (y - \sqrt{k}) \in \mathcal{P}$, so

$$2\sqrt{k} \cdot \sqrt{k} = 2k \in \mathcal{P}. \tag{8.18}$$

Given that

$$(y + \sqrt{k})(y - \sqrt{k}) = (y^2 - k) = (x^3) = (x)^3,$$

then by Corollary 2.5 again, since $(x)^3 \subseteq \mathcal{P}$, then $\mathcal{P} \mid (x)^3$. However, since $\mathcal{P}$ is prime $\mathcal{P} \mid (x)$, and once more by Corollary 2.5, we conclude that

$$x \in \mathcal{P}. \tag{8.19}$$

Now we invoke (8.17)–(8.19) to get that both rx and $2ks$ are in $\mathcal{P}$ so $1 = rx + 2ks \in \mathcal{P}$, a contradiction that establishes the claim.

By Theorem 2.9 on page 73, $\mathfrak{O}_F$ is a Dedekind domain, so by Claim 8.17 and Exercise 8.18 on page 285, there exists an integral $\mathfrak{O}_F$-ideal $\mathfrak{I}$ such that $(y + \sqrt{k}) = \mathfrak{I}^3$. In other words, $\mathfrak{I}^3 \sim 1$, but $h_{\mathfrak{O}_F} \not\equiv 0 \,(\mathrm{mod}\, 3)$, so by Exercise 8.19, $\mathfrak{I} \sim 1$. Thus, by Theorem 1.3 on page 6, there exist integers u, v such that $\mathfrak{I} = (u + v\sqrt{k})$. Hence, $(y + \sqrt{k}) = (u + v\sqrt{k})^3 = \left([u + v\sqrt{k}]^3\right)$.

By Exercise 8.20, there is a unit w in $\mathfrak{O}_F$ such that

$$y + \sqrt{k} = w(u + v\sqrt{k})^3, \tag{8.20}$$

and by Theorem 1.4 on page 8, $w = \pm 1$. Now we conjugate (8.20) to get

$$y - \sqrt{k} = w(u - v\sqrt{k})^3. \tag{8.21}$$

Hence,

$$x^3 = y^2 - k = (y - \sqrt{k})(y + \sqrt{k}) = w^2(u + v\sqrt{k})^3(u - v\sqrt{k})^3 = (u^2 - v^2 k)^3.$$

Therefore,

$$x = u^2 - v^2 k. \tag{8.22}$$

Now by adding (8.20)–(8.21), we get

$$2y = w\left[(u + v\sqrt{k})^3 + (u - v\sqrt{k})^3\right] = 2w(u^3 + 3uv^2 k), \tag{8.23}$$

and by subtracting (8.21) from (8.20), we get

$$2\sqrt{k} = w\left[(u + v\sqrt{k})^3 - (u - v\sqrt{k})^3\right] = 2w\sqrt{k}(3u^2 v + v^3 k). \tag{8.24}$$

Hence, from (8.23)–(8.24), we get, respectively, that

$$y = w(u^3 + 3uv^2 k) \tag{8.25}$$

and
$$1 = w(3u^2 v + v^3 k) = wv(3u^2 + v^2 k). \tag{8.26}$$

From (8.26), we get that $v = \pm w$, so from (8.22), (8.25)–(8.26), we have,

$$x = u^2 - k, \ y = w(u^3 + 3uk), \text{ and } 1 = \pm(3u^2 + k).$$

It follows that $k = \pm 1 - 3u^2$, $x = 4u^2 \mp 1$, and $y = \varepsilon(3u \mp 8u^2)$, where $\varepsilon = \pm 1$ is allowed in either case. Therefore, the two cases are encapsulated in the following, $(k, x, y) = (\pm 1 - 3u^2, 4u^2 \mp 1, \varepsilon u(3 \mp 8u^2))$, and

$$x^3 + k = (4u^2 \mp 1)^3 \pm 1 - 3u^2 = 64u^6 \mp 48u^4 + 9u^2 = (\varepsilon u(3 \mp 8u^2))^2 = y^2,$$

as required.   □

Remark 8.5 Note that in Theorem 8.4, u is odd when $k = 1 - 3u^2$ and u is even when $k = -1 - 3u^2$ by the hypothesis that $k \not\equiv 1 \pmod 4$, and the fact that k is a radicand, which precludes that $k \equiv 0 \pmod 4$—see Definition 3.11 on page 121.

Example 8.6 We may now easily achieve a result that we proved about Bachet's equation in Chapter 1 via Theorem 8.4 as follows. If $k = -2$, then we have $(x, y) = (3, \pm 5)$ are the only solutions of (8.15), which is Theorem 1.19 on page 47.

Example 8.7 We may also invoke some results from §8.2 to illustrate Theorem 8.4 as follows. In Example 8.4 on page 281, we looked at $y^2 + 2 = p^{3m}$, changing the notation to suit our current situation, when p is a prime of the form $p = a^2 + 2b^2$. We saw that the only solutions are for $b = m = 1$, $y = 5$, and $p = 3$. In terms of Theorem 8.4, $k = -2$, $x = p^m = 3 = 4u^2 - 1$, where $u = 1$. This brings us back to Example 8.6 for yet another interpretation.

Example 8.8 Corollary 8.1 on page 280 in §8.2 may be illustrated here as well. That result told us that, in our current notation,

$$y^2 = p^{3d} + k$$

for a prime $p = u^2 - kv^2$ and $k < 0$ has a solution if and only if $v = d = 1$, $k = \pm 1 - 3u^2$, $y = 8u^3 \pm 3u$, so

$$p = 4u^2 \pm 1,$$

which we see is the conclusion of Theorem 8.4 with the relevant sign associations.

See Exercises 8.21–8.24 for more examples. Also, see Exercise 8.25 for results similar to Theorem 8.4 on page 282 for the case where $k > 0$.

Exercises

8.18. Suppose that I, J are nonzero integral R-ideals where R is a Dedekind domain with I and J relatively prime—see Definition 2.15 on page 79. Prove that if K is an R ideal and $n \in \mathbb{N}$ such that $IJ = K^n$, then there exist R ideals $\mathfrak{I}, \mathfrak{J}$ such that $I = \mathfrak{I}^n$, $J = \mathfrak{J}^n$, and $K = \mathfrak{I}\mathfrak{J}$.

(*Hint: use Theorem 2.12 on page 77.*)

8.19. Let $\mathfrak{O}_F$ be the ring of integers of an algebraic number field F with class number $h_{\mathfrak{O}_F}$. Prove that if I is an integral $\mathfrak{O}_F$-ideal such that $I^n \sim 1$ for some $n \in \mathbb{N}$ with $\gcd(h_{\mathfrak{O}_F}, n) = 1$, then $I \sim 1$.

8.20. Let α, β be nonzero elements in a Dedekind domain R. Prove that the principal R-ideals $(\alpha) = (\beta)$ if and only if $\alpha = \beta u$ where u is a unit in R.

8.21. Suppose that p is a prime of the form $p = u^2 + 13v^2$ for some $u, v \in \mathbb{N}$. Find all solutions to $y^2 = p^{3m} - 13$, for $m \in \mathbb{N}$ if any exist.

(*Note that -13 is the smallest value of $|k|$ of the form $k = -1 - 3u^2$ such that the hypothesis of Theorem 8.4 is satisfied. Also, $h_{\mathbb{Z}[\sqrt{-13}]} = 2$.*)

8.22. Find all solutions of $y^2 = x^3 - 193$ if they exist.

(*With reference to Exercise 8.21, the next smallest $|k|$ of the form $k = -1 - 3u^2$ such that the hypothesis of Theorem 8.4 is satisfied is $k = -193$. Also, $h_{\mathbb{Z}[\sqrt{-193}]} = 4$.*)

8.23. Find all solutions of $y^2 = x^3 - 47$ if they exist. (Note that $h_{\mathbb{Z}[\sqrt{-47}]} = 5$.)

8.24. Find all solutions of $y^2 = x^3 - 57$ if they exist. (Note that $h_{\mathbb{Z}[\sqrt{-57}]} = 4$.)

8.25. Suppose that $k \in \mathbb{N}$ is a radicand of a real quadratic field $F = \mathbb{Q}(\sqrt{k})$ and $k \not\equiv 1 \pmod 4$, such that $h_{\mathfrak{O}_F} \not\equiv 0 \pmod 3$, with F having fundamental unit ε_k—see page 259. Let $\varepsilon = \varepsilon_k$ if ε_k has norm 1, let $\varepsilon = \varepsilon_k^2$ otherwise, and set $\varepsilon = T + U\sqrt{k}$. Prove that (8.15) has no solutions if $k \equiv 4 \pmod 9$ and $U \equiv 0 \pmod 9$.

(*Hint: Assume there is a solution (x, y) to (8.15). Then you may assume that $y + \sqrt{k} = w(u + v\sqrt{k})^3$ for a unit $w \in \mathfrak{O}_F$ and some $u, v \in \mathbb{Z}$, since the argument is the same as in the proof of Theorem 8.4.*)

(*Note that more results for $k > 0$ of this nature, which typically involve congruences on T and U, may be found, for instance, in Mordell's classic text [73] on Diophantine equations.*)

8.4   The Fermat Equation

> *There are no such things as applied sciences, only applications of science.*
>   from an address given on the inauguration of the Faculty of Science,
> University of Lille, France on December 7, 1854.
>   **Louis Pasteur (1822–1895)**
>   **French chemist and bacteriologist**

In this section, we look at FLT, and its related *prime Fermat equation*

$$x^p + y^p + z^p = 0, \tag{8.27}$$

solved for the case of $p = 3$ in Theorem 1.18 on page 41. It suffices to solve
(8.27) in order to solve the general Fermat equation (1.44) on page 41. The
following uses our techniques from Chapters 1 and 2, including factorization in
prime cyclotomic fields $F = \mathbb{Q}(\zeta_p)$, where ζ_p is a primitive p-th root of unity
for a prime $p > 2$ when $p \nmid h_{\mathfrak{O}_F}$, in which case p is called a *regular* prime. The
proof is due to Kummer—see Biography 3.2 on page 124. Some of the following
is adapted from [64].

Theorem 8.5 | **Kummer's Proof of FLT for Regular Primes**

   *If $p > 2$ is prime and $p \nmid h_{\mathfrak{O}_F}$ for $F = \mathbb{Q}(\zeta_p)$, then (8.27) has no solutions
with $p \nmid xyz \neq 0$.*

Proof. Assume that (8.27) has a solution $xyz \neq 0$ for $x, y, z \in \mathbb{Z}$. Without
loss of generality, we may assume that x, y, z are pairwise relatively prime.
Furthermore, we may write (8.27) as the *ideal* equation

$$\prod_{j=0}^{p-1}(x + \zeta_p^j y) = (z)^p. \tag{8.28}$$

Claim 8.5 $(x + \zeta_p^j y)$ *and* $(x + \zeta_p^k y)$ *are relatively prime for* $0 \leq j \neq k \leq p - 1$.

   Let $\mathcal{P}$ be a prime $\mathfrak{O}_F$-ideal dividing both of the above ideals. Therefore, $\mathcal{P}$
divides

$$(x + \zeta_p^k y) - (x - \zeta_p^j y) = y\zeta_p^k(1 - \zeta_p^{j-k}).$$

By Exercise 8.26 on page 291, $\lambda = 1 - \zeta_p$ and $1 - \zeta_p^{j-k}$ are associates for $j \neq k$,
and clearly ζ_p^k is a unit, so $\mathcal{P} \mid (y\lambda)$. By primality, $\mathcal{P} \mid (y)$ or $\mathcal{P} \mid (\lambda)$. If $\mathcal{P} \mid (y)$,
then $\mathcal{P} \mid (z)$ from (8.28). Since $\gcd(y, z) = 1$, there exist $u, v \in \mathbb{Z}$ such that
$uy + vz = 1$. Since $y, z \in \mathcal{P}$, then $1 \in \mathcal{P}$, a contradiction. Hence, $\mathcal{P} \mid (\lambda)$. By
Theorem 2.3 on page 58 and Exercise 8.26, (λ) is a prime $\mathfrak{O}_F$-ideal. Therefore,
$\mathcal{P} = (\lambda)$, so $(\lambda) \mid (z)$. By Exercise 2.29 on page 96, $N_F(\lambda) \mid N_F(z)$. However,
by Exercise 8.27,

$$N_F(z) = z^{p-1},$$

so $p = N_F(\lambda) \mid z$, contradicting the hypothesis. This completes Claim 8.5.

By Claim 8.5 and Theorem 2.12 on page 77,

$$(x + \zeta_p y) = I^p,$$

for some $\mathfrak{O}_F$-ideal I. Since $p \nmid h_F$, then by Exercise 8.19 on page 285, $I \sim 1$. Hence, there exists an $\alpha \in \mathfrak{O}_F$ such that

$$x + \zeta_p y = u_1 \alpha^p, \tag{8.29}$$

where $u_1 \in \mathcal{U}_F$. Our next task is to show that $u_1 \zeta_p^s \in \mathbb{R}$ for some $s \in \mathbb{Z}$. This first requires establishing the following.

Claim 8.6 $\mathfrak{O}_F = \mathbb{Z}[\zeta_p]$.

Clearly $\mathbb{Z}[\zeta_p] \subseteq \mathfrak{O}_F$. If $\alpha \in \mathfrak{O}_F$, by Theorem 1.5 on page 10, there exist $q_j \in \mathbb{Q}$ for $j = 0, 1, \ldots, q_{p-2}$ such that

$$\alpha = \sum_{j=0}^{p-2} q_j \zeta_p^j. \tag{8.30}$$

Now we show that $pq_j \in \mathbb{Z}$ for each such j. Let T_F be as given in Definition 2.19 on page 91. Then $T_F(\zeta_p^k) = -1$ for any k relatively prime to p by Exercise 1.54 on page 46. Therefore, for any $k = 0, 1, \ldots, p-2$,

$$T_F(\alpha \zeta_p^{-k}) = \sum_{j=0}^{p-2} q_j T_F(\zeta_p^{j-k}) = -\sum_{j=0}^{k-1} q_j + (p-1)q_k - \sum_{j=k+1}^{p-2} q_j = -\sum_{j=0}^{p-2} q_j + pq_k.$$

Hence,

$$T_F(\alpha \zeta_p^{-k} - \alpha \zeta_p) = T_F(\alpha \zeta_p^{-k}) - T_F(\alpha \zeta_p) = -\sum_{j=0}^{p-2} q_j + pq_k + \sum_{j=0}^{p-2} q_j = pq_k,$$

for any such k. Since

$$\alpha \zeta_p^{-k} - \alpha \zeta_p \in \mathfrak{O}_F,$$

then by Exercise 2.25 on page 96, $pq_k \in \mathbb{Z}$. Thus, from (8.30),

$$p\alpha = \sum_{j=0}^{p-2} pq_j \alpha_p^j$$

with $pq_j \in \mathbb{Z}$ for all such j. However, since $\zeta_p = 1 - \lambda$, then using the binomial theorem, we may write

$$p\alpha = \sum_{j=0}^{p-2} z_j \lambda^j \tag{8.31}$$

with $z_j \in \mathbb{Z}$ for all such j. However, by Exercise 8.26, $\lambda \mid p$, since

$$p = \prod_{j=1}^{p-1}(1 - \zeta_p^j),$$

so from (8.31), $\lambda \mid z_0$. However,

$$p = N_F(\lambda) \mid N_F(z_0) = z_0^{p-1}$$

so $p \mid z_0$ as well.

Now, by Exercise 8.26, $1 - \zeta_p^j$ are associates for $j = 1, 2, \ldots, p-1$, so the following equation involving principal $\mathfrak{O}_F$-ideals holds,

$$(\lambda)^{p-1} = \prod_{j=1}^{p-1}(1 - \zeta_p^j) = \left(\prod_{j=1}^{p-1}[1 - \zeta_p^j]\right) = (N_F(\lambda)) = (p), \qquad (8.32)$$

where the last equality also holds by Exercise 8.26. Hence, this implies that $\lambda^{p-1} \mid z_0$.

Now considering (8.31) modulo λ^2, we get that $\lambda^2 \mid z_1\lambda$, so $\lambda \mid z_1$, and as above $p \mid z_1$. Continuing in this fashion, we see that $p \mid z_j$ for $j = 0, 1, \ldots, p-2$. Then dividing (8.31) by p yields

$$\alpha \in \mathbb{Z}[\lambda] = \mathbb{Z}[\zeta_p],$$

so $\mathfrak{O}_F \subseteq \mathbb{Z}[\zeta_p]$. We have shown that $\mathfrak{O}_F = \mathbb{Z}[\zeta_p]$ thereby securing Claim 8.6. In the following, the reader is reminded of the notion of congruence modulo an ideal, explored in Exercises 8.32–8.39.

Claim 8.7 *If z is a unit in $\mathbb{Z}[\zeta_p]$, then $z\zeta_p^s \in \mathbb{R}$ for some $s \in \mathbb{Z}$.*

If z is a unit in $\mathbb{Z}[\zeta_p]$, then so is its complex conjugate $\bar{z}$, and

$$\tau = z/\bar{z} \in \mathbb{Z}[\zeta_p]. \qquad (8.33)$$

By Exercise 8.27, and Theorem 2.19 on page 88, the only roots of unity in $\mathbb{Q}(\zeta_p)$ are ζ_p^t for $t \in \mathbb{Z}$. Also, since for any F-monomorphism ρ,

$$\rho(\tau) = \rho(z)/\rho(\bar{z}) = \rho(z)/\overline{\rho(z)},$$

so $|\rho(\tau)| = 1$. By Exercises 2.23 on page 96 and 8.27 on page 291, $\mathbb{Q}(\zeta_p)$ can have only finitely many complex units, and $|\tau^k| = 1$ for all $k \in \mathbb{N}$, so $\tau^k = \tau^\ell$ for some $k < \ell$. Thus, $\tau^{\ell-k} = 1$, which implies that τ is a root of unity. Set

$$\tau = \pm\zeta_p^t.$$

Since

$$\zeta_p^j \equiv 1 \pmod{\lambda} \text{ for all } j, \qquad (8.34)$$

then letting

$$z = \sum_{j=0}^{p-2} a_j \zeta_p^j$$

and using the fact that $\rho(\zeta_p) = \zeta_p^k$ for some k we get that

$$z \equiv \rho(z) \pmod{\lambda}.$$

In particular, $z \equiv \bar{z} \pmod{\lambda}$. In the case that

$$\tau = -\zeta_p^t, \text{ which implies that } z = -\zeta_p^t \bar{z}, \text{ by (8.33)},$$

then by (8.34),

$$z \equiv -\bar{z} \pmod{\lambda},$$

which implies that $2z \equiv 0 \pmod{\lambda}$, an impossibility. Therefore,

$$z = \zeta_p^t \bar{z} = \zeta_p^{-2s} \bar{z},$$

where $-2s \equiv t \pmod{p}$. Hence,

$$\zeta_p^s z = \overline{\zeta_p^s z},$$

which says that $z\zeta_p^s \in \mathbb{R}$, which is the claim.

Now returning to (8.29) on page 287, using Claim 8.7, there is an $k \in \mathbb{Z}$ and $w \in \mathbb{R} \cap \mathfrak{U}_F$, with

$$x + \zeta_p y = w \zeta_p^k \alpha^p. \tag{8.35}$$

By Exercise 8.39 on page 293 there exists a $z_1 \in \mathbb{Z}$ such that

$$\alpha \equiv z_1 \pmod{(\lambda)}.$$

By taking norms on the latter, we get

$$\alpha^p - z_1^p = \prod_{j=0}^{p-1} (\alpha - \zeta_p^j z_1).$$

Since $\zeta_p \equiv 1 \pmod{(\lambda)}$, then for each $j = 0, 1, \ldots, p-1$,

$$\alpha - \zeta_p^j z_1 \equiv \alpha - z_1 \pmod{(\lambda)}.$$

Hence,

$$\alpha^p \equiv z_1^p \pmod{(\lambda)^p},$$

so (8.35) becomes

$$x + \zeta_p y \equiv w z_1^p \zeta_p^k \pmod{(\lambda)^p}.$$

However, $(p) \mid (\lambda)^{p-1}$ by (8.32), so

$$x + \zeta_p y \equiv w z_1^p \zeta_p^k \pmod{(p)}.$$

Since ζ_p^k is a unit, then

$$\zeta_p^{-k}(x + \zeta_p y) \equiv wz_1^p \pmod{(p)}. \tag{8.36}$$

By taking complex conjugates in (8.36), we get

$$\zeta_p^k(x + \zeta_p^{-1}y) \equiv wz_1^p \pmod{(p)}. \tag{8.37}$$

Subtracting (8.37) from (8.36), we get

$$\zeta_p^{-k}x + \zeta_p^{1-k}y - \zeta_p^k x - \zeta_p^{k-1}y \equiv 0 \pmod{(p)}. \tag{8.38}$$

Claim 8.8 $2k \equiv 1 \pmod{p}$.

If $p \mid k$, then $\zeta_p^k = 1$, so (8.38) becomes

$$0 \equiv y(\zeta_p - \zeta_p^{-1}) \equiv y\zeta_p^{-1}(\zeta_p^2 - 1) \equiv y\zeta_p^{-1}(\zeta_p - 1)(\zeta_p + 1) \equiv y\zeta_p^{-1}\lambda(\zeta_p + 1) \pmod{(p)}.$$

However, by setting $x = -1$ in

$$\sum_{j=0}^{p-1} x^j = \prod_{j=1}^{p-1}(x - \zeta_p^j),$$

we get that $1 + \zeta_p \in \mathcal{U}_F$, so

$$y\lambda \equiv 0 \pmod{(p)}.$$

Also, from (8.32), and the fact that $p \geq 3$, we get that $\lambda \mid y$. By Exercise 2.29 again, $N_F(\lambda) \mid N_F(y)$, so we get that $p \mid y$, contradicting the hypothesis. Therefore, $k \not\equiv 0 \pmod{p}$. By (8.38) there exists an $\alpha_1 \in \mathfrak{O}_F$ such that

$$\alpha_1 p = x\zeta_p^{-k} + y\zeta_p^{1-k} - x\zeta_p^k - y\zeta_p^{k-1}. \tag{8.39}$$

If $k \equiv 1 \pmod{p}$, then (8.38) becomes

$$x(\zeta_p^{-1} - \zeta_p) \equiv 0 \pmod{(p)}.$$

In the same fashion as in the elimination of the case $k \equiv 0 \pmod{p}$, we get that $p \mid x$, contradicting the hypothesis. Since $k \not\equiv 0, 1 \pmod{p}$, then

$$\alpha_1 = \frac{x}{p}\zeta_p^{-k} + \frac{y}{p}\zeta_p^{1-k} - \frac{x}{p}\zeta_p^k - \frac{y}{p}\zeta_p^{k-1}. \tag{8.40}$$

By Claim 8.6,

$$\{1, \zeta_p, \ldots, \zeta_p^{p-1}\}$$

is a $\mathbb{Z}$-basis of $\mathfrak{O}_F$. Thus, if all exponents $-k$, $1-k$, k, and $k-1$ are incongruent modulo p, then $x/p \in \mathbb{Z}$, contradicting the hypothesis. Thus, two of the aforementioned exponents are congruent modulo p. The only possibility remaining after excluding $k \equiv 0, 1 \pmod{p}$ is

$$2k \equiv 1 \pmod{p}.$$

This establishes Claim 8.8.

Hence, (8.39) becomes

$$\alpha_1 p \zeta_p^k = x + y\zeta_p - x\zeta_p^{2k} - y\zeta_p^{2k-1} = (x - y)\lambda.$$

By taking norms, we get $p \mid (x - y)$, namely $x \equiv y \,(\mathrm{mod}\ p)$. Thus, by (8.27) $y \equiv z \,(\mathrm{mod}\ p)$ as well. Therefore, since $p \nmid x$,

$$0 \equiv x^p + y^p + z^p \equiv 3x^p \quad (\mathrm{mod}\ p).$$

Thus, $p = 3$, which was eliminated in Theorem 1.18, so we have completed the proof. $\qquad\qquad\square$

Remark 8.6 The case where $p \nmid xyz$ for a regular prime is called *case I* in FLT. Kummer conjectured that there exist infinitely many regular primes, but this problem remains open to this day. However, it is possible to show that there are infinitely many primes $p \mid h_{\mathfrak{O}_F}$ for $F = \mathbb{Q}(\zeta_p)$, called *irregular primes*—see [64, §3.6]. This is done using Bernoulli numbers and polynomials—see §5.1. For Kummer's proof of FLT for regular primes $p \mid xyz$, called *Case II for FLT*, see [64, Theorem 4.124, p. 251].

In §8.5, we look at a related equation to the Fermat equation, which has also been relatively recently solved, the *Catalan equation* and the combined equations for the *Fermat–Catalan conjecture* and the impact of the *ABC conjecture* on the latter, which remains an open problem, as of course does the ABC conjecture.

Exercises

8.26. Prove that for a prime $p > 2$ and $F = \mathbb{Q}(\zeta_p)$, $N_F(1 - \zeta_p) = p$. Also, show that $1 - \zeta_p$ and $1 - \zeta_p^i$ are associates for $i = 1, 2, \ldots, p - 1$.

8.27. Prove that if $n \in \mathbb{N}$ with $n > 2$, then

$$|\mathbb{Q}(\zeta_n) : \mathbb{Q}| = \phi(n).$$

(*Hint: Use Exercise 2.24 on page 96 in conjunction with Theorem 1.7 and Definition 1.9 on page 11.*)

For the following exercises, the reader should be familiar with the basics concerning "actions on rings" such as presented in [68, Appendix A, pp. 303–306].

8.28. Prove that if R is a Dedekind domain and I, J are R-ideals, then

$$\frac{R}{I} \cong \frac{J}{IJ}.$$

as additive groups.

(*Hint: Use Exercises 2.10–2.11 on page 85, and employ the* Fundamental Isomorphism Theorem for Rings *which says:* If $\mathcal{R}$ and $\mathcal{S}$ are commutative rings with identity and $\psi : \mathcal{R} \mapsto \mathcal{S}$ is a homomorphism of rings, then $\mathcal{R}/\ker(\psi) \cong \operatorname{img}(\psi)$.)

8.29. Let $\mathfrak{O}_F$ be the ring of integers of a number field F, $\mathcal{P}$ a prime $\mathfrak{O}_F$-ideal, and $n \in \mathbb{N}$. Prove that
$$\left| \frac{\mathfrak{O}_F}{\mathcal{P}^n} \right| = \left| \frac{\mathfrak{O}_F}{\mathcal{P}} \right|^n.$$

(*Hint: Use Exercise 8.28.*)

8.30. Let R be a Dedekind domain, and let I be an R-ideal with
$$I = \prod_{j=1}^{r} \mathcal{P}_j^{a_j}, \tag{8.41}$$

for distinct prime R-ideal $\mathcal{P}_j$. Prove that
$$\left| \frac{R}{I} \right| = \prod_{j=1}^{r} \left| \frac{R}{\mathcal{P}_j} \right|^{a_j}.$$

8.31. Let R be a commutative ring with identity, and let I be an R-ideal. Prove that the additive abelian group R/I is a ring with identity, and whose multiplication is given by $(a + I)(b + I) = ab + I$.

8.32. Let F be a number field and I a nonzero $\mathfrak{O}_F$-ideal. If $\alpha, \beta \in \mathfrak{O}_F$, we say that α and β are *congruent modulo* I if $\alpha - \beta \in I$, denoted by
$$\alpha \equiv \beta \pmod{I}.$$

The set of those $\alpha \in \mathfrak{O}_F$ which are congruent to each other modulo I is called a *residue class modulo* I. Prove that the number of residue classes is equal to the *norm of* I, defined by $N(I) = |\mathfrak{O}_F/I|$.

(*Note that by Exercise 8.30, we know that* $|\mathfrak{O}_F/I|$ *is finite. Also, if I is given by* (8.41), *then Exercise 8.30 tells us that*
$$N(I) = \prod_{j=1}^{r} \left| \frac{\mathfrak{O}_F}{\mathcal{P}_j} \right|^{a_j}.$$

It follows that if I, J are R-ideals, then $N(IJ) = N(I)N(J)$.)

The balance of the exercises are in reference to Exercise 8.32. The reader should recall the developments in Chapter 2 for the terminology used in what follows.

8.33. Let R be a Dedekind domain. Prove that if $\gcd((\alpha), I) = 1$, then for any $\beta \in R$, there is a $\gamma \in R$, uniquely determined modulo I, such that $\alpha\gamma \equiv \beta$ (mod I). Furthermore, prove that this congruence is solvable for some $\gamma \in \mathfrak{O}_F$ if and only if $\gcd((\alpha), I) \mid (\beta)$.

8.34. In view of Exercise 8.33, two elements of $\mathfrak{O}_F$ that are congruent modulo I have the same gcd with I. Hence, this is an invariant of the class, since it is a property of the whole residue class. We denote the *number of residue classes relatively prime to* I, by the symbol $\Phi(I)$. Let I, J be relatively prime $\mathfrak{O}_F$-ideals. Prove that

$$\Phi(I) = N(I) \prod_{\mathcal{P} \mid I} \left(1 - \frac{1}{N(\mathcal{P})} \right),$$

where the product runs over all distinct prime divisors of I. Conclude that if I, J are relatively prime $\mathfrak{O}_F$-ideals, then $\Phi(IJ) = \Phi(I)\Phi(J)$.

8.35. Suppose that $I = \prod_{j=1}^{r} \mathcal{P}_j^{a_j}$, where the $\mathcal{P}_j$ are distinct $\mathfrak{O}_F$-ideals. Prove that

$$\Phi(I) = N(I) \prod_{j=1}^{r} \left(1 - \frac{1}{N(\mathcal{P}_j)} \right).$$

Note that when $F = \mathbb{Q}$, then Φ is the ordinary Euler totient function ϕ.

8.36. Let $\alpha_j \in \mathfrak{O}_F$ for $j = 1 \ldots, d$, and let $\mathcal{P}$ be a prime $\mathfrak{O}_F$-ideal. Prove that the polynomial congruence

$$f(x) = x^d + \alpha_1 x^{d-1} + \cdots + \alpha_{d-1} x + \alpha_d \equiv 0 \pmod{\mathcal{P}}$$

has at most d solutions $x \in \mathfrak{O}_F$ that are incongruent modulo $\mathcal{P}$, or else $f(\alpha) \equiv 0 \pmod{\mathcal{P}}$ for all $\alpha \in \mathfrak{O}_F$. (We also allow the case where $\deg(f) = 0$, in which case $f(x) = \alpha_0 \equiv 0 \pmod{\mathcal{P}}$ means that $\alpha_0 \in \mathcal{P}$.)

8.37. Prove that the residue classes modulo I, relatively prime to I, form an abelian group under the multiplication given in Exercise 8.31 on the preceding page. Prove that this group has order $\Phi(I)$. In particular, show that if I is a prime $\mathfrak{O}_F$-ideal, then the group is cyclic.

8.38. Suppose that I is a nonzero $\mathfrak{O}_F$-ideal and $\alpha \in \mathfrak{O}_F$ is relatively prime to I. Prove that

$$\alpha^{\Phi(I)} \equiv 1 \pmod{I},$$

called *Euler's Theorem for Ideals.* Conclude that if $I = \mathcal{P}$ is a prime $\mathfrak{O}_F$-ideal, then

$$\alpha^{N(\mathcal{P})-1} \equiv 1 \pmod{\mathcal{P}},$$

called *Fermat's Little Theorem for Ideals.*

8.39. Let $\mathcal{P}$ be a nonzero prime $\mathfrak{O}_F$-ideal, and let $\alpha \in \mathfrak{O}_F$. Prove that there exists a $z \in \mathbb{Z}$ such that

$$\alpha \equiv z \pmod{\mathcal{P}} \text{ if and only if } \alpha^p \equiv \alpha \pmod{\mathcal{P}}, \text{ where } (p) = \mathcal{P} \cap \mathbb{Z}.$$

8.5   Catalan and the ABC Conjecture

> *The last thing one knows in constructing a work is what to put the first.*
>   translated from section 1, no. 19 of *Pensées* (1670) ed. L. Brunschvicg
> (1909)
>
>   **Blaise Pascal (1623–1662)**
>   **French mathematician, physicist, and moralist**

In 1844, Charles Catalan conjectured that

$$a^b - c^d = 1 \tag{8.42}$$

with all integers a, b, c, d bigger than 1 has solutions for only $(a, b, c, d) = (3, 2, 2, 3)$. In an elementary course in number theory, one may look at this equation for special cases and solve it via congruence conditions and other such techniques—see [68, Biography 3.1, p. 144]. Indeed, in the Middle Ages, Hebraeus solved (8.42) for $(a, c) = (3, 2)$. In 1738, Euler solved it for $(b, d) = (2, 3)$, and in 1850, Lebesgue solved it for $d = 2$. Moving into the twentieth century, Nagell solved it in 1921 for $(b, d) = (3, 3)$, and C. Ko for the case $d = 2$ in 1967. In 1976, R. Tijdeman proved that (8.42) has solutions only for

$$c^d < \exp(\exp(\exp(\exp(730)))),$$

a monster of a bound, but this shows that it can have solutions for only finitely many values. Not long later, M. Langevin proved that the bounds for solutions to (8.42) must satisfy

$$b, d < 10^{110}.$$

Then Mignotte improved this to

$$\max\{b, d\} < 7.78 \cdot 10^{16}.$$

In the other direction, in 1997, Y. Roy and Mignotte, proved that a *lower* bound on such solutions must satisfy

$$\min\{b, d\} > 10^5.$$

It seemed, therefore, that the bounds were closing in. As with the Fermat equation, discussed in §8.4, a proof was eventually found. In 2002, Preda Mihăilescu discovered a proof, which employs wide use of cyclotomic fields and Galois modules. In 2004, it was published in [58]. Thus, the Catalan conjecture is now known as *Mihăilescu's theorem.*

Now that both the Fermat equation and the Catalan equation have been resolved, we may look at a problem that combines them both, and is still open.

> ### The Fermat–Catalan Conjecture
>
> There are only finitely many powers x^p, y^q, z^r satisfying
>
> $$x^p + y^q = z^r, \tag{8.43}$$
>
> where $x, y, z \in \mathbb{N}$ and are relatively prime, and $p, q, r \in \mathbb{N}$ with
>
> $$\frac{1}{p} + \frac{1}{q} + \frac{1}{r} < 1. \tag{8.44}$$

By the 1995 results of Darmon and Granville [20] it is known that for *fixed* p, q, r with (8.44) satisfied, (8.43) has only finitely many solutions.

Although the Fermat–Catalan conjecture remains unresolved, there is a means of proving it under the assumption of yet another unresolved conjecture, a process that has become "fashionable" in the literature. In order to properly present these ideas, let us set the stage by looking at the very foundations of solving Diophantine equations from a historical perspective.

In 1900, Hilbert posed a list of 23 problems—see Biography 3.5 on page 127. Among them was the problem, which we would understand today as asking: *Is there a comprehensive algorithm which can determine whether a given Diophantine polynomial equation* (with integral coefficients) *has a solution in integers?* The very interpretation of this query and the resulting search for an answer ultimately was resolved in 1970 by Matiyasevich [55] who provided a rather definitive negative answer, to what we now call *Hilbert's tenth problem.* What this means for the modern mathematician is that we can never find an algorithm for the decision problem: *Does a given Diophantine equation have a solution or not?* However, this does not deter us from looking at certain classes of Diophantine equations, or as was done with the Catalan equation above, finding bounds on the number of solutions to determine whether or not such solutions exist.

Matiyasevich's aforementioned proof is based upon the notion of *Diophantine sets.* Without getting embroiled in the definitions and technical aspects of this phenomenon, suffice it to say that in 1960, Putnam established that a set is Diophantine if and only if it coincides with the sets of positive values of a suitable polynomial taken at nonnegative integers. Putting this together with the Matiyasevich result, we achieve that there exists a polynomial $f(x_1, x_2, \cdots, x_n)$ whose positive values at integers $n_j \geq 0$ are primes, and every prime is representable in this fashion. Indeed, in 1976, Jones, Sato, Wada, and Wiens [44] explicitly found a polynomial of degree 25 in 26 variables which produces all prime numbers. It also takes on negative values and a given prime may be repeated. Yet it is open as to what the minimal possible degree and minimal number of variables for such a polynomial happen to be. Moreover, and perhaps more striking, is the fact that the above implies that *the set of prime numbers is Diophantine.*

With the previous discussion in mind, it would be valuable to have a general methodology for solving Diophantine equations employing a theory that applies to some certain selected sets of Diophantine equations. There is a conjecture, if proved, that would apply to a wide variety of such equations, and arguably one of the most important unsolved problems in number theory, first posed, independently, by David Masser and Joseph Oesterlé in 1985. In what follows, for any $n \in \mathbb{Z}$, $\mathcal{S}(n)$ denotes the largest squarefree divisor of n, also known as the *squarefree kernel* of n, as well as the *radical* of n.

The ABC Conjecture

If a, b, c are relatively prime integers which satisfy the equation

$$a + b = c,$$

then for any $\kappa > 1$, with finitely many exceptions, we have that

$$c < \mathcal{S}(abc)^{\kappa}.$$

To illustrate the power of this conjecture, if resolved affirmatively, the following list shows several results that would fall to the ABC conjecture.

❖ Consequences of the ABC Conjecture

The ABC conjecture implies each of (1)–(8):

(1) The Fermat–Catalan conjecture—see (8.43) on page 295.

(2) FLT–see (1.44) on page 41.

(3) The Thue–Siegel–Roth Theorem—see (4.2) on page 160.

(4) The Diophantine equation $y^m = x^n + k$ for $x, y, m, n, k \in \mathbb{Z}$ with $m > 1$ and $n > 1$ has only finitely many solutions. This is a generalization of *Tijdeman's theorem*, which is the case $k = 1$.

(5) *Hall's conjecture*, which says that if there are integer solutions x, y to the Bachet equation

$$y^2 = x^3 - k,$$

then for any $\varepsilon < 1/2$, there exists a constant $K(\varepsilon) > 0$ such that

$$|x^3 - y^2| > K(\varepsilon)x^{\varepsilon}.$$

In other words, the nonzero difference in absolute value, $x^3 - y^2$, cannot be less than $x^{1/2}$. This was posed by Marshall Hall in [37] in 1971 for any $k \neq 0$.

(6) The existence of infinitely many *non*-Wieferich primes, where a *Wieferich prime* p is one that satisfies

$$2^{p-1} \equiv 1 \pmod{p^2}.$$

(7) The *Erdös–Woods Conjecture* which says: There exists an integer k such that, for $m, n \in \mathbb{N}$, the conditions

$$S(m + j) = S(n + j) \text{ for } 0 \leq j \leq k - 1$$

imply $m = n$. This arose from [24], where Erdös asked how many pairs of products of consecutive integers have the same prime factors.

(8) There are only finitely many triples of consecutive powerful numbers. (A *powerful number* $n \in \mathbb{N}$ satisfies that $p^2 \mid n$ whenever a prime $p \mid n$.) The above is a weak form of the *Erdös–Mollin–Walsh* conjecture, which states that there are *no* consecutive triples of powerful numbers—see Granville [35], as well as Mollin–Walsh [71].

The ABC conjecture is *equivalent* to

(9) the *Granville–Langevin conjecture*, which says that if

$$f(x, y) \in \mathbb{Z}[x, y]$$

is a square-free binary quadratic form of degree $n > 2$, then for every $\beta > 2$, there exists a constant $C(f, \beta) > 0$ such that

$$S(f(x, y)) \geq C(f, \beta) \max\{|x|, |y|\}^{n-\beta} \text{ for every } x, y \in \mathbb{Z}$$

$$\text{with } \gcd(x, y) = 1, f(x, y) \neq 0.$$

The above list is by no means exhaustive, since numerous other results follow from, or are equivalent to, the ABC conjecture. However, we see that there is ample reason to believe that this is one of the most important outstanding problems in number theory. Now we are in a position to prove what we asserted earlier, namely number (1) on the above list.

Theorem 8.6 | **Fermat–Catalan Follows From ABC**

The ABC conjecture implies the Fermat–Catalan conjecture.

Proof. Considering the right-hand side of (8.44) on page 295, we note that the largest possible choices for p, q, r are given by

$$\frac{1}{2} + \frac{1}{3} + \frac{1}{7} = \frac{41}{42},$$

so replacing < 1 by $\leq 41/42$, and applying the ABC conjecture with $\kappa = 1.01$, observing that,

$$\mathcal{S}(x^p y^q z^r) = \mathcal{S}(xyz) \leq xyz = (x^p)^{\frac{1}{p}}(y^q)^{\frac{1}{q}}(z^r)^{\frac{1}{r}} \leq (z^r)^{\frac{1}{p}+\frac{1}{q}+\frac{1}{r}},$$

we have, with finitely many possible exceptions,

$$z^r < z^{\kappa(r/p+r/q+1)}.$$

Hence,

$$r < \kappa(r/p + r/q + 1),$$

which in turn implies

$$1 < \kappa(1/p + 1/q + 1/r).$$

However, $(1/p + 1/q + 1/r) \leq 41/42$ and $\kappa = 1.01$, so

$$1 < \kappa(1/p + 1/q + 1/r) < 1.01\frac{41}{42} = 0.9859523819\cdots,$$

a contradiction. Hence, there can only be finitely many solutions to (8.43). □

The following are the only known examples of solutions to the Fermat–Catalan equation (8.43), the last five of which were discovered by F. Beukers and D. Zagier—see [19, pp. 382–383]:

$$1^p + 2^3 = 3^2$$

$$2^5 + 7^2 = 3^4$$

$$13^2 + 7^3 = 2^9$$

$$2^7 + 17^3 = 71^2$$

$$3^5 + 11^4 = 122^2$$

$$33^8 + 1549034^2 = 15613^3$$

$$1414^3 + 2213459^2 = 65^7$$

$$9262^3 + 15312283^2 = 113^7$$

$$17^7 + 76271^3 = 21063928^2$$

$$43^8 + 96222^3 = 30042907^2$$

We now show how item (8) in the list on page 297 follows from ABC. The reader should solve Exercises 8.42–8.43 on page 300, which we will use in the following.

Theorem 8.7 $\boxed{\textbf{ABC Implies Weak Erdös–Mollin–Walsh Conjecture}}$

The ABC conjecture implies there exist only finitely many consecutive triples of powerful numbers.

Proof. By Exercises 8.42–8.43, if $(n-1, n, n+1)$ are powerful, then

$$n = x^2 y^3$$

is even and $n^2 - 1$ is powerful. Let

$$a = 1, b = n^2 - 1, \text{ and } c = n^2$$

in the ABC conjecture. Then since $c = a + b$,

$$S(abc) \leq \sqrt{bn} < n^{3/2},$$

so for any $\kappa > 1$, with finitely many possible exceptions, we have

$$n^2 = c < S(abc)^{\kappa} < n^{3\kappa/2}.$$

In particular, if $\kappa = 1.01$, then

$$1 < n^{0.485} = n^{2 - 3.03/2} < 1,$$

which is a contradiction. We have shown there are at most finitely many consecutive triples of powerful numbers. □

$\boxed{\clubsuit \text{ Concluding comments}}$

In 1994, Bombieri [9] proved that that ABC conjecture implies the Thue–Siegel–Roth Theorem, (3) in the list on page 297. A more far-reaching result was proved in 1999 by Frankenhuysen [26] that included not only Bombieri's conclusion from ABC, but also Elkies' [23] derivation of *Mordell's conjecture* from ABC. In 1922, Mordell posed that any curve of genus bigger than 1 defined over a number field F has only finitely many rational points in F. It is beyond the scope of this book to go into any depth on this topic. Suffice it to say that Elkies' proof was based upon recasting the ABC conjecture in terms of a specified rational point in the one-dimensional projective line. Then the Mordell conjecture is boiled down to the ABC conjecture via the Riemann–Hurwitz formula which describes the relationship of what is known as the Euler characteristic of two surfaces when one is a ramified covering of the other. For a nice explanation including terminology and methodology, see [42]. Of course, there is an unconditional proof of Mordell's conjecture for which Faltings [25] won the Fields medal in 1983 using techniques from algebraic geometry. But Bombieri [10] provided an elementary proof in 1990, which the reader may also find presented in [42].

There are numerous other applications of the ABC conjecture upon which we have not touched such as that proved by Granville and Stark in [36], which establishes that the ABC conjecture implies that there do not exist any *Siegel zeros*, also called called *Landau–Siegel zeros*, of Dirichlet L-functions for characters of complex quadratic fields, where a Siegel zero is a potential counterexample to the Riemann hypothesis in that it is a value $s \in \mathbb{C}$ with $\Re(s) \neq 1/2$ such that $L(s, \chi) = 0$—see §7.2. There are also generalizations of the ABC conjecture to number fields which was introduced by Vojta in [99]. However, we have covered a sufficient amount to demonstrate that the ABC conjecture is indeed one of the main open problems in number theory and may remain so well into the future.

Exercises

8.40. Prove that for sufficiently large $n \in \mathbb{N}$ the ABC conjecture implies FLT. In other words, there exists an $N \in \mathbb{N}$ such that

$$x^n + y^n = z^n$$

has no nontrivial integer solutions for all $n > N$.

8.41. Prove that the ABC conjecture implies that the Erdös–Woods conjecture holds for $k = 3$, with finitely many possible exceptions. This is (7) of the list on page 297.

8.42. With reference to item (8) on the list on page 297, prove that the conjecture is equivalent to the following statement. There are only finitely many even powerful numbers n such that

$$n^2 - 1$$

is also powerful (with $\gcd(n - 1, n + 1) = 1$.)

8.43. With reference to Exercise 8.42, prove that $n \in \mathbb{N}$ is powerful if and only if

$$n = x^2 y^3$$

for some $x, y \in \mathbb{N}$.

8.44. Show that the ABC conjecture implies that the largest prime factor of $1 + x^2 y^3$ goes to infinity as $x + |y|$ goes to infinity.

8.45. Given any even $a \in \mathbb{N}$ prove that the ABC conjecture implies the existence of infinitely many $m \in \mathbb{N}$ such that

$$a^{2m} - 1$$

is *not* powerful.

(*Hint: Use Exercise 8.42.*)

Chapter 9

Elliptic Curves

> *Is it so bad, then, to be misunderstood? Pythagoras was misunderstood, and Socrates, and Jesus, and Luther, and Copernicus, and Galileo, and Newton, and every pure and wise spirit that ever took flesh. To be great is to be misunderstood.*
>
> from **Self-Reliance** in **Essays 1841**
> **Ralph Waldo Emerson (1803–1882)**
> **American philosopher and poet**

Although the history of elliptic curves is well over a century old and was initially developed in the context of classical analysis, these essentially algebraic constructs have found their way into other areas of mathematics in the modern day. Elliptic curves have had impact, at a deep level, on both applied mathematics, for instance in the area of cryptology, as well as in pure mathematics, such as in the proof of FLT. Indeed, a key ingredient in the resolution of Fermat's equation, (1.44) on page 41, involved certain elliptic curves, which we will explore in §10.3. Moreover, as we shall see later in this chapter, elliptic curves are used in factoring algorithms, primality testing, as well as the discrete log problem, upon which certain elliptic curve ciphers base their security. In fact, elliptic curve methods are widely considered to be some of the most powerful and elegant tools available to the cryptographic community. To see the beauty, complexity, and power of this topic, we must begin with foundational material. Some of what follows is adapted from [64].

9.1 The Basics

In Chapter 8, we explored numerous applications of our methods, developed in earlier chapters, to a variety of Diophantine equations including the generalized Ramanujan-Nagell equation (8.5) in §8.2, Bachet's equation (8.15) in §8.3, the Fermat equation in §8.4, as well as the Catalan and related equations in

§8.5. In particular, Bachet's Equation motivates the very definition of elliptic curves since it is a special case.

Definition 9.1 | Elliptic Curves

Let F be a field with char$(F) \neq 2, 3$. If $a, b \in F$ are given such that

$$4a^3 + 27b^2 \neq 0$$

in F, then the *elliptic curve of*

$$y^2 = x^3 + ax + b$$

over F, denoted by $E(F)$, is the set of points (x, y) with $x, y \in F$ such that the equation

$$y^2 = x^3 + ax + b \tag{9.1}$$

holds in F together with a point $\mathfrak{o}$, called the *point at infinity*. The value

$$\Delta(E(F)) = -16(4a^3 + 27b^2)$$

is called the *discriminant of the elliptic curve* E. (*Elliptic curves can also be defined for* char$(F) = 2, 3$ *by an equation slightly different from* (9.1), *but we will not need those cases herein.* We assume throughout that char$(F) \neq 2, 3$.)

Remark 9.1 In order to understand the term *point at infinity*, we look at how *projective geometry* comes into play. Projective geometry studies the properties of geometric objects invariant under projection. For instance, projective 2-space over a field F, denoted by $\mathbb{P}^2(F)$, is the set

$$\{(x, y, z) : x, y, z \in F\} - \{(0, 0, 0)\}$$

of all equivalence classes of *projective points*

$$(tx, ty, tz) \sim (x, y, z)$$

for nonzero $t \in F$. So, if $z \neq 0$, then there exists a unique projective point in the class of (x, y, z) of the form $(x, y, 1)$, namely $(x/z, y/z, 1)$. Thus, $\mathbb{P}^2(F)$ may be identified with all points (x, y) of the *ordinary*, or *affine*, plane together with points for which $z = 0$. The latter are the points on the *line at infinity*, which one may regard as the *horizon* on the plane. With this definition, one sees that the point at infinity in Definition 9.1 is $(0, 1, 0)$ in $\mathbb{P}^2(F)$. This is the intersection of the y-axis with the line at infinity.

Remark 9.2 The historical significance of the very term "elliptic curve" is also worth exploring. The term *elliptic curve* is somewhat of a misnomer since the elliptic curves are not ellipses. The term comes from the fact that elliptic curves made their initial appearance during attempts to calculate the arc length of an ellipse. The most appropriate name for *elliptic curves* comes from an area of mathematical inquiry called *algebraic geometry*. There they are classified as *abelian varieties of dimension one*. Furthermore, (9.1) is used rather than the seemingly more general

$$Y^2 = X^3 + AX^2 + BX + C$$

since we may make the translation

$$X \mapsto x - A/3$$

to get (9.1) with

$$a = B - A^2/3 \text{ and } b = A^3/9 - AB/3 - A^3/27 + C.$$

Moreover, once the translation is made, we may find a root of

$$x^3 + ax + b = 0$$

from the formula:

$$x = \sqrt[3]{-\frac{b}{2} + c} + \sqrt[3]{-\frac{b}{2} - c},$$

where

$$c = \sqrt{\frac{b^2}{4} + \frac{a^3}{27}},$$

called *Cardano's Formula*. Also see (10.26) on page 353 for another standard form of equations for elliptic curves.

We now motivate the discussion of the group structure arising from elliptic curves by discussing some connections between elliptic curves and Diophantine equations that we studied earlier. As noted above, Bachet's Equation is an example of an elliptic curve. However, there are other, not so obvious, ones such as the Fermat Equation

$$x^3 + y^3 = z^3,$$

which is an elliptic curve after the transformations

$$X = 12z/(x + y) \text{ and } Y = 36(x - y)/(x + y),$$

which yield

$$Y^2 = X^3 - 432, \tag{9.2}$$

having no rational solutions, except $X = 12, |Y| = 36$, by Exercise 9.1 on page 309, in view of Theorem 1.18 on page 41 (see also Exercise 9.2). Hence, in his proof of Theorem 1.18, Gauss was essentially dealing with points in $F = \mathbb{Q}(\sqrt{-3})$ on elliptic curves over F. In fact, it was through such connections that Andrew Wiles used elliptic curves to motivate his solution of FLT for the general case. Essentially, Wiles showed that the existence of a solution to the Fermat Equation (1.44) would imply the existence of an elliptic curve which would exhibit a special property called a *modularity pattern*. In 1990, Ken Ribet, whose work inspired Wiles, had already shown that such a curve *cannot* be modular, and FLT fell to the contradiction—see §10.4 for a more detailed explanation of the proof of FLT and the involvement of these contributors. Hence, we cannot have a greater motivator for looking at such curves than the felling of a century's old problem. But this is not a sole motivator since, as mentioned at the outset, there are modern-day cryptographic applications, which are one of the main topics of this chapter.

Biography 9.1 Girolamo Cardano (1501–1576) *was born in Pavia, Duchy of Milan, now Italy, on September 24, 1501. In his early years, Cardano assisted his father, who was a lawyer and lecturer of mathematics primarily at the Platti foundation in Milan. Then he entered his father's alma mater, Pavia University, to study medicine. The university was closed when war erupted, so Cardano went to the University of Padua to continue his studies. Shortly after the death of his father, Cardano squandered his small inheritance, and became addicted to gambling, where his knowledge of probability fared him well. However, the company he kept is told by the fact that he always carried a knife, and once slashed the face of an opponent over a question of cheating. Despite the time wasted in these endeavors, he achieved his doctorate in 1525. After a series of attempts at medical practice and gambling, Cardano obtained his father's former post at the Platti foundation.*

In 1541, Niccolo Tartaglia (ca. 1500–1557) gained fame for solving the cubic equation. However, he was not the first to do so. That honour goes to Scipione del Ferro (ca. 1465–1526), a name absent from many historical accounts of the matter. When Cardano learned of the solution, he invited Tartaglia to his home and extracted the solution from him after Cardano promised, under oath, not to disclose it. In 1543, Cardano learned of Ferro's solution, and felt that he could therefore publish it despite his oath. In his book Ars Magna, *published in 1545, he did that along with a solution of the quartic equation. The latter had been solved by Ludovico Ferrari (1522–1569).*

Cardano became a respected professor at Bologna and Milan, and a prolific writer. He contributed to probability theory, hydrodynamics, mechanics, and geology. He died on September 21, 1576, ostensibly at his own hand, having correctly predicted the date of his demise some time earlier.

Example 9.1 Consider the elliptic curve

$$y^2 = x^3 + 3x + 4.$$

By observation we see that $P = (-1, 0)$ and $Q = (0, 2)$ are points on the intersection of the curve with a line. Let us find the third. Since

$$(2 - 0)/((0 - (-1)) = 2$$

is the slope of the line through P and Q, then the equation of the line is $y = 2(x + 1)$. The combined graphs are given in Diagram 9.1. To find the third point of intersection with the curve, we put

$$y = 2(x + 1) \tag{9.3}$$

into $y^2 = x^3 + 3x + 4$ to get

$$4(x + 1)^2 = x^3 + 3x + 4,$$

which simplifies to $x(x + 1)(x - 5) = 0$, so $x = 5$ and by plugging this into *(9.3)* we get $y = 12$.

In Example 9.1 we used the geometry of the situation to find a third point from two given points. We observe that if we can indeed find two rational points on a curve, then the third must also be rational since two of the three points (intersecting a straight line, possibly repeated) are roots of a quadratic equation, which is

Diagram 9.1

$$y^2 = x^3 + 3x + 4$$

and

$$y = 2(x + 1)$$

$$x^2 - 4x - 5 = 0$$

in Example 9.1. If we know only one rational point, then we cannot guarantee that the other two points on a line through that point, intersecting the curve, will be rational. For instance, if

$$y^2 = x^3 + x + 4, \tag{9.4}$$

then $(0, 2)$ is a point on the curve. However, if we take a line through this point with slope 1 say, then the equation of that line is

$$y = x + 2.$$

If we plug this into (9.4), we get

$$(x + 2)^2 = x^3 + x + 4,$$

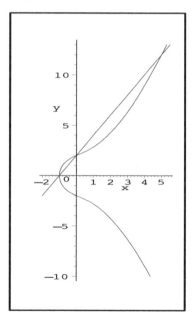

which simplifies to $x(x^2 - x - 3) = 0$. By the quadratic formula, $x^2 - x - 3 = 0$ has the solutions $x = (1 \pm \sqrt{13})/2$, which are not rational. Thus, in our quest to find rational points on elliptic curves, we should choose a straight line that goes through two rational points on an elliptic curve, since then the third point is guaranteed to be rational by the quadratic formula. This process is illustrated by Example 9.1.

Figure 9.1: $y^2 = x^3 - 4x$

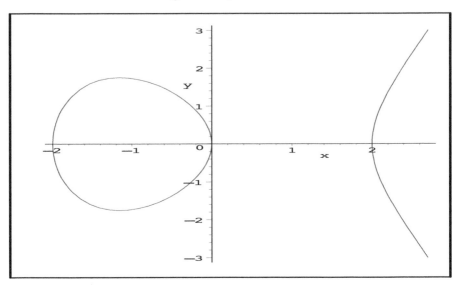

As seen earlier in (9.2) on page 303, there are elliptic curves with no non-trivial rational points, arising out of Diophantine problems. The following diagram illustrates another elliptic curve with no nontrivial rational points by Exercise 9.2 on page 309.

If one wishes to form a group out of the points of an elliptic curve, one must have a well-defined operation, such as addition. Let us look at adding two points P and Q on an elliptic curve $E(F)$. If $P \neq \mathfrak{o}$, and $P \neq \pm Q$ where $-Q$ is the reflection of Q in the x-axis, then there must be a third point R on $E(F)$, uniquely determined as the intersection point $E(F)$ of the line through P and Q. Note that $-Q$ is just the third point on the line joining Q and $\mathfrak{o}$, namely the vertical line through Q. This means that if

$$Q = (x, y),$$

then

$$-Q = (x, -y).$$

Observe, as well, that if $P = (x, z)$, then necessarily $y = \pm z$, namely

$$P = \pm Q.$$

As discussed above, if we require that all points be rational, then the existence of two distinct rational points P and Q guarantees that the third point must be rational.

Now the issue is to define the meaning of $P + Q$. It is tempting to set $P + Q = R$. However, suppose that we do this, namely we define the sum of two distinct points P and Q on an elliptic curve E to be the third point R of intersection of E with the line joining the P and Q. Suppose that this definition of addition leads to a group structure. Then in order to get $P + 0 = P$, where 0 is the additive identity, the line through any point P and 0 must intersect the curve as a tangent at P. However, by definition, this means that $P + P = P$, since this is the only point of intersection. Hence, given the existence of additive inverses $-P$, we get $P = 0$ for all P. Hence, the assumption of two distinct points on the curve leads to a contradiction. Instead, we define $P + Q = -R$, the reflection of R about the x-axis. The following figure illustrates this discussion.

Figure 9.2: Addition of distinct points on $y^2 = x^3 - 5x + 2$

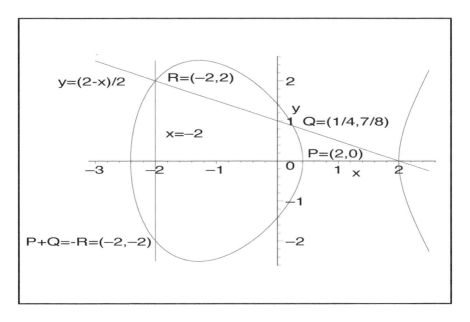

On the other hand, if $P = Q \neq \mathfrak{o}$ and $P \neq -Q$, then we take the tangent line at P, which gives rise to a third point $R = (x_3, y_3)$, uniquely determined as the intersection point of $E(F)$ with the tangent line. Then the reflection about the x-axis gives us:
$$P + P = 2P = -R.$$
Thus, $2P$ is the reflection of the point R about the x-axis, namely the other intersection $-R$ of the line $x = x_3$ with $E(F)$. Lastly, if $P = -Q$, then the line through P and $-Q$ is vertical, so $\mathfrak{o}$ is the third point of intersection, in which case
$$P + Q = \mathfrak{o}.$$
In the above fashion, $E(F)$ becomes an additive abelian group with identity $\mathfrak{o}$. This is an easy exercise, except for proving the associativity, for which the reader may want to use some mathematical software package. The following illustrates

the discussion for addition of nondistinct points $P = Q$, but $P \neq -Q$, namely P is not on a vertical tangent line.

Figure 9.3: Addition of a point to itself on $y^2 = x^3 - 4x + 1$

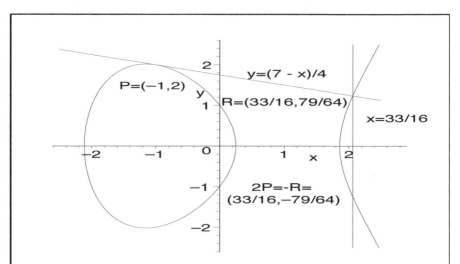

The following definition, motivated by the preceding discussion, provides a summary by giving the addition of points in parametric form.

Definition 9.2 (Addition of Points on Elliptic Curves)

Let $E(F)$ be an elliptic curve with $\text{char}(F) \neq 2, 3$. For any two points $P = (x_1, y_1)$ and $Q = (x_2, y_2)$ on $E(F)$, define

$$P + Q = \begin{cases} \mathfrak{o} & \text{if } x_1 = x_2 \text{ and } y_1 = -y_2, \\ Q & \text{if } P = \mathfrak{o}, \\ (x_3, y_3) & \text{otherwise,} \end{cases}$$

where

$$x_3 = m^2 - x_1 - x_2, \tag{9.5}$$

$$y_3 = m(x_1 - x_3) - y_1, \tag{9.6}$$

and

$$m = \begin{cases} (y_2 - y_1)/(x_2 - x_1) & \text{if } P \neq Q, \\ (3x_1^2 + a)/(2y_1) & \text{if } P = Q. \end{cases} \tag{9.7}$$

The preamble to Definition 9.2 provided a motivation for that definition in geometric terms. Now we have an algebraic explanation to supplement the geometry. Let $E(F)$ be given by

$$y^2 = x^3 + ax + b. \tag{9.8}$$

If $P = (x_1, y_1)$, $Q = (x_2, y_2)$ on $E(F)$ with $x_1 \neq x_2$, so $P \neq \pm Q$, then $-(P+Q)$ is the third point of intersection, $R = (x_3, -y_3)$, of $E(F)$ with the line joining P and Q. The equation of this line has slope $m = (y_1 - y_2)/(x_1 - x_2)$, which is (9.7) for the case $P \neq Q$. This may be rewritten as $y = m(x - x_1) + y_1$, and plugged into (9.8) to get:

$$m^2(x - x_1)^2 + 2m(x - x_1)y_1 + y_1^2 = x^3 + ax + b,$$

which simplifies to

$$x^3 - m^2 x^2 + Ax + B = 0, \tag{9.9}$$

where $A = a + 2m^2 x_1 - 2m y_1$ and $B = b - y_1^2 + 2m_1 x_1 - m^2 x_1^2$. However, by Exercise 2.25 on page 96, $m^2 = x_1 + x_2 + x_3$, or by rewriting,

$$x_3 = m^2 - x_1 - x_2,$$

which is (9.5). Thus $P + Q = (x_3, y_3)$, where

$$y_3 = m(x_1 - x_3) - y_1,$$

which is (9.6). If $P = Q = (x_1, y_1)$ and $P \neq -Q$, namely $y_1 \neq 0$, then the slope of the tangent at P is given by $2yy' = 3x^2 + a$, namely by

$$m = \frac{3x_1^2 + a}{2y_1},$$

which is the case (9.7) for $P = Q$. Lastly, if $P = -Q$, then the line through P and $-Q$ is vertical, so the third point of intersection is $\mathfrak{o}$, as noted above, and $P + Q = -Q + Q = \mathfrak{o}$.

Remark 9.3 All of the above can be summarized in a single equation that covers all cases including the possibility that $P = \mathfrak{o}$, and the possibility that the points are nondistinct. It is that if P, Q, R are three collinear points (all in the same straight line) on $E(F)$, then

$$P + Q + R = \mathfrak{o}.$$

Exercises

9.1. Prove that $x^3 + y^3 = z^3$ has solutions $x, y, z \in \mathbb{Z}$ with $xyz \neq 0$ if and only if $Y^2 = X^3 - 432$ has solutions $X, Y \in \mathbb{Q}$ with $|Y| \neq 36$.

9.2. Prove that $Y^2 = X^3 - 4X$ has nonzero solutions $X, Y \in \mathbb{Q}$ if and only if $x^4 + y^4 = z^2$ has nonzero solutions $x, y, z \in \mathbb{Z}$.

9.2 Mazur, Siegel, and Reduction

> *Mathematics, the non-empirical science par excellence...the science of sciences, delivering the key to those laws of nature and the universe which are concealed by appearances.*
>
> from contributions to **The New Yorker**
> **Hannah Arendt (1906–1975)**
> **Geman-born American political philosopher**

The principal thrust of this section is the presentation of the celebrated results by Mazur on torsion points, of Siegel on the finiteness of integer points on elliptic curves, and Mordell's result on elliptic curves over $\mathbb{Q}$ being finitely generated. First, we need to define some terms.

If we consider rational points on an elliptic curve $E(\mathbb{Q})$, then they are classified into two types as follows, with Definition 9.2 on page 308 in mind.

Definition 9.3 | **Torsion Points on Elliptic Curves**

If $E(\mathbb{Q})$ is an elliptic curve over $\mathbb{Q}$, and P is a point on $E(\mathbb{Q})$ such that

$$nP = \underbrace{P + P + \cdots + P}_{n \text{ summands}} = \mathfrak{o}$$

for some $n \in \mathbb{N}$, then P is called a *torsion point* or a *point of finite order*. The smallest such value of n is called the *order of P*. We call $\mathfrak{o}$ the *trivial torsion point*. If P is not a torsion point, then P is said to be a *point of infinite order*.

Remark 9.4 In 1922, Mordell proved that if $E(\mathbb{Q})$ is an elliptic curve over $\mathbb{Q}$, then $E(\mathbb{Q})$ is finitely generated—see Biography 9.2 on page 315. This remarkable result had been assumed without proof by Poincaré in 1901—see Biography 3.8 on page 147. Essentially this result says that the points of infinite order can be represented as an integral linear combination of some finite set of points $\{P_j\}_{j=1}^n$ on $E(\mathbb{Q})$. The value n is called the *rank of $E(\mathbb{Q})$*. The study of the rank of elliptic curves is one of the most active research areas in modern mathematics. In 1928, Weil generalized the Mordell result to elliptic curves $E(F)$, where F is an arbitrary number field—see Biography 9.3 on page 316. Thus, today we call the generalized result the *Mordell–Weil Theorem*. For a proof of this celebrated result see [88, Theorem 6.7, p. 220]. There are many deep results such as this, which we will state without proof in this section in order to give the reader some flavour of the richness of the subject. There is a vast literature on the subject for the interested reader to pursue.

Example 9.2 Let $E(\mathbb{Q})$ be defined by

$$y^2 = x^3 + 1,$$

illustrated in Figure 9.4 on the next page. Consider the point $P = (2, 3)$. By Definition 9.2, we calculate that

$$2P = (0, 1), 3P = (-1, 0), 4P = (0, -1), 5P = (2, -3), \text{ and } 6P = \mathfrak{o}.$$

These points are illustrated in Figure 9.4. Notice that we begin with the tangent line T at P, which intersects the curve at $(0,-1)$, so $2P = (0,1)$, the reflection of $(0,-1)$ about the x-axis. Then the line L through P and $(0,1)$ intersects the curve at $(-1,0)$, which is $3P$ since it is its own reflection in the x-axis. The intersection of L with $E(\mathbb{Q})$ is $(0,1)$, so $4P = (0,-1)$, the reflection of $(0,1)$ about the x-axis. Since $(0,-1)$ is on T, then the intersection of T with $E(\mathbb{Q})$ is $(2,3)$, so $5P = (2,-3)$, again the reflection of $(2,3)$ about the x-axis. Since P and $5P$ lie on the vertical line V, then $6P = \mathfrak{o}$. Thus, P is a torsion point of order 6.

Figure 9.4: Multiples of Torsion Points on $y^2 = x^3 + 1$

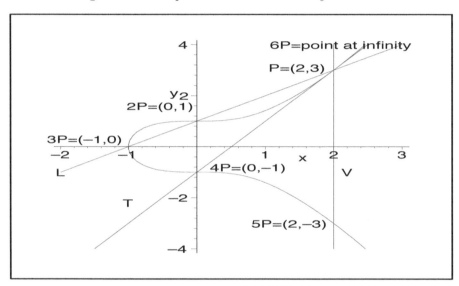

Example 9.2 illustrates a microcosm of a fact that is contained in the Mordell–Weil Theorem described in Remark 9.4 on the facing page, namely that every rational point on $E(\mathbb{Q})$ can be obtained from a finite set of points by repeatedly taking lines through pairs of them, intersecting with $E(\mathbb{Q})$, and reflecting about the x-axis to create new points. The torsion points, such as those given in Example 9.2, form a finite subgroup $E(\mathbb{Q})_t \subseteq E(\mathbb{Q})$, called the *torsion subgroup*. Thus, by the method illustrated in Example 9.2, we have an effective method for computing $E(\mathbb{Q})_t$. A pertinent result proved by Lutz and Nagell in the mid 1930's is given in the following.

Theorem 9.1 | **Nagell–Lutz Theorem**

If $P = (x_1, y_1) \in E(\mathbb{Q})_t$, *where* $E(\mathbb{Q})$ *is given by* $y^2 = x^3 + ax + b$, $a, b \in \mathbb{Z}$, *then* $x_1, y_1 \in \mathbb{Z}$ *and either* $y_1 = 0$ *or* $y_1^2 \mid (4a^3 + 27b^2)$.

Proof. See [88].   □

Theorem 9.1 says that all elements of $E(\mathbb{Q})_t$ must have rational integer coordinates, called *integer points*, where the ordinate (y-value) divides the discriminant of $E(\mathbb{Q})$. Thus, the Nagell–Lutz Theorem determines all integer points P such that $2P$ is also an integer point. (Nagell was the first to prove the result. Then Lutz later refined the proof.) Thus, we may conclude that if a multiple of an integer point is *not* an integer point, then that point has infinite order. For instance, in Figure 9.3 on page 308 the integer point $(-1, 2)$ must be of infinite order since

$$2P = (33/16, -79/64).$$

The deeper problem of actually determining the cardinality $|E(\mathbb{Q})_t|$ as $E(\mathbb{Q})$ varies over all elliptic curves over $\mathbb{Q}$ was solved by B. Mazur who proved in 1976:

Theorem 9.2 | **Mazur's Theorem**

If $E(\mathbb{Q})$ *is an elliptic curve over* $\mathbb{Q}$, *then either*

$$E(\mathbb{Q})_t \cong \mathbb{Z}/n\mathbb{Z},$$

for some $n \in \{1, 2, 3, 4, 5, 6, 7, 8, 9, 10, 12\}$ *or*

$$E(\mathbb{Q})_t \cong \mathbb{Z}/2\mathbb{Z} \oplus \mathbb{Z}/2n\mathbb{Z},$$

where $n \in \{1, 2, 3, 4\}$.

Thus, the torsion group cannot have order bigger than 16 for elliptic curves over $\mathbb{Q}$. For instance, by Exercise 9.3 on page 315, $E(\mathbb{Q})_t$ for the elliptic curve

$$y^2 = x^3 + 1$$

in Example 9.2 is made up of

$$(2, \pm 3), (0, \pm 1), (-1, 0), \text{ and } \mathfrak{o},$$

so $|E(\mathbb{Q})_t| = 6$ in that case. Figure 9.1 on page 306 provides an instance where $|E(\mathbb{Q})_t| = 2$ since $(0, 0)$ is the only nontrivial torsion point and it has order 2. An example of the case where $n = 1$ is given by the elliptic curve $E(\mathbb{Q})$ given by

$$y^2 = x^3 - 2$$

since $\mathfrak{o}$ is the only torsion point by Exercise 9.4. The problem of determining $|E(F)_t|$, as $E(F)$ varies over all elliptic curves for an arbitrary number field F, remains open. However, in 1996 L. Merel [57] proved what is called the *strong uniform boundedness conjecture (UBC)* of Mazur and Kamienny, namely that for an elliptic curve $E(F)$ over F, $|E(F)_t| \leq B_F$, where B_F is a constant depending only on $|F : \mathbb{Q}|$. For instance, Mazur's Theorem tells us that

$$|E(\mathbb{Q})_t| \leq B_{\mathbb{Q}} = 16.$$

We have seen that the number of torsion points is finite, in fact quite small for a given elliptic curve by Mazur's Theorem. However, we have also seen instances where an integer point is not a torsion point. Thus the question naturally arises: Are there infinitely many integer points on a given elliptic curve? In 1926, C.L. Siegel solved the problem by proving the following—see Biography 4.4 on page 170.

Theorem 9.3 | **Siegel's Theorem**

The equation $y^2 = x^3 + ax + b$, *with* $a, b, c \in \mathbb{Z}$ *and* $4a^3 + 27b^2 \neq 0$, *has only finitely many solutions* $x, y \in \mathbb{Z}$.

Remark 9.5 The nonvanishing condition on the discriminant in the hypothesis of Theorem 9.3 is necessary since, for instance, $y^2 = x^3$ has infinitely many integer solutions, which may be seen by letting $n \in \mathbb{N}$ and setting $y = n^3$, $x = n^2$.

Now that we have some basic knowledge of elliptic curves, we may turn our attention to elliptic curves over finite fields, since this is the gateway to the applications of elliptic curves to factoring and primality testing. To do this, the canonical approach is to begin with an elliptic curve $E(\mathbb{Q})$ over $\mathbb{Q}$ and reduce it modulo a prime p. To understand how this is done, we must first make precise what we mean by reduction of rational points.

Definition 9.4 | **Reduction of Rationals on Elliptic Curves**

Let $n \in \mathbb{N}$ and $x_1, x_2 \in \mathbb{Q}$ with denominators prime to n. Then

$x_1 \equiv x_2 \pmod{n}$ means $x_1 - x_2 = a/b$ where $\gcd(a, b) = 1, a, b \in \mathbb{Z}$, and $n | a$.

For any $x = c/d \in \mathbb{Q}$ with $\gcd(d, n) = 1 = \gcd(c, d)$, there exists a unique $r \in \mathbb{Z}$, with $0 \leq r \leq n - 1$, such that $x \equiv r \pmod{n}$, denoted by

$$r = \overline{x} \pmod{n}.$$

Note that we may take
$$r \equiv \overline{cd^{-1}} \pmod{n},$$

where d^{-1} is the unique multiplicative inverse of d modulo n. Hence, if $P = (x, y)$ is a point on an elliptic curve $E = E(\mathbb{Q})$ over $\mathbb{Q}$, with denominators of x and y prime to n, then

$$\overline{P} \pmod{n} \text{ means } (\overline{x} \pmod{n}, \overline{y} \pmod{n}).$$

Also, $\overline{E} \pmod{n}$ denotes the curve reduced modulo n, namely the curve defined by
$$y^2 = x^3 + \overline{a} \pmod{n} x + \overline{b} \pmod{n},$$

with
$$x = \overline{x} \pmod{n}, \text{ and } y = \overline{y} \pmod{n}.$$
The cardinality of the set $\overline{E} \pmod{n}$ is denoted by
$$|\overline{E} \pmod{n}|.$$

It turns out that $\overline{E} \pmod{n}$ in Definition 9.4 may not be a group, since certain elements may not be invertible. However, we may still use it for practical computational purposes, as illustrated below.

Example 9.3 If $x = 5/4$ and $n = 7$, then $\overline{x} \pmod{7} = 3 = r$ is the unique integer (least positive residue) modulo 7 such that $x \equiv r \pmod{7}$, since $5/4 - 3 = -7/4$. Note that $5/4 \equiv 5 \cdot 4^{-1} \equiv 5 \cdot 2 \equiv 3 \pmod{7}$.

The following result tells us how to add and reduce points on rational elliptic curves, and will be the chief tool in the description of the elliptic curve factoring method in §9.3.

Theorem 9.4 (Addition and Reduction of Points on Elliptic Curves)
   Let $n \in \mathbb{N}$, $\gcd(6, n) = 1$, and let $E = E(\mathbb{Q})$ be an elliptic curve over $\mathbb{Q}$ with equation
$$y^2 = x^3 + ax + b, \quad a, b \in \mathbb{Z},$$

and
$$\gcd(4a^3 + 27b^2, n) = 1.$$

Let P_1, P_2 be points on E where
$$P_1 + P_2 \neq \mathfrak{o},$$

and the denominators of P_1, P_2 are prime to n. Then $P_1 + P_2$ is on E with coordinates having denominators prime to n if and only if there does not exist a prime $p|n$ such that
$$\overline{P_1} \pmod{p} + \overline{P_2} \pmod{p} = \overline{\mathfrak{o}} \pmod{p}$$

on the elliptic curve $\overline{E} \pmod{p}$ over $\mathbb{F}_p$, with equation
$$y^2 = x^3 + \overline{a} \pmod{p}x + \overline{b} \pmod{p}.$$

Proof. See [47, Proposition VI.3.1, pp. 172–174].   □

Biography 9.2 Louis Joel Mordell (1888–1972) *was born in Philadelphia, Pennsylvania on January 28, 1888. He was educated at Cambridge, and lectured at Manchester College of Technology from 1920 to 1922. In 1922, he went to Manchester University where he remained until 1945 when he held the Sadleirian Chair at the College of St. John's in Cambridge. The topic for his inaugural lecture to the chair was the equation $y^2 = x^3 + k$. Although he retired from the chair in 1953, his mathematical output remained high. Indeed, roughly half of his 270 publications were published after he left the chair. In 1971, he was still traveling and lecturing, including an extensive tour of Asia after he attended a number theory conference in Moscow. Yet, he fell ill a few months later and died in Cambridge on March 12, 1972.*

   Among his honours were being elected as a member of the Royal Society in 1924, winning the De Morgan Medal in 1941, being president of the London Mathematical Society from 1943 to 1945, and winning the Sylvester Medal in 1949.

Exercises

9.3. Prove that the torsion points computed in Example 9.2 on page 310 are all of the points in $E(\mathbb{Q})_t$. (*Hint: Use the Nagell–Lutz Theorem.*)

9.4. Prove that there are no nontrivial torsion points on the elliptic curve $E(\mathbb{Q})$ given by $y^2 = x^3 - 2$. (*Hint: Look at Theorem 1.19 on page 47, and use the Nagell–Lutz Theorem.*)

9.5. Suppose that the equation defining an elliptic curve $E(\mathbb{F}_{p^k})$ over $\mathbb{F}_{p^k}$, p a prime, is
$$y^2 = x^3 + ax + b, \quad a, b \in \mathbb{Z}.$$
Prove that the number of elements on E, counting the point at infinity, is
$$p^k + 1 + \sum_{x \in \mathbb{F}_{p^k}} \chi(x^3 + ax + b),$$
where χ is a quadratic Dirichlet character modulo p^k. In other words, $\chi(y) = -1, 0, 1$ according as y is a quadratic nonresidue, 0, or a quadratic residue respectively for $y \in \mathbb{F}_{p^k}$.

   In Exercises 9.6–9.9, use the Nagell–Lutz Theorem 9.1 and Mazur's Theorem 9.2 both on page 312 to do the calculations.

9.6. If $y^2 = x^3 - 432$ defines the elliptic curve $E(\mathbb{Q})$, calculate $E(\mathbb{Q})_t$.

9.7. If $E(\mathbb{Q})$ is given by $y^2 = x^3 - 2x + 1$, determine $E(\mathbb{Q})_t$.

9.8. If $E(\mathbb{Q})$ is given by $y^2 = x^3 - x$, determine $E(\mathbb{Q})_t$.

9.9. If $E(\mathbb{Q})$ is given by $y^2 = x^3 + 1$, determine $E(\mathbb{Q})_t$.

Biography 9.3 André Weil, pronounced *vay* (1906–1998), *was born on May 6, 1906 in Paris, France. As he said in his autobiography,* The Apprenticeship of a Mathematician, *he was passionately addicted to mathematics by the age of ten.* He was also interested in languages, as evidenced by his having read the Bhagavad Gita *in its original Sanskrit at the age of sixteen. After graduating from the École Normal in Paris, he eventually made his way to Göttingen, where he studied under Hadamard. His doctoral thesis contained a proof of the Mordell–Weil Theorem, namely that the group of rational points on an elliptic curve over* $\mathbb{Q}$ *is a finitely generated abelian group. His first position was at Aligarh Muslim University, India (1930–1932), then the University of Strasbourg, France (1933–1940), where he became involved with the controversial* Bourbaki project, *which attempted to give a unified description of mathematics. The name* Nicholas Bourbaki *was that of a citizen of the imaginary state of Poldavia, which arose from a spoof lecture given in 1923. Weil tried to avoid the draft, which earned him six months in prison. It was during this imprisonment that he created the Riemann hypothesis—see Conjecture 5.1 on page 223. In order to be released from prison, he agreed to join the French army. Then he came to the United States to teach at Haverford College in Pennsylvania. He also held positions at São Paulo University, Brazil (1945–1947), the University of Chicago (1947–1958), and thereafter at the Institute for Advanced Study at Princeton. In 1947 at Chicago, he began a study, which eventually led him to a proof of the Riemann hypothesis for algebraic curves. He went on to formulate a series of conjectures that won him the Kyoto Prize in 1994 from the Inamori Foundation of Kyoto, Japan. His conjectures provided the principles for modern algebraic geometry. His honours include an honorary membership in the London Mathematical Society in 1959, and election as a Fellow of the Royal Society of London in 1966. However, in his own official biography he lists his only honour as* Member, Poldavian Academy of Science and Letters. *He is also known for having said,* "In the future, as in the past, the great ideas must be the simplifying ideas," *as well as,* "God exists since mathematics is consistent, and the devil exists since we cannot prove it." *This is evidence of his being known for his poignant phrasing and whimsical individuality, as well as for the depth of his intellect. He died on August 6, 1998 in Princeton, and is survived by two daughters, and three grandchildren. His wife Eveline died in 1986.*

9.3 Applications: Factoring & Primality Testing

> *In mathematics you don't understand things. You get used to them.*
> from **The Dancing Wu Li Masters**—see [106]
> **John von Neumann (1903–1945)**
> **Hungarian-born American mathematician and computer pioneer**

§9.1 and §9.2 put us in a position to describe Lenstra's factorization method using elliptic curves—see Biography 9.4 on the next page.

◆ **Lenstra's Elliptic Curve Factoring Method**

The following is the algorithm for factoring an odd composite $n \in \mathbb{N}$.

(1) In some random fashion, we generate a pair (E, P), where $E = E(\mathbb{Q})$ is an elliptic curve over $\mathbb{Q}$ with equation

$$y^2 = x^3 + ax + b, \quad a, b \in \mathbb{Z},$$

and P is a point on E.

(2) Check that $\gcd(n, 4a^3 + 27b^2) = 1$. If so, go to step (3). If not, then we have a factor of n, unless $\gcd(n, 4a^3 + 27b^2) = n$, in which case we choose a different pair (E, P).

(3) Choose $M \in \mathbb{N}$ and bounds $A, B \in \mathbb{N}$ such that the canonical prime factorization of M is

$$M = \prod_{j=1}^{\ell} p_j^{a_{p_j}},$$

for small primes $p_1 < p_2 < \ldots < p_\ell \leq B$, where

$$a_{p_j} = \lfloor \log_e A / \log_e p_j \rfloor$$

is the largest exponent such that $p_j^{a_{p_j}} \leq A$.

(4) For a sequence of divisors s of M, compute

$$\overline{sP} \pmod{n}$$

as follows. First compute

$$\overline{sP} = \overline{p_1^k P} \pmod{n},$$

for $1 \leq k \leq a_{p_1}$, then

$$\overline{sP} = \overline{p_2^k p_1^{a_{p_1}} P} \pmod{n},$$

for $1 \leq k \leq a_{p_2}$, and so on, until all primes p_j dividing M have been exhausted or the following occurs.

(5) If the calculation of either $(x_2 - x_1)^{-1}$ or $(2y_1)^{-1}$ in (9.7) on page 308, for some $s|M$ in step (4), shows that one of them is *not* prime to n, then there is a prime $p|n$ such that

$$\overline{sP} = \mathfrak{o} \pmod{p}, \tag{9.10}$$

by Theorem 9.4 on page 314. This will give us a nontrivial factor of n unless (9.10) occurs for all primes $p|n$. In that case $\gcd(s, n) = n$, and we go back and try the algorithm with a different (E, P) pair.

The value of B in step (3) of the above algorithm is the upper bound on the prime divisors of s, from which we form $\overline{sP}$. If B is large enough, then we increase the probability that $\overline{sP} = \mathfrak{o} \pmod{p}$ for some prime $p \mid n$. On the other hand, the larger the value of B, the longer the computational time. Hence, we must also choose B to minimize running time. Moreover, A is an upper bound on the prime powers that divide s, so similar considerations apply. Lenstra has some convincing conjectural evidence that $n \in \mathbb{N}$ can be factored by his algorithm in expected running time

$$O\left(e^{\sqrt{(2+\epsilon)\log_e p(\log_e \log_e p)}}(\log_e n)^2\right),$$

where p is the smallest prime factor of n and ϵ goes to zero as p gets large. (A corollary of this fact is that the elliptic curve method can be used to factor n in expected time

$$O(e^{\sqrt{(1+\epsilon)(\log_e n)(\log_e \log_e n)}}),$$

with ϵ as above.)

Biography 9.4 Hendrik Willem Lenstra Jr. (1949–) *was born in Zaandam, Netherlands. His father was a mathematician, and his brothers, Arjen and Jan, are also well-known mathematicians. Hendrik studied at the University of Amsterdam. He was an extraordinary student whose brilliance was demonstrated by his solution of a problem of Emmy Noether which he published in* Inventiones Mathematicae—*see Biography 2.1 on page 73. In 1977, he obtained his doctorate under the direction of Frans Oort. Then, when only twenty-eight, he was appointed full professor at the University of Amsterdam. In 1987, he went to the United States, where he was appointed a full professor at Berkeley. In 2003, he retired from Berkeley to take a full-time position at the University of Leiden, the oldest university in the Netherlands.*

Among his honours include the Fulkerton Prize in 1985, plenary lecturer at the International Congress of Mathematicians in 1986 at Berkeley, an honourary doctorate at the Université de Franche-Comté, Besançon in 1995, and Kloosterman-lecturer at the University of Leiden in 1995. Also, he received the Spinozapremie (*Spinoza Prize*) *in 1998. The latter is an annual award by the Netherlands Research Council of 1.5 million Euros, to be spent on new research. The award, named after the philosopher Baruch Spinoza, is the highest scientific award in the Netherlands—see the quote on page 331.*

In the next example, which illustrates the Lenstra's algorithm, we will make use of the following renowned result proved by Hasse.

Theorem 9.5 | **Hasse's Bound for Elliptic Curves Over $\mathbb{F}_p$**

If E is an elliptic curve over $\mathbb{F}_{p^k}$ for a prime $p > 3$, and $k \in \mathbb{N}$, then

$$\left| |\overline{E} \pmod{p^k}| - p^k - 1 \right| \leq 2\sqrt{p^k}.$$

Note that Exercise 9.5 on page 315 is related to the following inequality emanating in Theorem 9.5 for the case where $k = 1$,

$$(\sqrt{p} - 1)^2 = p + 1 - 2\sqrt{p} < |\overline{E} \pmod{p}| < p + 1 + 2\sqrt{p} = (\sqrt{p} + 1)^2. \quad (9.11)$$

Indeed, (9.11) represents the order of magnitude of the distance from p for the possible orders of $\overline{E} \pmod{p}$. Statistically speaking, the distance from the origin after addition over p elements of the Legendre symbol, the $k = 1$ case of Exercise 9.5, is proportional to $\sqrt{p}$, so Theorem 9.5 gives an expected statistical result:

$$\frac{\left| |\overline{E} \pmod{p}| - p - 1 \right|}{\sqrt{p}} \leq 2.$$

Based upon Hasse's Theorem 9.5, for $k = 1$ and the above expected running time, Lenstra concludes that if we take

$$A = p + 1 + 2\sqrt{p}, \text{ and } B = e^{\sqrt{(\log_e p)(\log_e \log_e p)/2}},$$

where p is the smallest prime factor of n, then about one out of every B iterations will be successful in factoring n. Of course, we do not know a prime divisor p of n in advance, so we replace p by $\lfloor \sqrt{n} \rfloor$ and look at incremental values up to that bound.

Once the values of A and B have been chosen, then for a given prime p, the set $\overline{E} \pmod{p}$ is a finite abelian group, since this is an elliptic curve over a finite field. Also, if the order g of $\overline{E} \pmod{p}$ is not divisible by any primes larger than B, and if p is a prime such that

$$p + 1 + 2\sqrt{p} < A,$$

then Hasse's Theorem 9.5 tells us that $g \mid m$ in the algorithm, so

$$\overline{mP} = \mathfrak{o} \pmod{p}.$$

When $\overline{E} \pmod{n}$ is not a group, then this is not a problem in the algorithm. The reason is that, even if P_1 and P_2 were points on such a curve and if $P_1 + P_2$ were not defined, then n must be composite! The noninvertibility that would result in step 5 of the algorithm would then give us a factor of n. This is indeed the underlying key element in the elliptic curve algorithm.

Remark 9.6 There is also the following valuable result on the group structure of $E = E(\mathbb{F}_p)$. If $p > 3$ is prime, then there are $m, n \in \mathbb{N}$ such that E is isomorphic to the product of a cyclic group of order m with one of order n, where $m| \gcd(n, p - 1)$. See [47].

The following example is chosen to best illustrate the algorithm for pedagogical purposes, wherein we choose relatively small values of n to factor. Even though modular reduction at each stage keeps the size of the rational points to a minimum, the larger the number, the higher the likelihood of a large number of stages before the algorithm terminates. Thus, we keep the value small so that the process may be illustrated without filling pages with calculations.

Example 9.4 Let $n = 3551$. Choose a family of elliptic curves

$$y^2 = x^3 + ax + 1,$$

each of which has the point $P = (0, 1)$ on it. We now choose successive natural numbers a until the process described above is successful in factoring n. We take $B = 3$, and since

$$\lfloor \sqrt{n} \rfloor = 59 \geq p,$$

then by Hasse's Theorem 9.5 on the preceding page, we may choose

$$A = 59 + 1 + 2\lfloor \sqrt{n} \rfloor = 178.$$

Thus,

$$M = 2^7 \cdot 3^4,$$

where

$$7 = \lfloor \log_e 178 / \log_e 2 \rfloor,$$

and

$$4 = \lfloor \log_e 178 / \log_e 3 \rfloor.$$

Using (9.5)–(9.7), we tabulate the following for $a = 1$. First we verify that the discriminant of E is prime to n. We have

$$\Delta(E(\mathbb{Q})) = -16(4 \cdot 1^3 + 27 \cdot 1^2) = -16 \cdot 31,$$

which is prime to n, so we may proceed. We therefore begin with the (E, P) pair

$$(y^2 = x^3 + x + 1, (0, 1)).$$

In Table 9.1, the value m is given by (9.7) on page 308.

s	$\overline{m}$	$s\overline{P}$
1	$--$	$(0,1)$
2	1776	$(888, 3106)$
2^2	2860	$(3422, 796)$
2^3	1218	$(3015, 1341)$
2^4	704	$(3099, 3441)$
2^5	3396	$(72, 3208)$
2^6	2022	$(1139, 1877)$
2^7	1977	$(151, 1900)$
$2^6 3$	1700	$(148, 3200)$
$2^7 3$	1085	$(1548, 1179)$
$2^6 3^2$	3476	$(525, 218)$
$2^7 3^2$	2939	$(639, 2081)$
$2^6 3^3$	3287	$(2932, 3152)$
$2^7 3^3$	117	$(723, 3180)$
$2^6 3^4$	3297	$(2612, 792)$
$2^7 3^4$	11	$(1999, 2400)$

Table 9.1

We now abandon the above (E, P) pair since we have exhausted all divisors of M without achieving a point at infinity modulo any prime p dividing n. Notice that on line nine of the column for s, we have

$$s = 2^6 3 = (2+1) \cdot 2^6 = 2^7 + 2^6.$$

We are adding the two distinct points, the ones on lines seven and eight. Then on line ten, $s = 2^7 3$ is twice $s = 2^6 3$ on the previous line. Similarly, this also occurs for

$$s = 2^6 3^2 = 3 \cdot 2^6 + 3 \cdot 2^7, \ s = 2^6 3^3 = 3^2 \cdot 2^7 + 3^2 \cdot 2^6, \ 2^6 3^4 = 3^3 \cdot 2^7 + 3^3 \cdot 2^6.$$

This natural process of doubling and reduction signifies the method in the algorithm that we are illustrating. (This method of repeated doubling is a method of multiplying a point P on an elliptic curve E by a given $s \in \mathbb{N}$. This is the analogue of raising an element of a finite field $\mathbb{F}_q$ to the power s. It is known that this can be accomplished in $O((\log_e s)(\log_e q)^3)$ bit operations.)

The reader may now go to Exercise 9.10 on page 325 which verifies that we also exhaust all divisors of M for each (E, P) pair

$$(y^2 = x^3 + ax + 1, (0, 1)) \text{ with } 2 \le a \le 8.$$

We now move to the next (E, P) pair which is

$$(y^2 = x^3 + 9x + 1, (0, 1)).$$

Observe that

$$\gcd(\Delta(E), n) = \gcd(-2^4 3^3 109, 3551) = 1,$$

so we may proceed.

Table 9.2

s	$\overline{m}$	$(\overline{x_3}, \overline{y_3})$
1	$--$	$(0, 1)$
2	1780	$(908, 3015)$
2^2	$--$	$--$

We terminate the calculations at $m = 2476943/6030$ since $\gcd(6030, 3551) = 67$. This gives us the factorization $3551 = 53 \cdot 67$. Thus, we have reached step (5) of the algorithm where $y_1^{-1} = 3015^{-1}$ does not exist modulo n for the pair $(\overline{x_1}, \overline{y_1}) = (908, 3015)$, so we cannot use (9.7) to compute the $(\overline{x_3}, \overline{y_3})$ pair for $2^2 P$, and the algorithm terminates with a factorization.

Example 9.4 provides ample illustrations of one reason for having to choose a new elliptic curve from the family, namely running out of divisors of M. The other reason for having to choose another such curve is the obtaining of the trivial factorization during the implementation of the algorithm. In other words, before exhaustion of the divisors of M, we could encounter a value whose gcd with n *is* n, as indicated in step (5) of the algorithm.

Lenstra's algorithm is exceptional at finding small prime factors (those with no more that forty digits) of large composite numbers. However, since it requires relatively little storage space, it can be used as a subroutine in conjunction with other methods. For this reason, among many others, the elliptic curve methods enjoy great favour among modern-day cryptographers.

We now show how Lenstra's algorithm may be modified to obtain a primaility testing algorithm. The primality test is based upon the following result.

Theorem 9.6 $\boxed{\textbf{Elliptic Curve Primality Test}}$

Let $n \in \mathbb{N}$ with $\gcd(n, 6) = 1$, and let $E = E(\mathbb{Q})$ be an elliptic curve over $\mathbb{Q}$. Suppose that

(a) $n + 1 - 2\sqrt{n} \leq |\overline{E}(\mathrm{mod}\ n)| \leq n + 1 + 2\sqrt{n}$.

(b) $|\overline{E}(\mathrm{mod}\ n)| = 2p$, where $p > 2$ is prime.

If $P \neq \mathfrak{o}$ is a point on E and $\overline{pP} = \mathfrak{o}$ on $\overline{E}(\mathrm{mod}\ n)$, then n is prime.

Proof. See [18, Lemma 14.23, p. 324]. $\qquad\qquad\qquad\qquad\qquad\qquad\square$

Theorem 9.6 is employed by picking in some random fashion points P_j for $j = 1, 2, \ldots, m \in \mathbb{N}$ on an elliptic curve E and, for a given prime p, calculating $\overline{pP_j}$ for each such j. If the outcome is that $\overline{pP_j} = \mathfrak{o}$ for some $j = 1, 2, \ldots, m$, then n is prime. For instance, a suitable choice for P_1 is $2Q_1$, where Q_1 is randomly chosen. If $P_1 \neq \mathfrak{o}$, but $pP_1 = \mathfrak{o}$, then n is prime. If $P_1 \neq \mathfrak{o} \neq pP_1$, then n is composite.

The following illustration is again chosen for pedagogical reasons. A "realistic" value of n cannot be chosen, given the depth of calculations that would be involved.

Example 9.5 Let $n = 1231$. Since we enjoyed success in Example 9.4 on page 320 with the elliptic curve E given by $y^2 = x^3 + 9x + 1$, we use it here. First we observe that

$$\gcd(n, 6) = \gcd(\Delta(E), n) = \gcd(2^4 3^3 109, 1231) = 1.$$

Now we proceed to check n for primality. If n were prime, then Exercise 9.5 on page 315 tells us that $|\overline{E}(\bmod n)| = 2 \cdot 619$. Also,

$$1161 < n + 1 - 2\sqrt{n} < |\overline{E} \ (\bmod n)| < 1302 < n + 1 + 2\sqrt{n}.$$

Therefore, conditions (a)–(b) of Theorem 9.6 are satisfied. To test n for primality, we begin with a primitive element that has a chance of generating enough points on E. Let $P = (0, 1)$ and observe that

$$619 = 2^9 + 2^6 + 2^5 + 2^3 + 2^1 + 2^0,$$

so we calculate up to 2^9 and test $\overline{619P}$. Again, in what follows, m is the value in (9.7) on page 308.

Table 9.3

s	$\overline{m}$	$\overline{sP}$
1	$--$	$(0, 1)$
2	620	$(328, 985)$
2^2	1213	$(899, 676)$
2^3	1156	$(134, 1037)$
2^4	226	$(337, 1094)$
2^5	302	$(667, 188)$
2^6	996	$(958, 492)$
2^7	1173	$(217, 846)$
2^8	1201	$(466, 469)$
2^9	457	$(1109, 1120)$
$576 = 2^9 + 2^6$	852	$(9, 520)$
$608 = 576 + 2^5$	557	$(592, 964)$
$616 = 608 + 2^3$	954	$(912, 275)$
$618 = 616 + 2$	3	$(0, 1230)$
$619 = 618 + 1$	$--$	$\mathfrak{o}$

Observe that via (9.7), $\overline{(0, 1)} + \overline{618P}$ has a zero denominator so we cannot invert in $\mathbb{Z}/n\mathbb{Z}$, thereby yielding that $619P = \mathfrak{o}$, so 1231 is prime by Theorem 9.6.

We observe that if part (a) of Theorem 9.6 fails to hold, then we have a compositeness test by Hasse's Theorem 9.5 on page 319. Also, part (b) of Theorem 9.6 is very special and does not hold for many elliptic curves. The reader may get a sense of this by checking a few examples via Exercise 9.5 on page 315. Moreover, our n in Example 9.5 was sufficiently small such that we were able to calculate $|\overline{E}(\bmod n)|$ with relative ease. However, as n gets large, $|\overline{E}(\bmod n)|$ gets large, so we may not be able to determine its value. In fact, calculating this cardinality may be as difficult as proving that n is prime. These problems were overcome in a primality test by Goldwasser and Kilian [34]. In order to discuss it, a primality proving algorithm upon which Goldwasser and Killian based their primality test is within our reach and provides a basis for discussing the latter. Recall that a *primality proving algorithm* is one that given an input n, verifies the hypothesis of a theorem whose conclusion is "n is prime"—see [68, §1.8].

Theorem 9.7 | **Goldwasser–Killian Primality Proving Algorithm** |.

Let $n > 1$ be an integer with $\gcd(6, n) = 1$, and let $m, r \in \mathbb{N}$ with $r \mid m$. Furthermore, assume $E = E(\mathbb{Q})$ is an elliptic curve over $\mathbb{Q}$. If there exists a point P on E such that $mP = \mathfrak{o}$, and for every prime $p \mid r$ we have that

$$\left(\frac{m}{p}\right) P \neq \mathfrak{o},$$

then for every prime $q \mid n$ we have

$$|\overline{E} \ (\mathrm{mod} \ q)| \equiv 0 \ (\mathrm{mod} \ r). \qquad (9.12)$$

Also, if

$$r > (n^{1/4} + 1)^2,$$

then n is prime.

Proof. Let q be a prime divisor of n and let d be the order of P on $\overline{E} \, (\mathrm{mod} \ q)$. It follows that $r \mid d$, so (9.12) follows. Now assume that $r > (n^{1/4} + 1)^2$. However, by Hasse's Theorem 9.5,

$$|\overline{E} \ (\mathrm{mod} \ q)| < (q^{1/2} + 1)^2.$$

Hence,

$$(q^{1/2} + 1) > \sqrt{|\overline{E} \ (\mathrm{mod} \ q)|} > r^{1/2} > (n^{1/4} + 1),$$

so $q > \sqrt{n}$. Yet, $n = qt$ for some $t \in \mathbb{N}$, so if $t \geq 2$, then $q^2 \geq 2q$, a contradiction, which yields that n is prime. $\qquad\qquad\qquad\qquad\qquad\qquad\qquad\qquad\qquad\qquad\square$

Goldwasser and Killian employed Theorem 9.7 to provide a primality test where an input $n \in \mathbb{N}$ could be tested in an expected number of operation $O(\log_e^C n)$ for a constant C. The kernel of the idea in their test comes in two parts. One is to randomly select elliptic curves modulo n for a large number of $n \in \mathbb{N}$. Then whenever we get

$$|\overline{E} \ (\mathrm{mod} \ n)| = 2p,$$

where p is a *probable prime*, then use Theorem 9.6 on page 322 to check for primality of p. If this test succeeds in demonstrating that p is indeed prime, then it follows from probabilistic compositeness tests that n is *provably prime*—see [68, §2.7, pp. 121–126].

The second idea is to make the above process recursive. They do this by proving p is prime using Theorem 9.6 on an elliptic curve over $\mathbb{Z}/p\mathbb{Z}$ of order $2r$, where r is a probable prime in Theorem 9.7. In this fashion, the primality of r implies the primality of p. Moreover, since each iteration reduces the size by a half, since $p \approx n/2$, then it follows that the numbers will get sufficiently small so that trial division may be used to prove it to be prime. Then by this process, the original n may be shown to be, in the last iteration, (provably) prime. If, in any iteration, the probable prime is shown to be composite,then one goes back to the initial iteration with another candidate—see [49] for more details. Also, see [18] for other interesting and deep connections.

In §9.4, we will look at applications of elliptic curves to cryptography as a fitting close to this chapter where we may employ what we have learned herein thus far.

Exercises

9.10. Perform the calculations in Lenstra's Elliptic Curve Factoring Method for each (E, P) pair $(y^2 = x^3 + ax + 1, (0, 1))$ where $3 \le a \le 8$. This shows that all divisors of M are exhausted in each case without achieving a nontrivial factor of 3551.

9.11. Use Lenstra's Elliptic Curve Factoring Algorithm to factor each of the following.

    (a) 16199   (b) 13261

    (c) 53059   (d) 10403

9.12. Use Lenstra's Elliptic Curve Method to factor each of the following.

    (a) 2201   (b) 16199

    (c) 9073   (d) 32107

9.13. Use the Elliptic Curve Primality Test to test each of the following for primality.

    (a) 7489   (b) 8179

    (c) 9533   (d) 26869

9.4   Elliptic Curve Cryptography (ECC)

Quod gratis assertiur, gratis negatur—What is asserted without reason (or proof), may be denied without reason (or proof).

Latin Maxim

For this section, the reader should be familiar with the basics on cryptology as set out for instance in [68, §2.8, pp. 127–138]. Part of the following is adapted from [66].

In the 1980s, there was a development of the notion of public-key cryptography in the realm of elliptic curves. In particular, in 1985, Miller (see [59]) and Koblitz (see [46]) independently proposed using elliptic curves for public-key cryptosystems. However, they did not invent a cryptographic algorithm for use with elliptic curves, but rather implemented then-existing public-key algorithms in elliptic curves over finite fields. These types of cryptosystems are more appealing than cryptosystems over finite fields since, rather than just the group of a finite field $\mathbb{F}_p^*$, one has many *elliptic curves over $\mathbb{F}_p$* from which to choose. Also, whenever the elliptic curve is properly chosen, there is no known *subexponential time algorithm* for cryptanalyzing such cryptosystems, where such an algorithm is defined as one for which the complexity for input $n \in \mathbb{N}$ is

$$O(\exp((c + o(1))(\log_e n)^r ((\log_e n)(\log_e \log_e n))^{1-r}),$$

where $r \in \mathbb{R}$ with $0 < r < 1$ and c is a constant

–see [68, Appendix B: Complexity]. Such algorithms are faster than exponential-time algorithms and slower than polynomial time algorithms. An example of a pioneer subexponential time algorithm is the Brillhart–Morrison continued fraction factoring method—see [68, §5.4, pp. 240–242].

The security of Elliptic Curve Cryptosystems depends upon the intractability of the following problem.

Definition 9.5 (Elliptic Curve Discrete Log Problem (ECDL))

If E is an elliptic curve over a field F, then the *Elliptic Curve Discrete Log Problem* to base $Q \in E(F)$ is the problem of finding an $x \in \mathbb{Z}$ (if one exists) such that $P = xQ$ for a given $P \in E(F)$.

Currently, the Discrete Log Problem in elliptic curve groups is several orders of magnitude more difficult than the Discrete Log Problem in the multiplicative group of a finite field (of similar size)—see [68, §3.5, p. 167]. What this means explicitly is that for a suitably chosen elliptic curve E over $\mathbb{F}_q$, the discrete log problem for the group of $E(\mathbb{F}_q)$ appears to be (given our current state of knowledge) of complexity exponential in the size $\lceil \log_2 q \rceil$ of the field elements, whereas there exist subexponential algorithms in $\lceil \log_2 q \rceil$ for the discrete log problem in $\mathbb{F}_q^*$, where $\lceil * \rceil$ is the ceiling function—see [68, §2.5]. The canonical choices for F in ECC are $F = \mathbb{F}_p$ for a prime $p > 3$ or F_{2^k} for $k \in \mathbb{N}$. We focus upon the odd prime case.

Remark 9.7 In 1991, Menezes, Okamoto, and Vanstone found a new means of attacking the ECDL (appearing two years later in [56]). Their method, currently called the *MOV attack* in the literature, involves the use of what is called a *Weil Pairing*—see [88, Section 3.8, pp. 95–99], which embeds an elliptic curve over a finite field into the multiplicative group of some finite extension field of the given finite field. Hence, their method reduces the problem to the discrete log problem in that extension field, called an *MOV reduction*. To be of any use, the degree of the extension field must be small, and essentially the only elliptic curves for which this degree is small are of a special type called *supersingular*—see [88, p. 137]. They demonstrated that if we have a supersingular curve, then the discrete log problem in an elliptic curve group can be reduced in expected polynomial time to the discrete log problem in the extension field of degree no more than 6 over the finite field. However, the vast majority of elliptic curves are *not* supersingular, called *nonsupersingular* or *ordinary*. For the nonsupersingular curves, the MOV reduction virtually never leads to a subexponential time algorithm. What this suggests is that one of the basic open questions in ECC is whether or not we can find a subexponential time algorithm for the ECDL on some set of nonsupersingular elliptic curves—a difficult question at the present time. The MOV attack was generalized by Frey and Rück [28] in 1994. Also, there is a useful test for approximating the security level of an ECC, called the *MOV threshold*—see [90] which may be accessed online at *http://grouper.ieee.org/groups/1363/*.

Another attack on elliptic curves E with $|E| = p$ involves p-adic arithmetic, called the *Semaev–Smart–Satoh–Araki attack*—see [83], [86] and [89]. Also, there is the Silver–Pohlig–Hellman algorithm, which reduces the problem to subgroups of prime order—see [67, §D.2, p. 530]. Other attacks include Shanks' baby-step-giant-step method—see [67, §D.3, p. 533]; Pollards's methods including his rho method—see [68, §4.3, pp. 206–208]; and the Frey–Rück attack using the Weil Pairing, described above. Of all of these, only the Semaev-Smart-Satoh-Araki attack runs in polynomial time, while the others are, at best, subexponential. Up to the modern day, the ECDL remains a very hard computational problem. Indeed, evidence of the power of ECC is the fact that the NSA had adopted ECC, saying that it "provides greater security and more efficient performance than the first generation public key techniques (RSA and Diffie-Hellman) now in use. As vendors look to upgrade their systems they should seriously consider the elliptic curve alternative for the computational and bandwidth advantages they offer at comparable security."—see *http://www.nsa.gov/business/programs/elliptic_curve.shtml*.

Now we are in a position to present an explicit ECC whose security is based upon the assumption that the ECDL is intractable, in particular, in the cyclic subgroup of the elliptic curve group.

◆ Menezes–Vanstone Elliptic Curve Cryptosystem

Let E be an elliptic curve over $\mathbb{F}_p$ where $p > 3$ is prime and let H be a subgroup of $E(\mathbb{F}_p)$ generated by a point $P \in E(\mathbb{F}_p)$. Assume that randomly chosen $k \in \mathbb{Z}/|H|\mathbb{Z}$ and $a \in \mathbb{N}$ are secret. If entity A wants to send message

$$m = (m_1, m_2) \in (\mathbb{Z}/p\mathbb{Z})^* \times (\mathbb{Z}/p\mathbb{Z})^*,$$

then A does the following.

Enciphering stage:

(1) $\beta = aP$, where P and β are public.

(2) $(y_1, y_2) = k\beta$.

(3) $c_0 = kP$.

(4) $c_j \equiv y_j m_j \pmod{p}$ for $j = 1, 2$.

Then A sends the following enciphered message to B,

$$\mathfrak{E}_k(m) = (c_0, c_1, c_2) = c,$$

and upon receipt, B calculates the following to recover m.

Deciphering stage:

(1) $ac_0 = (y_1, y_2)$.

(2) $\mathfrak{D}_k((c_1, c_2)) = (c_1 y_1^{-1} \pmod{p}, c_2 y_2^{-1} \pmod{p}) = m$.

Example 9.6 Let E be the elliptic curve given by

$$y^2 = x^3 + 4x + 4$$

over $\mathbb{F}_{13}$, and let $P = (1, 3)$. Then by Exercise 9.5 on page 315, $|E(\mathbb{F}_p)| = 15$, which is necessarily cyclic. Also, $P = (1, 3)$ is a generator of E. If the private keys are $k = 5$ and $a = 2$, then given a message

$$m = (12, 7) = (m_1, m_2),$$

entity A computes

$$\beta = aP = 2(1, 3) = (12, 8),$$

$$(y_1, y_2) = k\beta = 5(12, 8) = (10, 11),$$

$$c_0 = kP = 5(1, 3) = (10, 2),$$

$c_1 \equiv y_1 m_1 = 10 \cdot 12 \equiv 3 \pmod{13}$, and $c_2 \equiv y_2 m_2 = 11 \cdot 7 \equiv 12 \pmod{13}$.

Then A sends

$$\mathfrak{E}_k(m) = \mathfrak{E}_5(12, 7) = (c_0, c_1, c_2) = ((10, 2), 3, 12) = c$$

to B. Upon receipt, B computes

$$ac_0 = 2(10, 2) = (10, 11) = (y_1, y_2)$$

and

$$\mathfrak{D}_k((c_1, c_2)) = \mathfrak{D}_5(3, 12) = (3 \cdot 10^{-1} \pmod{13}, 12 \cdot 11^{-1} \pmod{13}) = (12, 7) = m.$$

(See Exercise 9.18.)

Exercises

9.14. A given $n \in \mathbb{N}$ is called a *congruent number* or simply *congruent* if it is the area of a right-angled triangle. Prove that the following are equivalent.

  (1) $n = ab/2$ is congruent, where (a, b, c) is a *Pythagorean triple*. (Recall that such triples are solutions $(x, y, z) \in \mathbb{N}^3$ to

$$x^2 + y^2 = z^2.$$

  Furthermore, such a solution with $\gcd(x, y, z) = 1$, called a *primitive Pythagorean triple*, exists with x even, if and only if

$$(x, y, z) = (2uv, v^2 - u^2, v^2 + u^2)$$

  for relatively prime natural numbers u and v of opposite parity—see [68, Theorem 7.6, p. 281].)

  (2) There exists an integer x such that x, $x - n$, and $x + n$ are all perfect squares of rational numbers.

9.15. Let E be an elliptic curve over $\mathbb{Q}$ given by

$$y^2 = (x - \alpha_1)(x - \alpha_2)(x - \alpha_3),$$

where $\alpha_j \in \mathbb{Q}$ for $j = 1, 2, 3$. Assume that for a given point $(x_2, y_2) \neq \mathfrak{o}$ on E, there exists a point (x_1, y_1) on E such that

$$2(x_1, y_1) = (x_2, y_2).$$

Prove that $x_2 - \alpha_j$ are squares of rational numbers for $j = 1, 2, 3$.

9.16. Let E be an elliptic curve over $\mathbb{Q}$ defined by

$$y^2 = x^3 - n^2 x$$

for some squarefree $n \in \mathbb{N}$. Prove that the conditions in Exercise 9.14 are equivalent to E having a rational point other than $(\pm n, 0)$, $(0, 0)$, and $\mathfrak{o}$.

In other words, n is congruent if and only if E has a rational point other than $(\pm n, 0)$, $(0,0)$, and $\mathfrak{o}$.

(*It can be shown (see* [45, Theorem 5.2, p. 134]) *that when E is given by*

$$y^2 = x^3 + Ax$$

with $A \in \mathbb{Z}$ assumed to be fourth-power free, then

$$E(\mathbb{Q})_t = \mathbb{Z}/2\mathbb{Z} \oplus \mathbb{Z}/2\mathbb{Z}$$

if $-A$ is a perfect square,

$$E(\mathbb{Q})_t = \mathbb{Z}/4\mathbb{Z}$$

when $A = 4$, and

$$E(\mathbb{Q})_t = \mathbb{Z}/2\mathbb{Z}$$

otherwise. Thus, for the case given in this exercise, n is congruent if and only if E has a point of infinite order.)

9.17. Let $n \in \mathbb{N}$ be squarefree. Prove that the following are equivalent.

(1) n is a congruent number.

(2) The simultaneous (homogeneous Diophantine) equations

$$x^2 + ny^2 = z^2 \text{ and } x^2 - ny^2 = t^2$$

have a solution in integers x, y, z, t with $y \neq 0$. (A polynomial of degree d is said to be *homogeneous* if each term has degree d. For example,

$$x^3 + xyz = z^3$$

is a homogeneous polynomial of degree $d = 3$ and $x + y = z$ is one of degree $d = 1$.)

9.18. Given the same curve E and point P as in Example 9.6, decipher

$$c = ((12, 8), 2, 8)$$

assuming that it was enciphered using the Menezes-Vanstone Elliptic Curve Cryptosystem with $k = 2$ and $a = 5$.

Chapter 10

Modular Forms

10.1   The Modular Group

In Remark 3.1 on page 98, we discussed unimodular transformations in the context of binary quadratic forms involving $\mathrm{SL}(2, \mathbb{Z})$. Also, in Exercise 2.5 on page 66, the content therein is that two $\mathbb{Z}$-modules having the same basis are connected by a unimodular transformation, namely via those $A \in \mathrm{GL}(2, \mathbb{Z})$ with $\det(A) = \pm 1$.

In order to discuss modular forms, and their connection with elliptic curves studied in Chapter 9, we need to expand this discussion into the analytic realm.

First, we let

$$\mathrm{SL}(2, \mathbb{R})$$

be the generalization of $\mathrm{SL}(2, \mathbb{Z})$ to $\mathbb{R}$, namely the group of 2×2-matrices with coefficients in $\mathbb{R}$ and determinant 1.

Then we let

$$\tilde{\mathbb{C}} = \mathbb{C} \cup \{\infty\},$$

called the *Riemann sphere*.

Definition 10.1 | **Möbius Transformations**

Define an action of $\mathrm{SL}(2, \mathbb{R})$ on $\tilde{\mathbb{C}}$ via the *fractional linear transformation*, also called a *Möbius transformation*, where $\alpha = \begin{pmatrix} a & b \\ c & d \end{pmatrix} \in \mathrm{SL}(2, \mathbb{R})$:

$$\sigma : z \mapsto \alpha z = \sigma(z) = \begin{cases} (az + b)/(cz + d) & \text{if } z \in \mathbb{C} \text{ and } z \neq -d/c, \\ \infty & \text{if } z = -d/c \\ a/c & \text{if } z = \infty \text{ and } c \neq 0, \\ \infty & \text{if } z = \infty \text{ and } c = 0. \end{cases}$$

A value $\sigma(\infty) = a/c \neq \infty$ is called a *cusp* of α.

By Exercise 10.1 on page 335, the imaginary part of $\alpha z \in \mathbb{C}$ is given by

$$\Im(\alpha z) = \frac{\Im(z)}{|cz + d|^2}. \tag{10.1}$$

Now set

$$\mathfrak{H} = \{z \in \mathbb{C} : \Im(z) > 0\},$$

namely the upper half plane. Thus, by (10.1), the Möbius transformation σ maps $\mathfrak{H} \mapsto \mathfrak{H}$, which says that $\mathfrak{H}$ is *stable*, meaning $\mathfrak{H}$ is *preserved* under the action of $\mathrm{SL}(2, \mathbb{R})$. Also, since

$$\sigma(z) = \alpha z = -\alpha z,$$

namely α and $-\alpha$ represent the same transformation, then

$$-1 = \begin{pmatrix} -1 & 0 \\ 0 & -1 \end{pmatrix}$$

acts trivially on $\mathfrak{H}$, so the group

$$\mathrm{PSL}(2, \mathbb{R}) = \mathrm{SL}(2, \mathbb{R})/\{\pm 1\},$$

called the *projective special linear group*, is actually isomorphic to the group of fractional linear transformations. When we specialize to $\mathbb{Z}$, we have the topic in this section's header.

Definition 10.2 | **The Modular Group**

The group

$$\Gamma = \mathrm{PSL}(2, \mathbb{Z}) = \mathrm{SL}(2, \mathbb{Z})/\{\pm 1\}$$

is called the *modular group*.

Note that Γ in Definition 10.2 is the image of $\mathrm{SL}(2, \mathbb{Z})$ in $\mathrm{PSL}(2, \mathbb{R})$. Moreover, the following describes properties of the modular group in detail.

Theorem 10.1 | **Generation of the Modular Group**

Let Γ be the modular group given in Definition 10.2, and set

$$T = \begin{pmatrix} 1 & 1 \\ 0 & 1 \end{pmatrix} \text{ and } S = \begin{pmatrix} 0 & -1 \\ 1 & 0 \end{pmatrix}.$$

Then Γ is generated by S and T. In other words, every $\alpha \in \Gamma$ may be expressed (not uniquely) in the following form

$$\alpha = T^{a_1} S T^{a_2} S \cdots S T^{a_n},$$

for integers a_j, $j = 1, 2, \ldots, n$.

Proof. Suppose that $\alpha = \begin{pmatrix} a & b \\ c & d \end{pmatrix} \in \Gamma$. If $c < 0 \le |a|$, then

$$\begin{pmatrix} a & b \\ c & d \end{pmatrix} = S^2 \begin{pmatrix} -a & -b \\ -c & -d \end{pmatrix}, \tag{10.2}$$

so we may assume that $c \ge 0$, since the right-hand side of (10.2), with $-c \ge 0$, tells us that this case suffices. If $c = 0$, then

$$1 = ad - bc = ad,$$

so $a = d = \pm 1$. Hence,

$$\begin{pmatrix} a & b \\ c & d \end{pmatrix} = \begin{pmatrix} \pm 1 & b \\ 0 & \pm 1 \end{pmatrix} = \begin{pmatrix} 1 & \pm b \\ 0 & 1 \end{pmatrix} = T^{\pm b}.$$

Now we use induction on $c > 0$. If $c = 1$, then

$$1 = ad - bc = ad - b,$$

so $b = ad - 1$. Thus,

$$\alpha = \begin{pmatrix} a & ad - 1 \\ 1 & d \end{pmatrix} = \begin{pmatrix} 0 & a \\ 1 & 0 \end{pmatrix} \begin{pmatrix} 0 & -1 \\ 1 & 0 \end{pmatrix} \begin{pmatrix} 1 & d \\ 0 & 1 \end{pmatrix} = T^a S T^d.$$

so we may now assume that the result holds for all $\alpha \in \Gamma$ with lower left-hand element $< c$ for some $c > 1$. Since $ad - bc = 1$, we have $\gcd(c, d) = 1$, so with

$$q = b/a, \text{ and } r = 1/a,$$

then $d = cq + r$ where $0 < r < c$, with

$$\alpha T^{-q} = \begin{pmatrix} a & b \\ c & d \end{pmatrix} \begin{pmatrix} 1 & -q \\ 0 & 1 \end{pmatrix} = \begin{pmatrix} a & -aq + b \\ c & r \end{pmatrix},$$

where we note that $a \ne 0$ since $c > 1$. Also,

$$\alpha T^{-q} S = \begin{pmatrix} a & -aq + b \\ c & r \end{pmatrix} \begin{pmatrix} 0 & -1 \\ 1 & 0 \end{pmatrix} = \begin{pmatrix} -aq + b & -a \\ r & -c \end{pmatrix}. \tag{10.3}$$

The right-hand side of (10.3) is now available to the induction hypothesis since $r < c$, so this completes the induction. $\qquad\square$

Remark 10.1 We have shown that Γ has generators S and T with relations $(ST)^3 = (TS)^3 = 1$. One can show that Γ is the product of the cyclic group of order 2 generated by S and the cyclic group of order 3 generated by ST—see Exercise 10.4 on the next page. Indeed, T and S are matrix representations of the linear transformations

$$T : z \mapsto z + 1$$

and

$$S : z \mapsto -\frac{1}{z},$$

where clearly

$$S^2 = 1 \text{ and } (ST)^3 = 1.$$

Thus, the argument to prove the above comment is essentially a topological argument that shows Γ has a *presentation* of the form

$$\Gamma = \langle S, T; S^2, (ST)^3 \rangle,$$

which is another way of stating that it is a free product of the cyclic groups mentioned above. Recall that a "presentation" of a group is defined to be a group G, generated by a subset $\mathcal{S}$ and some collection of relations $R_1, R_2, \ldots, R_n$, where R_j is an equation in the elements from $\mathcal{S} \cup \{1\}$, and is denoted by

$$G = \langle \mathcal{S}; R_1, R_2, \ldots, R_n \rangle.$$

Also, a "free product" is a product of two or more groups G and H such that, given presentations of G and of H, we take the generators of G and of H, from the disjoint union of those, and adjoin the corresponding relations for G and for H. This is a presentation of the product of G and H, with the property that there should be no "interaction" between G and H, justifying the term "free product."

Also, there is a correspondence between positive definite binary quadratic forms and points of $\mathfrak{H}$ as follows. If

$$f(x, y) = ax^2 + bxy + cy^2$$

is a positive definite binary quadratic form, then

$$f(x, y) = a(x - \omega y)(x - \overline{\omega} y)$$

with $\omega \in \mathfrak{H}$. Hence, the association

$$f \mapsto \omega$$

is a one-to-one correspondence between the positive definite binary quadratic forms with fixed discriminant $D = b^2 - 4ac$ and the points of $\mathfrak{H}$. Moreover, two forms are equivalent if and only if the points lie in the same $\mathrm{SL}(2, \mathbb{Z})$ orbit, where an orbit means the equivalence relation given in Definition 3.1 on page 98 for properly equivalent forms. As well, Theorem 10.1 on the previous page implies that every positive definite binary quadratic form is equivalent to a reduced form, and two reduced forms are equivalent if and only if they are equal—see Theorem 3.1 on page 100.

Exercises

10.1. Verify equality (10.1) on page 332.

10.2. Let Γ be the modular group give in Definition 10.2 on page 332, and set

$$D = \{z \in \mathbb{C} : |z| \geq 1 \text{ and } |\Re(z)| \leq 1/2\}.$$

Prove that for every $z \in \mathfrak{H}$, there exists an $\gamma \in \Gamma$ such that $\gamma z \in D$.

(*Hint: Use Theorem 10.1 and Equation (10.1).*)

10.3. With reference to Exercise 10.2, prove that if $z \in D$ and $\alpha \in \Gamma$, with α not the identity, such that $\alpha z \in D$, then either $|\Re(z)| = 1/2$ and $\alpha z = z \pm 1$, or else $|z| = 1$ and $\alpha z = -1/z$.

(*Note that D is called a* fundamental domain *for the action of Γ on $\mathfrak{H}$, with the properties in Exercises 10.2–10.3 being the two main properties that a fundamental domain must satisfy. Typically, the approach to proving Theorem 10.1 is the use of facts concerning D. However, the more elementary approach provided herein is more constructive and informative. Exercises 10.2–10.4 are designed to provide information on fundamental domains for the edification of the reader, since we will be using these facts in §10.2.*)

10.4. With reference to Exercise 10.2, prove that if $z \in D$, then $\alpha \in \Gamma$ satisfies $\alpha z = z$ if and only if one of the following holds, where S, T are given in Theorem 10.1 on page 333.

(a) α is the identity.

(b) $z = \sqrt{-1}$, in which case $\alpha = S$.

(c) $z = \zeta_3^2 = ((-1 + \sqrt{-3})/2)^2$, in which case $\alpha = (ST)^j$ for $j \in \{1, 2\}$.

(d) $z = \zeta_3$, in which case $\alpha = (TS)^j$ for $j \in \{1, 2\}$.

10.2   Modular Forms and Functions

> *The Answer to the Great Question of...Life, the Universe, and Everything...*
> *is forty-two.*
>   from Chapter 27 of **The Hitchhiker's Guide to the Galaxy (1979)**
>   **Douglas Adams (1951–2001)**
>   **English science fiction writer**

We now build upon the modular group Γ introduced in §10.1 by presenting and studying forms related to it. The reader will need to have solved Exercises 10.2–10.4 before proceeding.

Definition 10.3 $\boxed{\text{Modular Forms and Functions}}$

A function $f(z)$ defined for $z \in \mathfrak{H}$ is called a *modular function of weight* $k \in \mathbb{Z}$ *associated with the modular group* Γ if the following properties hold.

(a) f is analytic in $\mathfrak{H}$.

(b) f satisfies the *functional equation*:

$$f(z) = (cz + d)^{-k} f \left(\frac{az + b}{cz + d} \right) = (cz + d)^{-k} f(\gamma z),$$

$$\text{with } z \in \mathfrak{H} \text{ and } \gamma = \begin{pmatrix} a & b \\ c & d \end{pmatrix} \in \Gamma.$$

(c) The Fourier series of f in the variable $q = \exp(2\pi i z)$ is given by:

$$f(z) = \sum_{n=n_0(f)}^{\infty} c_n q^n, \tag{10.4}$$

where $n_0(f) \in \mathbb{Z}$ —see §5.1.

A modular function of weight k is called a *modular form of weight* k if, in addition, $n_0(f) = 0$. In this case, we say that f is analytic at ∞ and write $f(\infty) = c_0$. In the case where $f(\infty) = c_0 = 0$, we say that f is a *cusp form*.

In the literature modular functions of weight k are sometimes called *weakly modular functions of weight* k or an *unrestricted modular form of weight* k. However, the definition of *modular form* or *cusp form* of weight k appears to be uniform. Sometimes the cusp form is called a *parabolic* form.

Remark 10.2 If $\gamma = \begin{pmatrix} -1 & 0 \\ 0 & -1 \end{pmatrix}$ in Definition 10.3, then $\gamma z = z$ for all $z \in \mathfrak{H}$. Therefore, if f is a modular form of weight $k = 2m + 1$ for $m \in \mathbb{Z}$, then

$$f(z) = (-1)^{-k} f(\gamma z) = -f(z),$$

so if $f(z) \neq 0$, then dividing through the equation by $f(z)$, we get $1 = -1$, a contradiction. Thus, f is just the zero map, sometimes called *identically zero*. Hence, a nontrivial modular form on Γ must necessarily be of even weight. Also, by taking $\gamma = \begin{pmatrix} 1 & 1 \\ 0 & 1 \end{pmatrix} = T$ in Definition 10.3, we obtain that

$$f(z+1) = f(z), \tag{10.5}$$

namely f is invariant under the transformation $z \mapsto z+1$. This is what allows us to expand f into the expansion *(10.4)*, which is called the *q-expansion of* f. (*If we went into the details, we could invoke the Cauchy integral theorem using (10.5) to show symmetry in a certain line integral on* $f(z)\exp(-2\pi iz)$, *and the interested reader with knowledge of this area can derive the q-expansion in this fashion.*) Note that condition (c) implies that if $z = x + yi$ and $y \to \infty$, then $q \to 0$ as $y \to \infty$. Thus the q-expansion (10.4) may be considered as an expansion about $z = \infty$, which justifies the reference to f being called *holomorphic at* ∞. The condition above for a cusp form tells us, therefore, that f vanishes as $y \to \infty$.

Example 10.1 The Eisenstein series of weight $k \geq 2$ are defined by the infinite series

$$G_{2k}(z) = \sum_{m,n \in \mathbb{Z}-(0,0)} (nz+m)^{-2k}, \text{ for } \Im(z) > 0 \tag{10.6}$$

where the notation $m, n \in \mathbb{Z} - (0,0)$ means that m and n run over all integers except that $m = n = 0$ is not allowed. The Eisenstein series of even weight are the first nontrivial examples of modular forms on Γ. Indeed, the following, which establishes this fact, is of interest from the viewpoint of arithmetic functions studied in Chapter 5.

Theorem 10.2 | **Eisenstein Series as Modular Forms**

For $q = \exp(2\pi iz)$ *and* $\Im(z) > 0$, *the* Eisenstein series *given in* (10.6) *has Fourier expansion given by*

$$G_{2k}(z) = 2\zeta(2k) + 2\frac{(2\pi i)^{2k}}{(2k-1)!} \sum_{n=1}^{\infty} \sigma_{2k-1}(n)q^n,$$

where $k \geq 2$, $\zeta(s)$ *is the Riemann* ζ-function, *and* $\sigma_a(n) = \sum_{d|n} d^a$ *is a sum of* a-th powers of positive divisors of n. *Accordingly,* $G_{2k}(z)$ *is a modular form of weight* $2k$.

Proof. We know from elementary calculus that the following identity holds

$$\pi \cot(\pi z) = \frac{1}{z} + \sum_{m=1}^{\infty} \left(\frac{1}{z+m} + \frac{1}{z-m} \right) \tag{10.7}$$

see [101, p. 344]. For $\Im(z) > 0$ (so $|q| < 1$) we get

$$\pi \cot(\pi z) = \pi \frac{\cos(\pi z)}{\sin(\pi z)} = i\pi \frac{q+1}{q-1} = i\pi - \frac{2\pi i}{1-q} = i\pi - 2\pi i \sum_{c=0}^{\infty} q^c, \qquad (10.8)$$

where the second equality comes from the fact that

$$\cot(\pi z) = \frac{i(e^{2\pi iz} + 1)}{e^{2\pi iz} - 1}, \qquad (10.9)$$

and the last equality follows from the standard geometric formula

$$\lim_{N \to \infty} \sum_{c=0}^{N} q^c = \lim_{N \to \infty} \frac{q^{N+1} - 1}{q - 1} = \frac{1}{1-q},$$

where the last equality follows from the fact that $|q| < 1$—see [68, Theorem 1.2, p. 2]. Therefore, (10.7)–(10.8) imply that

$$\frac{1}{z} + \sum_{m=1}^{\infty} \left(\frac{1}{z+m} + \frac{1}{z-m} \right) = i\pi - 2\pi i \sum_{c=0}^{\infty} q^c. \qquad (10.10)$$

Now differentiating (10.10) $2k - 1$ times with respect to z we get

$$(-1)^{2k-1}(2k-1)! z^{-2k} + (-1)^{2k-1}(2k-1)! \sum_{m=1}^{\infty} \left(\frac{1}{(z+m)^{2k}} + \frac{1}{(z-m)^{2k}} \right)$$

$$= -(2\pi i)^{2k} \sum_{c=1}^{\infty} c^{2k-1} q^c,$$

which implies that

$$z^{-2k} + \sum_{m=1}^{\infty} \left(\frac{1}{(z+m)^{2k}} + \frac{1}{(z-m)^{2k}} \right) = \frac{(2\pi i)^{2k}}{(2k-1)!} \sum_{c=1}^{\infty} c^{2k-1} q^c,$$

so

$$\sum_{m=-\infty}^{\infty} \frac{1}{(z+m)^{2k}} = \frac{(2\pi i)^{2k}}{(2k-1)!} \sum_{c=1}^{\infty} c^{2k-1} q^c. \qquad (10.11)$$

However, since

$$G_{2k}(z) = \sum_{m,n \in \mathbb{Z}-(0,0)} (nz+m)^{-2k} = \sum_{m \neq 0} \frac{1}{m^{2k}} + \sum_{n \neq 0} \sum_{m=-\infty}^{\infty} \frac{1}{(nz+m)^{2k}}, \qquad (10.12)$$

and we know from (5.26) on page 218 that

$$\sum_{m=-\infty}^{1} \frac{1}{m^{2k}} = \sum_{m=1}^{\infty} \frac{1}{m^{2k}} = \zeta(2k),$$

as well as the fact that the sum over nonzero values of n is twice the sum over positive values of n in the second summand of (10.12), then

$$G_{2k}(z) = 2\zeta(2k) + 2\sum_{n=1}^{\infty}\sum_{m=-\infty}^{\infty}\frac{1}{(m+nz)^{2k}}. \tag{10.13}$$

Hence, by replacing z by nz in (10.11) and applying it to the last summand in (10.13), we achieve that

$$G_{2k}(z) = 2\zeta(2k) + \frac{2(2\pi i)^{2k}}{(2k-1)!}\sum_{c=1}^{\infty}\sum_{a=1}^{\infty}c^{2k-1}q^{ac} = 2\zeta(2k) + \frac{2(2\pi i)^{2k}}{(2k-1)!}\sum_{n=1}^{\infty}\sigma_{2k-1}(n)q^{n}.$$

For the last statement, we note that it follows that

$$G_{2k}(\gamma z) = (cz+d)^{2k}G_{2k}(z),$$

for

$$\gamma = \begin{pmatrix} a & b \\ c & d \end{pmatrix} \in \Gamma,$$

so $G_{2k}(z)$ is a modular form of weight $2k$. $\square$

Corollary 10.1 $G_{2k}(\infty) = 2\zeta(2k)$.

Proof. We have

$$\lim_{z\to\infty}G_{2k}(z) = 2\zeta(2k) + 2\frac{(2\pi i)^{2k}}{(2k-1)!}\sum_{n=1}^{\infty}\sigma_{2k-1}(n)\lim_{z\to\infty}q^{n},$$

but by Remark 10.2 on page 336, $\lim_{z\to\infty}q = 0$, which is the result. $\square$

Example 10.2 From Theorem 10.2 on page 337, we get

$$G_{2k}(z) = 2\zeta(2k)E_{2k}(z),$$

with

$$E_{2k}(z) = \frac{G_{2k}(z)}{2\zeta(z)} = 1 + \alpha_k \sum_{n=1}^{\infty}\sigma_{2k-1}(n)q^{n}$$

where, via (5.4)–(5.5) on pages 197–198,

$$\alpha_k = (-1)^k\frac{4k}{|B_{2k}|},$$

and B_k is the k-th Bernoulii number given in Definition 5.1 on page 192. The modular form E_{2k} is called the *weight k Eisenstein series*, which is *not* a cusp form.

Thus, for $k = 2$,

$$E_4(z) = 1 + 240 \sum_{n=1}^{\infty} \sigma_3(n)q^n,$$

and for $k = 3$,

$$E_6(z) = 1 - 504 \sum_{n=1}^{\infty} \sigma_5(n)q^n.$$

A few more examples are for $k = 4$,

$$E_8(z) = 1 + 480 \sum_{n=1}^{\infty} \sigma_7(n)q^n,$$

for $k = 5$,

$$E_{10}(z) = 1 - 264 \sum_{n=1}^{\infty} \sigma_9(n)q^n,$$

and for $k = 6$,

$$E_{12}(z) = 1 + \frac{65520}{691} \sum_{n=1}^{\infty} \sigma_{11}(n)q^n.$$

Remark 10.3 The first two cases in Example 10.2 motivate a basic notion which we now develop. The weight k Eisenstein series are foundational elements for the development of all modular forms in the sense that any modular form can be expressed as a polynomial in E_4 and E_6. For instance, $|\mathbb{C} : M_8(\Gamma)| = 1$, by Remark 10.4, so M_8 is one-dimensional space spanned by E_8. Moreover, E_4^2 has weight 8 and constant term 1 by Example 10.2, so $E_4^2 = E_8$—see Exercise 10.15 on page 346, as well as more information in Example 10.4 on page 342.

First we let

$$g_2 = 60G_4 \text{ and } g_3 = 140G_6,$$

where the need for the coefficients will become clear when we link modular forms to elliptic curves in §10.3, as will the contents of the following.

Definition 10.4 | **Modular Discriminant Function and j-Invariant**

The function $\Delta : \mathfrak{H} \mapsto \mathbb{C}$ given by

$$\Delta = g_2^3 - 27g_3^2$$

is called the *discriminant function*, and the *j-invariant* is given by

$$j(\Delta) = \frac{1728g_2^3}{\Delta}.$$

Example 10.3 The discriminant function given in Definition 10.4 was proved by Jacobi to be of the form

$$\Delta(q) = (2\pi)^{12} q \sum_{n=1}^{\infty} (1 - q^n)^{24},$$

with $q \in \mathbb{C}$ with $|q| < 1$—see Exercise 10.16 on page 346. Indeed, the n-th coefficients of the cusp form $F(z) = (2\pi)^{-12}\Delta(z)$ are values of $\tau(n)$, the distinguished *Ramanujan's τ-function*:

$$\sum_{n=1}^{\infty} \tau(n) q^n = q \sum_{n=1}^{\infty} (1 - q^n)^{24}$$

where $\tau : \mathbb{N} \mapsto \mathbb{Z}$. Note that since $g_2(\infty) = 120\zeta(4)$ and $g_3(\infty) = 280\zeta(6)$, then using Exercise 10.14 on page 346,

$$g_2(\infty) = \frac{4\pi^4}{3}, \text{ and } g_3(\infty) = \frac{8\pi^6}{27}.$$

Thus,

$$\Delta(\infty) = \left(\frac{4\pi^4}{3}\right)^3 - 27\left(\frac{8\pi^6}{27}\right)^2 = 0,$$

which means that Δ is a cusp form and by Exercise 10.16, it is of weight 12.

Another formula for the discriminant function that lends itself more readily to computations than that given above is in terms of the *Dedekind-η function* defined by:

$$\eta(z) = q^{1/24} \prod_{n=1}^{\infty} (1 - q^n),$$

where $q = \exp(2\pi i z)$ and $q^{1/24} = \exp(\pi i/12)$. Thus,

$$\Delta(z) = (2\pi)^{12} \eta(z)^{24},$$

where by Exercise 10.18,

$$\eta(z + 1) = \exp(\pi i/12)\eta(z) \text{ and } \eta(-z^{-1}) = (-iz)^{1/2}\eta(z), \qquad (10.14)$$

where we take the branch of the square root is chosen to be positive on the imaginary axis. Also, by Exercise 10.17, the j-invariant is a modular function of weight 0, namely a modular function, which has q-expansion given by

$$j(z) = \frac{1}{q} + 744 + \sum_{n=1}^{\infty} c_n q^n,$$

where $z \in \mathfrak{H}$ and $q = \exp(2\pi i z)$. It can be shown that

$$j : \frac{\mathfrak{H}}{\Gamma} \mapsto \mathbb{C}$$

is an isomorphism (*of Riemann surfaces*) and that any modular function of weight 0 must be a rational function of j—see [87, Propositions 5–6, p. 89].

We now look at spaces of forms and how they fit into the picture we have been painting.

Definition 10.5 | **Space of Modular Forms**

The set of modular forms of weight k on Γ forms a complex vector space denoted by $M_k(\Gamma)$. The subspace of cusp forms is denoted by $M_k^0(\Gamma)$.

Remark 10.4 It can be shown that the following dimensions hold—see [87].

$$|\mathbb{C} : M_k(\Gamma)| = \begin{cases} \lfloor k/12 \rfloor + 1 & \text{if } k \not\equiv 2 \pmod{12}, \\ \lfloor k/12 \rfloor & \text{if } k \equiv 2 \pmod{12}. \end{cases}$$

Also,

$$|\mathbb{C} : M_k^0(\Gamma)| = \begin{cases} \lfloor k/12 \rfloor & \text{if } k \not\equiv 2 \pmod{12}, \\ \lfloor k/12 \rfloor - 1 & \text{if } k \equiv 2 \pmod{12}. \end{cases}$$

Example 10.4 With reference to Theorem 10.2 and Remark 10.4, for $k = 14$,

$$M_{14} = \mathbb{C}E_{14}.$$

Moreover, in terms of Eisenstein series and cusp forms we have the following direct sum for k even $k \geq 4$,—see [68, p. 305],

$$M_k = \mathbb{C}E_k \oplus M_k^0.$$

Observe, by Remark 10.4 that $M_{14}^0(\Gamma) = 0$. Further, with reference to Remark 10.3 on page 340, it may be shown that the space M_k has for basis the family of monomials $G_2^\alpha G_3^\beta$ for all nonnegative integers α, β with $2\alpha + 3\beta = k$—see [87, Corollary 2, p. 89]. Moreover, it can be shown that multiplication by the discriminant function Δ defines an isomorphism of M_{k-12} onto M_k^0, which is equivalent to the following. If $M = \sum_{k=0}^\infty M_k$, called a *graded algebra*, the direct sum of the M_k, and $h : \mathbb{C}[x, y] \mapsto M$ is the homomorphism sending x to G_2 and y to G_3, then h is an isomorphism—see [87, Theorem 4, p. 88ff].

Remark 10.5 In the area of algebraic geometry,[10.1] most of the interesting entities come into view when we look at arithmetically defined subgroups of finite index in Γ. One such class of groups is called *Hecke congruence subgroups* denoted by $\Gamma_0(n)$ for any $n \in \mathbb{N}$, defined by

$$\Gamma_0(n) = \left\{ \begin{pmatrix} a & b \\ c & d \end{pmatrix} \in \Gamma : c \equiv 0 \pmod{n} \right\}.$$

[10.1] Algebraic geometry is a branch of mathematics combining methods in use in abstract algebra, especially commutative algebra, with the language of geometry. It has interconnections with complex analysis, topology, and number theory. At its most basic level, algebraic geometry deals with *algebraic varieties*, which are geometric manifestations of solutions of polynomial equations. For instance, plane algebraic curves, which include circles and parabolas for instance, comprise one of the most investigated classes of algebraic varieties.

It is known that the *index* of $\Gamma_0(n)$ in Γ is given by

$$|\Gamma : \Gamma_0(n)| = n \prod_{\substack{p \mid n \\ p=\text{prime}}} \left(1 + \frac{1}{p}\right),$$

the product over *distinct* primes divding n. See Exercises 10.6–10.8 on page 344 for applications of this fact.

An example of a modular form related to $\Gamma_0(n)$ is given by

$$f(z) = \eta(z)^2 \eta(11z)^2, \tag{10.15}$$

which is a cusp form of weight 2 related to the group $\Gamma_0(11)$. Here η is the Dedekind-η function introduced in Example 10.3 on page 341.

Hecke groups defined in Remark 10.5 allow us to add another "level" to the notion of a modular form.

Definition 10.6 $\boxed{\textbf{Levels of Modular Forms}}$

If f is an analytic function on $\mathfrak{H}$ with

$$f(\gamma z) = (cz + d)^k f(z) \text{ for all } \gamma \in \Gamma_0(n),$$

and has a q-expansion

$$f(z) = \sum_{j=n_0(f)}^{\infty} a_j(f) q^j \text{ where } q = \exp(2\pi i z) \text{ with } n_0(f) \in \mathbb{Z}, \tag{10.16}$$

then f is called a *modular function of weight k and level n*. A modular function of weight k and level n is called a *modular form of weight k and level n* if $n_0(f) = 0$. Moreover, if $a_0(f) = 0$, we call f a *cusp form of weight k and level n*. When $a_1(f) = 1$, and $a_0(f) = 0$, we say that f is a *normalized cusp form of weight k and level n*.

Spaces of modular and cusp forms of weight k and level n are denoted by $M_k(\Gamma_0(n))$, respectively $S_k(\Gamma_0(n))$.

Example 10.5 It can be shown that $S_2(\Gamma_0(11))$ is a one-dimensional space spanned by Equation (10.15)—see [88, Remark 12.17, p. 351]. This example will have significant implications for a celebrated conjecture—see Example 10.9 on page 360. Also, $S_2(\Gamma_0(2))$ is the zero space and this too will have implications for the proof of FLT—see Theorem 10.4 on page 365.

Note that Definition 10.1 on page 332 and Exercises 10.6–10.8 tell us (10.16) implies that a modular function of weight k and level n is *holomorphic at the cusps*.

In §10.4, we will see that, roughly speaking, all rational elliptic curves arise from modular functions of a certain level and weight, and explore the interconnections, including critical implications for the proof of Fermat's Last Theorem. We begin in §10.3 with linking elliptic curves and modular forms.

Biography 10.1 Erich Hecke (1887–1947) *was born in Buk, Posen, Germany (now Pozan, Poland) on September 20. His studies at university included the University of Breslau, the University of Berlin where he studied under Landau, and finally Göttingen, where Hilbert was his supervisor—see Biographies 3.1 on page 104 and 3.5 on page 127. In 1910, he was awarded his doctorate, and remained at Göttingen as assistant to Hilbert and Klein. After a brief stint at Basel, he returned to a chair of mathematics at Göttingen, but left again, this time for a chair at Hamburg in 1919. One of the reasons for leaving was that the university at Hamburg was founded in that year and he felt he could influence its development. Indeed he did and remained there for the rest of his professional life.*

Hecke is probably best remembered for his work in analytic number theory, where he proved results that simplified theorems in class field theory, a branch of algebraic number theory that deals with abelian extensions of number fields, namely those with an abelian Galois group—see [64]. He studied Riemann's ζ-function and its generalization to any number field. He also introduced the concept of a Grössencharakter *and its corresponding L-series. He then used the properties of analytic continuation he had proved for the ζ-function and extended them to his L-series. One of his most renowned results was achieved in 1936 when he introduced the algebra of what we now call Hecke operators and the Euler products associated with them.*

Hecke died of cancer in Copenhagen, Denmark on February 13, 1947 in his fifty-ninth year.

Exercises

10.5. Let f be a function that is analytic on $\mathfrak{H}$. Prove that condition (b) of Definition 10.3 on page 336 is equivalent to the conditions

(1) For all $z \in \mathfrak{H}$, $f(z+1) = f(z)$.

(2) For all $z \in \mathfrak{H}$, and some $k \in \mathbb{Z}$, $f(-1/z) = (-z)^k f(z)$.

(*Hint: Prove that conditions (1)–(2) imply that the subset of Γ generated by the elements for which* (b) *hold is a subgroup of Γ. Consequently, this subgroup must be all of Γ since S and T are in this subgroup. Do this by defining*

$$\mathfrak{d}(\gamma, z) = cz + d,$$

for $\gamma = \begin{pmatrix} a & b \\ c & d \end{pmatrix} \in \Gamma$. Then prove that $\mathfrak{d}(\alpha\gamma, z) = \mathfrak{d}(\alpha, \gamma z)\mathfrak{d}(\gamma, z)$ and $\mathfrak{d}(\gamma^{-1}, z) = (\mathfrak{d}(\gamma, \gamma^{-1}z))^{-1}$ for all $\alpha, \gamma \in \Gamma$, and $z \in \mathfrak{H}$. The converse is straightforward given Remark 10.2 on page 336.)

10.6. In Remark 10.5 on page 343, the index of the congruence subgroup $\Gamma_0(n)$ in Γ was given. If $n = p$ a prime, find left coset representatives γ_j for

$j = 0, 1, 2, \ldots, p$ such that

$$\Gamma = \cup_{j=0}^{p} \gamma_j \Gamma_0(p).$$

10.7. With reference to Exercise 10.6, find coset representatives γ_j for

$$\Gamma = \cup_{j=0}^{p^a + p^{a-1}} \gamma_j \Gamma_0(p^a),$$

where p is prime and $a > 1$.

10.8. With reference to Exercise 10.2 on page 335 and Remark 10.5, let

$$\Gamma = \cup_{j=0}^{n} \gamma_j \Gamma_0(n) \tag{10.17}$$

be a left coset decomposition of $\Gamma_0(n)$ in Γ. Then

$$D_n = \cup_{j=0}^{n} \gamma_j D$$

is a fundamental domain for $\Gamma_0(n)$, where D is a fundamental domain for Γ. Find the decomposition for D_2.

10.9. Let Γ have a decomposition as in (10.17) above. Prove that every $\gamma_j(\infty)$ represents a cusp as given in Definition 10.1 on page 332.

10.10. With reference to Exercise 10.9, prove that if $i \neq j$ and $b \in \mathbb{Z}$, then

$$\gamma_j \gamma_i^{-1} = \pm \begin{pmatrix} 1 & b \\ 0 & 1 \end{pmatrix}$$

implies that both $\gamma_i^{-1}(\infty)$ and $\gamma_j^{-1}(\infty)$ represent the same cusp, namely for some $\alpha \in \Gamma_0(n)$, we have that $\gamma_j^{-1}(\infty) = \gamma_i^{-1}\alpha(\infty)$. Apply the condition to the case in Exercise 10.6.

10.11. Is the condition in Exercise 10.10 necessary?

(*Hint: Look at the case $n = 8$ in Exercise 10.7.*)

10.12. Prove that the function f, defined by $f(x) = \Gamma(x)\Gamma(1-x)\sin(\pi x)$, satisfies $f(x) = f(x+1)$, where the Gamma function is given in Definition 5.6 on page 224.

(*Hint: Use Formula (5.34) on page 224.*)

10.13. Prove that

$$\sin x = x \prod_{j=1}^{\infty} \left(1 - \frac{x^2}{j^2 \pi^2} \right).$$

(*Hint: Use Exercise 10.12. Also, you may use the formula*

$$\Gamma(x)\Gamma(1-x) = \frac{\pi}{\sin(\pi x)} \tag{10.18}$$

—see [101, *Formula* (25), p. 697], *as well as the Weierstrass product formula for the Gamma function,*

$$\Gamma(x) = e^{-\gamma x}\frac{1}{x}\prod_{j=1}^{\infty}\frac{e^{x/j}}{1+x/j},\tag{10.19}$$

where γ is Euler's constant given by (4.13) *on page 172—see Biography 4.6 on page 179.*)

10.14. Establish (5.5) on page 198.

(*Hint: Use Exercise 10.13 by differentiating and compare the result with the formula*

$$z\cot z = 1 - \sum_{n=1}^{\infty} B_n \frac{2^{2n}z^{2n}}{(2n)!},$$

which follows from Definition 5.1 on page 192 by putting $x = 2iz$.)

10.15. With reference to Remark 10.3 on page 340, prove that

$$\sigma_7(n) = \sigma_3(n) + 120\sum_{j=1}^{n-1}\sigma_3(n)\sigma_3(n-j).$$

10.16. Prove that Δ given in Example 10.3 on page 341 is a modular form of weight 12, namely that:

$$\Delta(q) = (2\pi)^{12}q\sum_{n=1}^{\infty}(1-q^n)^{24}.$$

10.17. Prove that the j-invariant of Definition 10.4 on page 340 is a modular function of weight 0 with q-expansion

$$j(z) = \frac{1}{q} + 744 + \sum_{n=1}^{\infty}c_n q^n,$$

where $z \in \mathfrak{H}$ and $q = \exp(2\pi i z)$.

10.18. Establish (10.14) on page 341.

10.3 Applications to Elliptic Curves

> *I believe that if mathematicians on any other planet, anywhere in the universe, have sufficiently advanced knowledge of arithmetic and geometry, they will know the Pythagorean theorem, that pi is 3.14+, and that 113 is prime. Of course, they will express these truths in their own language and symbols. Within formal systems, mathematical theorems, unlike a culture's folkways and mores, and even its laws of science, are absolutely certain and eternal.*
>
> see [**22, pp. 274–275**]
> **Martin Gardner (1914–)**
> **American science writer specializing in recreational mathematics**

In this section, we apply the knowledge gained in Chapter 9 and in §10.1–10.2 to elliptic curves to show the wealth of results emanating from our journey. We begin with a link between elliptic curves and modular functions.

Definition 10.7 | Elliptic Modular Functions |

If f is a function analytic on $\mathbb{C}$ such that for $n \in \mathbb{N}$ and $z \in \mathbb{C}$,

$$f(\gamma z) = f(z) \text{ for all } \gamma \in \Gamma(n),$$

then f is called an *elliptic modular function*, where

$$\Gamma(n) = \left\{ \begin{pmatrix} a & b \\ c & d \end{pmatrix} \in \Gamma : b \equiv c \equiv 0 \pmod{n} \right\}$$

is called the *principal congruence subgroup* of Γ.

Note that

$$\Gamma(n) \subseteq \Gamma_0(n) \subseteq \Gamma.$$

In general, any analytic function that is invariant under a group of linear transformations is called an *automorphic function*. The classic elliptic modular function has already been encountered in §10.2.

Example 10.6 The j-invariant

$$j(\Delta) = \frac{1728 g_2^3}{\Delta} = \frac{1}{q} + 744 + \sum_{n=1}^{\infty} c_n q^n,$$

where $z \in \mathfrak{H}$ and $q = \exp(2\pi i z)$ is an elliptic modular function.

The j-invariant is linked to elliptic curves in a natural way as follows.

Definition 10.8 | **Weierstrass Equations for Elliptic Curves**

If F is a field of characteristic different from 2 or 3 and $E(F)$ is an elliptic curve over F, then

$$y^2 = 4x^3 - g_2 x - g_3$$

where $g_2, g_3 \in F$, and

$$\Delta = g_2^3 - 27g_3^2 \neq 0$$

is called the *Weierstrass equation for E*.

In order to give our first example of Weierstrass equations, we need the following concept. We encountered real lattices in Definition 4.4 on page 182. We now look at a complex version. Recall, for the following that, in general, a *singularity* of a complex function is a point at which the function is not defined. Also, an *isolated* singularity z_0 is one for which there are no other singularities of the function "close" to it, which means that there is an open disk

$$D = \{z \in \mathbb{C} : |z - z_0| < r \in \mathbb{R}^+\}$$

such that f is holomorphic on $D - \{z_0\}$.

Definition 10.9 | **Lattices in $\mathbb{C}$ and Elliptic Functions**

A *lattice in* $\mathbb{C}$ is an additive subgroup of $\mathbb{C}$ which is generated by two complex numbers ω_1 and ω_2 that are linearly independent over $\mathbb{R}$, denoted by

$$L = [\omega_1, \omega_2].$$

Then an *elliptic function for L* is a function f defined on $\mathbb{C}$, except for isolated singularities, satisfying the following two conditions:

(a) $f(z)$ is meromorphic on $\mathbb{C}$.

(b) $f(z + \omega) = f(z)$ for all $\omega \in L$.

Remark 10.6 Condition *(b)* in Definition 10.9 is equivalent to

$$f(z + \omega_1) = f(z + \omega_2) = f(z),$$

for all z, a property known as *doubly periodic*. Hence, an elliptic function for a lattice L is a doubly periodic meromorphic function and the elements of L are called *periods*.

Definition 10.10 Lattice Discriminant and Invariant

The *j-invariant of a lattice* L is the complex number

$$j(L) = \frac{1728 g_2(L)^3}{g_2(L)^3 - 27 g_3(L)^2},\tag{10.20}$$

where

$$g_2(L) = 60 \sum_{w \in L - \{0\}} \frac{1}{w^4},$$

and

$$g_3(L) = 140 \sum_{w \in L - \{0\}} \frac{1}{w^6}.$$

The *discriminant of a lattice* L is given by

$$\Delta(L) = g_2(L)^3 - 27 g_3(L)^2.$$

One of the most celebrated of elliptic functions is the following.

Definition 10.11 Weierstrass $\wp$-Functions

Given $z \in \mathbb{C}$ such that $z \notin L = [\omega_1, \omega_2]$, the function

$$\wp(z; L) = \frac{1}{z^2} + \sum_{w \in L - \{0\}} \left(\frac{1}{(z - w)^2} - \frac{1}{w^2} \right)\tag{10.21}$$

is called the Weierstrass $\wp$-function for the lattice L.

Remark 10.7 The Weierstrass $\wp$-function is an elliptic function for L whose singularities can be shown to be double poles at the points of L. This is done by showing that $\wp(z)$ is holomorphic on $\mathbb{C} - L$ and has a double point at the origin. Then one may demonstrate that since

$$\wp'(z) = -2 \sum_{w \in L} \frac{1}{(z - w)^3},$$

which can be shown to converge absolutely, then $\wp'(z)$ is an elliptic function for

$$L = [\omega_1, \omega_2].$$

Since $\wp(z)$ and $\wp(z + \omega_j)$ have the same derivative, given that $\wp'(z)$ is periodic, then they differ by a constant which can be shown to be zero by the fact that $\wp(z)$ is an even function. This demonstrates the periodicity of $\wp(z)$ from which it follows that the poles of $\wp(z)$ are double poles and lie in L—see [18, Theorem 10.1, p. 200].

Example 10.7 By Exercise 10.22 on page 352, the *Laurent series expansion* (generally one of the form $\sum_{n=-\infty}^{\infty} a_n z^n$) for $\wp(z)$ about $z = 0$ is given by

$$\wp(z) = \frac{1}{z^2} + \sum_{n=1}^{\infty} (2n+1)G_{2n+1}(L)z^{2n}, \qquad (10.22)$$

where for a lattice L, and an integer $r > 2$,

$$G_r(L) = \sum_{\omega \in L - \{0\}} \frac{1}{\omega^r}.$$

From this, by Exercise 10.23, it follows that if $x = \wp(z; L)$ and $y = \wp'(z; L)$,

$$y^2 = 4x^3 - g_2(L)x - g_3(L), \qquad (10.23)$$

where $g_j(L)$ for $j = 2, 3$ are given in Definition 10.10 on the preceding page.

Remark 10.8 If E is an elliptic curve over $\mathbb{C}$ given by the Weierstrass equation

$$y^2 = 4x^3 - g_2x - g_3,$$

with $g_1, g_2 \in \mathbb{C}$ and $g_2^3 - 27g_3^2 \neq 0$, then there is a unique lattice $L \subseteq \mathbb{C}$ such that

$$g_2(L) = g_2 \text{ and } g_3(L) = g_3$$

—see [18, Proposition 4.3, p. 309].

The j-invariant may be used with elliptic curves as follows.

Definition 10.12 $\boxed{\textbf{j-Invariants for Elliptic Curves}}$

If E is an elliptic curve defined by the Weierstrass equation in Definition 10.8 on page 348, then

$$j(E) = 1728\frac{g_2^3}{g_2^3 - 27g_3^2} = 1728\frac{g_2^3}{\Delta} \in F$$

is called the j-invariant of E.

In Definition 10.12, $\Delta \neq 0$ and $1728 = 2^6 \cdot 3^3$. Since we are not in characteristic 2 or 3, then $j(E)$ is well defined. If $F = \mathbb{C}$, then when E is the elliptic curve defined by the lattice $L \subseteq \mathbb{C}$,

$$j(L) = j(E). \qquad (10.24)$$

By Exercise 10.19 on the facing page isomorphic elliptic curves have the same j-invariant. Also, Definition 10.12 provides a means of looking at classes of elliptic curves.

Definition 10.13 | **Weierstrass and Elliptic Curves**

Suppose that
$$E_j = E_j(F) \text{ for } j = 1, 2$$
are elliptic curves over F defined by Weierstrass equations
$$y^2 = 4x^3 - g_2^{(j)}x - g_3^{(j)} \text{ for } j = 1, 2.$$

Then E_1 and E_2 are isomorphic over F if there is a nonzero $\alpha \in F$ such that
$$g_2^{(2)} = \alpha^4 g_2^{(1)} \text{ and } g_3^{(2)} = \alpha^6 g_3^{(1)}.$$

This is denoted by
$$E_1 \cong E_2,$$
induced by the map
$$(x, y) \mapsto (\alpha^2 x, \alpha^3 y).$$

In §10.4, we will be able to use the concepts developed thus far to be able to state the *Shimura–Taniyama–Weil conjecture* that was proved in the last century and whose solution implies Fermat's Last Theorem. The proof of this conjecture is arguably the most striking and important mathematical development of the twentieth century and it will be a fitting conclusion to the main text of this book.

Exercises

10.19. Prove that isomorphic elliptic curves have the same j-invariant.

10.20. Prove that the discriminant of a lattice L satisfies
$$\Delta(L) = 16(e_1 - e_2)^2(e_1 - e_3)^2(e_2 - e_3)^2,$$
where the e_j for $j = 1, 2, 3$ are the roots of
$$4x^3 - g_2(L)x - g_3(L).$$

(*Hint: Use Exercise 2.25 on page 96. Then compare the coefficients of*
$$(g_2^3 - 27g_3^2)/16$$
with those of
$$\prod_{1 \le e_i < e_j \le 3} (e_i - e_j)^2.)$$

10.21. Prove that for $|x| < 1$, we have that

$$\frac{1}{(1-x)^2} - 1 = \sum_{n=1}^{\infty} (n+1)x^n.$$

(*Hint: You may use the fact from standard geometric series that*

$$\sum_{n=0}^{\infty} x^n = (1-x)^{-1}.)$$

10.22. Establish (10.22) on page 350.

(*Hint: Use Exercise 10.21 to get a series expansion for*

$$(\omega - z)^{-2} - \omega^{-2},$$

then plug this into the representation for $\wp$ given in Definition 10.11 on page 349.)

10.23. Establish (10.23) on page 350.

(*Hint: Use Exercise 10.22. Then employ what is called* Liouville's theorem for elliptic functions *which says:* An elliptic function with no poles (or no zeros) is constant. *This theorem may be found in any standard text on complex analysis. More generally,* Liouville's theorem *is often stated as:* A bounded entire function on $\mathbb{C}$ is constant, *often called* Liouville's boundedness theorem *from which the fundamental theorem of algebra follows as a simple consequence.*)

10.24. Prove that the discriminant of a lattice given in Definition 10.10 on page 349 is nonzero.

(*Hint: Use Exercise 10.23 and the fact given in Remark 10.7 on page 349, that $\wp'(z)$ is an odd elliptic function.*)

10.25. Two lattices L_j for $j = 1, 2$ are called *homothetic* if there exists a $\lambda \in \mathbb{C}$ such that $L_1 = \lambda L_2$. Prove that if E_j are elliptic curves with respect to L_j for $j = 1, 2$, respectively, then

$$E_1 \cong E_2 \text{ if and only if } L_1 \text{ and } L_2 \text{ are homothetic.}$$

(*Hint: Use Exercise 10.19.*)

10.4   Shimura–Taniyama–Weil & FLT

> *Casually in the middle of a conversation this friend told me that Ken Ribet*
> *had proved a link between Taniyma–Shimura and Fermat's Last Theorem. I*
> *was electrified. I knew that moment that the course of my life was changing*
> *because this meant that to prove Fermat's Last Theorem all I had to do was*
> *to prove the Taniyama–Shimura conjecture. ... Nobody had any idea how to*
> *approach Taniyma–Shimura but at least it was mainstream mathematics. ...*
> *So the romance of Fermat, which had held me all my life, was now combined*
> *with a problem that was professionally acceptable. ... It was one morning in*
> *late May. ... I was sitting around thinking about the last stage of the proof.*
> *... I forgot to go down for lunch. ... My wife, Nada, was very surprised that*
> *I'd arrived so late. Then I told her I'd solved Fermat's Last Theorem.*
>
> from an interview with *NOVA*
> —for the full interview see *http://www.pbs.org/wgbh/nova/proof/wiles.html*
> **Andrew Wiles (1953–)**—see Biography 5.5 on page 225
> **British mathematician living in the U.S.A.**

In order to display the force of the Shimura–Taniyama–Weil (STW) conjecture, it is an important motivator to set the stage by briefly outlining the events leading to its proof and the connections with FLT. We begin with the latter. FLT would seem on the face of it to have no connections with elliptic curves since $x^n + y^n = z^n$ is not a cubic equation. However, in 1986 Gerhard Frey published [27], which associated, for a prime $p > 5$, the elliptic curve

$$y^2 = x(x - a^p)(x + b^p) \qquad (10.25)$$

with nontrivial solutions to $a^p + b^p = c^p$. We call elliptic curves, given by equation (10.25), *Frey curves*. It turns out that this curve is of the type mentioned in the STW conjecture. In other words, existence of a solution to the Fermat equation would give rise to elliptic curves which would contradict STW. Now we need to describe the technical details.

In general, an elliptic curve E defined over a field F may be given by the *global* Weierstrass equation

$$y^2 + a_1 xy + a_3 y = x^3 + a_2 x^2 + a_4 x + a_6, \qquad (10.26)$$

where $a_j \in F$ for $1 \le j \le 6$. Then when F has characteristic different from 2, we may complete the square, replacing y by $(y - a_1 x - a_3)/2$ to get the more familiar Weierstrass equation

$$y^2 = 4x^3 + b_2 x^2 + 2b_4 x + b_6 \qquad (10.27)$$

with

$$b_2 = a_1^2 + 4a_2,$$

$$b_4 = 2a_4 + a_1 a_3,$$

and
$$b_6 = a_3^2 + 4a_6.$$
In this case the discriminant $\Delta(E) = \Delta$ is given by
$$\Delta(E) = -b_2^2 b_8 - 8b_4^3 - 27b_6^2 + 9b_2 b_4 b_6, \tag{10.28}$$
where
$$b_8 = a_1^2 a_6 + 4a_2 a_6 - a_1 a_3 a_4 + a_2 a_3^2 - a_4^2.$$
Also, the j-invariant is given by
$$j(E) = c_4^3/\Delta(E), \tag{10.29}$$
where
$$c_4 = b_2^2 - 24b_4 \tag{10.30}$$
and
$$j(E) = 1728 + c_6^2/\Delta \tag{10.31}$$
where
$$c_6 = -b_2^3 + 36b_2 b_4 - 216b_6. \tag{10.32}$$
By Exercise 10.26 on page 366, these definitions for $\Delta(E)$ and $j(E)$ coincide with Definition 10.12 on page 350 for the special case of the Weierstrass equation covered in §10.3. We may further simplify Equation (10.27) by replacing (x,y) with $((x - 3b_2)/36, y/108)$ to achieve
$$y^2 = x^3 - 27c_4 x + 54c_6. \tag{10.33}$$
By Exercise 10.27,
$$\Delta(E) = \frac{c_4^3 - c_6^2}{1728}. \tag{10.34}$$

Remark 10.9 Note, however, that if we *begin* with Equation (10.33), then the discriminant is
$$\Delta(E) = 2^6 \cdot 3^9 (c_4^3 - c_6^2),$$
which differs from (10.34) by a factor of $2^{12} \cdot 3^{12}$, and this is explained by the scaling introduced in change of variables in going from (10.26) to (10.27), then to (10.33).

Remark 10.9 shows that a change of variables may "inflate" a discriminant with new factors. Thus, for our development, we need to find a "minimal discriminant." In order to proceed with this in mind, we need the following concept.

Definition 10.14 | **Admissible Change of Variables**

If $E = E(\mathbb{Q})$ is an elliptic curve over $\mathbb{Q}$, given by (10.26) where we may assume that $a_j \in \mathbb{Z}$ for $j = 1, 2, 3, 4, 6$, then an *admissible* change of variables is one of the form

$$x = u^2 X + r \text{ and } y = u^3 Y + su^2 X + t,$$

where $u, r, s, t \in \mathbb{Q}$ and $u \neq 0$ with resulting equation

$$Y^2 + a_1' XY + a_3' Y = X^3 + a_2' X^2 + a_4' X + a_6' \qquad (10.35)$$

where

$$a_1' = \frac{a_1 + 2s}{u}, \qquad a_2' = \frac{a_2 - sa_1 + 3r - s^2}{u^2},$$

$$a_3' = \frac{a_3 + ra_1 + 2t}{u^3}, \qquad a_4' = \frac{a_4 - sa_3 + 2ra_2 - (t + rs)a_1 + 3r^2 - 2st}{u^4},$$

and

$$a_6' = \frac{a_6 + ra_4 + r^2 a_2 + r^3 - ta_3 - t^2 - rta_1}{u^6}.$$

Remark 10.10 From the projective geometry viewpoint discussed in Remark 9.1 on page 302, considering equivalence classes of points (x, y, z), an admissible change of variables fixes the point at infinity $(0, 1, 0)$ and carries the line for which $z = 0$ to the same line. The original Weierstrass form (10.26) gets sent to the same curve in Weirstrass form (10.35). Modulo a constant, admissible changes of variables are the most general linear transformations satisfying these properties.

In the special case where $r = s = t = 0$, the admissible change of variables multiplies the a_i by u^{-i} for $i = 1, 2, 3, 4, 6$. In this case, we say a_i has *weight* i. Indeed, Definition 10.13 on page 351 is just this special case of an admissible change of variables. *In general, we may define two elliptic curves to be isomorphic if they are related by an admissible change of variables.* Hence, by Exercise 10.19 on page 351, two elliptic curves over $\mathbb{Q}$ are related by an admissible change of variables if and only if they have the same j-invariant. There is another term in the literature used to describe this phenomenon as well. *Two elliptic curves over $\mathbb{Q}$ having the same j-invariant are said to be* twists *of one another.*

Since the discriminant Δ is given by (10.34) in terms of c_4 and c_6, then Δ is unaffected by r, s, t in an admissible change of variables given that the new variables for (10.35) are related by

$$c_4' = c_4 / u^4 \text{ and } c_6' = c_6 / u^6.$$

Hence, the triple (Δ, c_4, c_6) is a detector for curves that are equivalent under an admissible change of variables. In fact, by the above discussion, two elliptic curves E_1 and E_2 with discriminant Δ_1 and Δ_2, respectively, related by an admissible change of variables, must satisfy

$$\Delta_1/\Delta_2 = u^{\pm 12}.$$

This now sets the stage for looking at elliptic curves with minimal discriminants.

For the ensuing development, the reader should be familiar with the notation and topics covered in §6.2, especially Theorem 6.1 on page 236 and the notation $\nu_p(x) \geq 0$ that characterizes the p-adic integers $x \in \mathcal{O}_p$. Also, the notation of Definition 10.14 remains in force.

Definition 10.15 | **Minimal Equations for Elliptic Curves**

If $E = E(\mathbb{Q})$ is an elliptic curve over $\mathbb{Q}$, given by (10.26) where $a_j \in \mathbb{Z}$ for $j = 1, 2, 3, 4, 6$, with discriminant Δ, then (10.26) is called *minimal* at the prime p if the power of p dividing Δ cannot be decreased by making an admissible change of variables with the property that the new coefficients $a'_j \in \mathcal{O}_p$. If (10.26) is minimal for all primes p with $a_j \in \mathbb{Z}$ for $j = 1, 2, 3, 4, 6$, then it is called a *global minimal Weierstrass equation*.

Remark 10.11 Since an equation for $E(\mathbb{Q})$ given in Definition 10.15 can be assumed, without loss of generality, to have integral coefficients, then $|\Delta|_p \leq 1$ where $|\cdot|_p$ is the p-adic absolute value given in Definition 6.3 on page 233. Hence, in only finitely many steps $|\Delta|_p$ can be increased and still maintain $|\Delta|_p \leq 1$. Hence, it follows that in finitely many admissible changes of variables, we can get an equation minimal for E at p. In other words, there always exists a global minimal Weierstrass equation for $E(\mathbb{Q})$.

Note that

$$|\Delta|_p = 1 \text{ if and only if } p \nmid \Delta.$$

Also, by Exercise 10.30 on page 366, if any of

(1) $|\Delta|_p > p^{-12}$.

(2) $|c_4|_p > p^{-4}$.

(3) $|c_6|_p > p^{-6}$.

holds then (10.26) is minimal for p. Moreover, if $p > 3$, $|\Delta|_p \leq p^{-12}$, and $|c_4|_p \leq p^{-4}$, then (10.26) is *not* minimal for p.

For the following, the reader is reminded, via Exercise 9.5 on page 315, of

$$N_p = p + 1 + \sum_{x \in \mathbb{F}_p} \chi(x^3 + ax + b), \tag{10.36}$$

being the number of points on the elliptic curve $E(\mathbb{F}_p)$, including the point at infinity, over a field of p elements for a prime p.

Definition 10.16 | **The Reduction Index for Elliptic Curves**

Suppose that E is an elliptic curve over Q given by a minimal Weierstrass equation. If the $\overline{E} \,(\mathrm{mod}\ p) \neq 0$ for a prime p, then p is said to be a prime of *good reduction* for E. Furthermore, if N_p for a prime p is given by (10.36), then let

$$a_p(E) = p + 1 - N_p.$$

If p is a prime of good reduction, then $a_p(E)$ is called the *good reduction index for E at p*, and the sequence $\{a_p(E)\}_p$ indexed over the primes of good reduction is called the *good reduction sequence for E*. Primes that are *not* of good reduction are called primes of *bad reduction* for E, and $a_p(E)$ is called the *bad reduction index for E*.

Note that there are only finitely many primes of bad reduction since these are the primes dividing Δ. Also, by Theorem 9.5 on page 319, we know that $|a_p(E)| < 2\sqrt{p}$. There is much more of interest in the reduction index.

Example 10.8 Consider the elliptic curve given by

$$y^2 + y = x^3 - x^2.$$

Via the formulas in (10.26)–(10.34) on pages 353–354, we have

$$a_1 = 0, a_3 = 1, a_2 = -1, a_4 = 0 = a_6, b_2 = -4, b_4 = 0, b_6 = 1, \text{ and } b_8 = -1.$$

Therefore,

$$\Delta(E) = -b_2^2 b_8 - 8b_4^3 - 27b_6^2 + 9b_2 b_4 b_6$$
$$= -(-4)^2(-1) - 8(0)^3 - 27 \cdot 1^2 + 9(-4)(0)(1) = -11,$$

so E has good reduction at all primes $p \neq 11$. Now we compute the good reduction index for this curve at various primes $p \neq 11$, which we call a *good reduction table* for E.

p	2	3	5	7	13	17	19	23	29	31	37	41
N_p	5	5	5	10	10	20	20	25	30	25	35	50
$a_p(E)$	-2	-1	-2	1	4	-2	0	-1	0	7	3	-8

See Exercise 10.28 on page 366 for more related illustrations. Also, see Example 10.9 on page 360.

Remark 10.12 To say that p is a prime of good reduction for E is to say that E is *nonsingular* over $\mathbb{F}_p$, meaning that $\Delta(\overline{E} \,(\mathrm{mod}\ p))$ is not divisible by p. We now explain this in detail. A point $P = (x_0, y_0)$ on an elliptic $E(F) = E$ curve over a field F is called a *singular point* if P satisfies the equation, defining E, given by

$$f(x, y) = y^2 + a_1 xy + a_3 y - x^3 - a_2 x^2 - a_4 x - a_6 = 0 \qquad (10.37)$$

with the partial derivatives satisfying

$$\partial f/\partial x(P) = \partial f/\partial y(P) = 0.$$

Thus, to say that P is a singular point of E is to say that E is a singular curve at P. To say that E is *nonsingular* over F is to say that the curve has *no singular points*. By Exercise 10.31 on page 366, E is nonsingular if and only if $\Delta(E) \neq 0$. Note that E never has a singular point at infinity also shown in that exercise.

Singular points may be classified as follows. By Exercise 10.32 on page 367,

$$E \text{ has a } node \text{ if and only if } \Delta(E) = 0 \text{ and } c_4 \neq 0;$$

and

$$E \text{ has a } cusp \text{ if and only if } \Delta(E) = 0 = c_4.$$

More explicitly, we may view the Taylor expansion of $f(x, y)$ at P via

$$f(x, y) - f(x_0, y_0) = [(y - y_0) - \alpha(x - x_0)][(y - y_0) - \beta(x - x_0)] - (x - x_0)^3.$$

Then P is a *node* if $\alpha \neq \beta$ having tangent lines at P given by $y - y_0 = \alpha(x - x_0)$ and $y - y_0 = \beta(x - x_0)$. An example is the curve given by $y^2 = x^3 + x^2$, for which $\Delta = 0$ and $c_4 = 16$. Here $P = (x_0, y_0) = (0, 0)$, and the two tangents are $y = x$ and $y = -x$ as in Figure 10.1.

Figure 10.1: $y^2 = x^3 + x^2$

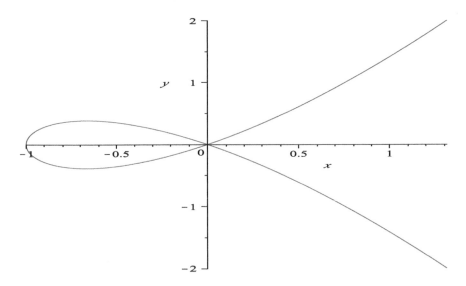

Also, P is a cusp if $\alpha = \beta$ with tangent line at P given by

$$y - y_0 = \alpha(x - x_0).$$

An example is the curve given by $y^2 = x^3$, where $\Delta = c_4 = 0$. The single tangent is $y = 0$ at $P = (x_0, y_0) = (0, 0)$. See Figure 10.2.

Figure 10.2: $y^2 = x^3$

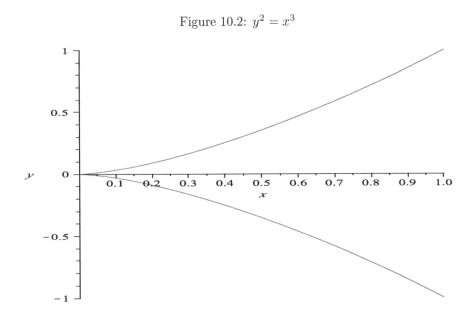

Remark 10.13 The good reduction index is a mechanism for representing arithmetic data about E that is captured in patterns of the good reduction sequence $\{a_p(E)\}_p$. How it does this is contained in the subtext of the Shimura–Taniyama–Weil conjecture. The pattern involves the normalized modular cusp forms of weight 2 and level $n \in \mathbb{N}$ that we introduced in Definition 10.6 on page 343.

Definition 10.17 |Modular Elliptic Curves|

Let $E(\mathbb{Q})$ be an elliptic curve over $\mathbb{Q}$ with good reduction sequence $\{a_p(E)\}_p$. If there exists an $n \in \mathbb{N}$ and a normalized weight 2 cusp form of level n

$$f(z) = q + \sum_{j=2}^{\infty} a_j(f) q^j, \text{ where } q = \exp(2\pi i z),$$

such that

$$a_p(E) = a_p(f),$$

then E is called a *modular elliptic curve*.

Now we may state the celebrated conjecture.

Conjecture 10.1 | **The Shimura–Taniyama–Weil (STW) Conjecture**

If E is an elliptic curve over $\mathbb{Q}$, then E is modular.

Example 10.9 By Example 10.5 on page 343, the function given in (10.15) spans $S_2(\Gamma_0(11))$ and is explicitly given by

$$f(z) = \eta(z)^2 \eta(11z)^2 = \sum_{n=1}^{\infty} c_n q^n = q \prod_{n=1}^{\infty} (1 - q^n)^2 \cdot (1 - q^{11n})^2 =$$

$$q - \mathbf{2q^2} - \mathbf{q^3} + 2q^4 + \mathbf{q^5} + 2q^6 - \mathbf{2q^7} - 2q^9 - 2q^{10} + \mathbf{q^{11}} - 2q^{12} + \mathbf{4q^{13}}$$

$$+ 4q^{14} - q^{15} - 4q^{16} - \mathbf{2q^{17}} + 4q^{18} + 2q^{20} + 2q^{21} - 2q^{22} - \mathbf{q^{23}} - 4q^{25}$$

$$- 8q^{26} + 5q^{27} - 4q^{28} + 2q^{30} + \mathbf{7q^{31}} + \cdots + \mathbf{3q^{37}} + \cdots - \mathbf{8q^{41}} + \cdots .$$

We have highlighted the prime powers of q and their coefficients to show that these coefficients are exactly the nonzero values of the good reduction index $a_p(E)$ in Example 10.8 on page 357, thereby illustrating that E is a modular function.

Remark 10.14 The notion of a *conductor* of an elliptic curve must now come into play for our discussion. The technical definition involves a cohomological description that we do not have the tools to describe. However, we can talk about it in reference to the discriminant and related prime divisors in order to understand what it means. Given an elliptic curve $E(\mathbb{Q}) = E$ with global minimal Weierstrass equation and discriminant $\Delta(E) = \Delta$, the conductor n divides Δ and has the same prime factors as Δ. The power to which a given prime appears in n is determined as follows. The power of a prime p dividing n is 1 if and only if $E(\mathbb{F}_p)$ has a *node*, which is characterized by having two candidate tangents at the point, which in turn, means that (10.35) has a double root. See Exercise 10.33 on page 367 for an illustration. Also, see Remark 10.12 on page 357. If $p > 3$, then the power of p dividing n is 2 if and only if $E(\mathbb{F}_p)$ has a cusp. In the case where $p = 2$ or $p = 3$, which we selectively have ignored for the sake of simplicity of presentation, the conductor can be computed using Tate's algorithm, which is uncomplicated, although the process of using it can be somewhat protracted, see [94]. For $p \neq 2, 3$, the power of p dividing the conductor n is at most 2, so for our purposes, the above discussion suffices.

From the above, we conclude that the conductor of E is not divisible by any primes of good reduction, also called *stable* reduction. In other words, only primes of bad reduction divide the conductor. Moreover, a prime p to the first power exactly divides the conductor precisely when $E(\mathbb{F}_p)$ has a node, in which case E is said to have *multiplicative* or *semi-stable* reduction at p. Hence, E has semi-stable reduction at *all* primes, in which case E is called semi-stable, precisely when the conductor n is squarefree. For instance, the curve in Example 10.8 on page 357 has conductor 11, an instance of a semi-stable elliptic curve. The conductor of E is exactly divisible by p^2 precisely when $E(\mathbb{F}_p)$ has a cusp, in which case we say that E has *additive* or *unstable* reduction.

By Exercise 10.38 on page 367, the conductor is an "isogeny invariant," as well. The STW conjecture implies that we have the conductor n equal to the level n in $\Gamma_0(n)$ of weight 2 cusp forms—see the reformulation of STW in terms of L-functions on page 364.

Now we illustrate the modularity theorem in different terms that will bring more of the structure and interconnections to light. To do this, we concentrate upon the example $n = 11$, which will be a template for the general theory.

Example 10.10 From Example 10.5 on page 343, for $n = 11$, the group $\Gamma_0(11)$ can be shown to be generated by

$$T = \begin{pmatrix} 1 & 1 \\ 0 & 1 \end{pmatrix}, U = \begin{pmatrix} 8 & 1 \\ -33 & -4 \end{pmatrix}, V = \begin{pmatrix} 9 & 1 \\ -55 & -6 \end{pmatrix},$$

and if $\gamma \in S_2(\Gamma_0(11))$, then we map $\Gamma_0(11)$ to $\mathbb{C}$, additively via $\phi_\gamma(U) = \omega_1$, $\phi_\gamma(V) = \omega_2$, and $\phi_\gamma(T) = 0$. Hence, $L = [\omega_1, \omega_2]$ is a lattice in $\mathbb{C}$. It can be shown that $\mathbb{C}/L$, called a *complex torus*, is analytically isomorphic to an elliptic curve $E(\mathbb{C})$, where L is determined up to homothety by E—see Exercise 10.25. For our purposes the "analytic isomorphism"

$$\mathbb{C}/L \mapsto E(\mathbb{C})$$

is explicitly given by

$$z \mapsto \begin{cases} (\wp(z), \wp'(z), 1) & \text{if } z \notin L, \\ (0, 1, 0) & \text{if } z \in L \end{cases}$$

—see Remark 10.10. This is a holomorphic map carrying $\mathbb{C}/L$ one-to-one onto the elliptic curve $E = E(\mathbb{C})$ where E is given by the form

$$y^2 = 4x^3 - g_2 x - g_3,$$

with g_2 and g_3 given in Definition 10.10 on page 349. Altogether, we get a holomorphic map from $X_0(11)$ onto $\mathbb{C}/L$, then onto $E(\mathbb{C})$. Thus, it can be shown that this provides a holomorphic surjection

$$X_0(11) = \frac{\Gamma_0(11)}{\mathfrak{H}^*} \mapsto E(\mathbb{C}) \text{ where } \mathfrak{H}^* = \mathfrak{h} \cup \mathbb{Q} \cup \{\infty\},$$

where $X_0(11)$ is called a *compact Riemann surface*, which is a complex one-dimensional manifold. *Think of a Riemann surface as a "deformed" complex plane, which looks like the complex plane locally near a given point, but the global topology may be different. The complex plane may be described as the most basic Riemann surface.* $\mathbb{C}/L$ is also a complex manifold and the principal feature of such surfaces is that holomorphic maps can be defined between them as we have done above—see [88] for more details.

One may actually calculate the j-invariant via (10.20) to get

$$j(L) = -\frac{(2^4 \cdot 31)^3}{11^5}, \qquad (10.38)$$

which demonstrates that E is defined over $\mathbb{Q}$ and gives more meaning to the above mapping involving $X_0(11)$ and E over $\mathbb{Q}$. However, from (10.31) on page 354 via (10.24) on page 350, we have

$$j(L) = \frac{c_4^3}{\Delta} = 1728 + \frac{c_6^2}{\Delta}, \qquad (10.39)$$

so by Exercise 10.39 on page 367, there is an integer $k \neq 0$ such that

$$c_4 = 2^4 \cdot 31k^2, c_6 = 2^3 \cdot 2501k^3, \text{ and } \Delta = -11^5 k^6. \qquad (10.40)$$

By Exercise 10.40, (10.40) yields a global minimal Weierstrass equation exactly when k has no odd square factor, and

$$k \equiv r \pmod{16} \text{ where } r \in \{1, 2, 5, 6, 9, 10, 12, 13, 14\}. \qquad (10.41)$$

We call the association of $X_0(11)$ and $E = E(\mathbb{Q})$ given by (10.41), with global minimal Weierstrass equation provided by (10.40), a $\mathbb{Q}$-*structure of* E. The simplest $\mathbb{Q}$-structure occurs when $k = 1$ in which case we get the global minimal equation given by

$$E(\mathbb{C}) : y^2 + y = x^3 - x^2 - 10x - 20, \qquad (10.42)$$

which is the curve in Exercise 10.28 on page 366. What we have accomplished is a mapping of $X_0(11)$ onto $E(\mathbb{C})$.

Now, if we define

$$\begin{pmatrix} \omega_2' \\ \omega_1' \end{pmatrix} = \begin{pmatrix} 1 & 3 \\ 0 & 5 \end{pmatrix} \begin{pmatrix} \omega_2 \\ \omega_1 \end{pmatrix}$$

and we let

$$L' = [\omega_1', \omega_2'],$$

it can be shown that

$$j(L') = -16^3/11,$$

so a corresponding elliptic curve E' can be defined over $\mathbb{Q}$, and this curve is given by

$$E' : y^2 + y = x^3 - x^2, \qquad (10.43)$$

which is the curve in Example 10.8 on page 357, with discriminant -11, and as we saw above the discriminant of (10.42) is -11^5. In Exercise 10.38, this was shown to be *isogenous* to the curve in (10.42). In Remark 10.14 on page 360, we saw that the conductor is an isogeny invariant, in this case $n = 11$.

We may reformulate the STW conjecture now in terms of the above, which we have illustrated for the case $n = 11$.

◆ STW Conjecture in Terms of Modular Parametrizations

Given an elliptic curve E over $\mathbb{Q}$, there exists an $n \in \mathbb{N}$ for which there is a nonconstant surjective holomorphic map $F : X_0(n) \mapsto E$, defined over $\mathbb{Q}$, in which case E is said to have a *modular parametrization modulo* n, and E is called a *Weil curve*.

Remark 10.15 We have illustrated the above for the case $n = 11$ in Example 10.10 on page 361, but the theory, called *Eichler–Shimura theory*, holds for any of the compact Reimann surfaces $X_0(n)$ where n is the level of the weight 2 cusp forms, so given the aforementioned proof of STW, the above is a statement of the *modularity theorem*.

The phrase "defined over $\mathbb{Q}$" in the above interpretation of the STW conjecture is important in that we may have holomorphic surjections without the rationality property but for which the L-functions of the curves and the cusp forms do not agree. Now we must explain this comment by introducing the notions of L-functions for elliptic curves and forms. Note that the construction of the map from $X_0(11)$ to $E(\mathbb{C})$ in Example 10.10 on page 361 is indeed defined over $\mathbb{Q}$. In the literature, such maps are rational maps defined at every point, called *morphisms*—see [88].

We turn our attention to L-functions, a concept we introduced in §7.2, but have not yet linked with elliptic curves. Elliptic curves that are isogenous over $\mathbb{Q}$ have the same L-functions which we now define and discuss.

Let $E(\mathbb{Q})$ be an elliptic curve over $\mathbb{Q}$ given by a global minimal Weierstrass equation, which is no loss of generality by Remark 10.11 on page 356. Then the L-function for E, having discriminant Δ, is given by

$$L(E, s) = \prod_{p \mid \Delta} \left[\left(1 - a_p(E)p^{-s}\right)^{-1} \right] \prod_{p \nmid \Delta} \left[\left(1 - a_p(E)p^{-s} + p^{1-2s}\right)^{-1} \right].$$

It can be shown that $L(E, s)$ converges for $\Re(s) > 2$, and is given by an absolutely convergent Dirichlet series—see §5.3. Thus, we may write

$$L(E, s) = \sum_{n=1}^{\infty} \frac{c_n}{n^s}.$$

Now by Definition 10.6 on page 343, a normalized cusp form $f \in S_2(\Gamma_0(n))$ of weight 2 and level n satisfies

$$f(z) = q + \sum_{n=2}^{\infty} a_n(f)q^n.$$

Thus, we may define the *L*-function of f by

$$L(f,s) = \sum_{n=1}^{\infty} \frac{a_n(f)}{n^s}.$$

Now the STW conjecture may be reformulated in terms of L functions:

◆ STW Conjecture in Terms of L-Functions

For every elliptic curve E defined over $\mathbb{Q}$, there exists a normalized cusp form of weight 2 and level n, $f \in S_2(\Gamma_0(n))$, such that

$$L(f,s) = L(E,s),$$

and n is the conductor of E.

We have concentrated upon $X_0(11)$ in Example 10.10 on page 361 since it is the simplest case, namely having what is called *genus one* with corresponding $S_2(\Gamma_0(11))$ having dimension one as we saw above. In general, the dimension of $S_2(\Gamma_0(n))$ is called the *genus* of $X_0(n)$. To see the intimate connection with FLT, we return to the discussion of Frey curves (10.25) introduced on page 353. Suppose that

$$a^p + b^p = c^p \tag{10.44}$$

is a counterexample to FLT for a prime $p \geq 5$. The Frey curve is given by

$$E : y^2 = x(x - a^p)(x - c^p), \tag{10.45}$$

for which

$$\Delta = 16a^{2p}b^{2p}c^{2p}, \tag{10.46}$$

and

$$c_4 = 16(a^{2p} - a^pc^p + c^{2p}). \tag{10.47}$$

Then when a, b, c are pairwise relatively prime, by Exercise 10.41 on page 367, the conductor of E is the product of all primes dividing abc, which tells us, by Remark 10.14, that E is semi-stable.

Now we are in a position to return to a discussion of the STW conjecture and FLT. In 1995, Wiles and Taylor published papers [95] and [103], which proved that every semi-stable elliptic curve is modular. In 1998, Conrad, Diamond, and Taylor [16] proved the STW conjecture for all elliptic curves with conductor not divisible by 27. Then in 2001, Breuil, Conrad, Diamond, and Taylor published a proof of the full STW conjecture, which we now call the *modularity theorem* [11]. However, in 1990, Ribet proved the following, which via the affirmative verification of the STW conjecture, allowed a proof of FLT as follows.

Theorem 10.3 | **Ribet's Theorem**

Suppose that E is an elliptic curve over $\mathbb{Q}$ given by a global minimal Weierstrass equation and having discriminant $\Delta = \prod_{p \mid \Delta} p^{f_p}$ and conductor $n = \prod_{p \mid \Delta} p^{g_p}$, both canonical prime factorizations. Furthermore, if E has a modular parametrization of level n with $f \in S_2(\Gamma_0(n))$ having normalized expansion

$$f(z) = q + \sum_{n=2}^{\infty} a_j(f) q^n,$$

then for a fixed prime p_0, set

$$n' = \frac{n}{\displaystyle\prod_{\substack{p \\ p_0 \mid f_p \\ g_p = 1}} p}. \tag{10.48}$$

Then there exists an $f' \in S_2(\Gamma_0(n'))$ such that $f' = \sum_{n=1}^{\infty} b_j(f') q^n$ with $b_j(f') \in \mathbb{Z}$ satisfying $a_j(f) \equiv b_j(f') \pmod{p_0}$ for all $n \in \mathbb{N}$.

Proof. See [81]. □

Now we may state our target result, which follows [45, Corollary 12.13, p. 399], where it is cited as a Frey–Serre–Ribet result.

Theorem 10.4 | **Proof of Fermat's Last Theorem**

The STW conjecture implies FLT.

Proof. Assume that FLT is false. Then by Theorem 10.3, the Frey curve given in (10.45) has conductor $n = \prod_{p \mid abc} p$, which when compared to the coefficients in (10.48), yields $n' = 2$. However, by Example 10.5 on page 343, $S_2(\Gamma_0(2))$ is the zero space, so $b_j(f') = 0$ for all $n \in \mathbb{N}$. Yet,

$$b_j(f') \equiv a_j(f) \pmod{p_0}$$

for all $n \in \mathbb{N}$. In particular,

$$0 = b_1(f') \equiv a_1(f) = 1 \pmod{p_0},$$

a contradiction. □

With the above, this completes the main text and demonstrates the power of the tools we developed herein. It is an appropriate juncture to leave since the proof of FLT to the extent we have been able to demonstrate herein shows the accomplishments of centuries of mathematical exploration.

Exercises

10.26. Prove that, in (10.28) and (10.29) on page 354, the definitions for discriminant and j-invariant agree with those given in §10.3, namely when $b_2 = 0$, $2b_4 = -g_2$, $b_6 = -g_3$, and $c_4 = 12g_2$.

10.27. With reference to (10.30) and (10.32) on page 354, prove that the discriminant of E given by (10.28) is equal to $\Delta(E) = (c_4^3 - c_6^2)/1728$.

10.28. By a suitable transformation, show that $y^2 + y = x^3 - x^2 - 10x - 20$ is of the form

$$y^2 = x^3 - 27c_4 x - 54c_6 \tag{10.49}$$

with $\Delta(E) = -11^5$. Conclude that E has good reduction for all primes $p \neq 11$.

(*Hint: Use Exercise 10.27.*)

10.29. For the elliptic curve given in Exercise 10.28, provide a good reduction table for the same primes as given for the curve in Example 10.8.

10.30. Let E be an elliptic curve given by (10.26) on page 353, where $|a_j|_p \leq 1$ for $j = 1, 2, 3, 4, 6$. With reference to Remark 10.11 on page 356, prove that (10.26) is minimal for E at p if any of the following hold.

(1) $|\Delta|_p > p^{-12}$.

(2) $|c_4|_p > p^{-4}$.

(3) $|c_6|_p > p^{-6}$.

Moreover, if $p > 3$, prove that (10.26) is *not* minimal for E at p if both of the following hold

(1) $|\Delta|_p \leq p^{-12}$.

(2) $|c_4|_p \leq p^{-4}$.

10.31. Prove that an elliptic curve $E = E(F)$ over a field F is always nonsingular at infinity. Then prove that E is nonsingular over F if and only if $\Delta(E) \neq 0$.

(*Hint: To prove that E never has a singular point at infinity, consider the homogeneous equation*

$$\mathcal{F}(X, Y, Z) = Y^2 Z + a_1 XYZ + a_3 Y Z^2 - X^3 - a_2 X^2 Z - a_4 X Z^2 - a_6 Z^3,$$

so the point at infinity is $\mathfrak{o} = (0, 1, 0)$. Then show $\partial \mathcal{F}/\partial Z(\mathfrak{o}) \neq 0$—see Remark 9.1 on page 302. Recall that we are always assuming characteristic not 2 or 3, which simplifies computations although the result still holds in the latter cases.)

10.32. Prove that an elliptic curve $E = E(\mathbb{F}_p)$ has a node if and only if $c_4 \neq 0$ and $\Delta(E) = 0$; and E has a cusp if and only if $\Delta = 0 = c_4$.

(*Hint: Prove that you may assume, without loss of generality, that the singular point occurs at the origin. Then consider* (10.37) *at the origin with respect to partial derivatives.*)

10.33. Prove that the curve in Exercise 10.29 has a node over $\mathbb{F}_{11}$ by displaying a graph, reduced modulo 11 to display the node. This illustrates Remark 10.14 on page 360 from which we may conclude that this curve has conductor 11.

(*Hint: Use Exercise 10.31.*)

10.34. Prove that the curve in Example 10.8 on page 357 has conductor 11 by reducing it modulo 11 and graphing its node there using Remark 10.14.

10.35. Prove that if $p > 3$ is prime then the elliptic curve given by $y^2 = x^3 + px^2 + 1$ has good reduction at p.

10.36. Prove that the elliptic curve $E = E(\mathbb{F}_p)$ given by $y^2 = x^3 + x^2 + p$ for a prime $p > 3$ has a node.

10.37. Prove that the elliptic curve $E = E(\mathbb{F}_p)$ given by $y^2 = x^3 + p$ for a prime $p > 3$ has a cusp.

10.38. Given elliptic curves $E_j = E_j(\mathbb{C})$ for $j = 1, 2$, an *isogeny* is defined to be an analytic map $h : E_1 \mapsto E_2$, where the identity gets mapped to the identity. Show that there is an isogeny between the curve given in Exercise 10.28, which we will call E_2 and the curve given in Example 10.8, which we will call E_1. Two curves E_1 and E_2 are said to be *isogenous* if there is a nonconstant isogeny h between them.

10.39. Verify (10.40) on page 362.

(*Hint: Use* (10.38) *on page 362.*)

10.40. In Example 10.10, show that k has no odd square factor and verify (10.41) on page 362.

(*Hint: Look at the elliptic curve that one gets from the elliptic curve of the form* $y^2 = x^3 + a_2x^2 + a_4x + a_6$, *satisfying* (10.40) *by an admissible change of variables with* $u = 2$.)

10.41. Prove that the Frey curve has conductor $n = \prod_{p|abc} p$, where a, b, c are given in (10.44).

(*Hint: First verify that* (10.46)–(10.47) *on page 364 hold. Then prove that* (10.45) *is minimal Weierstrass for any prime* $p \mid \Delta$. *Also, check the odd primes* p *dividing* ac *and* b *separately, as well as* $p = 2$. *Then find an admissible change of variables at* $p = 2$ *for* (10.45) *so that the new equation is global minimal.*)

Appendix

Sieve Methods

> *Work without hope draws nectar in a sieve,*
> *And hope without an object cannot live.*
>
> from **Work without Hope (1828)**
> **Samuel Taylor Coleridge (1772–1834)**
> **English poet, critic, and philosopher**

The purpose of this appendix is to provide an overview of sieve methods used in factoring, recognizing primes, finding natural numbers in arithmetic progressions whose common difference is prime, or generally to estimate the cardinalities of various sets defined by the use of multiplicative properties. Recall that use of a sieve or *sieving* is a process whereby we find numbers via searching up to a prescribed bound and eliminate candidates as we proceed until only the desired solution set remains. In other words, sieve theory is designed to estimate the size of sifted sets of integers. For instance, sieves may be used to attack the following *open* problems, for which sieve methods have provided some advances.

(a) **(The Twin Prime Conjecture)**

There are infinitely many primes p such that $p + 2$ is also prime.

(b) **(The Goldbach Conjecture)**

Every even integer $n > 2$ is a sum of two primes.

(c) **(The p $= n^2 + 1$ Conjecture)**

There are infinitely many primes p of the form $p = n^2 + 1$.

(d) **(The q $= 4p + 1$ Conjecture)**

There are infinitely many primes p such that $q = 4p + 1$ is also prime.

(e) **(Artin's Conjecture)**

For any nonsquare integer $a \notin \{-1, 0, 1\}$, there exist infinitely many primes p such that a is a primitive root modulo p.

Indeed, in 1986, Heath-Brown [39] used sieving methods to advance the Artin conjecture to within a hair of a solution when he proved that, for a given prime p, with the possible exception of at most two primes, there are infinitely many primes q such that p is a primitive root modulo q. Thus, sieve methods are important to review for their practical use in number theory and the potential for solutions of outstanding problems such as the above. The methodology used to prove these results could serve as a course in itself, so we have relegated these facts to an appendix without proofs.

The fundamental goal of sieve theory is to produce upper and lower bounds for cardinalities of sets of the type,

$$S(\mathcal{S}, \mathcal{P}, y) = \{n \in \mathcal{S} : p \mid n \text{ implies } p > y \text{ for all } p \in \mathcal{P}\}, \qquad \text{(A.1)}$$

where $\mathcal{S}$ is a finite subset of $\mathbb{N}$, $\mathcal{P}$ is a subset of $\mathbb{P}$, the set of all primes, and y is a positive real number.

Example A.1 Let

$$\mathcal{S} = \{n \in \mathbb{N} : n \le x\} \text{ and } \sqrt{x} < y \le x.$$

Then

$$|S(\mathcal{S}, \mathcal{P}, y)| = \left|\{n \le x : p \mid n \text{ implies } p > y\}\right| = \pi(x) - \pi(y) + 1,$$

one more than the number of primes between x and y.

To illustrate (A.1) more generally, we begin with what has been called "the oldest nontrivial algorithm that has survived to the present day." From antiquity, we have the Sieve of Eratosthenes, which is covered in a first course in number theory—see [68, Example 1.16, p. 31, Biography 1.6, p. 32], which sieves to produce primes to a chosen bound. However, as discussed therein, this sieve is highly inefficient. Indeed, since in order to determine the primes up to some bound using this sieve for $n \in \mathbb{N}$, one must check for divisibility by all primes not exceeding $\sqrt{n}$, then the sieve of Eratosthenes has complexity $O((n \log_e n)(\log_e \log_e n))$, which even using the world's fastest computers, is beyond hope for large integers as a method for recognizing primes. Yet there is a formulation of this sieve that fits nicely into the use of arithmetic functions, and has applications as a tool for modern sieves, so we present that here for completeness and interests sake.

Recall the definition of the Möbius function $\mu(d)$, given by (5.22) on page 214. Also, let $\omega(d)$ denote the number of distinct prime divisors of d.

Theorem A.1 | **Eratosthenes' Sieve** |

Suppose that

$$\mathcal{P} = \{p_1, p_2, \ldots, p_n\} \subseteq \mathbb{P}$$

is a set of distinct primes and

$$\mathcal{S} \subseteq \mathbb{N} \text{ with } |\mathcal{S}| < \infty.$$

Denote by S the number of elements of $\mathcal{S}$ not divisible by any of the p_j's and by $\mathcal{S}_d$ the number of elements of $\mathcal{S}$ divisible by a given $d \in \mathbb{N}$. Then

$$S = \sum_{d \mid p_1 p_2 \cdots p_n} \mu(d) \mathcal{S}_d.$$

Moreover, for $m = 1, 2, \ldots, \lfloor n/2 \rfloor$, we have

$$\sum_{\substack{d \mid p_1 p_2 \cdots p_n \\ \omega(d) \leq 2m-1}} \mu(d) \mathcal{S}_d \leq S \leq \sum_{\substack{d \mid p_1 p_2 \cdots p_n \\ \omega(d) \leq 2m}} \mu(d) \mathcal{S}_d, \qquad (A.2)$$

where (A.2) *is called Eratosthenes' sieve.*

Proof. See [75, Corollary 2, p. 147]. $\qquad\square$

For instance, an application of Theorem A.1 is that it may be used to prove the following result on the number of primes less than a certain bound, first proved in 1919, by the Norwegian mathematician Viggo Brun (1882–1978).

Theorem A.2 $\boxed{\textbf{Brun's Theorem}}$

If $n \in \mathbb{N}$ and $A_{2n}(x)$ denotes the number of primes $p \leq x$ for which $|p + 2n|$ is also prime, then

$$A_{2n}(x) = O(x(\log_e \log_e x)^2 \log_e^{-2} x).$$

Proof. See [75, Theorem 4.3, p. 148]. $\qquad\square$

Theorem A.2 has, as a special case, implications for the twin prime conjecture as follows. Recall that the symbol $<<$ is synonymous with the "big Oh" notation.

Corollary A.1 $\boxed{\textbf{Brun's Constant}}$

Let $\mathcal{Q}$ be the set of all primes p such that $p + 2$ is also prime, then

$$\sum_{\substack{p \in \mathcal{Q} \\ p \leq x}} 1 << \frac{x(\log_e \log_e x)^2}{\log_e^2 x},$$

and the series

$$\sum_{p \in \mathcal{Q}} \frac{1}{p} = B \qquad (A.3)$$

is convergent, where (A.3) *is called Brun's constant.*

Proof. See [75, Corollary, p. 152]. $\qquad\square$

Remark A.1 We do not know if $\mathcal{Q}$ in Corollary A.1 is finite or not since its infinitude would be the twin prime conjecture. We *do know* that the sum of the reciprocals of *all* primes diverges, but since the series (A.3) converges, this is not a proof of the conjecture since we would need divergence to get the infinitude. The behaviour of the two series does tell us that, although the twin prime conjecture may be true, the twin primes must be appreciably less dense than the entire set of primes. Brun's result, that the reciprocals of twin primes converges, is one of the centerpiece achievements of sieve theory.

The value of Brun's constant is

$$B \approx 1.9021605824,$$

with an error within ± 0.000000003, computed by Thomas R. Nicely in 1999. It is worth noting the now famous fact that, in 1995, Nicely was doing computations on Brun's constant which led him to discover a flaw in the floating-point arithmetic of the Pentium computer chip, costing literally millions of dollars to its manufacturer Intel—see *http://www.trnicely.net/twins/twins2.html*.

Theorem A.1 on page 370 tells us that the sieve of Eratosthenes investigates the function

$$|S(\mathcal{S}, \mathcal{P}, x)| = \sum_{\substack{n \in \mathcal{S} \\ \gcd(n, \Pi)=1}} 1, \text{ where } \Pi = \prod_{\substack{p \in \mathcal{P} \\ p < x}} p$$

via the equality

$$|S(\mathcal{S}, \mathcal{P}, x)| = \sum_{n \in \mathcal{S}} \sum_{\substack{d \mid n \\ d \mid \Pi}} \mu(d) = \sum_{d \mid \Pi} \mu(d) \mathcal{S}_d.$$

The general basic sieve problem emanates from this, namely find arithmetic functions $\lambda_\ell(d) : \mathbb{N} \mapsto \mathbb{R}$ and $\lambda_u(d) : \mathbb{N} \mapsto \mathbb{R}$ with

$$\sum_{\substack{d \mid n \\ d \mid \Pi}} \lambda_\ell(d) \leq \begin{cases} 1 & \text{if } \gcd(n, \Pi) = 1, \\ 0 & \text{if } \gcd(n, \Pi) > 1, \end{cases}$$

and

$$\sum_{\substack{d \mid n \\ d \mid \Pi}} \lambda_u(d) \geq \begin{cases} 1 & \text{if } \gcd(n, \Pi) = 1, \\ 0 & \text{if } \gcd(n, \Pi) > 1, \end{cases}$$

such that

$$\sum_{d \mid \Pi} \lambda_\ell(d) \mathcal{S}_d = \sum_{n \in \mathcal{S}} \sum_{\substack{d \mid n \\ d \mid \Pi}} \lambda_\ell(d) \leq |S(\mathcal{S}, \mathcal{P}, x)| \leq \sum_{n \in \mathcal{S}} \sum_{\substack{d \mid n \\ d \mid \Pi}} \lambda_u(d) = \sum_{d \mid \Pi} \lambda_u(d) \mathcal{S}_d.$$

$$\text{(A.4)}$$

Now we interpret the above in terms of what Selberg did to create his famous sieve and how Theorem A.1 comes into play—see [68, Biography 1.21, p. 67]. With the notation of Theorem A.1 still in force, we add that P denotes the product of the primes in $\mathcal{P}$, $|\mathcal{S}| = N$, and call the following *Selberg's condition* on $\mathcal{S}$.

There exists a multiplicative function $f(d)$ such that if $d \mid P$, then

$$\mathcal{S}_d = \frac{f(d)}{d} N + R(d), \tag{A.5}$$

where $|R(d)| \leq f(d)$ and $d > f(d) > 1$. With the Selberg condition plugged into the right-hand side of (A.4), we have

$$|S(S, \mathcal{P}, x)| \leq \sum_{d \mid \Pi} \frac{\lambda_u(d) f(d) N}{d} + \sum_{d \mid \Pi} \lambda_u(d) R(d)$$

$$= N \sum_{d \mid \Pi} \frac{\lambda_u(d) f(d)}{d} + O\left(\sum_{d \mid \Pi} |\lambda_u(d) R(d)|\right). \qquad (A.6)$$

Selberg's sieve arose from his attempts to minimize (A.6) subject to Selberg's condition (A.5). Theorem A.1 on page 370 comes into play again in that it is used in the proof of the following, first proved by Selberg [85] in 1947. The following is considered to be the fundamental theorem concerning Selberg's sieve, which for the above-cited reasons, is often called *Selberg's upper bound sieve*.

Theorem A.3 │Selberg's Sieve│

Let $\mathcal{P}$ be a finite set of primes, P denoting their product, $S \subseteq \mathbb{N}$ with $|S| = N \in \mathbb{N}$, where the elements of S satisfies Selberg's condition (A.5), and let

$$S = |S(S, \mathcal{P}, x)|$$

be the number of elements of S not divisible by primes $p \in \mathcal{P}$ with $p \leq x$ where $x > 1$. If for $p \mid P$, we have that $f(p) > 1$,

$$g(n) = \prod_{d \mid n} \frac{\mu(n/d) d}{f(d)},$$

and

$$Q_x = \sum_{\substack{d \mid P \\ d \leq x}} g^{-1}(d),$$

then

$$S \leq \frac{N}{Q_x} + x^2 \prod_{\substack{p \in \mathcal{P} \\ p \leq x}} \left(1 - \frac{f(p)}{p}\right)^{-2}.$$

Proof. See [75, Theorem 4.4, p. 158]. ☐

An application of Theorem A.3 is the following, where $\pi(x; k, \ell)$ denotes the number of primes $p \leq x$ such that $p \equiv \ell \pmod{k}$. In the notation of Theorem A.3, we have that

$$\mathcal{P} = \{p \in \mathbb{P} : p \nmid k \text{ and } p \leq \sqrt{x}\}.$$

Also,

$$S = \{y = kn + \ell : n \in \mathbb{N} \text{ and } y \leq x\}.$$

Then $N = \lfloor x/k \rfloor$,

$$S(\mathcal{S}, \mathcal{P}, x) = \pi(x; k, \ell) - \pi(\sqrt{x}; k, \ell) = \pi(x; , k, \ell) + O(\sqrt{x}).$$

It follows that $f(d) = 1$, $S_d = \lfloor N/d \rfloor + R_d$ with $|R_d| \leq 1$, $g(n) = \phi(n)$, and $Q_x = \sum_{x \geq d|P} \phi^{-1}(d)$.

Theorem A.4 $\boxed{\text{The Brun–Titchmarsh Theorem}}$

There exists a $C = C(\varepsilon) \in \mathbb{R}^+$ such that for $1 \leq q < x$ and $\gcd(k, \ell) = 1$, we have

$$\pi(x; k, \ell) \leq \frac{Cx}{\phi(k) \log_e(x/q)}.$$

Proof. See [75, Corollary, p. 161] and see Biography A.2 on page 376. □

Remark A.2 Theorem A.4 is known to hold when the constant $c = 2$. Moreover, if $1 \leq q \leq x^{1-\varepsilon}$ for $\varepsilon > 0$, then the upper bound is at the expected order of magnitude.

Another interpretation of Theorem A.4 is that if x, y are positive reals, and $k, \ell \in \mathbb{Z}$ with $y/k \to \infty$, then

$$\pi(x + y, k, \ell) - \pi(x, k, \ell) < \frac{(2 + o(1))y}{\phi(k) \log_e(y/k)}.$$

Yet another formulation is given as follows. There exists an effective constant $k > k_0(\varepsilon)$ such that

$$\pi(x + ky, k, \ell) - \pi(x, k, \ell) < \frac{(2 + \varepsilon)y}{\phi(k) \log_e y},$$

for all y, x, ℓ with $y > k$. The amazing aspect of Brun–Titchmarsh is that if we could replace 2 by $2 - \delta$ for any $\delta > 0$, then Landau–Siegel zeros cannot exist—see page 300.

Selberg's sieve also has applications to some other classical problems. For instance, the twin-prime conjecture may be interpreted as follows. Suppose that $f(d)$ represents the number of elements of

$$\{n(n + 2) : d \mid n(n + 2) \text{ where } 1 \leq n \leq d\}$$

which are divisible by d and for some $m \in \mathbb{N}$,

$$\mathcal{S} = \{j(j + 2) : j = m, m + 1, \ldots, m + N - 1\}.$$

Let $\pi_2(N)$ be the number of twin primes less than N, from which it follows that

$$\pi_2(N) \leq |S(\mathcal{S}, \mathcal{P}, N^{1/3})| + N^{1/3}$$

because if $p \leq N$ has a twin prime, then either $p \leq N^{1/3}$ or else $p(p+2)$ has no prime factor $\leq N^{1/3}$. Thus, using Selberg's sieve to estimate $|S(\mathcal{S}, \mathcal{P}, N^{1/3})|$, we have $f(2) = 1$ and $f(p) = 2$ for odd primes p. We claim that

$$\prod_{p \leq N^{1/3}} \left(1 - \frac{f(p)}{p}\right)^{-1} << (\log_e N)^2.$$

This follows from the fact that for $p > 3$,

$$\left(1 - \frac{2}{p}\right)^{-1} \leq \left(1 - \frac{1}{p}\right)^{-2} \left(1 - \frac{2}{p^2}\right)^{-1}$$

and the fact that

$$\prod_{p \leq N^{1/3}} \left(1 - \frac{1}{p}\right)^{-1} << \log_e N^{1/3},$$

which, in turn, follows from Merten's Theorem 5.12 on page 222, keeping in mind that $\prod_{p \leq N^{1/3}} (1 - 2p^{-2})^{-1}$ converges. One may also deduce a lower bound as follows,

$$\sum_{\substack{d \leq N^{1/3} \\ d \text{ odd}}} \frac{f(d)}{d} \geq (\log_e N)^2.$$

Putting this all together via Theorem A.3 on page 373, we get the following.

Theorem A.5 | Selberg's Sieve on Twin Primes

The number $\pi_2(N)$ of twin primes less than N satisfies

$$\pi_2(N) << \frac{N}{(\log_e N)^2}.$$

Remark A.3 With the above application of Selberg's sieve, it is certainly worth mentioning another highlight of sieve theory with respect to the twin-prime conjecture, namely *Chen's Theorem*, which shows that there are infinitely many primes p such that $p+2$ is either prime or a product of two primes. Again, sieve methods allowed a result that is within a hair of the affirmation of another classical conjecture.

Another of the list of conjectures from our discussion at the outset is the Goldbach conjecture. Now we look at applications of Selberg's sieve to this classical problem. To this end, let $N = 2m$ for $m \in \mathbb{N}$, and for some $k \in \mathbb{N}$,

$$\mathcal{S} = \{j(N - j) : j = k, k+1, \ldots, k+N-1\},$$

and let $\mathfrak{r}(N)$ be the number of representations of N as a sum of two primes. Also, $f(d)$ is the number of elements of

$$\{n(N - n) : n = 1, 2, \ldots, d\}$$

divisible by d. It follows that

$$\mathfrak{r}(N) \leq |S(\mathcal{S}, \mathcal{P}, N^{1/3})| + 2N^{1/3}.$$

Thus, $f(p) = 2$ if $p \nmid N$ and $f(p) = 1$ if $p \mid N$. Applying Theorem A.3, and arguing in a similar fashion to the above, we get the following, a complete proof of which may be found in [75, Theorem 4.6, p. 162].

Theorem A.6 | **Selberg's Sieve on the Goldbach Conjecture**

For $N \in \mathbb{N}$,

$$\mathfrak{r}(N) << \frac{N}{(\log_e N)^2} \prod_{p \mid d} \left(1 + \frac{2}{p}\right).$$

Biography A.2 Edward Charles Titchmarsh (1899–1963) *was born in Newbury, Berkshire, England on June 1, 1899. At the early age of seventeen, he won an Open Mathematical Scholarship to Balliol College, Oxford. In October of 1917, he began his studies at that college. When he turned eighteen, he was inducted into the service in World War I, becoming a dispatch rider in France. He served until after the war, and returned to his studies at Oxford in October of 1919. While there he was taught by G.H. Hardy, who had a profound influence on Titchmarsh, including their shared passion for cricket. He graduated in 1922, and, in the following year, won the Prize Fellowship at Magdalen College Oxford. He also held a Senior Lecturer position at University College in London at the same time. Eventually, he was appointed to succeed Hardy for the Savilian chair at Oxford when Hardy left for Cambridge. All of Titchmarsh's work was in analysis, including work on the Riemann ζ-function. Arguably, his most important, and certainly his most popular, book was published in 1932,* The Theory of Functions. *His work had influence on diverse areas including quantum mechanics, via his work on series expansions of eigenfunctions of differential equations. Indeed, the latter topic consumed a quarter century of his professional life. He published a significant amount of that work in* Eigenfunction expansions associated with second-order differential equations *in the late 1940s and 1950s. Among the honours received in his lifetime were: election to the Royal Society in 1931, being awarded the De Morgan Medal in 1953, winning the Sylvester Medal in 1955, and although he did not formally study to receive a doctorate, he was awarded an honourary one by the University of Sheffield in 1953. He died in Oxford, Oxfordshire on January 18, 1963.*

We have amply illustrated the applications of Selberg's sieve to a variety of classical problems. It is now time to look at other sieves and their contributions. One of these is due to Linnik [52] first produced in 1941—see Biography A.6 on page 384. To understand what it says, we provide a preamble that takes into account what we have learned thus far. Brun's result, Theorem A.2 on page 371, may be interpreted as a generalization of Eratosthenes' sieve as follows. Take

$1, 2, \ldots, n$ and for each prime $p \leq \sqrt{n}$, we eliminate k residue classes modulo p, then the number remaining does not exceed $C(k)N/(\log_e^k n)$, where $C(k) > 0$ depends on k. Linnik considered a more general situation by considering for each prime $p \leq \sqrt{n}$, and eliminating $f(p)$ classes modulo p where $f(p)$ gets large as p does. Linnik called this the *large sieve*. This is formalized in terms of the notation we have developed herein as follows.

Theorem A.7 $\boxed{\textbf{The Large Sieve Inequality}}$

Suppose that $N \in \mathbb{N}$ and for every prime $p \leq \sqrt{N}$, let $f(p)$ residue classes modulo p be given, where $0 \leq f(p) < p$. If I_N is any interval of natural numbers of length N, then in I_N there are at most

$$\frac{(1 + \pi)N}{\sum_{p \leq \sqrt{N}} f(p)/(p - f(p))}$$

integers not lying in any of the given residue classes.

Proof. See [75, Corollary 2, p. 170]. $\qquad\square$

The large sieve can be applied to Artin's conjecture, one of the classical problems from our list at the outset. From the large sieve Theorem A.7, we have the following.

Theorem A.8 $\boxed{\textbf{The Large Sieve on Artin's Conjecture}}$

Let I_N be an interval of natural numbers of length $N \in \mathbb{N}$ and let

$$\mathfrak{C}(N) = \left| \left\{ n \in I_N : n \text{ is not a primitive root modulo for any prime } p \leq \sqrt{N} \right\} \right|.$$

Then

$$\mathfrak{C}(N) << \sqrt{N} \log_e(N).$$

Proof. See [75, Theorem 4.8, p. 171]. $\qquad\square$

Corollary A.2 *Almost every $n \in \mathbb{N}$ is a primitive root for some prime.*

Using the large sieve, Bombieri [6] and Vinogradov [98] independently found a result on distribution of primes in arithmetic progression that is quite pleasant—see Biography A.3 on page 380. In the next result, we use the following. The (basic) *Mangoldt function* is given by

$$\Lambda(n) = \log_e p \text{ if } n = p^a \text{ for some prime and } p, a \in \mathbb{N}, \text{ and } \Lambda(n) = 0 \text{ otherwise.}$$

In the

Theorem A.9 | **The Bombieri–Vinogradov Theorem**

For any real number $A > 0$, there is a constant $B = B(A)$ such that, for $Q = \sqrt{x}(\log_e x)^{-B}$,

$$\sum_{q \leq Q} \max_{y \leq x} \max_{a \in (\mathbb{Z}/q\mathbb{Z})^*} \left| \psi(y; q, a) - \frac{y}{\phi(q)} \right| << \frac{x}{(\log_e x)^A}, \tag{A.7}$$

where

$$\psi(x; q, a) = \sum_{\substack{n \leq x \\ n \equiv a \,(\text{mod } q)}} \Lambda(n).$$

In keeping with the above, we now show how some classical problems can be tackled with Theorem A.9. If $\tau(x)$ is the number of divisors function, and $n \in \mathbb{N}$, is fixed, then the *Titchmarsh divisor problem* is to compute the order of the function

$$S(x) = \sum_{p \leq x} \tau(p + n)$$

—see page 208. Theorem A.9 can be applied to this problem to get the following—see [75, Theorem 5.11, p. 202] for a related result.

Theorem A.10 | **Bombieri–Vinogradov Applied to Titchmarsh**

For any $n \in \mathbb{N}$, there exists a constant $c \in \mathbb{R}^+$ such that

$$S(x) = cx + O\left(\frac{x \log_e \log_e x}{\log_e x} \right).$$

This establishes more than that proved by Titchmarsh [96] in 1930, wherein he showed that $S(x) = O(x)$.

Bombieri also provided a sieve, essentially generalizing the Selberg sieve, that was highly useful in establishing another highlight of sieve theory. To describe this and the application, we need the following notions. If (A.7) holds for any $A > 0$ and any $\varepsilon > 0$ with $Q = x^{\nu - \varepsilon}$, then we say the primes have *level of distribution* ν. Thus, according to Theorem A.9, the primes are known to have level of distribution $\nu = 1/2$. The *Elliott–Halberstam conjecture* says the primes have level of distribution $\nu = 1$. This remains open.

The *generalized Mangoldt function* is given by

$$\Lambda_k(n) = \sum_{d|n} \mu(d) \log_e^k(n/d).$$

Now let $\{a_n\}_{n=1}^\infty$ be a sequence of positive real numbers,

$$A(x) = \sum_{n \leq x} a_n, \text{ and } H = \prod_p (1 - f(p))(1 - 1/p)^{-1},$$

for a multiplicative function f, then the following, proved by Bombieri in 1976—see [8]—under the assumption of the validity of the Elliott–Halberstam conjecture, is called the *asymptotic sieve*, where $k \geq 2$

$$\sum_{n \leq x} a_n \Lambda_k(n) \sim k H A(x) (\log_e x)^{k-1}. \tag{A.8}$$

The case $k = 2$ and $a_n = 1$ for all n is essentially Selberg's sieve.

The most striking application to date of (A.8) was achieved by Friedlander and Iwaniec in 1998—see [29]–[30]—when they proved the following—see Biographies A.4 on page 381 and A.5 on page 382.

Theorem A.11 $\boxed{\textbf{The Friedlander–Iwaniec Theorem}}$

There are infinitely many primes of the form $a^2 + b^4$.

We have covered an overview of some of the successes of sieve methods, but there are weaknesses. In particular, sieve methods cannot, in general, distinguish between numbers with an even number of prime factors and an odd number of prime factors, which is called the *parity problem*. Bombieri's sieve clarified some of this issue in [7]–[8], by showing that his sieve implies an asymptotic formula for

$$\sum_{n \leq x} a_n F(n)$$

precisely when a function F provides what is called *equal weight* to integers with an even number of prime factors and those with an odd number of prime factors. It turns out that the generalized Mangoldt functions have exactly this property for $k > 1$. Of course, the parity problem remains, but the above strides and applications are indicative of the power of sieve methods.

It is worth pointing out, before we turn to another topic, that the Elliot–Halberstram conjecture implies some fascinating recent results for gaps between primes as well as implications for the twin-prime conjecture. These were found by Goldston, Pintz, and Yildirim in 2005—see [31]–[33]. For the following statement recall that the *infimum of a set S* is the greatest lower bound of S and is denoted inf(S). Also, the *limit inferior*, denoted by lim inf, is given by

$$\liminf_{n \to \infty} a_n = \lim_{n \to \infty} \left(\inf_{m \geq n} a_m \right)$$

for a sequence $\{a_n\}$.

The first result is unconditional.

Theorem A.12 $\boxed{\textbf{Unconditional Goldston–Pintz–Yildirim}}$

If p_n denotes the n-th prime, then

$$\liminf_{n \to \infty} \frac{p_{n+1} - p_n}{\sqrt{\log_e p_n} (\log_e \log_e p_n)^2} < \infty.$$

Also, if $\{a_n\}$ is a sequence of natural numbers satisfying that

$$|\{a_n : n \leq N\}| > C(\log_e N)^{1/2}(\log_e \log_e N)^2$$

for all sufficiently large N, then infinitely many of the differences of two elements of $\{a_n\}$ can be expressed as the difference of two primes.

The following is the conditional result.

Theorem A.13 | **The Conditional Goldston–Pintz–Yildirim Theorem**

If the Elliott–Halberstam conjecture is true, then

$$\liminf_{n\to\infty} p_{n+1} - p_n \leq 16.$$

Remark A.4 It is worth noting that, in joint work with S. Graham, Goldston, Pintz, and Yildirim proved that if q_n is the n-th natural number with exactly two prime factors, then under the assumption of a generalized Elliot–Halberstram conjecture:

$$\liminf_{n\to\infty} q_{n+1} - q_n \leq 6$$

–see: http://www.math.boun.edu.tr/instructors/yildirim/yildirim.htm.

Biography A.3 Enrico Bombieri (1940–) *was born in Milan, Italy on November 26, 1940. He achieved his doctoral degree at the University of Milan in 1963. In 1966, he was appointed to a chair in mathematics at the University of Pisa. He also taught at the Scuola Normale Superiore at Pisa. He was awarded the Field's medal in Vancouver in 1974 for his work in the study of the theory of functions of several complex variables, the study of primes, as well as to partial differential equations and minimal surfaces. Bombieri's large sieve methods improved upon the methods of Rényi, who had in turn extended the sieve method developed by Linnik—see Biography A.6 on page 384. Theorems A.8–A.10 are a few examples of the applicability of Bombieri's large sieve method. In 1980, Bombieri was awarded the Balzan International Prize, and in 1984, he was elected as a foreign member of the French Academy of Sciences. He is also a foreign member of the Royal Swedish Academy, and the Academia Europea. In 1996, he was elected to be a member of the National Academy of Sciences. He currently works in the U.S.A. as the IBM Von Neumann Professor of Mathematics at the Institute for Advanced Study at Princeton, New Jersey, where he has been since 1977.*

Theorems A.12–A.13 are outcroppings of results on sieve methods that began with Selberg's sieve, which has been supplanted by other methods. Selberg's sieve applies to twin primes as we saw in Theorem A.5 on page 375. In 1997, Heath-Brown generalized Selberg's application to the problem of *almost primes*, which are natural numbers that are either prime or a product of two primes. The authors of Theorems A.12–A.13 used Heath-Brown's argument in ways that theretofore had not been applied to primes themselves and achieved these spectacular results. The description of the details of their method is described at the end of the paper [33].

Biography A.4 John Friedlander (1941–) *is a Canadian mathematician at the University of Toronto, who specializes in analytic number theory. In particular, he is considered to be a world leader in the theory of primes and Dirichlet L-functions. In 1965, he received his B.Sc. from the University of Toronto, his M.Sc. in 1966 from the University of Waterloo, and his Ph.D. in 1972 from Pennsylvania State University. He was a lecturer at M.I.T. from 1974 to 1976 and has been at the University of Toronto since 1977. He served as chair in the mathematics department from 1987 to 1991. He spent many years at the Institute for Advanced Study at Princeton and has collaborated with Bombieri among others—see Biography A.3 on the preceding page. He was elected as a member of the Royal Society in 1988. In 1997, he collaborated with Iwaniec to prove Theorem A.11 on page 379 using Bombieri's asymptotic sieve—see Biography A.5 on the next page. He has received the CRM-Fields Prize recognizing his achievements. In 1999, he was invited to give the Jeffery-Williams lecture to recognize his leadership in Canadian mathematics.*

We now turn to a powerful sieve that is used to great success in factoring. The following is adapted from [64].

In 1988, John Pollard circulated a manuscript that contained the outline of a new algorithm for factoring integers, which we studied in §2.3. In 1990, the first practical version of Pollard's algorithm was given in [51], published in 1993, the authors of which dubbed it the *number field sieve*. Pollard had been motivated by a discrete logarithm algorithm given in 1986, by the authors of [17], which employed quadratic fields. Pollard looked at the more general scenario by outlining an idea for factoring certain large integers using number fields. The special numbers that he considered are those large composite natural numbers that are "close" to being powers, namely those $n \in \mathbb{N}$ of the form $n = r^t - s$ for small natural numbers r and $|s|$, and a possibly much larger natural number t. Examples of such numbers, which the number field sieve had some successes factoring, may be found in tables of numbers of the form

$$n = r^t \pm 1, \text{ called } Cunningham \text{ } numbers.$$

However, the most noteworthy success was factorization of the ninth Fermat number $F_9 = 2^{2^9} + 1 = 2^{512} + 1$ (having 155 decimal digits), by the Lenstra brothers, Manasse and Pollard in 1990, the publication of which appeared in 1993 (see [50]).

To review some of the recent history preceding the number field sieve, we observe the following. Prior to 1970, a 25-digit integer was considered difficult to factor. In 1970, the power of the continued fraction method raised this to 50 digits (see [68, §5.4, pp. 240–242]). Once the algorithm was up and running in 1970, legions of 20- to 45-digit numbers were factored that could not be factored before. The first major success was the factorization of the seventh Fermat number

$$F_7 = 2^{2^7} + 1 = 2^{128} + 1,$$

a 39-digit number, which we described via Pollard's method in §2.3. By the mid 1980's, the quadratic sieve algorithm was felling 100-digit numbers. With the dawn of the number field sieve, 150-digit integers were now being tackled. The number field sieve is considered to be asymptotically faster than any known algorithm for the special class of integers of the above special form to which it applies. Furthermore, the number field sieve can be made to work for arbitrary integers. For details, see [13], where the authors refer to the number field sieve for the special number $n = r^t - s$ as the *special number field sieve*. The more general sieve has come to be known as the *general number field sieve*.

Biography A.5 Henryk Iwaniec (1947–) *is a Polish-American mathematician, who was born on October 9, 1947 in Elblag, Poland. He obtained his doctorate from the University of Warsaw in 1972 under the direction of Andrzej Schinzel. He was employed at the Institute of Mathematics of the Polish Academy of Sciences until 1983, when he left Poland for the U.S.A. He held visiting positions at the Institute for Advanced Study, the University of Michigan, and the University of Colorado at Boulder. Then he went to Rutgers University, where he has been a professor since 1987. In 1997, he and John Friedlander proved Theorem A.11 on page 379 using Bombieri's asymptotic sieve—see Biographies A.3 on page 380 and A.4 on page 381. For this he was awarded the Ostrowski Prize in 2001, where the citation mentioned his "profound understanding of the difficulties of the problem." In 2002 he was awarded the fourteenth Frank Nelson Cole Prize in number theory. He has contributed many results to analytic number theory, but in particular to modular forms on the general linear group and to sieve methods.*

Much older than any of the aforementioned ideas for factoring is that attributed to Fermat, namely the writing of n as a difference of two squares. However, this idea was enhanced by Maurice Kraitchik in the 1920's. Kraitchik reasoned that it might suffice to find a *multiple of n* as a difference of squares, namely,

$$x^2 \equiv y^2 \pmod{n}, \tag{A.9}$$

so that one of $x - y$ or $x + y$ *could* be divisible by a factor of n. We say *could* here since we fail to get a nontrivial factor of n when $x \equiv \pm y \pmod{n}$. However, it can be shown that if n is divisible by at least two distinct odd primes, then for at least half of the pairs x (modulo n), and y (modulo n), satisfying (A.9) with $\gcd(x, y) = 1$, we will have

$$1 < \gcd(x - y, n) < n.$$

This classical idea of Kraitchik had seeds in the work of Gauss, but Kraitchik introduced it into a new century in the pre-dawn of the computer age. This idea is currently exploited by many algorithms via construction of these (x, y)-pairs. For instance, the aforementioned continued fraction, and quadratic sieve algorithms use it. More recently, the number field sieve exploits the idea. To see how this is done, we give a brief overview of the methodology of the number field sieve. This will motivate the formal description of the algorithm.

For $n = r^t - s$, as above, we wish to choose a number field of degree d over $\mathbb{Q}$. The following choice for d is made for reasons (which we will not discuss here), which makes it the optimal selection, at least theoretically. (The interested reader may consult [51, Sections 6.2–6.3, pp. 31–32] for the complexity analysis and reasoning behind these choices.) Set

$$d = \left(\frac{(3 + o(1)) \log n}{2 \log \log n} \right)^{1/3}. \tag{A.10}$$

Now select $k \in \mathbb{N}$, which is minimal with respect to $kd \geq t$. Therefore,

$$r^{kd} \equiv sr^{kd-t} \pmod{n}.$$

Set

$$m = r^k, \text{ and } c = sr^{kd-t}. \tag{A.11}$$

Then

$$m^d \equiv c \pmod{n}.$$

Set

$$f(x) = x^d - c,$$

and let $\alpha \in \mathbb{C}$ be a root of f. Then this leads to a choice of a number field, namely $F = \mathbb{Q}(\alpha)$. Although the number field sieve can be made to work when $\mathbb{Z}[\alpha]$ is *not* a UFD, the assumption that it *is* a UFD simplifies matters greatly in the exposition of the algorithm, so we will make this assumption. Note that once made, this assumption implies that

$$\mathfrak{O}_F = \mathbb{Z}[\alpha].$$

See [51] for a description of the modifications necessary when it is not a UFD.

Now the question of the irreducibility of f arises. If f is reducible over $\mathbb{Z}$, we are indeed lucky, since then

$$f(x) = g(x)h(x), \text{ with } g(x), h(x) \in \mathbb{Z}[x],$$

where $0 < \deg(g) < \deg(f)$. Therefore,

$$f(m) = n = g(m)h(m)$$

is a nontrivial factorization of n, and we are done. Use of the number field sieve is unnecessary. However, the probability is high that f is irreducible since *most primitive polynomials over $\mathbb{Z}$ are irreducible*. Hence, for the description of the number field sieve, we may assume that f is irreducible over $\mathbb{Z}$.

Biography A.6 Yuri Vladimirovich Linnik (1915–1972) *was born in Belaya Tserkov, Ukraine on January 21, 1915. His university studies began in 1932 when he entered Leningrad University, from which he graduated in 1938. He began studying for his doctorate under the guidance of Vladimir Tartakovski, and produced a thesis on quadratic forms that earned him the higher degree of D.Sc. in Mathematics and Physics. In April of 1940, the Leningrad branch of Steklov Institute for Mathematics was formed and Linnik began working there from the outset. At this time the German army was approaching Leningrad and Linnik was involved in the fighting in Kazan. When the siege of Leningrad ended in 1944, he returned to the Steklov Institute. He was also appointed as professor of mathematics at Leningrad State University, and he stayed in Leningrad for the rest of his life, working on number theory, probability, and statistics. One of his contributions to the analytic theory of quadratic forms was to introduce ergodic methods into its study. In 1941, he published a paper* [52] *which introduced his* large sieve. *He used this term to describe the method of eliminating some residue classes modulo a prime from a given set of integers where the number of classes (possibly) increased when the prime increased. He was motived to create his sieve in order to tackle Vinogradov's hypothesis, which postulated that the size* n_p *of the smallest quadratic nonresidue modulo a given prime p is* $O(p^e)$ *for any* $e > 0$*. He was able to use his sieve to show that the number of primes* $p < x$ *for which* $n_p > p^e$ *is* $O(\log_e \log_e x)$*. Linnik's results using his sieve naturally led him to study Dirichlet L-functions, where he generalized density theorems to them. His interest in probability theory also led him to introduce the dispersion method into number theory. In 1959, he used his method to prove that any sufficiently large integer can be represented as the sum of a prime and two squares of integers—see* [53]*. He also solved problems in statistics and applied his methods to number-theoretic problems. He was highly talented outside of mathematics as well, speaking seven languages fluently and had interests in poetry and history. Among his honours were: election to the International Statistical Institute, the Academy of Sciences of the USSR in 1964, being awarded the State Prize in 1947, and the Lenin Prize in 1970. He was also awarded an honourary doctorate from the University of Paris. He died on June 30, 1972 in Leningrad, now St. Petersburg, Russia.*

Biography A.7 Maurice Kraitchik (1882–1957) *was born on April 21, 1882 in Minsk, capital of the former Belorussian Soviet Socialist Republic. From 1915 to 1948, he was an engineer in Brussels, Belgium and also held a directorship at the Mathematical Sciences section of the Mathematics Institute of Advanced Studies there. From 1941 to 1946, he was Associate Professor at the New School for Social Research in New York. He died on August 19, 1957 in Brussels.*

Since $f(m) \equiv 0 \pmod{n}$, we may define the natural homomorphism,

$$\psi : \mathbb{Z}[\alpha] \mapsto \mathbb{Z}/n\mathbb{Z},$$

given by
$$\alpha \mapsto \overline{m} \in \mathbb{Z}/n\mathbb{Z}.$$

Then

$$\psi\left(\sum_j a_j \alpha^j\right) = \sum_j a_j \overline{m}^j.$$

Now define a set $\mathcal{S}$ consisting of pairs of relatively prime integers (a, b), satisfying the following two conditions:

$$\prod_{(a,b)\in\mathcal{S}} (a + bm) = c^2, \quad (c \in \mathbb{Z}), \tag{A.12}$$

and

$$\prod_{(a,b)\in\mathcal{S}} (a + b\alpha) = \beta^2, \quad (\beta \in \mathbb{Z}[\alpha]). \tag{A.13}$$

Thus,

$$\psi(\beta^2) = \overline{c}^2,$$

so

$$\psi(\beta^2) \equiv c^2 \pmod{n}.$$

In other words, since $\psi(\beta^2) = \psi(\beta)^2$, then if we set $\psi(\beta) = h \in \mathbb{Z}$,

$$h^2 \equiv c^2 \pmod{n}.$$

This takes us back to Kraitchik's original idea, and we may have a nontrivial factor of n, namely $\gcd(h \pm c, n)$ (provided that $h \not\equiv \pm c \pmod{n}$).

The above overview of the number field sieve methodology is actually a special case of an algebraic idea, which is described as follows. Let R be a ring with homomorphism

$$\phi : R \mapsto \mathbb{Z}/n\mathbb{Z} \times \mathbb{Z}/n\mathbb{Z},$$

together with an algorithm for computing nonzero diagonal elements (x, x) for $x \in \mathbb{Z}/n\mathbb{Z}$. Then the goal is to multiplicatively combine these elements to obtain squares in R whose square roots have an image under ϕ not lying in $(x, \pm x)$ for nonzero $x \in \mathbb{Z}/n\mathbb{Z}$. The number field sieve is the special case

$$R = \mathbb{Z} \times \mathbb{Z}[\alpha], \text{ with } \phi(z, \beta) = (\overline{z}, \psi(\beta)).$$

Before setting down the details of the formal number field sieve algorithm, we discuss the crucial role played by *smoothness* introduced in Definition 2.21 on page 93. Recall that a smooth number is one with only "small" prime factors. In particular, $n \in \mathbb{N}$ is B-smooth for $B \in \mathbb{R}^+$, if n has no prime factor bigger than B. Smooth numbers satisfy the triad of properties:

(1) They are fairly numerous (albeit sparse).

(2) They enjoy a simple multiplicative structure.

(3) They play an essential role in discrete logarithm algorithms.

If $F = \mathbb{Q}(\alpha)$ is a number field, then by definition

an algebraic number $a + b\alpha \in \mathbb{Z}[\alpha]$ is B-smooth if $|N_F(a + b\alpha)|$ is B-smooth.

Hence, $a + b\alpha$ is B-smooth if and only if all primes dividing $|N_F(a + b\alpha)|$ are less than B. Thus, the idea behind the number field sieve is to look for small relatively prime numbers a and b such that both $a + \alpha b$ and $a + \overline{m}b$ are smooth. Since $\psi(a + \alpha b) = a + \overline{m}b$, then each pair provides a congruence modulo n between two products. Sufficiently many of these congruences can then be used to find solutions to $h^2 \equiv c^2 \pmod{n}$, which may lead to a factorization of n.

The above overview leaves open the demanding questions as to how we choose the degree d, the integer m, and how the set of relatively prime integers a, b such that Equations (A.12)–(A.13) can be found. These questions may now be answered in the following formal description of the algorithm.

✦ The Number Field Sieve Algorithm

Step 1. (Selection of a Factor Base and Smoothness Bound)
There is a consensus that smoothness bounds are best chosen empirically. However, there are theoretical reasons for choosing such bounds as

$$B = \exp((2/3)^{2/3}(\log n)^{1/3}(\log \log n)^{2/3}),$$

which is considered to be optimal since it is based upon the choice for d as above. See [51, Section 6.3, p. 32] for details. Furthermore, the reasons for this being called a smoothness bound will unfold in the sequel.

Define a set $\mathcal{S} = \mathcal{S}_1 \cup \mathcal{S}_2 \cup \mathcal{S}_3$, where the component sets $\mathcal{S}_j$ are given as follows. $\mathcal{S}_1 = \{p \in \mathbb{Z} : p \text{ is prime and } p \leq B\}$,

$$\mathcal{S}_2 = \{u_j : j = 1, 2, \ldots, r_1 + r_2 - 1, \text{ where } u_j \text{ is a generator of } \mathcal{U}_F\}.$$

(Here $\{r_1, r_2\}$ is the signature of F, and the generators u_j are the generators of the infinite cyclic groups given by Dirichlet's Unit Theorem—see [64, Theorem 2.78, p. 114].) Also,

$$\mathcal{S}_3 = \{\beta = a + b\alpha \in \mathbb{Z}[\alpha] : |N_F(\beta)| = p < B_2 \text{ where } p \text{ is prime }\},$$

where B_2 is chosen empirically. Now we set the factor base as

$$\mathcal{F} = \{a_j = \psi(j) \in \mathbb{Z}/n\mathbb{Z} : j \in \mathcal{S}\}.$$

Also, we may assume $\gcd(a_j, n) = 1$ for all $j \in \mathcal{S}$, since otherwise we have a factorization of n and the algorithm terminates.

Step 2. (Collecting Relations and Finding Dependencies)

We wish to collect relations (A.12)–(A.13) such that they occur simultaneously, thereby yielding a potential factor of n. One searches for relatively prime pairs (a, b) with $b > 0$ satisfying the following two conditions.

(i) $|a + bm|$ is B-smooth except for at most one additional prime factor p_1, with $B < p_1 < B_1$, where B_1 is empirically determined.

(ii) $a + b\alpha$ is B_2-smooth except for at most one additional prime $\beta \in \mathbb{Z}[\alpha]$ such that $|N_F(\beta)| = p_2$ with $B_2 < p_2 < B_3$, where B_3 is empirically chosen.

The prime p_1 in (i) is called the *large prime*, and the prime p_2 in (ii) is called the *large prime norm*. Pairs (a, b) for which p_1 and p_2 do not exist (namely when we set $p_1 = p_2 = 1$) are called *full relations*, and are called *partial relations* otherwise. In the sequel, we will only describe the full relations since, although the partial relations are more complicated, they lead to relations among the factor base elements in a fashion completely similar to the ones for full relations. For details on partial relations, see [50, Section 5].

First, we show how to achieve relations in Equation (A.12), the "easy" part (relatively speaking). (This is called the *rational part*, whereas relations in Equation (A.13) are called the *algebraic part*.) Then we show how to put the two together. To do this, we need the following notion from linear algebra.

Every $n \in \mathbb{N}$ has an exponent vector $v(n)$ defined by $n = \prod_{j=1}^{\infty} p_j^{v_j}$, where p_j is the jth prime, only finitely many of the v_j are nonzero, and

$$v(n) = (v_1, v_2, \ldots) = (v_j)_{j=1}^{\infty}$$

with an infinite string of zeros after the last significant place. We observe that n is a square if and only if each v_j is even. Hence, for our purposes, the v_j give *too much* information. Thus, to simplify our task, we reduce each v_j modulo 2. Henceforth, then $\overline{v_j}$ means v_j reduced modulo 2. We modify the notion of the exponent vector further for our purposes by letting $B_1 = \pi(B)$, where $\pi(B)$ is the number of primes no bigger than B. Then, with $p_0 = -1$, $a + bm = \prod_{j=0}^{B_1} p_j^{v_j}$ is the factorization of $a + bm$. Set

$$v(a + bm) = (\overline{v_0}, \ldots, \overline{v_{B_1}}),$$

for each pair (a, b) with $a + b\alpha \in S_3$. The choice of B allows us to make the assumption that $|S_3| > B_1 + 1$. Therefore, the vectors in $v(a+bm)$ for pairs (a, b) with $a + b\alpha \in S_3$ exceed the dimension of the $\mathbb{F}_2$-vector space $\mathbb{F}_2^{B_1+1}$. In other words, we have *more than* $B_1 + 1$ vectors in a $B_1 + 1$-dimensional vector space. Therefore, there exist nontrivial linear dependence relations between vectors. This implies the existence of a subset $\mathcal{J}$ of S_3 such that

$$\sum_{a+b\alpha \in \mathcal{J}} v(a + bm) = 0 \in \mathbb{F}_2^{B_1+1},$$

so

$$\prod_{a+b\alpha \in \mathcal{J}} (a + bm) = z^2 \quad (z \in \mathbb{Z}).$$

This solves Equation (A.12).

Now we turn to the algebraic relations in Equation (A.13). We may calculate the norm of $a + b\alpha$ by setting $x = a$ and $y = b$ in the homogeneous polynomial

$$(-y)^d f(-x/y) = x^d - c(-y)^d,$$

with $f(x) = x^d - c$. Therefore, $N_F(a + b\alpha) = (-b)^d f(-ab^{-1}) = a^d - c(-b)^d$. Let

$$R_p = \{r \in \mathbb{Z} : 0 \le r \le p - 1, \text{ and } f(r) \equiv 0 \pmod{p}\}.$$

Then for relatively prime pairs (a, b), we have

$$N_F(a + b\alpha) \equiv 0 \pmod{p} \text{ if and only if } a \equiv -br \pmod{p},$$

and this r is unique. Observe that by the relative primality of a and b, the multiplicative inverse b^{-1} of b modulo p is defined since, for $b \equiv 0 \pmod{p}$, there are no nonzero pairs (a, b) with $N_F(a + b\alpha) \equiv 0 \pmod{p}$.

The above shows that there is a one-to-one correspondence between those $\beta \in \mathbb{Z}[\alpha]$ with $|N_F(\beta)| = p$, a prime and pairs (p, r) with $r \in R_p$. Note that the kernel of the natural map

$$\psi : \mathbb{Z}[\alpha] \mapsto \mathbb{Z}/p\mathbb{Z} \text{ is } \ker(\psi) = \langle a + b\alpha \rangle,$$

the cyclic subgroup of $\mathbb{Z}[\alpha]$ generated by $a + b\alpha$. It follows that

$$|\mathbb{Z}[\alpha] : \langle a + b\alpha \rangle| = |N_F(a + b\alpha)| = p,$$

so $\mathbb{Z}[\alpha]/\langle a + b\alpha \rangle$ is a field.

This corresponds to saying that the $\mathbb{Z}[\alpha]$- ideal $\mathcal{P} = (a + b\alpha)$ is a principal, first-degree prime $\mathbb{Z}[\alpha]$-ideal, namely one for which $N_F(\mathcal{P}) = p^1 = p$. Hence, $\mathbb{Z}[\alpha]/\mathcal{P} \cong \mathbb{F}_p$, the finite field of p elements.

The above tells us that in Step 1 of the number field sieve algorithm , the set $\mathcal{S}_3$ essentially consists of the first-degree prime $\mathbb{Z}[\alpha]$-ideals of norm $N_F(\mathcal{P}) \le B_2$. These are the *smooth, degree one, prime* $\mathfrak{O}_F$-*ideals*, namely those ideals whose prime norms are B_2-smooth.

In part (ii) of Step 2 of the algorithm on page 387, the additional prime element $\beta \in \mathbb{Z}[\alpha]$ such that $|N_F(\beta)| = p_2$ with $B_2 < p_2 < B_3$ corresponds to the prime $\mathfrak{O}_F$-ideal $\mathcal{P}_2$ called the *large prime ideal*. Moreover, $\mathcal{P}_2$ corresponds to the pair $(p_2, c \pmod{p_2})$, where $c \in \mathbb{Z}$ is such that $a \equiv -bc \pmod{p_2}$, thereby enabling us to distinguish between prime ideals of the same norm. If the large prime in Step 2 does not occur, we write $\mathcal{P}_2 = (1)$. Now, since

$$|a + bm| = \prod_{p \in \mathcal{S}_1} p^{v_p},$$

and

$$|a + b\alpha| = \prod_{u \in \mathcal{S}_2} u^{t_u} \prod_{s \in \mathcal{S}_3} s^{v_s}, \tag{A.14}$$

for nonnegative $t_u, v_s \in \mathbb{Z}$, and since $\psi(a + bm) = \psi(a + b\alpha)$, then

$$\prod_{p \in S_1} \psi(p)^{v_p} = \prod_{u \in S_2} \psi(u)^{t_u} \prod_{s \in S_3} \psi(s)^{v_s},$$

in $\mathbb{Z}/n\mathbb{Z}$. Therefore, we achieve a relationship among the elements of the factor base $\mathcal{F}$, as follows

$$\prod_{u \in S_2} \psi(u)^{t_u} \prod_{s \in S_3} \psi(s)^{v_s} \equiv \prod_{p \in S_1} \psi(p)^{v_p} \pmod{n}. \tag{A.15}$$

Furthermore, we may translate (A.14) ideal-theoretically into the ideal product

$$|a + b\alpha| = \prod_{u \in S_2} u^{t_u} \prod_{\mathcal{P} \in S_3} \pi_{\mathcal{P}}^{v_{\mathcal{P}}}, \tag{A.16}$$

where $\mathcal{P}$ ranges over all of the first-degree prime $\mathbb{Z}[\alpha]$-ideals of norm less than B_2, and $\pi_{\mathcal{P}}$ is a generator of $\mathcal{P}$.

Thus, (A.15) gives rise to the identity

$$\prod_{p \in S_1} \psi(p)^{v_p} = \prod_{u \in S_2} \psi(u)^{t_u} \prod_{\mathcal{P} \in S_3} \psi(\pi_{\mathcal{P}})^{v_{\mathcal{P}}}.$$

If $|S_3| > \pi(B)$, then by applying Gaussian elimination for instance, we can find $x(a, b) \in \{0, 1\}$ such that simultaneously

$$\prod_{a + b\alpha \in S_3} (a + b\alpha)^{x(a,b)} = \left(\left(\prod_{u \in S_2} u^{\overline{t_u}} \right) \left(\prod_{s \in S_3} s^{\overline{v_s}} \right) \right)^2,$$

and

$$\prod_{a + b\alpha \in S_3} (a + bm)^{x(a,b)} = \left(\left(\prod_{p \in S_1} p^{\overline{v_p}} \right) \right)^2,$$

hold. From this a factorization of n may be gleaned, by Kraitchik's method.

Practically speaking, the number field sieve tasks consist of sieving all pairs (a, b) for $b = b_1, b_2 \ldots, b_n$ for short (overlapping) intervals $[b_1, b_2]$, with $|a|$ less than some given bound. All relations, full and partial, are gathered in this way until sufficiently many have been collected.

The big prize garnered by the number field sieve was the factorization of F_9, the ninth Fermat number, as described in [50]. In 1903, A.E. Western found the prime factor $2424833 = 37 \cdot 2^{16} + 1$ of F_9. Then in 1967, Brillhart determined that $F_9/2424833$ (having 148 decimal digits) is composite by showing that it fails to satisfy Fermat's Little Theorem. Thus, the authors of [50] chose

$$n = F_9/2424833 = \left(2^{512} + 1 \right)/2424833.$$

Then they exploited the above algorithm as follows. If we choose d as in Equation (A.10), we get that $d = 5$. The authors of [50] then observed that since $2^{512} \equiv -1 \,(\text{mod } n)$, then for $h = 2^{205}$, we get

$$h^5 \equiv 2^{1025} \equiv 2 \cdot \left(2^{512}\right)^2 \equiv 2 \quad (\text{mod } n).$$

This allowed them to choose the map

$$\psi : \mathbb{Z}[\sqrt[5]{2}] \mapsto \mathbb{Z}/n\mathbb{Z}, \text{ given by } \psi : \sqrt[5]{2} \mapsto 2^{205}.$$

Here $\mathbb{Z}[\sqrt[5]{2}]$ is a UFD. Then they chose m and c as in Equation (A.11), namely since $r = 2$, $s = -1$, and $t = 512$, then the minimal k with $5k = dk \geq t = 512$ is $k = 103$, and $m = 2^{103}$, so $c = -8 \equiv 2^{5 \cdot 103} \,(\text{mod } n)$. This gives rise to $f(x) = x^5 + 8$ with root $\alpha = -\sqrt[5]{2}^3$, and $\mathbb{Z}[\alpha] \subseteq \mathbb{Z}[\sqrt[5]{2}]$. Observe that

$$8F_9 = 2^{515} + 8 = \left(2^{103}\right)^5 + 8.$$

Thus,

$$\psi(\alpha) = m = 2^{103} \equiv -2^{615} \equiv -\left(2^{205}\right)^3 \quad (\text{mod } n).$$

Notice that 2^{103} is small in relation to n, and is in fact closer to $\sqrt[5]{n}$. Since

$$\psi(a + b\alpha) = a + 2^{103}b \in \mathbb{Z}/n\mathbb{Z},$$

we are in a position to form relations as described in the above algorithm. Indeed, the authors of [50] actually worked only in the subring $\mathbb{Z}[\alpha]$ to find their relations. The sets they chose from Step 1 are $\mathcal{S}_1 = \{p \in \mathbb{Z} : p \leq 1295377\}$,

$$\mathcal{S}_2 = \{-1, -1 + \sqrt[5]{2}, -1 + \sqrt[5]{2}^2 - \sqrt[5]{2}^3 + \sqrt[5]{2}^4\},$$

for units $u_1 = -1$, $u_2 = -1 + \sqrt[5]{2}$, and $u_3 = -1 + \sqrt[5]{2}^2 - \sqrt[5]{2}^3 + \sqrt[5]{2}^4$, and

$$\mathcal{S}_3 = \{\beta \in \mathbb{Z}[\alpha] : |N_F(\beta)| = p \leq 1294973, \ p \text{ a prime}\}.$$

The authors began sieving in mid-February of 1990 on approximately thirty-five workstations at Bellcore. On the morning of June 15, 1990 the first of the dependency relations that they achieved turned out to give rise to a trivial factorization! However, an hour later their second dependency relation gave way to a 49-digit factor. This and the 99-digit cofactor were determined by A. Odlyzko to be primes, on that same day. They achieved: $F_9 = q_7 \cdot q_{49} \cdot q_{99}$, where q_j is a prime with j decimal digits as follows:

$$q_7 = 2424833,$$

$$q_{49} = 7455602825647884208337395736200454918783366342657,$$

and $q_{99} = 741640062627530801524787141901937474059940781097519$

$$02390582131614441575950470500809281871169394 0737.$$

Fermat numbers have an important and rich history, which is intertwined with the very history of factoring itself. Euler was able to factor F_5. In 1880, Landry used an idea attributable to Fermat to factor F_6. As noted above, F_7 was factored by Pollard. Brent and Pollard used a version of Pollard's "rho"-method to factor F_8 (see [68, pp. 206–208] for a detailed description with examples of the rho-method). As we have shown above, F_9 was factored by the number field sieve. Lenstra's elliptic curve method was used by Brent to factor F_{10} and F_{11}—see §9.3. Several other Fermat numbers are known to have certain small prime factors, and the smallest Fermat number for which there is no known factor is F_{24}. For updates on the largest prime discoveries, see the website:

http://www.utm.edu/research/primes/largest.html.

We have covered several applications of sieve methods as well as their historical development. The power of the theory is clearly paramount, but the complete proofs of the results in this section would provide the foundation for a *third course in number theory*. Fittingly, we close our discussion here.

Bibliography

[1] D.J. Albers, *"Freeman Dyson: Mathematician, Physicist, and Writer"*: *Interview with Donald J. Albers*, College Math. J. **25** (1994), 3–21. (*Cited on page 155.*)

[2] R. Alter and K.K. Kubota, *The Diophantine equation $x^2 + D = p^n$*, Pacific J. Math. **46** (1973), 11–16. (*Cited on page 276.*)

[3] A. Baker, *Linear forms in logarithms of algebraic numbers*, Mathematica **13** (1966), pp. 204–216; **14** (1967), pp. 102–107; and **15** (1968), pp. 204–216. (*Cited on page 166.*)

[4] A. Baragar, *On the unicity conjecture for Markoff numbers*, Canad. Math. Bull. **39** (1996), 3–9. (*Cited on page 123.*)

[5] S. Beatty, *Problem 3173*, American Math. Monthly **33** (1926), 159. (*Cited on page 264.*)

[6] E. Bombieri, *On the large sieve*, Mathematika **12**, 201–225 (1965). (*Cited on page 377.*)

[7] E. Bombieri, *On twin-almost primes*, Acta Arith. **28** (1975), 177–193, 457–461. (*Cited on page 379.*)

[8] E. Bombieri, *The asymptotic sieve*, Mem. Acad. Naz. dei **XL** (1976), 243–269. (*Cited on page 379.*)

[9] E. Bombieri, *Roth's theorem and the abc-conjecture*, preprint ETH Zürich (1994). (*Cited on page 299.*)

[10] E. Bombieri, *The Mordell conjecture revisited*, Ann. Sc. Norm. Super. PisaCl. Sci **17** (1990), 615–640. (*Cited on page 299.*)

[11] C. Breuil, B. Conrad, F. Diamond, and R. Taylor, *On the modularity of elliptic curves over Q: Wild 3-adic exercises*, J. Amer. Math. Soc. **14** (2001), 843939. (*Cited on page 364.*)

[12] Y. Bugeaud and T.N. Shorey, *On the number of solutions of the generalized Ramanujan-Nagell equation*, J. für die Reine und Angew. Math. **539** (2001), 55–74. (*Cited on page 281.*)

[13] J.P. Buhler, H.W. Lenstra Jr., and C. Pomerance, *Factoring integers with the number field sieve*, in **The Development of the Number Field Sieve**, A.K. Lenstra and H. W. Lenstra Jr. (Eds.), Lecture Notes in Mathematics, Springer-Verlag, Berlin, Heidelberg, New York **1554** (1993), 50–94. (*Cited on page 382.*)

[14] H. Chatland and H. Davenport, *Euclid's algorithm in quadratic number fields*, Bulletin of the American Math. Society **55** (1949), 948–953. (*Cited on page 50.*)

[15] D.A. Clark, *A quadratic field which is Euclidean but not norm-Euclidean*, Manuscripta Mathematica **83** (1994), 327–330. (*Cited on page 50.*)

[16] B. Conrad, F. Diamond, and R. Taylor, *Modularity of certain potentially Barsotti-Tate Galois representations*, J. Amer. Math. Soc. **12** (1999), 521–567. (*Cited on page 364.*)

[17] D. Coppersmith, A. Odlyzko, and R. Schroeppel, *Discrete logarithms in $GF(p)$*, Algorithmica **I** (1986), 1–15. (*Cited on page 381.*)

[18] D.A. Cox, **Primes of the Form** $x^2 + ny^2$, Wiley, New York, (1989). (*Cited on pages 98, 100, 322, 325, 349–350.*)

[19] R. Crandall and C. Pomerance, **Prime Numbers: A Computational Perspective** Springer, New York, Berlin (2001). (*Cited on page 298.*)

[20] H. Darmon and A. Granville, *On the equations $z^m = F(x,y)$ and $Ax^p + By^q = Cz^r$*, Bull. London Math. Soc., **27** (1995), 513–543. (*Cited on page 295.*)

[21] H. Davenport, **The Work of K.F. Roth**, Proc. Int. Cong. Math. (1958), **LVII-LX** Cambridge University Press, 1960. (*Cited on page 160.*)

[22] J. dePhillis, **Mathematical Conversation Starters**, M.A.A., Washington, (2002). (*Cited on pages 67, 347.*)

[23] N.D. Elkies, *ABC implies Mordell*, Indagationes Math. **11** (2000), 197–200. (*Cited on page 299.*)

[24] P. Erdős, *How many pairs of products of consecutive integers have the same prime factors?*, Amer. Math. Monthly **87** (1980), 391–392. (*Cited on page 297.*)

[25] G. Faltings, *Diophantine approximations on abelian varieties*, Ann. Math. **133** (1991), 549–576. (*Cited on page 299.*)

[26] M. Van Frankenhuysen, *The abc-conjecture implies Roth's theorem and Mordell's conjecture*, Math. Contemp. **16** (1999), 45–72. (*Cited on page 299.*)

[27] G. Frey, *Links between stable elliptic curves and certain Diophantine equations*, Annales Universitatis Saraviensis, Series Mathematicae **1** (1986), 1–40. (*Cited on page 353.*)

[28] G. Frey and H.-G. Rück, *A remark concerning m-divisibility and the discrete logarithm problem in the divisor class group of curves*, Math. Comp. **62** (1994), 865–874. (*Cited on page 327.*)

[29] J. Friedlander and H. Iwaniec, *The polynomial $X^2 + Y^4$ captures its primes*, Annals of Math. **148** (1998), 945–1040. (*Cited on page 379.*)

[30] J. Friedlander and H. Iwaniec, *Asymptotic sieve for primes*, Annals of Math. **148** (1998), 1041–1065. (*Cited on page 379.*)

[31] D.A. Goldston, J. Pintz, and C.Y. Yildirim, *Primes in tuples I* (preprint (2005)-19 of http://aimath.org/preprints.html); to appear in Ann. of Math. (*Cited on page 379.*)

[32] D.A. Goldston, J. Pintz, and C.Y. Yildirim, *Primes in tuples II* (preprint, see:http://front.math.ucdavis.edu/author/D.Goldston). (*Cited on page 379.*)

[33] D.A. Goldston, J. Pintz, and C.Y. Yildirim, *The path to recent progress on small gaps between primes*, Clay Math. Proceed. **7** (2007). (*Cited on pages 379, 381.*)

[34] S. Goldwasser and J. Killian, *Almost all primes can be quickly certified*, Proceed. eighteenth annual ACM symp. on theory of computing (STOC), Berkely (1986), 316–329. (*Cited on page 324.*)

[35] A. Granville, *Some conjectures related to Fermat's last theorem* in **Number Theory** (R.A. Mollin, ed.) Walter de Gruyter, Berlin, New York (1990), 177-192. (*Cited on page 297.*)

[36] A. Granville and H. Stark, *abc implies no Siegel zeros for L-functions of characters with negative discriminant*, Invent. Math. **139** (2000), 509–523. (*Cited on page 300.*)

[37] M. Hall, *The Diophantine equation $x^3 - y^2 = k$*, in **Computers in Number Theory** (A. Atkin, B. Birch, eds.) Academic Press (1971). (*Cited on page 296.*)

[38] R. Harris, **Enigma**, Arrow Books, Random House, London (2001). (*Cited on page 47.*)

[39] D.R. Heath-Brown, *Artin's conjecture for primitive roots*, Quart. J. Math. Oxford **37** (1986), 27–38. (*Cited on page 369.*)

[40] K. Heegner, *Diophantische Analysis und Modulfunktionen*, Math. Zeitscr., **56** (1952), 227–253. (*Cited on page 141.*)

[41] H. Heilbronn, *On Euclid's algorithm in real quadratic fields*, Proc. Cambridge Philos. Soc. **34** (1938), 521–526. (*Cited on page 50.*)

[42] M. Hindy and J.H. Silverman, **Diophantine Geometry, an Introduction**, Springer, New York, (2000). (*Cited on page 299.*)

[43] N. Hofreiter, *Quadratische Körper mit und ohne Euklidischen Algorithmus*, Monatshefte für Mathematik und Physik **42** (1935), 397–400. (*Cited on page 50.*)

[44] J.P. Jones, D. Sato, H. Wada, and D. Wiens, *Diophantine representation of the set of prime numbers*, Amer. Math. Monthly **83** (1976), 449–464. (*Cited on page 295.*)

[45] A.W. Knapp, **Elliptic Curves**, Math. Notes **40**, Princeton University Press, Princeton, N.J. (1992). (*Cited on pages 330, 365.*)

[46] N. Koblitz, *Elliptic curve cryptosystems*, Math. Comp. **48** (1987), 203–209. (*Cited on page 326.*)

[47] N. Koblitz, **A Course in Number Theory and Cryptography**, Academic Press, New York, London (1988). (*Cited on pages 314, 320.*)

[48] E. Landau, *Über die Klassenzahl der binären quadratischen Formen von negativer Discriminante*, Math. Annalen **56** (1903), 671–676. (*Cited on page 102.*)

[49] H.W. Lenstra, *Factoring integers with elliptic curves*, Annals of Math. **126** (1987), 649–673. (*Cited on page 325.*)

[50] A.K. Lenstra, H.W. Lenstra, M.S. Manasse, and J.M. Pollard, *The factorization of the ninth Fermat number*, Math. Comp. **61** (1993), 319–349. (*Cited on pages 381, 387, 389–390.*)

[51] A.K. Lenstra, H.W. Lenstra Jr., M.S. Manasse, and J.M. Pollard, *The number field sieve*, in **The Development of the Number Field Sieve**, A.K. Lenstra, and H. W. Lenstra Jr. (Eds.), Lecture Notes in Mathematics, Springer-Verlag, Berlin, Heidelberg, New York **1554** (1993), 11–42. (*Cited on pages 381–383, 386.*)

[52] Yu. V. Linnik, *The large sieve*, Dokl. AN USSR **30** (1941), 290–292.[Russian] (*Cited on pages 376–384.*)

[53] Yu.V. Linnik, *The dispersion method in binary additive problems*, Amer. Math. Soc. (1963) (Translated from Russian). (*Cited on page 384.*)

[54] K. Mahler, *Lectures on transcendental numbers*, LNM **546**, Springer, Berlin, Heidelberg, New York, (1976). (*Cited on page 177.*)

[55] Y. Matiyasevich, *Enumerable sets are Diophantine*, Doklady Akad. Nauk SSSR **191** (1970), 279–282. [Russian] English translation in Soviet Mathematics, Doklady **11** (1970). (*Cited on page 295.*)

[56] A. Menezes, T. Okamoto, and S. A. Vanstone, *Reducing elliptic curve logarithms to logarithms in a finite field*, IEEE Trans. Inform. Theory, **39** (1993), 1639–1646. (*Cited on page 327.*)

[57] L. Merel, *Bornes pour la torsion des courbes elliptiques sur les corps de nombres*, Invent. Math. **124** (1996), 437–449. (*Cited on page 312.*)

[58] P. Mihăilescu, *Primary cyclotomic units and a proof of Catalan's conjecture*, J. Reine Angew. Math. **572** (2004), 167–195. (*Cited on page 294.*)

[59] V. Miller, *Use of elliptic curves in cryptography* in **Advances in Cryptography**—Crypto '85 Proceed., Springer-Verlag, Berlin, LNCS **218** (1987), 417–426. (*Cited on page 326.*)

[60] R.A. Mollin, **Number Theory and Applications**, Proceedings of the NATO Advanced Study Institute, Banff Centre, Canada, 27 April–5 May 1988, Kluwer Academic Publishers, Dordrecht (1989). (*Cited on page xiii.*)

[61] R.A. Mollin, **Number Theory**, Proceedings of the First Conference of the Canadian Number Theory Association, Banff Centre, Canada, April 17–27, 1988, Walter de Gruyter, Berlin (1990). (*Cited on page xiii.*)

[62] R.A. Mollin, **Quadratics**, CRC Press, Boca Raton, London, Tokyo (1995). (*Cited on pages 60, 65, 108, 256, 276.*)

[63] R.A. Mollin, *An elementary proof of the Rabinowitch-Mollin-Williams criterion for real quadratic fields*, J. Math. Sci. **7** (1996), 17–27. (*Cited on page 153.*)

[64] R.A. Mollin, **Algebraic Number Theory**, Chapman and Hall/CRC Press, Boca Raton, London, Tokyo (1999). (*Cited on pages 30, 63, 182, 189, 286, 291, 301, 344, 381, 386.*)

[65] R.A. Mollin, **Fundamental Number Theory with Applications**, *First Edition*, CRC, Boca Raton, London, New York (1998). (*Cited on pages 60, 153, 276.*)

[66] R.A. Mollin, **An Introduction to Cryptography**, First Edition (2001). (*Cited on page 326.*)

[67] R.A. Mollin, **Codes: The Guide to Secrecy from Ancient to Modern Times**, CRC, Taylor & Francis Group, Boca Raton, London, New York (2008). (*Cited on pages 205, 327.*)

[68] R.A. Mollin, **Fundamental Number Theory with Applications**, *Second Edition*, CRC, Taylor & Francis Group, Boca Raton, London, New York (2008). (*Cited on pages ix, 1, 10–12, 13, 15, 19, 21, 26–28, 40–41, 43, 47, 53, 55, 60, 63, 67, 79, 84, 88, 97–98, 102, 130, 132–133, 140, 152, 156, 159–160, 166, 167–168, 178, 182, 191, 198, 209, 213–214, 215, 221–222, 228–231, 236, 249, 260, 266, 271–272, 282, 291, 294, 324–327, 329, 338, 342, 370, 372, 381, 391, 429, 435.*)

[69] R.A. Mollin, *A note on the Diophantine equation* $D_1 x^2 + D_2 = ak^n$, Acta Math. Acad. Paed. Nyireg. **21** (2005), 21–24. (*Cited on page 281.*)

[70] R.A. Mollin, *Characterization of* $D = P^2 + Q^2$ *when* $\gcd(P, Q) = 1$ *and* $x^2 - Dy^2 = -1$ *has no integer solutions*, Far East J. Math. Sci. **32** (2009), 285–294 (*Cited on page 121.*)

[71] R.A. Mollin and P.G. Walsh, *A note on powerful numbers, quadratic fields, and the Pellian*, C.R. Math. Rep. Acad. Sci. Canada **8** (1986), 109–114. (*Cited on page 297.*)

[72] L.J. Mordell, *Reminiscences of an octogenarian mathematician*, Amer. Math. Monthly **78** (1971), 952–961. (*Cited on page 154.*)

[73] L.J. Mordell, **Diophantine Equations**, Academic Press, London and New york (1969). (*Cited on page 285.*)

[74] C.J. Moreno and S.S. Wagstaff, Jr., **Sums of Squares of Integers**. (*Cited on page 218.*)

[75] W. Narkiewicz, **Number Theory**,World Scientific Publishers, Singapore (1983). (*Cited on pages 371, 373–374, 376–377.*)

[76] A. Oppenheim, *Quadratic fields with and without Euclid's algorithm*, Math. Ann. **109** (1934), 349–352. (*Cited on page 50.*)

[77] O. Perron, *Quadratische Zahlkörper mit Euklidischen Algorithmus*, Math. Ann. **107** (1932), 489–495. (*Cited on page 50.*)

[78] J. M. Pollard, *Factoring with Cubic Integers* in **The Development of the Number Field Sieve**, A.K. Lenstra and H. W. Lenstra Jr. (Eds.), in LNM, Springer-Verlag, Berlin, Heidelberg, New York **1554** (1993), 4–10. (*Cited on page 92.*)

[79] G. Rabinowitsch, *Eindeutigkeit der Zerlegung in Primzahlfactoren in quadratischen Zahlkörpern*, J. Reine Angew. Math. **142** (1913), 153–164. (*Cited on pages 153–154.*)

[80] R. Remak, *Über den Euklidischen Algorithmus in reelquadratischen Zahlkörpern*, Jber. Deutschen Math. Verein **44** (1934), 238–250. (*Cited on page 50.*)

[81] K.A. Ribet, *On modular representations of $Gal(\overline{(\mathbb{Q})}/\mathbb{Q})$ arising from modular forms*, Invent. Math. **100** (1990), 431–476. (*Cited on page 365.*)

[82] J.P. Robertson and K.R. Matthews, *A Continued Fraction Approach to a Result of Feit*, American Math. Monthly, **115** (2008), 346–349. (*Cited on page 121.*)

[83] T. Satoh and K. Araki, *Fermat quotients and the polynomial time discrete logarithm for anomalous elliptic curves*, Comment. Math. Univ. St. Paul, (1998), 81–92. (*Cited on page 327.*)

[84] M.R. Schroeder, **Number Theory in Science and Communication**, Springer (1999). (*Cited on page 220.*)

[85] A. Selberg, *On an elementary method in the theory of primes*, Norske Vid. Selsk. Forh. Trondhjem **19**, 64-67, (1947). (*Cited on page 373.*)

[86] I. Semaev, *Evaluation of discrete logarithms in a group of p-torsion points of an elliptic curve in characteristic p*, Math. Comp. **67** (1998), 353–356. (*Cited on page 327.*)

[87] J.-P. Serre, **A Course in Arithmetic**, Springer-Verlag, New York, Heidelberg, Berlin (1973). (*Cited on pages 341–342.*)

[88] J.H. Silverman, **The Arithmetic of Elliptic Curves**, Springer, New York, Berlin, Heidelberg (1985). (*Cited on pages 310, 312, 327, 343, 361, 363.*)

[89] N. Smart, *The discrete logarithm problem on elliptic curves of trace one*, J. Cryptology **12** (1999), 193–196. (*Cited on page 327.*)

[90] J. Solinas, *Standard specifications for public key cryptography*, Annex A: Number-theoretic background. IEEE P1363 Draft (1998). (*Cited on page 327.*)

[91] B.K. Spearman and K.S. Williams, *Representing primes by binary quadratic forms*, American Math. Monthly, **99** (1992), 423–426. (*Cited on page 141.*)

[92] A. Srinavasan, *Markoff Numbers and Ambiguous Classes*, preprint. (*Cited on page 125.*)

[93] J. Steuding, **Diophantine Analysis**, Chapman and Hall/CRC Press, Boca Raton, London, Tokyo (2005). (*Cited on page 172.*)

[94] J. Tate, *Algorithm for determining the type of a singular fiber in an elliptic pencil* in **Modular Functions of One Variable IV**, LNM **476**, Springer-Verlag, (1975), 33–52. (*Cited on page 360.*)

[95] R. Taylor and A. Wiles, *Ring-theoretic properties of certain Hecke algebras*, Ann. of Math. **141** (1995), 553–572. (*Cited on page 364.*)

[96] E.C. Titchmarsh, *A divisor problem*, Rend. Circ. Mat. Palermo **54** (1930), 414–429. (*Cited on page 378.*)

[97] G.R. Veldekamp, *Remark on Euclidean rings*, Nieuw, Tid. Wisk, **48** (1960/61), 268–270 (Dutch). (*Cited on page 34.*)

[98] A.I. Vinogradov, *On the denseness hypothesis for Dirichlet L-series*, Izv. AN SSSR, Ser. Matem. **29** (1965), 903–934.[Russian] (*Cited on page 377.*)

[99] P. Vojta, **Diophantine Approximation and Value Distribution Theory**, LNM **1239**, Springer, Berlin, 1987. (*Cited on pages 300, 386.*)

[100] M. Waldschmidt, *Open Diophantine problems*, Moscow Math. J. **4** (2004), 245–305. (*Cited on page 179.*)

[101] E.W. Weisstein, **CRC Concise Encyclopedia of Mathematics**, CRC Press, Boca Raton, London, New York (1999). (*Cited on pages 227, 338, 346.*)

[102] H. Weyl, *A half-century of mathematics*, American Math. Monthly, **58** (1951), 523–553. (*Cited on page 18.*)

[103] A. Wiles, *Modular elliptic curves and Fermat's last theorem*, Ann. of Math. (1995), 443–551. (*Cited on page 364.*)

[104] A. Wintner, **The Theory of Measure in Arithmetical Semi-Groups**, Waverly Press, Baltimore (1944). (*Cited on page 216.*)

[105] P. Wolfskehl, *Beweis, dass der zweite Factor der Klassenzahl für die aus den elfen und dreizehnten Einheitswurzeln gebildeten Zahlen gleich Eins ist*, J. Reine Angew Math., **99** (1886) 173–178. (*Cited on page 224.*)

[106] G. Zukav, **The Dancing Wu Li Masters: An Overview of the New Physics**, Bantam Books, New York (1979). (*Cited on page 317.*)

Solutions to Odd-Numbered Exercises

Section 1.1

1.1 Since $a, b \in \mathbb{Q}$, then $\alpha a + b \in \mathbb{Q}(\alpha)$, so $\mathbb{Q}(a\alpha + b) \subseteq \mathbb{Q}(\alpha)$. However, $a \neq 0$, so a had an inverse a^{-1} in $\mathbb{Q}$, and $\alpha = a^{-1}(a\alpha + b) - ba^{-1} \in \mathbb{Q}(a\alpha + b)$, so $\mathbb{Q}(\alpha) \subseteq \mathbb{Q}(a\alpha + b)$. Hence, we have equality.

1.3 Since $(x^p - 1)/(x - 1) = x^{p-1} + x^{p-2} + \cdots + x + 1$ and ζ_p is a primitive pth root of unity, then this is the minimal polynomial $m_{\alpha,\mathbb{Q}}(x)$.

1.5 By Proposition 1.1 on page 13, $\alpha \in \mathfrak{U}_F$ if and only if $m_{\alpha,F}(0) = \pm 1$. However, since

$$m_{\alpha,F}(x) = \prod_{j=1}^{d}(x - \alpha_j),$$

then this occurs if and only if $\prod_{j=1}^{d} \alpha_j = \pm 1$. Hence, all α_j are units and the last statement is proved as well.

1.7 Since

$$x^n - 1 = \prod_{j=0}^{n-1}(x - \zeta_n^j) = \prod_{d|n} \prod_{\gcd(j,n)=d}(x - \zeta_n^j),$$

then it suffices to show that

$$\prod_{\gcd(j,n)=d}(x - \zeta_n^j) = \Phi_{n/d}(x),$$

since n/d runs over all divisors of n as d does. For $\gcd(j, n) = d$, let $j = dk$. Then $\zeta_n^j = \zeta_n^{dk} = \zeta_{n/d}^k$. Also, $\gcd(k, n/d) = 1$, so

$$\prod_{\gcd(j,n)=d}(x - \zeta_n^j) = \prod_{\gcd(k,n/d)=1}(x - \zeta_n^j) = \Phi_{n/d}(x).$$

Section 1.2

1.9 This is immediate from Corollary 1.3 since $m_{\alpha,F}(x)$ is irreducible over F.

1.11 Since

$$m_{\alpha.\mathbb{Q}}(x) = \prod_{j=1}^{d}(x - \alpha_j) = x^d + a_{d-1}x^{d-1} + \cdots + a_1 x + a_0 \in \mathbb{Q}[x],$$

then the coefficients of $m_{\alpha,\mathbb{Q}}(x)$ are sums of products of the α_j so by Exercise 1.10, $\alpha_j \in \mathbb{A}$ for all $j = 0, 1, 2, \ldots, d - 1$ if and only if $\alpha \in \mathbb{A}$. Hence,

$$m_{\alpha,\mathbb{Q}}(x) \in (\mathbb{Q} \cap \mathbb{A})[x] \text{ if and only if } \alpha \in \mathbb{A}$$

However, by Corollary 1.2 on page 4, $\mathbb{Q} \cap \mathbb{A} = \mathbb{Z}$, which proves the result.

1.13 $\alpha = \beta\sigma + \delta$ where:

(a) $\sigma = 1 + i$, $\delta = 1$. (b) $\sigma = 7 + i$, $\delta = -37i$. (c) $\sigma = 1 - 2i$, $\delta = 6i$. (d) $\sigma = 2 + i$, $\delta = -18i$.

1.15 $\alpha = 4x - 5y + (5x + 4y)i$ for any $x, y \in \mathbb{Z}$, since

$$\alpha = (4 + 5i)(x + yi) = \beta\sigma.$$

1.17 Suppose that γ is a greatest common divisor of α and β. If γ_j are associates of γ for $j = 1, 2$, then there are $u_j \in \mathfrak{U}_F$ for $j = 1, 2$ such that $\gamma = u_j\gamma_j$. Thus, $\gamma_j \mid \gamma$ which implies that γ_j divides both α and β for $j = 1, 2$. Now if δ divides both α and β, then $\delta \mid \gamma$ by Definition 1.14 on page 21. Therefore, since $\gamma_j = u_j^{-1}\gamma$, then $\delta \mid \gamma_j$ for $j = 1, 2$. Hence, by Definition 1.14, γ_j is a greatest common divisor of α and β for $j = 1, 2$. Conversely, if all associates of γ are greatest common divisors of α and β, then in particular γ is one.

For the last statement, if γ_j are gcds for $j = 1, 2$, then $\gamma_1 \mid \gamma_2$ and $\gamma_2 \mid \gamma_1$, so the result follows.

1.19 Let $\alpha = a + b\sqrt{D} \in \mathfrak{O}_F$. If $\alpha \in \mathfrak{U}_F$, then $1 \sim \alpha$ so $\alpha v = 1$ for some $v \in \mathfrak{U}_F$. Therefore, $N_F(\alpha v) = N_F(\alpha)N_F(v) = 1$, so $N_F(\alpha) = \pm 1$. Conversely, if $N_F(\alpha) = \pm 1$, then $(a + b\sqrt{D})(a - b\sqrt{D}) = \pm 1$ so by Definition 1.3 on page 2, $\alpha \in \mathfrak{U}_F$.

1.21 Since $\alpha \sim \beta$, then there exists $u \in \mathfrak{U}_F$ such that $\alpha = u\beta$ so

$$|N_F(\alpha)| = |N_F(u\beta)| = |N_F(u)||N_F(\beta)| = |N_F(\beta)|,$$

by Exercise 1.19.

1.23 Let $\alpha = 2 + i$ and $\beta = 2 - i$. Then $\gcd(N_F(\alpha), N_F(\beta)) = 5$. However, if $\delta \in \mathbb{Z}[i]$ such that $\delta \mid \alpha$ and $\delta \mid \beta$, then there exist $\sigma_1, \sigma_2 \in \mathbb{Z}[i]$ such that $\alpha = \delta\sigma_1$ and $\beta = \delta\sigma_2$. Thus, $N_F(\delta)N_F(\sigma_2) = N_F(\beta) = 5 = N_F(\alpha) = N_F(\delta)N_F(\sigma_1)$. Therefore, either $N_F(\delta) = 1$, in which case we have our counterexample since then $2 + i$ and $2 - i$ are relatively prime, or $N_F(\delta) = 5$ which implies $N_F(\sigma_1) = N_F(\sigma_2) = 1$. In the latter case, $\sigma_j \in \{\pm 1, \pm i\}$ for $j = 1, 2$, so $\alpha\sigma_1^{-1} = \beta\sigma_2^{-1}$, which implies $\alpha \sim \beta$ since $\alpha = \beta\sigma_1\sigma_2^{-1} = \beta u$, where $u \in \{\pm 1, \pm i\}$. However, all solutions of $\alpha = u\beta$ lead to contradictions. Hence, $N_F(\delta) = 1$ and we have our counterexample.

1.25 (a) $1 + 2i$ where $12 + 9i = (6 - 3i)(1 + 2i)$ and $2 + 69i = (28 + 13i)(1 + 2i)$ (b) $1 + i$ where $2 + 8i = (5 + 3i)(1 + i)$ and $21 + 9i = (15 - 6i)(1 + i)$

1.27 (a) $3 + 2i$ where $17 + 7i = (5 - i)(3 + 2i)$ and $71 + 4i = (17 - 10i)(3 + 2i)$ (b) 1

1.29 If α and β are relatively prime, then by Theorem 1.10 on page 21, there exist $\sigma, \tau \in \mathbb{Z}[i]$ such that

$$1 = \alpha\sigma + \beta\tau.$$

Thus, by taking conjugates over this equation, we get

$$1 = \alpha'\sigma' + \beta'\tau',$$

which implies that α' and β' are relatively prime since any common divisor of them must divide 1.

Conversely, if α' and β' are relatively prime, then as above, there exist $\sigma_1, \tau_1 \in \mathbb{Z}[i]$ such that $1 = \alpha'\sigma_1 + \beta'\tau_1$. Taking conjugates over this equation, we get

$$1 = (\alpha')'\sigma_1' + (\beta')'\tau_1' = \alpha\sigma_1' + \beta\tau_1',$$

so α and β are relatively prime.

1.31 If $a + bi$ is primary, then $a + b \equiv 1 \pmod 4$ where a is odd and b is even. Thus,

$$a + bi = 1 + \left(\frac{-1 + a + b}{4} + \left(\frac{1 - a + b}{4} \right) i \right) (2 + 2i) \equiv 1 \pmod{2 + 2i},$$

in $\mathbb{Z}[i]$.

Conversely, if $a + bi \equiv 1 \pmod{2 + 2i}$ in $\mathbb{Z}[i]$, then there exist $c, d \in \mathbb{Z}$ such that

$$a + bi = 1 + (c + di)(2 + 2i) = 1 + 2c - 2d + (2c + 2d)i.$$

By comparing coefficients,

$$a = 1 + 2c - 2d \equiv 1 \pmod 2, b = 2c + 2d \equiv 0 \pmod 2,$$

and

$$a + b = 1 + 4c \equiv 1 \pmod 4.$$

1.33 If $\alpha = a + bi$ is an odd Gaussian integer that is not primary, then one of the following holds, (a) a is even; (b) b is odd; or (c) $a + b \not\equiv 1 \pmod 4$. It remains to show that exactly one of its associates $-\alpha$, $i\alpha$, or $-i\alpha$ is primary.

If (a) holds, then $-\alpha$ cannot be primary since $-a$ is even. Also, $i\alpha = ai - b$. If b is even, then

$$\alpha = 2 \left(\frac{a}{2} + \frac{b}{2} i \right) = (1 + i)(1 - i) \left(\frac{a}{2} + \frac{b}{2} i \right),$$

which implies that $(1 + i) \mid \alpha \in \mathbb{Z}[i]$, contradicting that α is odd. Thus, b is odd. If $a - b \equiv 1 \pmod 4$, then $i\alpha$ is primary. However, $-i\alpha = b - ai$ is not primary since $b - a \equiv -1 \pmod 4$, so in this case exactly one of the associates, $i\alpha$, of α is primary, so we may assume that $a - b \not\equiv 1 \pmod 4$. Since $a - b$ is odd, then $a - b \equiv -1 \pmod 4$ which makes $-i\alpha$ the only primary associate. This takes care of case (a).

If (b) holds, and a is odd, then there exist $c, d \in \mathbb{Z}$ such that

$$\alpha = a + bi = 2c + 1 + (2d + 1)i = 2(c + di) + 1 + i = (1 + i)[c - ci + di + d + 1],$$

so $(1 + i) \mid \alpha$, contradicting that α is odd. Hence, a is even. However, this puts us back in case (a), with which we have already dealt.

If (c) holds, then given that we have already dealt with the cases where a is even and b is odd, we must have that a is *odd* and b is even. Since $a + b$ is odd, then $a + b \equiv -1 \pmod 4$, which makes $-\alpha = -a - bi$ primary, and neither of the other associates are primary.

This completes the analysis of the result for α not, itself, primary. If α is primary, then $-\alpha = -a - bi$ cannot be since $-a - b \equiv -1 \pmod 4$. Also,

$$\pm i\alpha = ai \mp b$$

cannot be primary since b is even.

1.35 (a) $(1 + i)(2 + 5i)^2(3 - 2i)^3$

(b) $(2 - i)(1 - 4i)^2(1 - 2i)^3$

(c) $(5 + 2i)(4 + 5i)^2(3 - 2i)^3$

(d) $(1 + i)(2 + 7i)^2(3 - 8i)$

1.37 If $p = (a+bi)(c+di)$ for $a, b, c, d \in \mathbb{Z}$ and neither right-hand factor is a unit, then $N_F(a+bi) = a^2 + b^2 = p = N_F(c+di) = c^2 + d^2$, since $N_F(p) = p^2$. However, as noted in Example 1.15 on page 28, it is not possible for a prime $p \equiv 3 \pmod 4$ to be a sum of two integer squares. Hence, one of the aforementioned factors must be a unit, so p is a Gaussian prime.

Section 1.3

1.39 Let $\alpha, \beta \in R$ be nonzero elements and set

$$S = \{\gamma \in R : \gamma = \rho\alpha + \eta\beta, \text{ for some } \rho, \eta \in R\}.$$

Since $1_R\alpha + 0 \in S$ and $0 + 1_R\beta \in S$, then S consists of more than just the zero element. If f is the Euclidean function on R, we may choose an element $\gamma_0 = \rho_0\alpha + \eta_0\beta \in S$ with $f(\gamma_0)$ as a minimum. Now let $\gamma = \rho\alpha + \eta\beta \in S$ be arbitrary. By condition (b) of Euclidean domains in Definition 1.17 on page 32, there are $\sigma, \delta \in R$ such that

$$\gamma = \sigma\gamma_0 + \delta, \text{ with either } \delta = 0, \text{ or } f(\delta) < f(\gamma_0).$$

Since

$$\delta = \gamma - \sigma\gamma_0 = \rho\alpha + \eta\beta - \sigma(\rho_0\alpha + \eta_0\beta) =$$
$$(\rho - \sigma\rho_0)\alpha + (\eta - \sigma\eta_0)\beta \in S,$$

then if $\delta \neq 0$, condition (b) of Euclidean domains tells us that

$$f(\delta) = f((\rho - \sigma\rho_0)\alpha + (\eta - \sigma\eta_0)) < f(\gamma_0),$$

a contradiction to the minimality of $f(\gamma_0)$. Thus, $\delta = 0$, and so $\gamma = \sigma\gamma_0$. In other words, $\gamma_0|\gamma$ for all $\gamma \in S$. In particular $\gamma_0|\alpha$ and $\gamma_0|\beta$. It remains to show that any common divisor of α and β in R must divide γ_0. Let $\gamma_1|\alpha$, and $\gamma_1|\beta$. Therefore, $\gamma_1|\sigma_0\alpha + \delta_0\beta = \gamma_0$. Hence, γ_0 is a gcd of α and β as required.

1.41 If the condition in the exercise holds and $\alpha\beta \neq 0$ for $\alpha, \beta \in R$, then $\alpha \mid \alpha\beta$ so $f(\alpha) \leq f(\alpha\beta)$, which is condition (a) in Definition 1.17. Conversely, if (a) holds and $\alpha \mid \beta$, then $\beta = \alpha\gamma$ for some $\gamma \in R$. Therefore, by (a), $f(\alpha) \leq f(\alpha\gamma) = f(\beta)$.

1.43 If $\alpha \in R$ is a unit, there exists a $u \in R$ such that $u\alpha = 1_R$. Thus, by Exercise 1.42 and condition (a) of Definition 1.17,

$$f(1_R) \leq f(\alpha) \leq f(u\alpha) = f(1_R),$$

so $f(\alpha) = f(1_R)$. Conversely, if $f(\alpha) = f(1_R)$, then for any $\beta \in R$, $\beta = \alpha\gamma + \delta$ for some $\gamma, \delta \in R$. If $\delta \neq 0$, then $f(\delta) < f(\alpha) = f(1_R) \leq f(\delta)$, a contradiction. Hence, for each $\beta \in R$, $\alpha \mid \beta$. In particular, $\alpha \mid 1_R$, which makes it a unit in R.

1.45 Since $a + b\sqrt{D} \in \mathbb{Z}[\sqrt{D}] \subseteq \mathfrak{O}_F$ for any quadratic field, and since $x^2 - Dy^2 = 1$ has infinitely many solutions for $D > 0$ by Pell's solutions, then we have our result.

1.47 If $2 + i = (a + bi)(c + di)$ for $a, b, c, d \in \mathbb{Z}$, then

$$N_F(2 + i) = 5 = N_F(a + bi)N_F(c + di) = (a^2 + b^2)(c^2 + d^2),$$

so either $a^2 + b^2 = 1$ or $c^2 + d^2 = 1$. Therefore by Exercise 1.19 on page 29, one of them is a unit. The argument for $2 - i$ is the same. Thus, $2 + i$ and $2 - i$ are irreducible.

1.49 If δ_1 and δ_2 are least common multiples of α and β, then by property (b) of Definition 1.21 on page 40, $\delta_1 \mid \delta_2$ and $\delta_2 \mid \delta_1$, so $\delta_1 \sim \delta_2$ by Exercise 1.16 on page 29.

1.51 The converse is false since 2 is irreducible in $\mathbb{Z}[\sqrt{10}]$ by Example 1.17, but $N_F(2) = 4$.

1.53 The converse is false since $\mathbb{Z}[i]$ is a UFD by Theorem 1.15 on page 34, and 3 is a Gaussian prime by Exercise 1.37 on page 30, but $N_F(3) = 9$.

Section 1.4

1.55 We may factor in the Gaussian integers $\mathbb{Z}[i]$ as follows.

$$(y + i)(y - i) = x^3.$$

By the same method as in the proof of Theorems 1.19–1.20 on pages 47 and 48 we have that $y + i$ and $y - i$ are relatively prime. Thus, by unique factorization ensured for the Gaussian integers, there is a $\beta = a + bi \in \mathbb{Z}[i]$ such that

$$y + i = \beta^3 = (a + bi)^3,$$

and

$$y - i = (a - bi)^3.$$

Subtracting the two equations and dividing by $2i$ we get

$$1 = b(3a^2 - b^2).$$

Therefore, $b = \pm 1$. However, $b = 1$ implies that $2 = 3a^2$, which is impossible, so $b = -1$. This forces $1 = -(3a^2 - 1)$. Thus, $a = 0$, so $y = 0$. Hence, $x = (-i)i = 1$, which secures the result.

1.57 Since $\alpha \mid N_F(\alpha) = \alpha\alpha' \in \mathbb{Z}$, then there is a least element $n \in \mathbb{N}$ such that $\alpha \mid n$. If $n = n_1 n_2$ for $n_j \in \mathbb{N}$, $j = 1, 2$, with $0 < n_1 \le n_2 < n$. Then $\alpha \mid (n_1 n_2)$, so $\alpha \mid n_1$ or $\alpha \mid n_2$, contradicting the minimality of n in this regard. Hence, n is a rational prime, say $n = p$. If $\alpha \mid q$ where q is a rational prime with $q \ne p$, then by the Euclidean algorithm for rational integers, there exist $a, b \in \mathbb{Z}$ such that $1 = ap + bq$, Since $\alpha \mid p$ and $\alpha \mid q$, then $\alpha \mid 1$, a contradiction, so p is the only rational prime divisible by α.

1.59 By Exercises 1.56–1.57, there is a unique rational prime p such that $N_F(\alpha) = \pm p$.

1.61 If 2 is not prime in $\mathfrak{D}_F$, then by Exercise 1.56, $2 = \alpha\alpha'$ where $\alpha = (a + b\sqrt{D})/2 \in \mathfrak{D}_F$. Thus,

$$\pm 2 = N_F(\alpha) = \alpha\alpha' = \frac{a^2 - b^2 D}{4},$$

where a, b have the same parity. If both are odd, then

$$\pm 8 = a^2 - b^2 D \equiv 1 - D \equiv -4 \pmod 8,$$

a contradiction. If both are even, then $\pm 2 = (a/2)^2 - (b/2)^2 D$, so both $a/2$ and $b/2$ have to be odd. Therefore, $\pm 2 \equiv 1 - D \equiv 4 \pmod 8$, a contradiction. Hence, 2 is prime in $\mathfrak{O}_F$ as required.

1.63 If $p \mid D$, then $|D| = pn$ for some $n \in \mathbb{N}$. If $n = 1$, then $p = \pm\sqrt{D} \cdot \sqrt{D}$, where $\sqrt{D}$ is a prime in $\mathfrak{O}_F$ by Exercise 1.52 on page 46, since $N_F(\sqrt{D}) = \pm p$. Thus, $p \sim \sqrt{D}^2$. If $n > 1$, then

$$D = p(D/p) = \sqrt{D} \cdot \sqrt{D}. \tag{S1}$$

However, p does not divide $\sqrt{D}$, since to do so would mean that $\sqrt{D} = p(a + b\sqrt{D})/2$ where $a, b \in \mathbb{Z}$ have the same parity, by Theorem 1.3 on page 6. However, this means that $a = 0$ and $pb/2 = 1$, where b must be even, a contradiction. Thus, p is not a prime in $\mathfrak{O}_F$. Therefore, by Exercise 1.56, $\alpha \mid p$ where α is prime in $\mathfrak{O}_F$ and $N_F(\alpha) = \pm p$. Now, by (S1), $\alpha \mid \sqrt{D}$, so $\alpha^2 \mid D$, which implies that $\alpha^2 \mid p$ since $\alpha \nmid D/p$. Thus, $p = \alpha^2 \beta$ where $\beta \in \mathfrak{O}_F$. However, $N_F(p) = p^2 = N_F(\alpha^2)N_F(\beta) = N_F(\alpha)^2 N_F(\beta) = p^2 N_F(\beta)$, so $N_F(\beta) = 1$, which means that $\beta \in \mathfrak{U}_F$. Therefore, $p \sim \alpha^2$.

Section 2.1

2.1 Let M be a $\mathbb{Z}$-module. If $r \in \mathbb{Z}$, and $m \in M$, then

$$r \cdot m = \underbrace{m + \cdots m}_{r},$$

so the properties of an additive abelian group are inherited from this action. Conversely, if M is an additive abelian group, then the addition within the group gives the $\mathbb{Z}$-module action as above.

2.3 We only prove this for $\sigma = 1$ since the other case is similar.

Suppose that I is an ideal. Therefore, $a\sqrt{D} \in I$, so $c|a$ by the minimality of c. We have

$$\sqrt{D}(b + c\sqrt{D}) = b\sqrt{D} + cD \in I,$$

so $c|b$. Moreover, since

$$\left(\frac{b}{c} - \sqrt{D}\right)(b + c\sqrt{D}) = \frac{b^2 - c^2 D}{c} \in I,$$

then

$$a|(b^2 - c^2 D)/c.$$

In other words,

$$ac|(b^2 - c^2 D).$$

Conversely, assume that I satisfies the conditions. To verify that I is an ideal, we need to show that $a\sqrt{D} \in I$ and $(b + \sqrt{D})\sqrt{D} \in I$. This is a consequence of the following identities, the details of which we leave to the reader for verification:

$$a\sqrt{D} = -(b/c)a + (a/c)(b + c\sqrt{D}),$$

and

$$b\sqrt{D} + cD = -(b^2 - c^2 D)/c + b(b + c\sqrt{D})/c,$$

so I is an ideal.

2.5 If $[\alpha, \beta] = [\gamma, \delta]$, there are integers $x, x_0, y, y_0, z, z_0, w, w_0$ such that

$$\alpha = x\gamma + y\delta, \quad \beta = w\gamma + z\delta,$$

and

$$\gamma = x_0 \alpha + y_0 \beta, \quad \delta = w_0 \alpha + z_0 \beta.$$

These two sets of equations translate into two matrix equations as follows.

$$\begin{pmatrix} \alpha \\ \beta \end{pmatrix} = X \begin{pmatrix} \gamma \\ \delta \end{pmatrix},$$

where

$$X = \begin{pmatrix} x & y \\ w & z \end{pmatrix},$$

and

$$\begin{pmatrix} \gamma \\ \delta \end{pmatrix} = X_0 \begin{pmatrix} \alpha \\ \beta \end{pmatrix},$$

where

$$X_0 = \begin{pmatrix} x_0 & y_0 \\ w_0 & z_0 \end{pmatrix}.$$

Hence,

$$\begin{pmatrix} \alpha \\ \beta \end{pmatrix} = X X_0 \begin{pmatrix} \alpha \\ \beta \end{pmatrix}.$$

Therefore, the determinants of X and X_0 are ± 1, so the result follows.

Conversely, assume that the matrix equation holds as given in the exercise. Then clearly

$$[\alpha, \beta] \subseteq [\gamma, \delta].$$

Since the determinant of X is ± 1, we can multiply both sides of the matrix equation by the inverse of X to get that γ and δ are linear combinations of α and β. Thus,

$$[\gamma, \delta] \subseteq [\alpha, \beta].$$

The result is now proved.

2.7 Let

$$J_i = (a_i, (b_i + \sqrt{\Delta})/2) \text{ for } i = 1, 2$$

be $\mathfrak{O}_F$-ideals such that $J_1 J_2 \subseteq \mathcal{P}$. Then by the multiplication formulas given on page 59, $J_1 J_2 = (a_3, (b_3 + \sqrt{\Delta})/2)$ where $a_3 = a_1 a_2 / g \equiv 0 \pmod{p}$ with $g = \gcd(a_1, a_2, (b_1 + b_2)/2))$. If $p \nmid a_2$ (which means that $J_2 \not\subseteq \mathcal{P}$), then $p \mid a_1$ since p cannot divide g given that it does not divide a_2. Thus, to show that $J_1 \subseteq \mathcal{P}$, it remains to show that $b_1 = 2pn + b$ for some $n \in \mathbb{Z}$, by Exercise 2.6. Now, by Exercise 2.4,

$$b_1^2 \equiv \Delta \pmod{4a_1} \text{ and } b^2 \equiv \Delta \pmod{4p},$$

so $b_1^2 \equiv b^2 \pmod{4p}$. Since p is prime, then $b_1 \equiv \pm b \pmod{2p}$. If

$$b_1 \equiv -b \pmod{2p}, \text{ then } J_1 \subseteq \mathcal{P}' = (p, (-b + \sqrt{\Delta})/2),$$

so if $(-b + \sqrt{\Delta})/2 \in \mathcal{P}$, then $J_1 \subseteq \mathcal{P}$ so we are done by Theorem 2.2 on page 57. If $(-b + \sqrt{\Delta})/2 \notin \mathcal{P}$, then $\mathcal{P} \cap \mathcal{P}' = (p)$, so $a_3 = 1$, and this forces $p \mid 1$, a contradiction. The remaining case is $b_1 \equiv b \pmod{2p}$, so $b_1 = 2pn + b$ for some $n \in \mathbb{Z}$, as required.

Section 2.2

2.9 Let $\mathcal{P}_j$ be distinct prime R-ideals with

$$I = \prod_{j=1}^{r} \mathcal{P}_j^{a_j} \text{ and } J = \prod_{j=1}^{r} \mathcal{P}_j^{b_j},$$

where $a_j, b_j \geq 0$. Choose $\alpha_j \in \mathcal{P}_j^{a_j} - \mathcal{P}_j^{a_j+1}$ for $j = 1, 2, \ldots, r$. By Theorem 2.18 on page 84, there exists an $\alpha \in R$ such that

$$\alpha - \alpha_j \in \mathcal{P}_j^{a_j+1} \text{ for all } j = 1, \ldots, r.$$

Thus,

$$\alpha \in \mathcal{P}_j^{a_j} \text{ and } \alpha \notin \mathcal{P}_j^{a_j+1} \text{ for } 1 \leq j \leq r.$$

Therefore,

$$\alpha \in \cap_{j=1}^{r} \mathcal{P}_j^{a_j} \subseteq I.$$

Therefore, by Remark 2.10 on page 81,

$$I \subseteq \gcd((\alpha), IJ) = (\alpha) + IJ \subseteq I,$$

so $\gcd((\alpha), IJ) = I$, as required.

2.11 By Exercise 2.11, there is an $\alpha \in I$ such that

$$(\alpha) + IJ = I. \tag{S2}$$

Since $(\alpha) \subseteq I$, then $I \mid (\alpha)$ by Corollary 2.5 on page 76, so there exists an R-ideal H such that $(\alpha) = HI$. Substituting this into (S2), we get

$$I = IH + IJ = I(H + J),$$

by Exercise 2.10. Hence, by Corollary 2.7 on page 77, $R = H + J = \gcd(H, J)$.

2.13 If R does not satisfy the DCC, there exists an infinite nonterminating descending sequence of ideals $\{I_j\}$, so there can exist no minimal element in this set. Conversely, if R satisfies the DCC, then any nonempty collection $\mathcal{S}$ of ideals has an element I. If I is not minimal, then it contains an element I_1. If I_1 is not minimal, then it contains an ideal I_2, and so on. Eventually, due to DCC, the process terminates, so the set contains a minimal element.

2.15 Since s is integral over R, there exists a monic polynomial $f(x) = \sum_{j=0}^{d} r_j s^j \in R[x]$ such that $f(s) = 0$. Thus,

$$s^d = -\sum_{j=0}^{d-1} r_j s^j,$$

so s^{d+k} for any nonnegative integer k can be expressed as an R-linear combination of s^i for $i = 0, 1, \ldots, d-1$. Hence,

$$R[s] = R + Rs + \cdots + Rs^{d-1},$$

which means that $R[s]$ is finitely generated as an R-module.

2.17 First we show that I^{-1} is unique for an invertible fractional R-ideal in the sense that if $IJ = R$ for some $J \in G$, then $J = I^{-1}$.

If $I \in G$ is invertible, and $J \in G$ with $IJ = R$, then

$$I^{-1} = I^{-1}R = I^{-1}(IJ) = (I^{-1}I)J = RJ = J,$$

so that $I^{-1} = J$.

If (a) holds, then, by the above, every nonzero fractional ideal I has a unique inverse given by I^{-1}. Since $I^{-1}H \in G$ for any $I, H \in G$ with I nonzero, this shows that G is a multiplicative group.

Conversely, if (b) holds, then every nonzero $I \in G$ has a unique inverse J, namely $IJ = R$. As above, $J = I^{-1}$, so I is invertible.

2.19 (a) By Theorem 2.12 on page 77,

$$I = \prod_{\mathcal{P}} \mathcal{P}^{\mathrm{ord}_{\mathcal{P}}(I)} \text{ and } J = \prod_{\mathcal{P}} \mathcal{P}^{\mathrm{ord}_{\mathcal{P}}(J)},$$

where there are only finitely many nonzero exponents. Moreover,

$$\prod_{\mathcal{P}} \mathcal{P}^{\mathrm{ord}_{\mathcal{P}}(IJ)} = IJ = \prod_{\mathcal{P}} \mathcal{P}^{\mathrm{ord}_{\mathcal{P}}(I)+\mathrm{ord}_{\mathcal{P}}(J)},$$

so via the uniqueness guaranteed by Theorem 2.12,

$$\mathrm{ord}_{\mathcal{P}}(IJ) = \mathrm{ord}_{\mathcal{P}}(I) + \mathrm{ord}_{\mathcal{P}}(J).$$

(b) Let $H = I + J$. Then it follows from Exercise 2.10 that

$$IH^{-1} + JH^{-1} = (I+J)H^{-1} = HH^{-1} = R,$$

where the last equality comes from Theorem 2.11 on page 76. Thus, we have that

both $IH^{-1} \subseteq IH^{-1} + JH^{-1} = R$ and $JH^{-1} \subseteq IH^{-1} + JH^{-1} = R$,

so both IH^{-1} and JH^{-1} are integral R-ideals. If both $IH^{-1} \subseteq \mathcal{P}$ and $JH^{-1} \subseteq \mathcal{P}$, then

$$R = IH^{-1} + JH^{-1} \subseteq \mathcal{P} + \mathcal{P} = \mathcal{P},$$

contradicting that $\mathcal{P}$ is prime. Thus, either $IH^{-1} \not\subseteq \mathcal{P}$ or $JH^{-1} \not\subseteq \mathcal{P}$. Therefore, by Corollary 2.5 on page 76, either $\mathcal{P} \nmid IH^{-1}$ or $\mathcal{P} \nmid JH^{-1}$. Thus,

$$\min(\mathrm{ord}_{\mathcal{P}}(IH^{-1}), \mathrm{ord}_{\mathcal{P}}(JH^{-1})) = 0. \tag{S3}$$

Also, by part (a),

$$\mathrm{ord}_{\mathcal{P}}(I) = \mathrm{ord}_{\mathcal{P}}(IH^{-1}H) = \mathrm{ord}_{\mathcal{P}}(IH^{-1}) + \mathrm{ord}_{\mathcal{P}}(H),$$

and

$$\mathrm{ord}_{\mathcal{P}}(J) = \mathrm{ord}_{\mathcal{P}}(JH^{-1}H) = \mathrm{ord}_{\mathcal{P}}(JH^{-1}) + \mathrm{ord}_{\mathcal{P}}(H).$$

Therefore, by (S3),

$$\min(\mathrm{ord}_{\mathcal{P}}(I), \mathrm{ord}_{\mathcal{P}}(J)) = \mathrm{ord}_{\mathcal{P}}(H) = \mathrm{ord}_{\mathcal{P}}(I + J).$$

(c) Suppose that $\mathrm{ord}_{\mathcal{P}}(I) = a$. Select an element $\alpha_{\mathcal{P}} \in \mathcal{P}^a - \mathcal{P}^{a+1}$. Then $\mathrm{ord}_{\mathcal{P}}((\alpha_{\mathcal{P}})) = a = \mathrm{ord}_{\mathcal{P}}(I)$. Then by induction, for any prime R-ideal $\mathcal{Q}$ dividing I, there exists an element

$$\alpha_{\mathcal{Q}} \in \mathcal{Q}^{\mathrm{ord}_{\mathcal{Q}}(I)} \prod_{\substack{\mathcal{P} \mid I \\ \mathcal{P} \neq \mathcal{Q}}} \mathcal{P}^{\mathrm{ord}_{\mathcal{P}}(I)+1} - \mathcal{Q}^{\mathrm{ord}_{\mathcal{Q}}(I)+1} \prod_{\substack{\mathcal{P} \mid I \\ \mathcal{P} \neq \mathcal{Q}}} \mathcal{P}^{\mathrm{ord}_{\mathcal{P}}(I)+1},$$

so $\mathrm{ord}_{\mathcal{Q}}((\alpha_{\mathcal{Q}})) = \mathrm{ord}_{\mathcal{Q}}(I)$. Hence, by selecting $\alpha = \sum_{\mathcal{Q} \mid I} \alpha_{\mathcal{Q}} \in F$, we have, by inductively extrapolating from part (b), that if $\mathcal{P}_j \mid I$ for $j = 1, 2, \ldots n$ are all the distinct prime R-ideals dividing I, then

$$\mathrm{ord}_{\mathcal{P}_1}((\alpha)) = \mathrm{ord}_{\mathcal{P}_1}\left(\sum_{j=1}^{n} \mathrm{ord}_{\mathcal{P}_j}(\alpha_{\mathcal{P}_j})\right) =$$

$$\min\left(\mathrm{ord}_{\mathcal{P}_1}((\alpha_{\mathcal{P}_1})), \mathrm{ord}_{\mathcal{P}_1}((\alpha_{\mathcal{P}_2})) \ldots, \mathrm{ord}_{\mathcal{P}_1}((\alpha_{\mathcal{P}_n}))\right) = \mathrm{ord}_{\mathcal{P}_1}((\alpha_{\mathcal{P}_1})),$$

namely $\mathrm{ord}_{\mathcal{P}_1}((\alpha)) = \mathrm{ord}_{\mathcal{P}_1}((\alpha_{\mathcal{P}_1}))$, as required.

Section 2.3

2.21 Let $F = \mathbb{Q}(\alpha)$ where $\alpha = \sqrt[3]{-2}$ for which

$$m_{\alpha,\mathbb{Q}}(x) = x^3 + 2 \text{ and } m_{\alpha^2,F}(x) = x^3 - 4.$$

Hence, $\alpha, \alpha^2 \in \mathfrak{O}_F$. Since

$$\deg(m_{\alpha,\mathbb{Q}}) = \deg(m_{\alpha^2,\mathbb{Q}}) = |F : \mathbb{Q}| = 3,$$

then $\{1, \alpha, \alpha^2\}$ provides a $\mathbb{Z}$-basis for $\mathbb{Z}[\sqrt[3]{-2}]$. Since $\alpha \in \mathbb{A} \cap F = \mathfrak{O}_F$, then $\mathbb{Z}[\alpha] \subseteq \mathfrak{O}_F$. It remains to show equality.

Since $|F : \mathbb{Q}| = 3$, then $|\mathfrak{O}_F : \mathbb{Z}| = 3$. However, $|\mathfrak{O}_F : \mathbb{Z}| = |\mathfrak{O}_F : \mathbb{Z}[\alpha]| \cdot |\mathbb{Z}[\alpha] : \mathbb{Z}|$, so either $|\mathfrak{O}_F : \mathbb{Z}[\alpha]| = 1$ or $|\mathfrak{O}_F : \mathbb{Z}[\alpha]| = 3$. In the former case, we are done since then $\mathfrak{O}_F = \mathbb{Z}[\alpha]$. In the latter case, $|\mathbb{Z}[\alpha] : \mathbb{Z}| = 1$ is forced and this means that $\mathbb{Z}[\alpha] = \mathbb{Z}$ so $\alpha \in \mathbb{Z}$ which is false.

2.23 Let $\{\alpha_i\}_{i \in \mathcal{I}}$ and $\{\beta_j\}_{j \in \mathcal{J}}$ be bases for K over F and E over K, respectively, where $\mathcal{I}$ and $\mathcal{J}$ are indexing sets, possibly infinite. We now show that the set of products

$$\{\alpha_i \beta_j\}_{(i,j) \in \mathcal{I} \times \mathcal{J}}$$

is a basis for E over F. If $\alpha \in E$, then it has a unique representation

$$\alpha = \sum_{j \in \mathfrak{J}} \gamma_j \beta_j, \text{ where } \gamma_j \in K \text{ for } j \in \mathfrak{J}.$$

Also, for each $j \in \mathfrak{J}$, there is a unique representation

$$\gamma_j = \sum_{i \in \mathfrak{I}} \delta_i \alpha_i, \text{ where } \delta_i \in F \text{ for } i \in \mathfrak{I}.$$

Hence, we have a unique representation

$$\alpha = \sum_{j \in \mathfrak{J}} \gamma_j \beta_j = \sum_{j \in \mathfrak{J}} \beta_j \sum_{i \in \mathfrak{I}} \delta_i \alpha_i = \sum_{j \in \mathfrak{J}} \sum_{i \in \mathfrak{I}} \delta_i \alpha_i \beta_j,$$

which yields the result.

2.25 By examining coefficients, we have

$$f_F(\beta) = \prod_{j=1}^{n} (x - \beta_j) = x^n - T_F(\beta) x^{n-1} + \cdots + (-1)^n N_F(\beta),$$

so by Exercise 2.24, $N_F(\beta), T_F(\beta) \in \mathbb{Q}$. If $\alpha \in \mathfrak{O}_F$, then by Corollary 1.4 on page 11, $m_{\alpha,\mathbb{Q}(x)} \in \mathbb{Z}[x]$, so by Exercise 2.24 again, $N_F(\beta), T_F(\beta) \in \mathbb{Z}$.

2.27 Since, for a primitive cube root of unity ζ_3, we have

$$N_F(\beta) = (a + b\alpha + c\alpha^2)(a + b\zeta_3\alpha + c\zeta_3^2\alpha^2)(a + b\zeta_3^2\alpha + c\zeta_3^4\alpha^2),$$

then using the fact that $\sum_{j=0}^{2} \zeta_3^j = 0$ we get

$$N_F(\beta) = (a + b\alpha + c\alpha^2)((a^2 + 2bc) - (ab + 2c^2)\alpha + (b^2 - ac)\alpha^2),$$

so, by simplifying,

$$N_F(\beta) = a^3 - 2b^3 + 4c^3 + 6abc.$$

2.29 Since $\beta \mid \gamma$, then there is a $\delta \in \mathfrak{O}_F$ such that $\gamma = \beta\delta$, so by Exercise 2.28,

$$N_F(\gamma) = N_F(\beta\delta) = N_F(\beta) N_F(\delta),$$

so $N_F(\beta) \mid N_F(\gamma)$.

2.31 Since $(5^7 - 1) \mid (5^{77} - 1)$ and $4 \mid (5^7 - 1)$, then $(5^7 - 1)/4 = 19531 \mid (5^{77} - 1)$.

2.33 Since $3(3^{239} - 1) = 3^{240} - 3 = x^3 - 3$, where $x = 3^{80}$, and $N_F(a + b\sqrt[3]{3}) = a^3 + 3b^3$, for $F = \mathbb{Q}(\sqrt[3]{3})$, then $N_F(x - \sqrt[3]{3}) = x^3 - 3$. An initial run shows that $\gcd(3^{240} - 3, a^3 + 3b^3) = 479$, for $a = 14$, and $b = 185$, so $479 \mid (3^{239} - 1)$.

Section 3.1

3.1 Clearly, since $f(x, y) = g(X, Y)$ for

$$X = px + qy \tag{S4}$$

and

$$Y = rx + sy, \tag{S5}$$

then equivalent forms represent the same integers by definition. Since $ps - qr = \pm 1$ and from (S4)–(S5), $x = \pm(sX - qY)$ and $y = \pm(rX - pY)$, so $\gcd(x, y) = 1$ if and only if $\gcd(X, Y) = 1$.

3.3 Suppose that $f(x,y) = g(X,Y)$ where $X = px+qy$, $Y = rx+sy$, and $ps-qr = 1$. If we set $x = X$ and $Y = y$, namely $p = s = 1$ and $q = r = 0$, then $f(x,y) = g(x,y)$ and we have the reflexive property. Also, since

$$g(X_1, Y_1) = f(x,y),$$

where $X_1 = sx - qy$ and $Y_1 = py - rx$, then we have the symmetry property. Lastly, for transitivity, assume that

$$g(X,Y) = h(PX + QY, RX + SY),$$

where $PS - QR = 1$. Then since

$$PX + QY = P(px+qy) + Q(rx+sy) = (Pp+Qr)x + (Pq+Qs)y = P_1x + Q_1y$$

and

$$RX + SY = R(px+qy) + S(rx+sy) = (Rp+Sr)x + (Rq+Ss)y = R_1x + S_1y$$

we have

$$P_1S_1 - Q_1R_1 = (Pp+Qr)(Rq+Ss) - (Pq+Qs)(Rp+Sr) =$$

$$PRpq + QRrq + PpSs + QrSs - PqRp - PqSr - QsRp - QsSr =$$
$$QR(rq - sp) + PS(ps - qr) = -QR + PS = 1,$$

so

$$f(x,y) = h(P_1x + Q_1y, R_1x + S_1y),$$

with $P_1S_1 - Q_1R_1 = 1$, which is the transitive property.

3.5 If $f \sim g$, $f = (a,b,c)$, $g = (a_1, b_1, c_1)$ with f primitive, then

$$ax^2 + bxy + cy^2 = a_1(px+qy)^2 + b_1(px+qy)(rx+sy) + c_1(rx+sy)^2 =$$

$$(a_1p^2 + b_1pr + c_1r^2)x^2 + (2pqa_1 + (ps+rq)b_1 + 2rsc_1)xy + (q^2a_1 + qsb_1 + c_1s^2)y^2,$$

so if $\gcd(a_1, b_1, c_1) = g$, then $g \mid \gcd(a,b,c) = 1$, and the result is secured.

3.7 Applying the substitution $x = pX + qY$ and $y = rX + sY$ to the form

$$f(x,y) = ax^2 + bxy + cy^2,$$

we get the form $AX^2 + BXY + CY^2$, where

$$A = ap^2 + bpr + cr^2,$$

$$B = 2apq + b(ps + qr) + 2crs,$$
$$C = aq^2 + bqs + cs^2.$$

A straightforward calculation shows that

$$B^2 - 4AC = (b^2 - 4ac)(ps - qr)^2,$$

which yields the result.

3.9 If the primitive form $f(x, y)$ properly represents $n \in \mathbb{Z}$, then

$$f(x, y) = nx^2 + bxy + cy^2$$

may be assumed by Exercise 3.2. Therefore, $D = b^2 - 4nc$. Thus, D is a quadratic residue modulo n. If n is even, then $D \equiv b^2 \pmod{8}$ where b is necessarily odd, so $D \equiv 1 \pmod{8}$. Conversely, if $D \equiv b^2 \pmod{|n|}$, where n is odd, we may assume that D and b have the same parity by replacing b by $b + n$, if necessary. Therefore, since $D \equiv 0, 1 \pmod{4}$, then $D \equiv b^2 \pmod{4|n|}$, which implies that there exists an integer m such that $D = b^2 - 4mn$. Hence, $nx^2 + bxy + my^2$ properly represents n and has discriminant D. Lastly, since $\gcd(D, n) = 1$, then $\gcd(n, b, m) = 1$, so $nx^2 + bxy + my^2$ is primitive. If n is even and $D \equiv b^2 \pmod{4|n|}$, then there exists an integer m such that $D = b^2 - 4mn$ and we proceed as above.

3.11 Let $f(x, y) = ax^2 + bxy + cy^2$ be a reduced form of discriminant $D < 0$. Thus, $b^2 \leq a^2$ and $a \leq c$. Therefore,

$$-D = 4ac - b^2 \geq 4a^2 - a^2 = 3a^2,$$

whence,

$$a \leq \sqrt{(-D)/3}.$$

For D fixed, $|b| \leq a$. This together with the latter inequality imply that there are only finitely many choices for a and b. However, since $b^2 - 4ac = D$, then there are only finitely many choices for c. We have shown that there are only finitely many reduced forms of discriminant D. By Theorem 3.1 on page 100, the number of equivalence classes of such forms is finite, which is the required result.

3.13 Since a reduced form has coefficients satisfying $b^2 \leq a^2 \leq ac$ and $b^2 - 4ac = D$, then

$$D = b^2 - 4ac \leq -3ac,$$

so $ac \leq -D/3$. When $D = -4n$, this means that

$$ac \leq 4n/3. \tag{S6}$$

We use (S6) to test for values up to the bound to prove the result.

    When $n = 1$, this means that $ac \leq 4/3$ so $a = c = 1$ is forced and $b = 0$. Hence, the only reduced form of discriminant -4 is $x^2 + y^2$. If $n = 2$, then $ac \leq 8/3$, so $c = 2$ and $a = 1$ is forced given that ac must be even since $b^2 - 4ac = -8$. Therefore, $b = 0$, and the only reduced form of discriminant -8 is $x^2 + 2y^2$. If $n = 3$, then $ac \leq 4$. Again, since ac must be even, $c \geq a$, and $\gcd(a, b, c) = 1$, then $c = 3$, $a = 1$, and $b = 0$ is forced. Thus $x^2 + 3y^2$ is the only primitive reduced form of discriminant -12. (There is one *imprimitive form*, namely $2x^2 + 2xy + 2y^2$, which we do not count.) If $n = 4$, then $ac \leq 16/3 < 6$. With the caveats as above, we must have $c = 4$, $a = 1$, $b = 0$, so $x^2 + 4y^2$ is the only primitive reduced form of discriminant -16. (There is one *imprimitive form*, namely $2x^2 + 2y^2$, which we do not count.)

    Lastly, if $n = 7$, then $ac \leq 28/3 < 9$, and $(b/2)^2 + 7 = ac$, so the only possibility is $c = 7$, $a = 1$, and $b = 0$, so $x^2 + 7y^2$ is the only primitive reduced form of discriminant -28. (There is one *imprimitive form*, namely $2x^2 + 2xy + 4y^2$, which we do not count.)

Section 3.2

3.15 If $\alpha \sim -\alpha$, then there exist $p, q, r, s \in \mathbb{Z}$ such that $ps - qr = 1$ and in the case where $\Delta_F \equiv 0 \pmod{4}$,

$$x^2 - \frac{\Delta_F}{4} y^2 = -(px + qy)^2 + \frac{\Delta_F}{4}(rx + sy)^2.$$

By comparing the coefficients of x^2, we get

$$p^2 - \frac{\Delta_F}{4} r^2 = -1,$$

so $p + r\sqrt{\Delta_F/4}$ is a unit of norm -1 in $\mathcal{O}_F = \mathbb{Z}[\sqrt{\Delta_F/4}]$.

When $\Delta_F \equiv 1 \pmod{4}$, then

$$x^2 + xy + \frac{1 - \Delta_F}{4} y^2 = -(px + qy)^2 - (px + qy)(rx + sy) - \frac{1 - \Delta_F}{4}(rx + sy)^2.$$

By comparing the coefficients of x^2 we get that

$$(2p + r)^2 - \Delta_F r^2 = -4,$$

so

$$p + \frac{1 + \sqrt{\Delta_F}}{2} r$$

is a unit of norm -1 in $\mathcal{O}_F = \mathbb{Z}[(1 + \sqrt{\Delta_F})/2]$.

3.17 Since we have that

$$\mathbf{C}_{\mathcal{O}_F}^+ = \frac{I_{\Delta_F}}{P_{\Delta_F}^+} \cong \frac{I_{\Delta_F}}{P_{\Delta_F}} \cdot \frac{P_{\Delta_F}}{P_{\Delta_F}^+},$$

then, when F is real, by Exercise 3.15, $\mathbf{C}_{\mathcal{O}_F}^+ = \mathbf{C}_{\mathcal{O}_F}$ if and only if $\mathcal{O}_F$ has a unit of norm -1. When F is complex, then $P_{\Delta_F} = P_{\Delta_F}^+$ since all norms are positive, so $\mathbf{C}_{\mathcal{O}_F}^+ = \mathbf{C}_{\mathcal{O}_F}$. This proves the assertion.

Section 3.3

3.19 Using the multiplication formulas as suggested in the hint, $g = a$, $a_3 = 1$, $b_3 = b$, $\delta = 1$, and $\mu = \nu = 0$, so

$$II' = (a) \left(1, \frac{b + \sqrt{\Delta_F}}{2}\right) \sim (a) \sim (1),$$

so $I' \sim I^{-1}$ in $\mathbf{C}_{\mathcal{O}_F}$.

3.21 Set $\alpha = 1 + u$ if $u \neq -1$, and $\alpha = \sqrt{\Delta_F}$ if $u = -1$. If $u \neq -1$, then

$$(1 + u')u = u + uu' = u + N_F(u) = u + 1.$$

Therefore,

$$\frac{\alpha}{\alpha'} = \frac{u + 1}{u' + 1} = u.$$

If $u = -1$, then

$$\frac{\alpha}{\alpha'} = \frac{\sqrt{\Delta_F}}{-\sqrt{\Delta_F}} = -1 = u,$$

as required.

Section 3.4

3.23 By Theorem 3.2 on page 102, we know that $h(D) = 1$ for

$$D \in \{-4, -8, -12, -16, -28\},$$

and indeed these are the only ones of the form $D = -4n$ with $h(D) = 1$. We now look at the remainder of the form $D \equiv 1 \pmod 4$. By the argument in the solution of Exercise 3.13 on page 413, a form $ax^2 + bxy + cy^2$ of discriminant $D = b^2 - 4ac$ must satisfy that $ac \leq -D/3$ and must satisfy the inequalities in Definition 3.4 on page 100. For $D = -7$ this says $ac \leq 7/3$ and the only values that satisfy these restrictions are $(a, b, c) = (1, 1, 2)$. For $D = -11$, $ac \leq 11/3$ and the only values satisfying our criteria are $(a, b, c) = (1, 1, 3)$. For $D = -19$, $ac \leq 19/3$ for which only $(a, b, c) = (1, 1, 5)$ works. Lastly for $D = -43$, only $(a, b, c) = (1, 1, 11)$ fits the inequalities. This completes the solution.

3.25 By Corollary 3.8 on page 138 $p = x^2 + 14y^2$ if and only if $p \equiv z^2 \pmod{56}$ or $p \equiv z^2 + 14 \pmod{56}$ for some integer z and this holds if and only if $p \equiv 1, 9, 15, 23, 25, 39$, where the values correspond to $z = 1, 3, 5$ in each case. Moreover, it is straightforward to check that $2x^2 + 7y^2$ represents the same congruence classes in $(\mathbb{Z}/56\mathbb{Z})^*$. Thus, they are in the same genus.

3.27 By Theorem 3.14 on page 142, the number of forms in each genus is $h_{\Delta_F}/2^{\tau-1}$. Thus, there is a single class of forms in each genus if and only if $h_{\Delta_F}/2^{\tau-1} = 1$.

3.29 Using the same argument as in the solution of Exercise 3.23, any reduced form $ax^2 + bxy + cy^2$ must satisfy

$$ac \leq -D/3 = 56/3 < 19.$$

Testing for this inequality together with the inequalities in Definition 3.4, the only solutions are for

$$(a, b, c) \in \{(1, 0, 14), (2, 0, 7), (3, 2, 5), (3, -2, 5)\}.$$

Thus, $h(-56) = 4$.

3.31 Using the hint, we see that when $b^2 - 4ac = \Delta_F \equiv 0 \pmod 4$, then b is even so

$$acx^2 + bxy + y^2 = (bx/2 + y)^2 - \frac{\Delta_F}{4}x^2$$

since comparing the coefficients of x^2, we get $b^2/4 - \Delta_F/4 = ac$, comparing the coefficients of xy we get $b = b/2 \cdot 2$, and the coefficients of y^2 are both 1. When $\Delta_F \equiv 1 \pmod 4$, then b is odd so

$$acx^2 + bxy + y^2 = \left(-\frac{b+1}{2}x - y\right)^2 + \left(-\frac{b+1}{2}x - y\right)x + \frac{1 - \Delta_F}{4}x^2,$$

since comparing the coefficients of x^2 we get

$$\left(\frac{b+1}{2}\right)^2 - \frac{b+1}{2} + \frac{1 - \Delta_F}{4} = \frac{b^2 + 2b + 1 - 2b - 2 + 1 - b^2 + 4ac}{4} = ac,$$

and comparing the coefficients of xy we get

$$2 \cdot \frac{b+1}{2} - 1 = b,$$

and the coefficients of y^2 are both 1.

3.33 If (a) holds, then $C_{\Delta_F}^2 = \{1\}$ by the hint, so every element in C_{Δ_F} has order 1 or 2. Thus, by Exercise 3.32, (b) holds. Conversely, if (b) holds, then by Exercise 3.32, every element in C_{Δ_F} has order 1 or 2, so the principal genus is $C_{\Delta_F}^2 = \{1\}$, a single class. However, every genus has the same numbers of classes of forms so we have the result.

3.35 They are $f = (a, b, c)$ for the values $(1, 0, 20)$, $(3, 2, 7)$, $(3, -2, 7)$, and $(4, 0, 5)$.

3.37 For each of the following values of z, and primes p we have

$$p \equiv z^2 + z - 57 \pmod{229}.$$

For $p = 643949$ we have $z = -803$ and $p = 803^2 - 803 - 57$. For $p = 17863$ we have $z = 113$ and $113^2 + 113 - 57 = 17863 - 22 \cdot 229$. For $p = 24733$ we have $z = 113$ and $113^2 + 113 - 57 = 24733 - 52 \cdot 229$.

Section 3.5

3.39 We have that $(-7/p) = (-1/p)(7/p) = 1$ if and only if $(-1/p) = (7/p) = -1$ or $(-1/p) = (7/p) = 1$. Thus, $(-7/p) = 1$ if and only if either $p \equiv -1 \pmod 4$ and $p \equiv \pm 1, 2, 4 \pmod 7$, or else $p \equiv 1 \pmod 4$ and $p \equiv \pm 1, 2, 4 \pmod 7$. In other words, $(-7/p) = 1$ if and only if either $p \equiv 11, 15, 23 \pmod{28}$ or $p \equiv 1, 9, 25 \pmod{28}$, which is to say if and only if $p \equiv 1, 9, 11, 15, 23, 25 \pmod{28}$.

3.41 Since $(-19/p) = (-1/p)(19/p) = 1$ if and only if $(-1/p) = (19/p) = -1$ or $(-1/p) = (19/p) = 1$, then $(-19/p) = 1$ if and only if either $p \equiv -1 \pmod 4$ and

$$p \equiv 1, 4, 5, 6, 7, 9, 11, 16, 17 \pmod{19},$$

or else $p \equiv 1 \pmod 4$ and

$$p \equiv 1, 4, 5, 6, 7, 9, 11, 16, 17 \pmod{19}.$$

This means that $(-19/p) = 1$ if and only if either

$$p \equiv 7, 11, 23, 35, 39, 43, 47, 55, 63, \pmod{76},$$

or

$$p \equiv 1, 5, 9, 17, 25, 45, 49, 61, 73 \pmod{76},$$

namely if and only if

$$p \equiv 1, 5, 7, 9, 11, 17, 23, 25, 35, 39, 43, 45, 47, 49, 55, 61, 63, 73 \pmod{76}.$$

By Example 3.10, Theorem 1.3 on page 6, and (3.6), we have that $h_{-19} = h_{\mathbb{Z}[(1+\sqrt{-19})/2]} = 1$. Thus, by Theorem 3.15, if $(\Delta_F/p) = (-19/p) = 1$, then $p = a^2 + ab + 5b^2$ for some integers a, b. Also $19 = 1^2 - 1 \cdot 2 + 5 \cdot 2^2$. Conversely, by Exercise 3.9 on page 104, if $p \neq 19$ and $p = a^2 + ab + 5b^2$, then $(-19/p) = 1$.

3.43 By the same methodology as in Exercise 3.41, we get that $(-67/p) = 1$ if and only if either

$$p \equiv 15, 19, 23, 35, 39, 47, 55, 59, 71, 83, 91, 103, 107, 123, 127, 131, 135, 143, 151, 155$$

$$159, 163, 167, 171, 183, 199, 207, 211, 215, 223, 227, 255, 263 \pmod{268}, \quad (S7)$$

or

$$p \equiv 1, 9, 17, 21, 25, 29, 33, 37, 49, 65, 73, 77, 81, 89, 93, 121, 129, 149, 153, 157, 169,$$

$$173, 181, 189, 193, 205, 217, 225, 237, 241, 257, 261, 265 \pmod{268}. \quad \text{(S8)}$$

Lastly, (S7)–(S8) hold if and only if

$$p \equiv 1, 9, 15, 17, 19, 21, 23, 25, 29, 33, 35, 37, 39, 47, 49, 55, 59, 65, 71, 73, 77, 81, 83,$$

$$89, 91, 93, 103, 107, 121, 123, 127, 129, 131, 135, 143, 149, 151, 153, 155, 157, 159,$$

$$163, 167, 169, 171, 173, 181, 183, 189, 193, 199, 205, 207, 211, 215, 217, 223, 225,$$

$$227, 237, 241, 255, 257, 261, 263, 265 \pmod{268}.$$

Now the result is established exactly as in Exercise 3.41.

3.45 The following are all of the prime values or 1 for each discriminant.

Δ_F	$x^2 + x + (1 - \Delta_F)/4$	> 0 values for $x = 1, 2, \ldots, \lfloor (\sqrt{\Delta_F} - 1)/2 \rfloor$
17	$x^2 + x - 4$	2
21	$x^2 + x - 5$	3
29	$x^2 + x - 7$	5, 1
37	$x^2 + x - 9$	7, 3
53	$x^2 + x - 13$	11, 7, 1
77	$x^2 + x - 19$	17, 13, 7
101	$x^2 + x - 25$	23, 19, 13, 5
173	$x^2 + x - 43$	41, 37, 31, 23, 13, 1
197	$x^2 + x - 49$	47, 43, 37, 29, 19, 7
293	$x^2 + x - 73$	71, 67, 61, 53, 43, 31, 17, 1
437	$x^2 + x - 109$	107, 103, 97, 89, 79, 67, 53, 37, 19
677	$x^2 + x - 169$	167, 163, 157, 149, 139, 127, 113, 97, 79, 59, 37, 13

Section 3.6

3.47 If $(1, 0, -\Delta_F) \sim (1, 0, -1)$, then there is a transformation $x = rX + sY$ and $y = tX + uY$ such that

$$p \nmid (ru - st) \quad \text{(S9)}$$

and

$$(rX + sY)^2 - \Delta_F(tX + uY)^2 \equiv X^2 - Y^2 \pmod{p}.$$

It follows that

$$r^2 - t^2 \Delta_F \equiv 1 \pmod{p}, \quad \text{(S10)}$$

$$s^2 - u^2 \Delta_F \equiv -1 \pmod{p}, \quad \text{(S11)}$$

and

$$rs \equiv tu \Delta_F \pmod{p}. \quad \text{(S12)}$$

Multiplying (S10) by u^2, we get

$$r^2 u^2 - t^2 u^2 \Delta_F \equiv u^2 \pmod{p}. \quad \text{(S13)}$$

Multiplying (S11) by t^2, we get

$$t^2 s^2 - t^2 u^2 \Delta_F \equiv -t^2 \pmod{p}. \tag{S14}$$

Now if $p \nmid ru$ and $p \nmid ts$, then we may multiply (S13) by $(ru)^{-1}$ modulo p and by employing (S12), we get

$$ru - ts \equiv ur^{-1} \pmod{p}. \tag{S15}$$

Similarly multiplying (S14) by $-(ts)^{-1}$, and using (S12),

$$ru - ts \equiv ts^{-1} \pmod{p}. \tag{S16}$$

From (S15)–(S16), we get

$$ur^{-1} \equiv ts^{-1} \pmod{p},$$

which implies that
$$tr \equiv us \pmod{p}. \tag{S17}$$

Multiplying (S12) by tu and employing (S17), we get

$$t^2 u^2 \Delta_F \equiv trus \equiv (us)^2 \pmod{p},$$

contradicting that Δ_F is a quadratic nonresidue modulo p. Hence, either $p \mid (ru)$ or $p \mid (ts)$ but not both due to (S9). If $p \mid (ur)$, and $p \nmid (ts)$, then either $p \mid u$ or $p \mid r$. If $p \mid u$, then $p \mid (tr)$ by (S17). Since $p \nmid t$, then $p \mid r$. Thus by (S10),

$$t^2 \Delta_F \equiv -1 \pmod{p},$$

which implies that $p \equiv 3 \pmod 4$ since Δ_F is a quadratic nonresidue modulo p. However, by (S11), $s^2 \equiv -1 \pmod{p}$ contradicting that $p \equiv 3 \pmod 4$. We have shown that $p \nmid u$. If $p \mid r$, then $p \mid (us)$ as above, but $p \nmid s$. We have shown that p cannot divide ur. Thus, $p \mid (ts)$, so $p \mid t$ or $p \mid s$. If $p \mid t$, then by (S17), p must divide s since it cannot divide u. Thus, by(S11), $u^2 \Delta_F \equiv 1 \pmod{p}$, contradicting that Δ_F is a quadratic nonresidue modulo p. This completes the proof that $(1, 0, -\Delta_F) \not\sim (1, 0, -1)$.

3.49 If $(0, 1, 0) \sim (1, 1, 1) \pmod 2$, then there is a transformation $x = rX + sY$ and $y = tX + uY$ with $ru - st$ odd, such that

$$(rX + sY)(tX + uY) \equiv X^2 + XY + Y^2 \pmod 2.$$

This implies that
$$rt = 1,$$
$$ru + st = 1,$$

and
$$su = 1.$$

However, the first and last equations imply that $r = t = 1$ or $r = t = -1$, and $s = u = 1$ or $s = u = -1$, and these do not solve the middle equation.

3.51 The existence of integers n_j with $\gcd(n_j, \Delta_{F_j}) = 1$ for $j = 1, 2$ is guaranteed by Lemma 3.1. Since $\gcd(\Delta_{F_j}, n_j) = 1$ and $p \mid \Delta_{F_j}$ for $j = 1, 2$, then $\gcd(n_j, p) = 1$. Also, there are integers x_j, y_j such that $n_j = a_j x_j^2 + b_j x_j y_j + c_j y_j^2$. Therefore,

$$\left(\frac{n_j}{p}\right) = \left(\frac{a_j x_j^2 + b_j x_j y_j + c_j y_j^2}{p}\right). \tag{S18}$$

However since

$$4a_j(a_j x_j^2 + b_j x_j y_j + c_j y_j^2) \equiv (2ax_j + b_j y_j)^2 \pmod{p},$$

given that

$$4a_j c_j \equiv b_j^2 \pmod{p},$$

because $p \mid \Delta_{F_j}$ for $j = 1, 2$, this implies that

$$\left(\frac{4a_j}{p}\right)\left(\frac{a_j x_j^2 + b_j x_j y_j + c_j y_j^2}{p}\right) = 1,$$

so by (S18),

$$\left(\frac{n_j}{p}\right) = \left(\frac{a_j}{p}\right) \text{ for } j = 1, 2. \tag{S19}$$

Suppose that $(a_1, b_1, c_1) \sim (a_2, b_2, c_2) \pmod{p}$. Then

$$\left(\frac{n_1}{p}\right) = \left(\frac{a_1}{p}\right) = \left(\frac{a_1 x_1^2 + b_1 x_1 y_1 + c_1 y_1^2}{p}\right) =$$

$$\left(\frac{a_2 x_2^2 + b_2 x_2 y_2 + c_2 y_2^2}{p}\right) = \left(\frac{a_2}{p}\right) = \left(\frac{n_2}{p}\right).$$

Conversely, if

$$\left(\frac{n_1}{p}\right) = \left(\frac{n_2}{p}\right),$$

then by (S18)–(S19),

$$\left(\frac{a_1}{p}\right) = \left(\frac{a_2}{p}\right).$$

Hence, there exists a $z \in \mathbb{Z}$ such that $a_1 \equiv z^2 a_2 \pmod{p}$. This implies

$$(a_1, b_1, c_1) \sim (n_1, 0, 0) \sim (a_1, 0, 0) \sim (a_2, 0, 0) \sim (n_2, 0, 0) \sim (a_2, b_2, c_2) \pmod{p}.$$

Section 4.1

4.1 Since we know from the hint that

$$\alpha - \frac{A_j}{B_j} = \frac{(-1)^j}{B_j(\alpha_{j+1}B_j + B_{j-1})},$$

then

$$\left| \alpha - \frac{A_j}{B_j} \right| = \left| \frac{1}{B_j((q_{j+1} + 1/\alpha_{j+2})B_j + B_{j-1})} \right| \leq \frac{1}{q_{j+1}B_j^2}.$$

4.3 For the first part, with $j = 1, 2$, let $d_j \geq 0$ with $f_j(x) = \sum_{i=0}^{d_j} a_i^{(j)} x^i$. Thus,

$$\gcd(f_1(x)f_2(x)) = \gcd\left(\sum_{i=0}^{d_1} a_i^{(1)} x^i \sum_{k=0}^{d_2} a_k^{(2)} x^k\right) = \gcd\left(\sum_{i=0}^{d_1} a_i^{(1)} \sum_{k=0}^{d_2} a_k^{(2)} x^{j+k}\right) =$$

$$\gcd\{a_i^{(1)} a_k^{(2)}\}_{\substack{1 \leq i \leq d_1 \\ 1 \leq k \leq d_2}} =$$

$$\gcd\left(\sum_{j=0}^{d_1} a_j^{(1)} x^j\right) \gcd\left(\sum_{k=0}^{d_2} a_k^{(2)} x^k\right) =$$

$$\gcd(f_1(x)) \gcd(f_2(x)).$$

Now, if $f(x) \in \mathbb{Z}[x]$, then we may assume, without loss of generality, that $\gcd(f(x)) = 1$ since we may otherwise just look at $F(x) = f(x)/\gcd(f(x)) \in \mathbb{Z}[x]$. If $f(x) = g(x)h(x)$ where $g(x), h(x) \in \mathbb{Q}[x]$, then we may find rational numbers ℓ_g and ℓ_h such that $\ell_g g(x) \in \mathbb{Z}[x]$, $\ell_h h(x) \in \mathbb{Z}[x]$, and $\gcd(\ell_g g(x)) = 1 = \gcd(\ell_h h(x))$, so from the above

$$\gcd(\ell_g \ell_h f(x)) = \gcd(\ell_g g(x)) \gcd(\ell_h h(x)) = 1.$$

Hence, $\ell_g \ell_h = \pm 1$. By setting

$$H(x) = \operatorname{sign}(\ell_h)\ell_h h(x) \text{ and } G(x) = \operatorname{sign}(\ell_g)\ell_g g(x),$$

where $\operatorname{sign}(\ell_g) = 1$, if $\ell_g > 0$, and $\operatorname{sign}(\ell_g) = -1$, if $\ell_g < 0$, and similarly for $\operatorname{sign}(\ell_h)$. Hence, we have that

$$f(x) = G(x)H(x),$$

as required.

4.5 Since the base-a expansion of the number is $(.100100001\ldots)_a$, which is infinitely nonrepeating, then we know that it is irrational.

Section 4.2

4.7 An easy check shows that

$$0 < (-1)^{n+1}\beta_n = \sum_{j=1}^{\infty} \frac{(-1)^{j+1}}{(n+j)!} < \frac{1}{(n+1)!}.$$

Thus,

$$0 < n!\beta_n(-1)^{n+1} < \frac{1}{n+1} < 1,$$

which implies that

$$n!e^{-1} = n!\alpha_n + n!\beta_n(-1)^{n+1} \notin \mathbb{Z}.$$

We have shown that $e^{-1} \notin \mathbb{Q}$ since $n!\alpha_n \in \mathbb{Z}$, so $e \notin \mathbb{Q}$.

4.9 Since $\pi = a/b \in \mathbb{Q}$ and

$$f(x) = f^{(0)}(x) = \frac{x^n(a - bx)^n}{n!},$$

then we may set

$$G(x) = \sum_{j=0}^{n} (-1)^j f^{(2j)}(x).$$

Since $f^{(2j)}(0)$ and $f^{(2j)}(\pi)$ are integers for all $j = 0, 1, \ldots, n$, then $G(0), G(\pi) \in \mathbb{Z}$. Also, since

$$\frac{d}{dx}\left(G'(x)\sin(x) - G(x)\cos(x)\right) = \left(G''(x) + G(x)\right)\sin(x)$$

$$= \left(f^{(0)}(x) + \sum_{j=0}^{n-1}(-1)^j f^{2(j+1)}(x) + \sum_{k=0}^{n-1}(-1)^{k+1}f^{2(k+1)}(x)\right)\sin(x) = f(x)\sin(x),$$

then

$$\int_0^\pi f(x)\sin(x)dx = G(\pi) + G(0) \in \mathbb{Z}. \tag{S20}$$

However, by selecting n large enough, we must have

$$0 < f(x)\sin(x) < \frac{\pi^n a^n}{n!} < \frac{1}{\pi},$$

so

$$0 < \int_0^\pi f(x)\sin(x)dx < 1,$$

contradicting (S20).

Section 4.3

4.11 Since every subgroup of a free abelian group of rank n is a free abelian group of rank at most n, set the rank of H to be $m \leq n$. Then G/H has $n - m$ infinite cyclic factors. Hence, G/H is finite if and only if $m = n$. If L is a lattice with free abelian subgroup H of rank n, then H is a full lattice in $\mathbb{R}^n$.

Section 5.1

5.1 Since $f(x) = x/(e^x - 1) + x/2$ is an even function, namely, $f(x) = f(-x)$, then $B_n = (-1)^n B_n$ for any $n > 1$, so for odd n, $B_n = 0$.

5.3 According to the hint, if $\sum_{j=1}^{\infty}(1/j) = d \in \mathbb{R}$. Then there is an $N \in \mathbb{N}$ such that $N \leq d < N + 1$. Also, note that

$$\sum_{j=1}^{\infty}\frac{1}{j} = 1 + \frac{1}{2} + \left(\frac{1}{3} + \frac{1}{4}\right) + \left(\frac{1}{5} + \frac{1}{6} + \frac{1}{7} + \frac{1}{8}\right) + \cdots > 1 + \frac{1}{2} + \frac{1}{2} + \frac{1}{2} + \cdots$$

so each block has a sum bigger than $1/2$. Let $M \in \mathbb{N}$ be chosen such that the number of blocks larger than $1/2$ satisfies $M \geq 2N$. Then

$$d = \sum_{j=1}^{\infty}\frac{1}{j} > 1 + \frac{2M}{2} \geq N + 1,$$

a contradiction.

5.5 Since $F(s,x) - F(s, x-1) = se^{s(x-1)}$, then

$$\frac{B_{n+1}(x) - B_{n+1}(x-1)}{n+1} = (x-1)^n. \tag{S21}$$

Adding (S21) for $x = 1, 2, \ldots k$, we get the result.

5.7 Since we know from Exercise 5.6 that $B'_{n+1}(x) = (n+1)B_n(x)$, then

$$\int_a^b B_n(t)dt = \frac{1}{n+1}\int_a^b B'_{n+1}(x) = \frac{1}{n+1}(B_{n+1}(b) - B_{n+1}(a)).$$

Section 5.2

5.9 By Theorem 5.9 on page 214 and the hint,

$$\frac{\sum_{j=1}^n \phi(j)}{n(n+1)/2} \approx \frac{6n^2}{\pi^2 n^2} = \frac{6}{\pi^2}.$$

5.11 By the definition of the Möbius function and Theorem 5.9, we have simply a restatement, namely

$$\sum_{n \le x} |\mu(n)| = \frac{6x}{\pi^2} + O(\sqrt{x}),$$

from which it follows that the mean value of μ^2 is $6/\pi^2$.

Section 5.3

5.13 Suppose that $|f(p)| \ge 1$ for some prime p. Then

$$\sum_{n=1}^\infty |f(n)| \ge \sum_{j=0}^\infty |f(p^j)| = \sum_{j=0}^\infty |f(p)|^j$$

and the latter series clearly diverges. This shows that $|f(p)| < 1$ for each prime p so

$$\sum_{j=0}^\infty f(p)^j - \sum_{j=0}^\infty f(p)^{j+1} = \lim_{n \to \infty}\left(\sum_{j=}^n f(p)^j - \sum_{j=0}^n f(p)^{j+1}\right)$$

$$= \lim_{n \to \infty}\left(\sum_{j=}^n (f(p)^j - f(p)^{j+1})\right) = \lim_{n \to \infty}\left(1 - f(p)^{n+1}\right) = 1,$$

so

$$\sum_{j=0}^\infty f(p^j) = \sum_{j=0}^\infty f(p)^j = \frac{1}{1 - f(p)}.$$

The result now follows.

5.15 Let $n = k + 1$ and $s = -k$ in Theorem 5.10. Thus,

$$\zeta(-k) = \frac{1}{-k-1} + \frac{1}{2} + \sum_{j=2}^{k+1} \frac{B_j}{j!}(-k)(-k+1)\cdots(-k+j-2)$$

$$= \frac{-1}{k+1}\left(1 + \left(-\frac{1}{2}\right)(k+1) + \sum_{j=2}^{k+1}\binom{k+1}{j}B_j\right)$$

$$= \frac{-1}{k+1}\sum_{j=0}^{k+1}\binom{k+1}{j}B_j = -\frac{B_{k+1}}{k+1},$$

if k is odd and equals 0 if k is even, by Exercise 5.5.

5.17 This is an immediate consequence of the answer provided in the solution of Exercise 5.15.

5.19 This is immediate from Theorem 5.10 on page 219.

5.21 The integral

$$\int_0^1 B_3(t - \lfloor t\rfloor)t^{-s-3}\,dt$$

is convergent for $Re(s) < -1$. Using Exercise 5.6 on page 206 and integration by parts (three times) we get

$$\int_0^1 B_3(t-\lfloor t\rfloor)t^{-s-3}\,dt = -\frac{1}{s+2}\int_0^1 B_3(t-\lfloor t\rfloor)\,dt^{-s-2} = -\frac{1}{s+2}B_3(t-\lfloor t\rfloor)t^{-s-2}\Big|_0^1$$

$$+\frac{1}{s+2}\int_0^1 t^{-s-2}\,dB_3(t-\lfloor t\rfloor) = \frac{3}{s+2}\int_0^1 t^{-s-2}B_2(t-\lfloor t\rfloor)\,dt$$

$$= -\frac{3}{(s+1)(s+2)}\int_0^1 B_2(t-\lfloor t\rfloor)\,dt^{-s-1} = -\frac{3}{(s+1)(s+2)}B_2(t-\lfloor t\rfloor)t^{-s-1}\Big|_0^1$$

$$+\frac{3}{(s+1)(s+2)}\int_0^1 t^{-s-1}\,dB_2(t-\lfloor t\rfloor) = -\frac{1}{2(s+1)(s+2)}$$

$$+\frac{6}{(s+1)(s+2)}\int_0^1 t^{-s-1}B_1(t-\lfloor t\rfloor)\,dt = -\frac{1}{2(s+1)(s+2)}$$

$$-\frac{6}{(s+1)(s+2)}\int_0^1 B_1(t-\lfloor t\rfloor)\,dt^{-s} = -\frac{1}{2(s+1)(s+2)}+$$

$$\frac{6}{s(s+1)(s+2)}\left(t^{-s}B_1(t-\lfloor t\rfloor)\right)\Big|_0^1 + \frac{6}{s(s+1)(s+2)}\int_0^1 t^{-s}\,dB_11(t-\lfloor t\rfloor)$$

$$= -\frac{1}{2(s+1)(s+2)} - \frac{3}{s(s+1)(s+2)} + \frac{6}{s(s+1)(s+2)}\int_0^1 t^{-s}B_0(t-\lfloor t\rfloor)\,dt$$

$$= -\frac{1}{2(s+1)(s+2)} - \frac{3}{s(s+1)(s+2)} + \frac{6}{s(s+1)(s+2)}\int_0^1 t^{-s}\,dt$$

$$= -\frac{1}{2(s+1)(s+2)} - \frac{3}{s(s+1)(s+2)} - \frac{6}{(s-1)s(s+1)(s+2)}t^{-s+1}\Big|_0^1$$

$$= -\frac{1}{2(s+1)(s+2)} - \frac{3}{s(s+1)(s+2)} - \frac{6}{(s-1)s(s+1)(s+2)}$$

$$= -\frac{(s+3)}{2s(s-1)(s+1)}.$$

Hence,

$$\frac{s(s+1)(s+2)}{6}\int_0^1 B_3(t - \lfloor t \rfloor)t^{-s-3}\,dt = -\frac{(s+2)(s+3)}{12(s-1)}$$

$$= -\frac{s}{12} - \frac{1}{2} - \frac{1}{s-1} = -\frac{B_2 s}{2} - \frac{1}{2} - \frac{1}{s-1},$$

as required.

5.23 By Exercise 5.22, the result is immediate since we let $x - 1 = -s$, then

$$\Gamma(1 - s) = \Gamma(x) = (x - 1)\Gamma(x - 1) = (-s)\Gamma(-s).$$

5.25 By the hint,

$$\Gamma(n) = (n-1)\Gamma(n-1) = (n-1)(n-2)\Gamma(n-3) = \cdots = (n-1)!.$$

Section 6.1

6.1 $x \equiv 20 \,(\mathrm{mod}\ 7^2)$.

6.3 $x \equiv 239, 1958, 2196 \,(\mathrm{mod}\ 13^3)$.

6.5 $5 + 3 \cdot 7 + 3 \cdot 7^2 + 3 \cdot 7^3 + 3 \cdot 7^4 + 3 \cdot 7^5 + 3 \cdot 7^6 + 3 \cdot 7^7 + 3 \cdot 7^8 + \cdots$.

6.7 $3 + 2 \cdot 5 + 2 \cdot 5^3 + 2 \cdot 5^4 + 4 \cdot 5^5 + 5^6 + 3 \cdot 5^7 + 5^8 + 5^9 + 2 \cdot 5^{11} + 2 \cdot 5^{12} + 2 \cdot 5^{15} + 5^{16} + 3 \cdot 5^{18} + 3 \cdot 5^{19} + + 5^{21} + 5^{24} + 5^{25} + 4 \cdot 5^{28} + 3 \cdot 5^{29} \cdots$

Section 6.2

6.9 First, parts (a)–(b) of Definition 6.2 follow immediately from Definition 6.3. Part (c) is, for any $x, y \in \mathbb{Q}$, that

$$|x + y|_p = p^{-(\nu_p(x+y))} \le p^{-\nu_p(x)} + p^{-\nu_p(y)},$$

since

$$v_p(x + y) \ge \min\{\nu_p(x), \nu_p(y)\}$$

from which the non-Archimedean property follows.

6.11 By Definition, for any $\varepsilon > 0$, there is an integer $n = n(\varepsilon)$ such that

$$|q_j - q_k|_p < \varepsilon \text{ for all } j, k > n.$$

Thus,

$$|q_j|_p - |q_k|_p \le |q_j - q_k|_p < \varepsilon \text{ for all } j, k > n.$$

By taking $k = n + 1$ and adding $|q_{n+1}|_p$ to both sides,

$$|q_j|_p < |q_{n+1}|_p + \varepsilon.$$

Hence, for all $j \in \mathbb{N}$,

$$|q_j|_p \le \max\{|q_1|_p, \ldots, |q_n|_p, |q_{n+1}|_p + \varepsilon\}.$$

By setting $M = \max\{|q_1|_p, \ldots, |q_n|_p, |q_{n+1}|_p + \varepsilon\}$, we have our result.

6.13 The reflexive property is clear. Also, since

$$\overset{(v)}{\underset{j\to\infty}{\lim}} (q_j - q_j') = 0 \text{ if and only if } \overset{(v)}{\underset{j\to\infty}{\lim}} (q_j' - q_j) = 0,$$

then $\overline{\{q_j\}} = \overline{\{q_j'\}}$ implies $\overline{\{q_j'\}} = \overline{\{q_j\}}$, which is symmetry. Lastly, if $\overline{\{q_j\}} = \overline{\{q_j'\}}$ and $\overline{\{q_j'\}} = \overline{\{q_j''\}}$, then

$$\overset{(v)}{\underset{j\to\infty}{\lim}} (q_j - q_j') = 0 \text{ and } \overset{(v)}{\underset{j\to\infty}{\lim}} (q_j' - q_j'') = 0,$$

so by symmetry,

$$\overset{(v)}{\underset{j\to\infty}{\lim}} (q_j'' - q_j') = 0.$$

Therefore,

$$\overset{(v)}{\underset{j\to\infty}{\lim}} (q_j - q_j' - (q_j'' - q_j')) = \overset{(v)}{\underset{j\to\infty}{\lim}} (q_j - q_j'') = 0,$$

so $\overline{\{q_j\}} = \overline{\{q_j''\}}$, which establishes transitivity. Hence, Cauchy sequences are partitioned into classes as an equivalence relation.

6.15 If $\lim_{j\to\infty} q_j = L \in \mathbb{R}$, then given $\varepsilon > 0$, select $N \in \mathbb{N}$ such that

$$|q_j - L| < \varepsilon \text{ for } j > N.$$

Then, if $j, k > N$, we have

$$|q_j - q_k| = |(q_j - L) - (q_k - L)| \leq |q_j - L| + |q_k - L| < 2\varepsilon,$$

from which it follows that the sequence is Cauchy.

6.17 Let $x = 5/4$ and $y = 5$. Then $x + y = 25/4$, so

$$|x + y|_5 = 5^{-2} < \max\{|x|_5, |y|_5\} = 5^{-1}.$$

6.19 Given three points x, y, z of a triangle, we have that

$$|x - y|_p + |y - z|_p = |x - z|_p,$$

so if $|x - y|_p \neq |y - z|_p$, then by Exercise 6.16,

$$|x - z|_p = |(x - y) + (y - z)|_p = \max\{|x - y|_p, |y - z|_p\},$$

so two of the sides must be equal.

Section 6.3

6.21 Let $k + \ell = m > j$, then by (6.8) on page 233,

$$|q_m - q_j|_p = |q_{k+\ell} - q_{k+\ell-1} + q_{k+\ell-1} - q_{k+\ell-2} \pm \cdots \pm q_{j+1} - q_j|_p$$

$$\leq \max\{|q_{k+\ell} - q_{k+\ell-1}|_p, \dots, |q_{j+1} - q_j|_p\},$$

which yields the result.

Section 6.4

6.23 By Theorem 6.4 on page 244,

$$\alpha = a/b = p^{-\ell}\left(\left(\sum_{j=0}^{m}c_jp^j\right) + \sum_{j=0}^{\infty}p^{m+1+jn}C\right),$$

where

$$C = \sum_{j=m+1}^{m+n}c_jp^{j-m-1}.$$

Thus, $|\alpha|_p \geq 0$ if and only if $\ell = 0$, namely $p \nmid b$.

6.25 If we let $\alpha \in \mathcal{O}_p$, then the polynomial $f(x) = \alpha x - 1$ has a root if and only if α is a unit in $\mathcal{O}_p$ and α^{-1} is its other root. Thus, $f(x) \equiv 0 \,(\text{mod } \mathcal{P})$ is solvable if and only if α is a unit, but

$$f'(x) = \alpha \not\equiv 0 \pmod{\mathcal{P}}$$

since no element of $\mathcal{P}$ can be invertible. By Lemma there exists a p-adic integer α^{-1} such that $f(\alpha^{-1}) = 0$ so $\alpha\alpha^{-1} = 1$. This shows that a p-adic integer is invertible in $\mathcal{O}_p$ if and only if $\alpha \in \mathcal{O}_p/\mathcal{P}$. By Theorem 2.7 on page 68, $\mathcal{P}$ is a maximal ideal.

6.27 If $|\alpha|_p = p^{-n}$, then $u = \alpha p^{-n} \in \mathcal{U}_p$, so $\alpha = up^n$. If

$$\alpha = up^n = vp^m,$$

where $u, v \in \mathcal{U}_p$, then

$$|\alpha|_p = p^{-n} = p^{-m},$$

so $m = n$ and $u = v$.

Section 7.1

7.1 Since $n \not\equiv 0, 1\,(\text{mod } D)$, then there is a prime p dividing D such that

$$n \not\equiv 1 \pmod{p^a} \text{ where } p^a \,\|\, D.$$

Also, as in the proof of Theorem 7.1 on page 249, there is a character χ^{p^a} with $\chi^{p^a}(n) \neq 1$ which is possible since there exist $\phi(p^a)$ distinct characters modulo p^a, and

$$\phi(p^a) = p^{a-1}(p-1) > 1,$$

since $D = p^a = 2$ is not possible given the existence of $n \not\equiv 0, 1\,(\text{mod } D)$. (For instance, if p is odd choose a primitive root g modulo p^a and the character $\chi^{p^a}(g) = g$. Since $n \equiv g^i \not\equiv 1\,(\text{mod } p^a)$ for some i with $1 \leq i < \phi(p^a)$, then $\chi(n) = \chi(g^i) = g^i \neq 1$.) If $D \neq p^a$, then select

$$\chi^{D/p^a} = \chi_0^{D/p^a},$$

then the product of these characters is a character χ for which $\chi(n) \neq 1$.

Section 7.2

7.3 By using Exercises 5.12–5.13, with $f(n) = \chi(n)n^{-s}$, the result is an immediate consequence in view of the absolute convergence given by Exercise 7.2.

7.5 We have for $0 < \Re(s) < 1$,

$$\Gamma(s)\Gamma(1-s) = \int_0^\infty e^{-t}t^{s-1}dt \int_0^\infty e^{-x}x^{-s}dx,$$

and by letting $t = xu$, we get that the latter equals

$$\int_0^\infty e^{-xu}(xu)^{s-1}xdu \int_0^\infty e^{-x}x^{-s}dx = \int_0^\infty (e^{-xu-x}dx)u^{s-1}du,$$

and now by letting $y = x(u+1)$, the latter equals

$$\int_0^\infty \frac{e^{-y}}{u+1}dy(u^{s-1}du) = \int_0^\infty [-e^{-y}]_0^\infty \left(\frac{u^{s-1}}{u+1}du\right) = \int_0^\infty \frac{u^{s-1}}{u+1}du,$$

and by the hint, this gives us the result. The last equality in the exercise follows from the formula from elementary calculus that

$$\sin(2\theta) = 2\sin\theta\cos\theta.$$

7.7 By Theorem 7.3,

$$L(s,\chi) = \prod_{p=\text{prime}} (1 - \chi(p)p^{-s})^{-1}.$$

By taking logs we get

$$\log_e L(s,\chi) = -\sum_{p=\text{prime}} \log_e(1 - \chi(p)p^{-s}) = \sum_{p=\text{prime}} \sum_{m=1}^\infty \frac{\chi(p^m)}{mp^{ms}}.$$

Since the latter is absolutely convergent for $\Re(s) > 1$, then we may interchange the order of summation to get that it equals

$$\sum_{m=1}^\infty \sum_{p=\text{prime}} \frac{\chi(p^m)}{mp^{ms}} = \sum_{p=\text{prime}} \frac{\chi(p)}{p^s} + R(s,\chi),$$

where

$$|R(s,\chi)| = \left|\sum_{m=2}^\infty \sum_{p=\text{prime}} \frac{\chi(p^m)}{mp^{ms}}\right| \leq \sum_{p=\text{prime}} \sum_{m=2}^\infty \frac{1}{mp^{m\Re(s)}} \leq \frac{1}{2}\sum_{p=\text{prime}} \sum_{m=2}^\infty \frac{1}{p^{m\Re(s)}}.$$

However, since we have the known geometric series

$$\sum_{m=2}^\infty \frac{1}{p^{m\Re(s)}} = \frac{1}{p^{\Re(s)}(p^{\Re(s)}-1)} \leq \frac{2}{p^{2\Re(s)}},$$

then it follows that

$$|R(s,\chi)| \leq \sum_{p=\text{prime}} \frac{1}{p^{2\Re(s)}}.$$

Also, for $\Re(s) > 1$, we have

$$\sum_{p=\text{prime}} \frac{1}{p^{2\Re(s)}} < \sum_{p=\text{prime}} \frac{1}{p^2} < \sum_{m=2}^{\infty} \frac{1}{m^2} = \frac{\pi^2}{6} - 1 < 1,$$

where the last equality comes from Remark 5.9 on page 220. We have shown that

$$\log_e L(s,\chi) = \sum_{p=\text{prime}} \frac{\chi(p)}{p^s} + O(1). \tag{S22}$$

Now if $a \in \mathbb{Z}$ with $\gcd(a,D) = 1$, then by part (b) of Corollary 7.2, via (S22),

$$\sum_{\chi \in G_{\text{char}}^D} \overline{\chi(a)} \log_e L(s,\chi) = \sum_{\chi \in G_{\text{char}}^D} \sum_{p=\text{prime}} \frac{\overline{\chi(a)}\chi(p)}{p^s} + O(\phi(D))$$

$$= \phi(D) \sum_{p \equiv a \pmod D} \frac{1}{p^s} + O\left(\phi(D)\right). \tag{S23}$$

But we also have

$$\sum_{\chi \in G_{\text{char}}^D} \overline{\chi(a)} \log_e L(s,\chi) = \log_e L(s,\chi_0) + \sum_{\substack{\chi \in G_{\text{char}}^D \\ \chi \neq \chi_0}} \overline{\chi(a)} \log_e L(s,\chi), \tag{S24}$$

so by equating (S23)–(S24), we get

$$\log_e L(s,\chi_0) + \sum_{\substack{\chi \in G_{\text{char}}^D \\ \chi \neq \chi_0}} \overline{\chi(a)} \log_e L(s,\chi) = \phi(D) \sum_{p \equiv a \pmod D} \frac{1}{p^s} + O\left(\phi(D)\right),$$

as required.

7.9 Assuming that $s > 1$, let $S(p) = \sum_{j=1}^{\infty} f(p^j)p^{-js}$. Thus,

$$S(p) < Kp^{-s} \sum_{j=0}^{\infty} p^{-js} = Kp^{-s}(1 - p^{-s})^{-1},$$

which implies that $S(p) < 2Kp^{-s}$. For a fixed bound $N \in \mathbb{N}$,

$$\sum_{p \leq N} S(p) < 2K \sum_{p} p^{-s} = B, \tag{S25}$$

say. Since f is multiplicative, then

$$\sum_{n=1}^{N} f(n)n^{-s} = \sum_{n=1}^{N} \prod_{p \leq N} f(p^j)p^{-js} < \prod_{p \leq N} \sum_{n=1}^{\infty} f(p^j)p^{-js}$$

$$= \prod_{p \leq N} S(p) < \prod_{p \leq N} (1 + S(p)) < \prod_{p \leq N} \exp(S(p)) = \exp\left(\sum_{p \leq N} S(p)\right),$$

where the last inequality follows from the fact that for any $x \in \mathbb{R}^+$, $1+x < \exp x$. Therefore, from (S25) it follows that

$$\sum_{n=1}^{N} f(n)n^{-s} < \exp B$$

for all N. Since f is nonnegative, this shows that $\sum_{n=1}^{\infty} f(n)n^{-s}$ converges. The last statement now follows immediately from Exercise 5.14.

Section 7.3

7.11 This follows from the definitions since the numerator is finite and the denominator goes to ∞.

7.13 Parts (a)–(b) are proved in the same way as given in Remark 7.1 on page 248. For part (c), we have

$$1 = \chi(1_p) = \chi(a \cdot a^{-1}) = \chi(a^{-1})\chi(a),$$

which implies $\chi(a^{-1}) = \chi(a)^{-1}$. Lastly, $\chi(a)^{-1} = \overline{\chi(a)}$ follows from the fact that $\chi(a) \in \mathbb{C}$ and $|\chi(a)| = |\zeta_{p-1}^j| = 1$ by part (b).

7.15 That $\chi\gamma$ and χ^{-1} are characters follows from the definition of the individual characters χ and γ. Therefore, if $\chi, \gamma \in G$, the set of multiplicative characters on $\mathbb{F}_p$, then $\chi\gamma^{-1} \in G$, which makes G into a group.

Now since $\mathbb{F}_p^*$ is cyclic—see [68, Theorem A.6, p. 300], let g be a generator of $\mathbb{F}_p^*$. Thus, if $a \in \mathbb{F}_p^*$, then $a = g^j$ for some $j = 0, 1, 2, \ldots, p-1$. Therefore, $\chi(a) = \chi(g^j) = \chi(g)^j$, so the value of $\chi(g)$ determines all other values. By part (b) of Exercise 7.13, $\chi(g)$ is a $(p-1)$-st root of unity. Hence, the order of the character group has order at most $p-1$. If we define for any $j = 0, 1, 2, \ldots, p-1$,

$$\alpha(g^j) = \zeta_{p-1}^j$$

for a primitive $p-1$-st root of unity ζ_{p-1}, then α is clearly a multiplicative character on $\mathbb{F}_p^*$. Suppose that $\alpha^k = \chi_0$. Therefore, $\alpha^k(g) = \chi_0(g) = 1$. However,

$$1 = \alpha^k(g) = \alpha(g)^k = \zeta_{p-1}^k,$$

and since ζ_{p-1} is a primitive $p-1$-st root of unity, then $(p-1) \mid k$. Moreover, since

$$\alpha^{p-1}(a) = \alpha(a^{p-1}) = \alpha(1) = 1,$$

then $\alpha^{p-1} = \chi_0$. This shows that α^j for $j = 0, 1, 2, \ldots, p-2$ are distinct. However, the order of G is at most $p-1$ from the above, so we have demonstrated that $|G| = p-1$ and G has generator α.

7.17 Let α and g be as in the solution of Exercise 7.15 above. Let $\chi = \alpha^{(p-1)/m}$. Therefore,

$$\chi(g) = \alpha^{(p-1)/m}(g) = \alpha(g)^{(p-1)/m} = \zeta_m.$$

In other words, $\chi(g)$ is a primitive m-th root of unity. Since $a = g^j$ for some j and since $x^m \neq a$ for any $x \in \mathbb{F}_p^*$, then $m \nmid j$. Hence,

$$\chi(a) = \chi(g)^j = \zeta_m^j \neq 1.$$

Lastly, $\chi^m = \alpha^{p-1} = \chi_0$.

7.19 If $a = x^2$ for $a \in \mathbb{F}_p$, then

$$N(2, a) = 2 = 1 + \left(\frac{a}{p}\right) = 1 + 1 = 2.$$

If $a \neq 0$, and if $a = 0$, then

$$N(2, a) = 1 + \left(\frac{a}{p}\right) = 1 + 0 = 1.$$

On the other hand, if $a = x^2$ is not solvable, then

$$N(2, a) = 1 + \left(\frac{a}{p}\right) = 1 - 1 = 0.$$

7.21 Since $a \neq 0$, then $\zeta_p^a \neq 1$ and

$$\sum_{j \in \mathbb{F}_p} \zeta_p^{aj} = \frac{\zeta_p^{ap} - 1}{\zeta_p^a - 1} = 0.$$

If $a = 0$, then $\zeta_p^a = 1$, so

$$\sum_{j \in \mathbb{F}_p} \zeta_p^{aj} = p.$$

7.23 This is virtually immediate from Exercise 7.21, since

$$p^{-1} \sum_{j \in \mathbb{F}_p} \zeta_p^{j(a-b)} = p^{-1}p = 1$$

if $a = b$ and is zero otherwise.

Section 8.1

8.1 By the quadratic formula, the solutions to Equation 8.1 on page 271 are

$$x = (\sqrt{R} \pm \sqrt{R - 4Q})/2.$$

Therefore,

$$\alpha + \beta = (\sqrt{R} + \sqrt{R - 4Q})/2 + (\sqrt{R} - \sqrt{R - 4Q})/2 = \sqrt{R},$$

and

$$\alpha\beta = (\sqrt{R} + \sqrt{R - 4Q})(\sqrt{R} - \sqrt{R - 4Q})/4 =$$
$$(R - (R - 4Q))/4 = Q.$$

Also,

$$\alpha - \beta = (\sqrt{R} + \sqrt{R - 4Q})/2 - (\sqrt{R} - \sqrt{R - 4Q})/2 = \sqrt{R - 4Q}.$$

8.3 (a)–(b) We use induction on n. The induction step is $U_1 = 1 \in \mathbb{Z}$, $U_2 = \sqrt{R}$. The induction hypothesis is

$$U_{2i+1} \in \mathbb{Z}, \text{ and } U_{2i} \text{ is an integer multiple of } \sqrt{R} \text{ for all } i < n.$$

Therefore, by part (a) of Theorem 8.1,

$$U_{2n} = \sqrt{R}U_{2n-1} - QU_{2n-2}$$

is an integer multiple of $\sqrt{R}$ by the induction hypothesis, which also implies that

$$U_{2n+1} = \sqrt{R}U_{2n} - QU_{2n-1} \in \mathbb{Z}.$$

The argument for the V_i's is similar.

8.5 We use induction on n.

Induction Step: For $n = 1$,

$$2^{n-1}U_n = 1 = \sum_{k=1}^{\lfloor (n+1)/2 \rfloor} \binom{n}{2k-1} V_1^{n-2k+1} \Delta^{k-1},$$

and

$$2^{n-1}V_n = V_1 = \sum_{k=0}^{\lfloor n/2 \rfloor} \binom{n}{2k} V_1^{n-2k} \Delta^k.$$

Induction hypothesis:

$$2^{n-2}U_{n-1} = \sum_{k=1}^{\lfloor n/2 \rfloor} \binom{n-1}{2k-1} V_1^{n-2k} \Delta^{k-1},$$

and

$$2^{n-2}V_{n-1} = \sum_{k=0}^{\lfloor (n-1)/2 \rfloor} \binom{n-1}{2k} V_1^{n-2k-1} \Delta^k.$$

We may assume that n is even since the other case is similar. By part (f) of Theorem 8.1, $2V_n = V_1V_{n-1} + \Delta U_{n-1}U_1$, and by the induction hypothesis,

$$2^{n-1}V_n = \sum_{k=0}^{n/2-1} \binom{n-1}{2k} V_1^{n-2k} \Delta^k + \sum_{k=1}^{n/2} \binom{n-1}{2k-1} V_1^{n-2k} \Delta^k =$$

$$V_1^n + \Delta^{n/2} + \sum_{k=1}^{n/2-1} \left(\binom{n-1}{2k} + \binom{n-1}{2k-1} \right) V_1^{n-2k} \Delta^k =$$

$$\sum_{k=0}^{n/2} \binom{n}{2k} V_1^{n-2k} \Delta^k.$$

Now we turn to the proof for U_n.

$2V_{n+1} = V_1V_n + \Delta U_nU_1$, by part (f) of Theorem 8.1. Thus, from what we have just proved we get

$$U_n = \frac{1}{\Delta}(2V_{n+1} - V_1 V_n) =$$

$$\frac{2}{2^n \Delta} \sum_{k=0}^{n/2} \binom{n+1}{2k} V_1^{n+1-2k} \Delta^k - \frac{V_1}{2^{n-1}\Delta} \sum_{k=0}^{n/2} \binom{n}{2k} V_1^{n-2k} \Delta^k.$$

Therefore,

$$2^{n-1} U_n = \sum_{k=0}^{n/2} \left(\binom{n+1}{2k} - \binom{n}{2k} \right) V_1^{n-2k+1} \Delta^{k-1} =$$

$$\sum_{k=1}^{n/2} \binom{n}{2k-1} V_1^{n-2k+1} \Delta^{k-1},$$

as required.

8.7 We may assume that Q is odd by Exercise 8.6. Also, by part (d) of Theorem 8.1, U_n is even if and only if V_n is even.

(a) In this case, $\sqrt{R} \equiv 0 \pmod 2$, which by definition means that $R \equiv 0 \pmod 4$. By part (a) of Theorem 8.1, $U_{n+2} \equiv U_n \pmod 2$. Since $U_0 = 0, U_1 = 1$, then $2|U_n$ if and only if n is even.

(b) Define: $U'_{2n} = U_{2n}/\sqrt{R}$, and $U'_{2n+1} = U_{2n+1}$. By part (a) of Theorem 8.1, $U'_{2n+2} \equiv U'_{2n+1} + U'_{2n} \pmod 2$, with $U'_0 = 0, U'_1 = 1$. Thus, $U'_n \equiv 0 \pmod 2$ if and only if $n \equiv 0 \pmod 4$.

(c) Since, $U'_{n+2} \equiv U'_{n+1} + U'_n \pmod 2$, with $U'_1 = U'_2 = 1$, then $U'_n \equiv 0 \pmod 2$ if and only if $n \equiv 0 \pmod 3$.

8.9 Let $n = mm_1$. Then

$$U_n/U_m = (\alpha^n - \beta^n)/(\alpha^m - \beta^m) = (\alpha^{mm_1} - \beta^{mm_1})/(\alpha^m - \beta^m) =$$

$$\alpha^{m(m_1-1)} + \alpha^{m(m_1-2)}\beta^m + \alpha^{m(m_1-3)}\beta^{2m} + \cdots + \alpha^m \beta^{m(m_1-2)} + \beta^{m(m_1-1)} =$$

$$V_{m(m_1-1)} + V_{m(m_1-3)}Q^m + V_{m(m_1-5)}Q^{2m} + \cdots + T,$$

where $T = Q^{m(m_1-2)/2}V_m$ if m_1 is even, and $T = Q^{m(m_1-1)/2}$ if m_1 is odd. In either case, U_n/U_m is an integral multiple of $\sqrt{R}$. Hence, $U_m|U_n$.

8.11 Let $d = \gcd(U_m, U_n)$. By Exercise 8.9, $U_g|U_m$ and $U_g|U_n$, so $U_g|d$. It remains to show that $d|U_g$. By Exercise 8.4,

$$2Q^m U_{n-m} = U_n V_m - V_n U_m \tag{S26}$$

and, by Exercise 8.6, $\gcd(U_m, Q) = 1 = \gcd(U_n, Q)$, so $d|2U_{n-m}$. If $2|d$, then V_m and V_n are even, so (S26) may be written $Q^m U_{n-m} = U_n(V_m/2) - (V_n/2)U_m$. Hence, $d|U_{n-m}$. By a reduction process that mimics the Euclidean algorithm, this shows that $d|U_g$.

Section 8.2

8.13 The equation has no solutions $x, d \in \mathbb{N}$ since $a^2 - D = 2^2 + 43 = 47$, but $D \neq -3a^2 \pm 1$.

8.15 The equation has no solution since $8^2 + 225 = 17^2$, so $a = 8$, but $D \neq -3a^2 \pm 1$.

8.17 $2^2 + 161047 = 11^5$.

Section 8.3

8.19 Since $I^{h_{\mathfrak{O}_F}} \sim 1$, $I^n \sim 1$, and $\gcd(h_{\mathfrak{O}_F}, n) = 1$, then there exist integers x, y such that $nx + h_{\mathfrak{O}_F} y = 1$. Therefore,

$$I = I^{nx + h_{\mathfrak{O}_F} y} = (I^n)^x (I^y)^{h_{\mathfrak{O}_F}} \sim 1,$$

as we sought to prove.

8.21 In Theorem 8.4, let $k = -13 = -1 - 3u^2$ with $u = 2$, for which $x = p^m = 4u^2 + 1 = 17$ with $m = 1$ and $y = \pm 2(3 + 8 \cdot 2^2) = \pm 70$. Thus, $p = 2^2 + 13$, and $70^2 = 17^3 - 13$. Thus, $(x, y) = (17, \pm 70)$.

8.23 By Theorem 8.4 there can be no solutions since $k = -47 \neq -3u^2 \pm 1$ for any integer u.

8.25 As per the hint, a solution (x, y) to (8.15) implies that

$$y + \sqrt{k} = w(u + v\sqrt{k})^3 \tag{S27}$$

for a unit $w \in \mathfrak{O}_F$ and some $u, v \in \mathbb{Z}$. Then $w = \pm \varepsilon_k^z$ for some $z \in \mathbb{Z}$. Since we may write $z = 3z_1 + r$ where $r \in \{0, \pm 1, \pm 2\}$, then we may absorb $(\pm \varepsilon_k^{z_1})^3$ into the cube $(u + v\sqrt{k})^3$, so we may assume, without loss of generality, that $w = \varepsilon_k^r$, where $r \in \{0, \pm 1, \pm 2\}$. Given the definition of ε and the fact that $(T + U\sqrt{k})^{-1} = T - U\sqrt{k}$, then we may assume $w \in \{\varepsilon_k^j : j = 0, 1, -1\}$ if ε_k has norm 1 and $w \in \{\varepsilon_k^j : j = 0, 2, -2\}$ if ε_k has norm -1. In either case, $w \in \{\varepsilon^j : j = 0, 1, -1\}$.

Case S.1 $w = 1$

From (S27),
$$y + \sqrt{k} = (u^3 + 3uv^2 k) + (3u^2 v + v^3 k)\sqrt{k},$$
so by comparing coefficients of $\sqrt{k}$, we have that

$$1 = 3u^2 v + v^3 k = v(3u^2 + v^2 k), \tag{S28}$$

so $v = \pm 1$. Hence, multiplying (S28) by v yields

$$\pm 1 = v = 3u^2 v^2 + v^4 k \geq k > 1,$$

a contradiction.

Case S.2 $w \in \{T \pm U\sqrt{k}\}$

From (S27) we have

$$y + \sqrt{k} = (T \pm U\sqrt{k})(u + v\sqrt{k})^3 = (T \pm U\sqrt{k})\left((u^3 + 3uv^2 k) + (3u^2 v + v^3 k)\sqrt{k}\right)$$

$$= (T(u^3 + 3uv^2 k) \pm (Uk(3u^2 + v^3 k)) + (T(3u^2 v + v^3 k) \pm U(u^3 + 3uv^2 k))\sqrt{k}.$$

Therefore, by comparing coefficients of $\sqrt{k}$ again yields

$$1 = T(3u^2 v + v^3 k) \pm U(u^3 + 3uv^2 k). \qquad (S29)$$

Since $k \equiv 4 \,(\mathrm{mod}\ 9)$ and $U \equiv 0 \,(\mathrm{mod}\ 9)$, then $1 = T^2 - kU^2$ implies that

$$T \equiv \pm 1 \quad (\mathrm{mod}\ 81).$$

Hence, by (S29),

$$1 \equiv \alpha(3u^2 + 4v^2)v \quad (\mathrm{mod}\ 9), \qquad (S30)$$

where $\alpha \equiv \pm 1 \,(\mathrm{mod}\ 9)$.

From (S30), $\alpha v \equiv \pm 1 \,(\mathrm{mod}\ 9)$, so

$$3u^2 + 4 \equiv \alpha v \equiv \pm 1 \quad (\mathrm{mod}\ 9).$$

Thus,

$$3u^2 \equiv 4, 6 \quad (\mathrm{mod}\ 9),$$

which are impossible. This completes all cases.

Section 8.4

8.27 By Exercise 2.24,

$$|\mathbb{Q}(\zeta_n) : \mathbb{Q}| = \phi(n) = \deg(m_{\zeta_n, \mathbb{Q}}(x)),$$

and by Theorem 1.7, $\Phi_n(x) = m_{\zeta_n, \mathbb{Q}}(x)$, so by Definition 1.9, the result follows.

8.29 By Exercise 8.28 with $I = \mathcal{P}$ and $J = \mathcal{P}^{m-1}$, where $m \in \mathbb{N}$ and $\mathcal{P}^0 = \mathfrak{D}_F$, we get

$$\frac{\mathfrak{D}_F}{\mathcal{P}} \cong \frac{\mathcal{P}^{m-1}}{\mathcal{P}^m},$$

and for any $n \in \mathbb{N}$,

$$\left(\frac{\mathfrak{D}_F}{\mathcal{P}} \right)^n \cong \frac{\mathfrak{D}_F}{\mathcal{P}} \times \frac{\mathcal{P}}{\mathcal{P}^2} \times \cdots \times \frac{\mathcal{P}^{n-1}}{\mathcal{P}^n} \cong \frac{\mathfrak{D}_F}{\mathcal{P}^n},$$

so

$$\left| \frac{\mathfrak{D}_F}{\mathcal{P}} \right|^n = \left| \frac{\mathfrak{D}_F}{\mathcal{P}^n} \right|,$$

which is what we sought to show.

8.31 The principal fact to establish is that the multiplication is well defined, namely that if $a + I = a' + I$, and $b + I = b' + I$, then $ab + I = a'b' + I$. Since $a' \in a' + I = a + I$, then $a' = a + j$ for some $j \in I$. Similarly, $b' = b + k$ for some $k \in I$. Thus,

$$a'b' = (a + j)(b + k) = ab + jb + ak + jk.$$

Therefore,

$$a'b' - ab = jb + ak + jk \in I,$$

since I is an ideal. However, a fundamental fact is that cosets are either equal, or have a trivial intersection. Thus,

$$ab + I = a'b' + I.$$

It now follows that R/I is a ring with the properties inherited by the well-defined operation of multiplication, and $1_R + I$ is the identity of R/I where 1_R is the multiplicative identity of R.

8.33 Suppose that γ runs through a system of $N(I)$ elements of R which are incongruent modulo I. Since

$$\alpha\gamma_1 \equiv \alpha\gamma_2 \pmod{I} \text{ for } \gamma_1, \gamma_2 \in R,$$

implies that
$$I \mid \alpha(\gamma_1 - \gamma_2),$$
then the relative primality of α and I implies that $I \mid (\gamma_1 - \gamma_2)$, namely

$$\gamma_1 \equiv \gamma_2 \pmod{I}.$$

Hence, $\alpha\gamma$ runs through *all* residue classes modulo I as γ runs over its system. Therefore, among the $\alpha\gamma$, there exists one residue class in which β sits. Moreover, it is clearly uniquely determined modulo I.

Now we prove the last assertion. Set $\gcd(\alpha, I) = G$. Assume first that there is a solution to the congruence $\alpha\gamma \equiv \beta \pmod{I}$. Then there exists a $\delta \in I$ such that $\alpha\gamma = \beta + \delta$. Hence, $G \mid I \mid (\delta)$. However, $G \mid (\alpha)$, so $G \mid (\beta) = (\alpha\gamma - \delta)$.

Conversely, if $G \mid (\beta)$, then $(\beta) \subseteq (\alpha) + I = \gcd((\alpha), I)$, so $\beta = \alpha\gamma + \delta$ for some $\gamma \in R$ and $\delta \in I$. Thus, $\beta \equiv \alpha\gamma \pmod{(\delta)}$, so since $I \mid (\delta)$, then $\beta \equiv \alpha\gamma \pmod{I}$.

8.35 This is immediate from Exercise 8.34.

8.37 If we are given $\alpha, \beta \in \mathfrak{O}_F$ both relatively prime to I, then $\alpha\beta + I$ is a class completely determined by α and β modulo I, and $\alpha\beta$ is relatively prime to I. Thus, the group is an abelian group. By definition, the order of the group is $\Phi(I)$. Moreover, if I is a prime $\mathfrak{O}_F$-ideal, then the group is isomorphic to the multiplicative subgroup of nonzero elements of the field $\mathfrak{O}_F/I$, and we are done, since it is known that the multiplicative subgroup of all nonzero elements in a field is cyclic—see [68, Theorem A.6, p. 300].

8.39 The classes of the group defined in Exercise 8.37, represented by a rational integer, form a subgroup thereof. These are the classes of the representatives $1, 2, \ldots, p-1$. Suppose that one of these integers z is not relatively prime to $\mathcal{P}$. Then since there exist $u, v \in \mathbb{Z}$ with

$$up + vz = 1,$$

and $p \in \mathcal{P}$, we would have $1 \in \mathcal{P}$, a contradiction. Hence, all of these representatives are relatively prime to $\mathcal{P}$, and they are distinct. Therefore, for any such class $\overline{z}$, we must have that $\overline{z}^{p-1} = \overline{1}$, the identity class of the group. By Exercise 8.37, the group is cyclic, so there are no more than $p-1$ classes $\overline{z}$ for which $\overline{z}^{p-1} = \overline{1}$. Thus, the subgroup of classes represented by a rational integer is identical with the group of classes whose elements raised to the power $(p-1)$ is the class $\overline{1}$. This yields the result.

Section 8.5

8.41 Applying the ABC-conjecture with

$$a = m(m+2),$$

$$b = 1,$$

and

$$c = (m+1)^2$$

yields that with finitiely many exceptions, for any $\kappa > 1$,

$$(m+1)^2 < \mathcal{S}(m(m+1)^2(m+2))^\kappa. \qquad \text{(S31)}$$

Now we assume that $m > n$ and prove that for $k = 3$, there are only finitely many such m for which

$$\mathcal{S}(m) = \mathcal{S}(n), \mathcal{S}(m+1) = \mathcal{S}(n+1), \text{ and } \mathcal{S}(m+2) = \mathcal{S}(n+2). \qquad \text{(S32)}$$

Now (S32) implies

$$m - n = (m+j) - (n+j) \equiv 0 \pmod{\mathcal{S}(m+j)}$$

for $0 \le j \le 2$. Given that

$$\gcd(\mathcal{S}(m), \mathcal{S}(m+1), \mathcal{S}(m+2)) \mid 2,$$

then

$$\mathcal{S}(m(m+1)^2(m+2)) \mid 2(m-n).$$

Using this in (S31) yields that with finitely many exceptions,

$$m^2 < (m+1)^2 < \mathcal{S}(m(m+1)^2(m+2))^\kappa < (2m)^\kappa,$$

which implies

$$m < 2^{\kappa/(2-\kappa)}$$

with finitely many exceptions. Hence, m is bounded by a constant. We have shown that for $k = 3$ in the Erdös–Woods Conjecture holds with finitely many exceptions, assuming the ABC-conjecture.

8.43 If n is powerful, then any prime $p \mid n$ has exponent $n(p) \ge 2$ in the canonical prime factorization of n. Let $\mathcal{S}$ denote the set of primes dividing n that appear to an odd exponent $n(p) \ge 3$. Then

$$n = \prod_{\substack{p \mid n \\ p \notin \mathcal{S}}} p^{n(p)} \prod_{p \in \mathcal{S}} p^{n(p)} = \prod_{\substack{p \mid n \\ p \notin \mathcal{S}}} p^{n(p)} \prod_{p \in \mathcal{S}} p^{n(p)-3} \prod_{p \in \mathcal{S}} p^3.$$

Letting

$$x = \prod_{\substack{p \mid n \\ p \notin \mathcal{S}}} p^{n(p)/2} \prod_{p \in \mathcal{S}} p^{(n(p)-3)/2}$$

and

$$y = \prod_{p \in \mathcal{S}} p$$

yields

$$n = x^2 y^3.$$

8.45 Let a be even and set $n = a^m$ in Exercise 8.42. Then there are only finitely many values such that $m > 1$ with $a^{2m} - 1$ being powerful. Hence, there cannot be infinitely many such values, which is what we sought to prove.

Section 9.1

9.1 If $x^3 + y^3 = z^3$ has nonzero integer solutions, then for

$$X = 12z/(x+y) \text{ and } Y = 36(x-y)/(x+y),$$

we get

$$Y^2 = X^3 - 432.$$

Since $xyz \neq 0$, then $|Y| \neq 36$. Conversely, assume that

$$Y^2 = X^3 - 432, \tag{S33}$$

for some $Y = A/B$ and $X = C/D$ with $A, B, C, D \in \mathbb{Z}$, $AD \neq 0$, and set

$$x = (36B + A)D, \ y = (36B - A)D, \text{ and } z = 6BC.$$

By (S33), $(36 + Y)^3 + (36 - Y)^3 = (6X)^3$. Therefore,

$$x^3 + y^3 = D^3[(36B + A)^3 + (36B - A)^3] = D^3(6XB)^3 =$$

$$D^3(6BC/D)^3 = (6BC)^3 = z^3.$$

Since $|Y| \neq 36$, then $xyz \neq 0$.

Section 9.2

9.3 Since $y^2 = x^3 + 1$, then $(y-1)(y+1) = x^3$. It is easy to see that

$$g = \gcd(y-1, y+1) \mid 2.$$

If $g = 1$, then there are $z_1, z_2 \in \mathbb{Z}$ such that $y - 1 = z_1^3$ and $y + 1 = z_2^3$. By subtracting, we get

$$2 = z_2^3 - z_1^3.$$

However, this is impossible since

$$z_2^3 - z_1^3 \equiv 0 \pmod{4},$$

given that z_1 and z_2 must have the same parity. Thus, $g = 2$. Therefore,

$$\left(\frac{y-1}{2}\right)\left(\frac{y+1}{2}\right) = 2\left(\frac{x}{2}\right)^3.$$

Hence, one of $(y+1)/2$ or $(y-1)/2$ is of the form z_1^3, for some $z_1 \in \mathbb{Z}$, and the other is of the form $2z_2^3$, for some $z_2 \in \mathbb{Z}$. Thus,

$$\pm 1 = z_1^3 - 2z_2^3.$$

One readily verifies that the only integer solutions to this last equation are $z_1 = \pm 1 = z_2$ and $z_2 = 0$. From these solutions emerge

$$(x, y) \in \{(2, \pm 3), (0, \pm 1), (-1, 0)\}.$$

The result now is a consequence of the Nagell-Lutz Theorem.

9.5 The number of incongruent solutions of

$$z^2 \equiv y \pmod{p^k}$$

is $1 + \chi(y)$, so the number of solutions of $y^2 = x^3 + ax + b$, counting the point at infinity, is

$$1 + \sum_{x \in \mathbb{F}_{p^k}} \left(1 + \chi(x^3 + ax + b)\right) = p^k + 1 + \sum_{x \in \mathbb{F}_{p^k}} \chi(x^3 + ax + b).$$

9.7 The discriminant is given by

$$\Delta(E(\mathbb{Q})) = 5.$$

Also, $(1, 0) = Q$ is clearly a point as is $P = (0, 1)$. Since $2P = (1, 0) = Q$, then P is a point of order 4. Moreover, $(0, -1) = -P$, so $(0, -1) = 3P$, and this implies that $E(\mathbb{Q})_t$ is generated by P, namely

$$E(\mathbb{Q})_t \cong \mathbb{Z}/4/\mathbb{Z}.$$

9.9 The discriminant is $\Delta(E(\mathbb{Q})) = -9$. Also, all of the points

$$\{\mathfrak{o}, (-1, 0), (0, \pm 1), (2, \pm 3)\}$$

are of finite order. Moreover, $(2, 3)$ is of order 6 and $E(\mathbb{Q})$ can be shown to be cyclic of order 6, so

$$E(\mathbb{Q}) \cong \mathbb{Z}/6\mathbb{Z}.$$

Section 9.3

9.11 (a) $97 \cdot 167$ (b) $89 \cdot 149$ (c) $97 \cdot 547$ (d) $101 \cdot 103$

9.13 (a)–(c) are prime and $26869 = 97 \cdot 277$

Section 9.4

9.15 Let $y = mx + b$ be the tangent line to E at x_1. Thus, since

$$2(x_1, y_1) = (x_2, y_2),$$

both (x_1, y_1) and $(x_2, -y_2)$ are on $y = mx + b$. Hence, both points satisfy

$$\prod_{j=1}^{3} (x - \alpha_j) = y^2 = (mx + b)^2.$$

Also, since $y = mx + b$ is tangent to E at x_1, then the three roots of

$$\prod_{j=1}^{3} (x - \alpha_j) - (mx + b)^2 = 0$$

are x_2 and x_1 repeated. In other words,

$$\prod_{j=1}^{3} (x - \alpha_j) - (mx + b)^2 = (x - x_2)(x - x_1)^2.$$

By setting $x = \alpha_j$ for each of $j = 1, 2, 3$ and observing that $x_1 \neq \alpha_j$ since $(x_2, y_2) \neq \mathfrak{o}$, then

$$(x_2 - \alpha_j) = \left(\frac{mx + b}{\alpha_j - x_1}\right)^2,$$

which is the square of a rational number for $j = 1, 2, 3$.

9.17 It is a trivial exercise to verify that

$$x^2 + ny^2 = z^2 \text{ and } x^2 - ny^2 = t^2 \tag{S34}$$

has a solution in integers with $y \neq 0$ if and only if it has a solution in rational numbers with $y \neq 0$.

Suppose first that n is a congruent number. Then by part (1) of Exercise 9.14,

$$b = 2n/a.$$

Thus,

$$c^2 = a^2 + b^2 = a^2 + 4n^2/a^2,$$

or via division by 4:

$$\left(\frac{c}{2}\right)^2 = \left(\frac{a}{2}\right)^2 + \left(\frac{n}{a}\right)^2. \tag{S35}$$

Adding $\pm n$ to each side of (S35) (to essentially complete the square), we get:

$$\left(\frac{c}{2}\right)^2 \pm n = \left(\frac{a}{2} \pm \frac{n}{a}\right)^2.$$

Setting

$$x = c/2, \ y = 1, \ z = a/2 + n/2, \text{ and } t = a/2 - n/2$$

yields a rational solution of (S34), so by the initial comment at the outset of this solution, it has an integral solution. This shows that (1) implies (2).

Now we assume that (2) holds. Without loss of generality, we may assume that $x, y, z, t \in \mathbb{N}$ and that these integers are pairwise relatively prime. If $y = 1$, then by adding the equations in (S34), we get that

$$2x^2 = z^2 + t^2.$$

Thus, both z and t have the same parity. If they are both even, then x is even contradicting the relative primality in pairs. Therefore, they are both odd. By subtracting the two equations in (S34), we get that

$$2n = z^2 - t^2 \equiv 0 \pmod 8$$

since

$$z^2 \equiv 1 \equiv t^2 \pmod 8.$$

Thus, 4 divides n contradicting the squarefreeness of n. We have shown that $y \neq 1$. By multiplying the equations in (S34), we get:

$$\left(\frac{xtz}{y^3}\right)^2 = \left(\frac{x^2}{y^2}\right)^3 - n^2 \frac{x^2}{y^2}.$$

This shows that

$$P = (X, Y) = (x^2/y^2, xtz/y^3)$$

is a rational (but not integral) point on E, so P has infinite order. Thus,

$$2P = (x_2, y_2) \neq \mathfrak{o}.$$

From Exercises 9.14–9.15, n is a congruent number. This shows that (2) implies (1) and we are done.

Section 10.1

10.1 We have, for $\bar{z} = e - fi$ being the complex conjugate of $z = e + fi$, that

$$\Im(\alpha z) = \Im\left(\frac{az+b}{cz+d}\right) = \Im\left(\frac{(az+b)(c\bar{z}+d)}{(cz+d)(c\bar{z}+d)}\right) = \Im\left(\frac{(az+b)(c\bar{z}+d)}{|cz+d|^2}\right),$$

where the denominator of the last equality comes from the fact that

$$(cz+d)(c\bar{z}+d) = c^2(e^2 + f^2) + 2cde + d^2 = (ce+d)^2 + c^2 f^2 = |cz+d|^2.$$

Hence,

$$\Im(\alpha z) = \frac{\Im[(az+b)(c\bar{z}+d)]}{|cz+d|^2},$$

so it remains to show that $\Im[(az+b)(c\bar{z}+d)] = \Im(z)$. However, this follows from the fact that $ad - bc = 1$ since

$$\Im[(az+b)(c\bar{z}+d)] = \Im[ac(e^2 + f^2) + ade + bce + bd + f(ad - bc)i]$$

$$= \Im(f(ad - bc)i) = f = \Im(z).$$

10.3 Assume that $\alpha = \begin{pmatrix} a & b \\ c & d \end{pmatrix} \in \Gamma$ such that $\alpha z \in D$. If $\Im(\alpha z) < \Im(z)$, then we may replace z by αz and α by α^{-1}, which tells us that $\Im(\alpha z) \geq \Im(z)$ may be assumed without loss of generality. Therefore, by (10.1) on page 332,

$$|cz+d|^2 = \frac{\Im(z)}{\Im(\alpha z)} \leq 1, \tag{S36}$$

so $|cz+d| \leq 1$. This means that $|c| \leq 1$, namely $c \in \{0, \pm 1\}$. If $c = 0$, then $d = \pm 1$, and $\alpha = \begin{pmatrix} \pm 1 & b \\ 0 & \pm 1 \end{pmatrix} \in \Gamma$, namely $\alpha z = z \pm b$. Given that $z, \alpha z \in D$, then $|b| = |z - \alpha z| \leq 1$, namely $b \in \{0, \pm 1\}$. If $b = 0$, then α is the identity, which contradicts the hypothesis. If $b = \pm 1$, then $\alpha z = z \pm 1$. Also, $|\Re(z)| \leq 1/2$ and $|\Re(z \pm 1)| \leq 1/2$, so $\Re(z) = \pm 1/2$ is forced.

Now consider the case where $c = \pm 1$. Then (S36) tells us that $|z + d| \leq 1$, which forces $d = 0$ *unless* $z = \zeta_3 = (-1 + \sqrt{-3})/2$ or $z = 1 + \zeta_3$, since this is the case where $\alpha z = z + 1 = -z^2$, $\Re(z) = -1/2$, and $d = 1$ (respectively $\alpha z = z - 1 = -(z-1)^2 - 1$, $\Re(z) = 1/2$, and $d = -1$)—see Exercise 1.54 on page 46.

When $d = 0$, since $ad - bc = 1$ and $bc = -1$, then either $b = 1 = -c$ or $c = 1 = -b$. Thus,

$$\alpha z = \pm a - 1/z = \pm a - \bar{z}, \tag{S37}$$

where $\bar{z}$ is the complex conjugate of z. If $a = 0$, then $\alpha z = -\bar{z} = -1/z$. Also, since $z \in D$, then $|z| \geq 1$, and since $|z + d| = |z| \leq 1$, then $|z| = 1$.

Now assume that $a \neq 0$. Since $z, \alpha z \in D$, then

$$|a| = |\Re(\alpha z) + \Re(z)| \leq 1, \tag{S38}$$

so $a = \pm 1$. Thus, by (S37), $\Re(\alpha z) = \Re(\pm 1 - \bar{z}) = \pm 1 - \Re(z)$, but $\Re(z) \leq 1/2$ and $\Re(\alpha z) \leq 1/2$, so it follows that if $a = -1$, then $\Re(z) = \Re(\alpha z) = -1/2$, while if $a = 1$, then $\Re(z) = \Re(\alpha z) = 1/2$. However, by (S37), $\Im(\alpha z) = \Im(z)$, so $\alpha z = z$, which forced α to be the identity contradicting the hypothesis. This completes all cases.

Section 10.2

10.5 If condition (b) is satisfied, then by (10.5) on page 337, $f(z+1) = f(z)$. Also, since $S = \begin{pmatrix} 0 & -1 \\ 1 & 0 \end{pmatrix}$ is a generator of Γ by Theorem 10.1 on page 333, then

$$f(\gamma z) = f(-1/z) = (-z)^k f(z).$$

Conversely, assume that conditions (1)–(2) hold. Given

$$\gamma = \begin{pmatrix} a & b \\ c & d \end{pmatrix} \in \Gamma, \tag{S39}$$

define

$$\eth(\gamma, z) = cz + d.$$

Now we show that for $\alpha, \gamma \in \Gamma$, we have

$$\eth(\alpha\gamma, z) = \eth(\alpha, \gamma z)\eth(\gamma, z). \tag{S40}$$

Let γ be given by (S39), and let $\alpha = \begin{pmatrix} a' & b' \\ c' & d' \end{pmatrix}$. Then

$$\alpha\gamma = \begin{pmatrix} a' & b' \\ c' & d' \end{pmatrix} \begin{pmatrix} a & b \\ c & d \end{pmatrix} = \begin{pmatrix} aa' + b'c & a'b + b'd \\ c'a + d'c & c'b + d'd \end{pmatrix},$$

so

$$\eth(\alpha\gamma, z) = (c'a + d'c)z + c'b + d'd$$

$$\eth(\alpha, \gamma z) = \eth\left(\alpha, \frac{az + b}{cz + d}\right) = c'\left(\frac{az + b}{cz + d}\right) + d',$$

and

$$\eth(\gamma, z) = cz + d.$$

Hence,

$$\eth(\alpha, \gamma z)\eth(\gamma, z) = \left[c'\left(\frac{az + b}{cz + d}\right) + d'\right] \cdot [cz + d]$$

$$= c'(az + b) + d'(cz + d) = (c'a + d'c)z + c'b + d'd = d(\alpha\gamma, z),$$

which establishes (S40).

Now we establish that

$$\eth(\gamma^{-1}, z) = (\eth(\gamma, \gamma^{-1}z))^{-1}. \tag{S41}$$

Since $\gamma^{-1} = \begin{pmatrix} d & -b \\ -c & a \end{pmatrix}$, then

$$\mathfrak{d}(\gamma^{-1}, z) = -cz + a,$$

and

$$\mathfrak{d}(\gamma, \gamma^{-1}z) = \mathfrak{d}\left(\gamma, \frac{dz-b}{-cz+a}\right) = c\left(\frac{dz-b}{-cz+a}\right) + d$$

$$= \frac{c(dz-b) + d(-cz+a)}{-cz+a} = \frac{1}{-cz+a} = (\mathfrak{d}(\gamma^{-1}, z))^{-1},$$

which is (S41).

Now assume that

$$f(\delta z) = \mathfrak{d}(\delta, z)^k f(z), \tag{S42}$$

where $z \in \mathfrak{H}$ and $\delta \in \Gamma$. Then (S40) tells us that (S42) holds for $\delta = \alpha\gamma$ and (S41) tells us that (S42) holds for $\delta = \gamma^{-1}$. Hence, the subset of Γ for which (S42) holds is a subgroup. However, conditions (1)–(2) tell us that this subgroup contains S and T, which generate all of Γ by Theorem 10.1 on page 333. Hence, (1)–(2) imply that (b) holds.

10.7 We have $|\Gamma : \Gamma_0(n)| = p^a + p^{a-1}$. For $0 \le \ell \le p^{a-1} - 1$ set $\gamma_\ell = \begin{pmatrix} 1 & 0 \\ p\ell & 1 \end{pmatrix}$ and

for $0 \le m \le p^a - 1$ set $\gamma_m = \begin{pmatrix} m & 1 \\ -1 & 0 \end{pmatrix}$. Therefore we have that

$$\cup_{m=0}^{p^a+p^{a-1}} \gamma_m \Gamma_0(p^a) \subseteq \Gamma,$$

so we merely have to show that these γ_m represent distinct cosets. If $0 \le \ell \le p^{a-1} - 1$ and $0 \le m \le p^a - 1$,

$$\gamma_\ell^{-1}\gamma_m = \begin{pmatrix} 1 & 0 \\ -p\ell & 1 \end{pmatrix}\begin{pmatrix} m & 1 \\ -1 & 0 \end{pmatrix} = \begin{pmatrix} m & 1 \\ -p\ell m - 1 & -p\ell \end{pmatrix} \notin \Gamma_0(p^a).$$

If $0 \le \ell, m \le p^{a-1} - 1$, then

$$\gamma_\ell^{-1}\gamma_m = \begin{pmatrix} 1 & 0 \\ -p\ell & 1 \end{pmatrix}\begin{pmatrix} 1 & 0 \\ pm & 1 \end{pmatrix} = \begin{pmatrix} 1 & 0 \\ p(m-\ell) & 1 \end{pmatrix},$$

which is in $\Gamma_0(p^a)$ if and only if $\ell = m$. Lastly, if $0 \le \ell, m \le p^a - 1$, then

$$\gamma_\ell^{-1}\gamma_m = \begin{pmatrix} 0 & -1 \\ 1 & \ell \end{pmatrix}\begin{pmatrix} m & 1 \\ -1 & 0 \end{pmatrix} = \begin{pmatrix} 1 & 0 \\ (m-\ell) & 1 \end{pmatrix},$$

which is in $\Gamma_0(p^a)$ if and only if $\ell = m$. Hence, all left cosets are distinct.

10.9 Let $a/c \in \mathbb{Q}$ with $\gcd(a, c) = 1$. Then by the Euclidean algorithm there exist $b, d \in \mathbb{Z}$ such that $ad - bc = 1$. Thus, $\gamma = \begin{pmatrix} a & b \\ c & d \end{pmatrix} \in \Gamma$. Select $\alpha \in \Gamma_0(n)$, as well as any γ_j and set $\gamma = \gamma_j\alpha$. Thus, $\gamma_j\alpha(\infty) = \gamma(\infty) = a/c$. Hence, a/c and $\gamma_j(\infty)$ represent the same cusp.

10.11 If $n = 8$ in Exercise 10.10, then taking

$$\gamma_j = \gamma_{\ell=1} = \begin{pmatrix} 1 & 0 \\ 2 & 1 \end{pmatrix}, \gamma_i = \gamma_{\ell=3} = \begin{pmatrix} 1 & 0 \\ 6 & 1 \end{pmatrix}, \text{ and } \alpha = \begin{pmatrix} 1 & 1 \\ 4 & 5 \end{pmatrix}$$

yields

$$\gamma_i^{-1}\alpha = \begin{pmatrix} 1 & 1 \\ -2 & -1 \end{pmatrix}.$$

Therefore, $\gamma_i^{-1}(\infty) = -1/2$, and since

$$\gamma_j^{-1} = \begin{pmatrix} 1 & 0 \\ -2 & 1 \end{pmatrix},$$

then $\gamma_j^{-1}(\infty) = -1/2$. Hence, both $\gamma_i^{-1}(\infty)$ and $\gamma_j^{-1}(\infty)$ represent the same cusp $-1/2$. However,

$$\gamma_j\gamma_i^{-1} = \begin{pmatrix} 1 & 0 \\ -4 & 1 \end{pmatrix},$$

which is not of the upper triangular form in the sufficient condition.

10.13 By Exercise 10.12 and (10.18) of the hint, we have that $f(x) = \pi$ for all x. Thus, since $f(x) = \Gamma(x)\Gamma(1-x)\sin(\pi x)$, then we deduce that

$$\Gamma(x)\Gamma(1-x) = \frac{\pi}{\sin(\pi x)}. \qquad (S43)$$

Also, by (5.34) on page 224, $\Gamma(x+1) = x\Gamma(x)$, so (S43) may be rewritten as

$$\sin(\pi x) = \frac{\pi}{-x\Gamma(x)\Gamma(-x)}.$$

Now (10.19) from the hint allows us to replace the gamma function to achieve

$$-x\Gamma(x)\Gamma(-x) = -x\left(\frac{e^{-\gamma x}}{x}\prod_{j=1}^{\infty}\frac{e^{x/j}}{1+x/j}\right)\left(\frac{e^{\gamma x}}{-x}\prod_{j=1}^{\infty}\frac{e^{-x/j}}{1-x/j}\right) = \frac{1}{x}\prod_{j=1}^{\infty}\frac{1}{1-\frac{x^2}{j^2}}.$$

Hence,

$$\sin(\pi x) = \pi x\prod_{j=1}^{\infty}\left(1 - \frac{x^2}{j^2}\right),$$

so by letting $z = \pi x$, we get the result,

$$\sin(z) = z\prod_{j=1}^{\infty}\left(1 - \frac{z^2}{\pi^2 j^2}\right).$$

10.15 By Remark 10.3, $E_4^2 = E_8$, so by Example 10.2 on page 339,

$$1 + 480\sum_{n=1}^{\infty}\sigma_7(n)q^n = \left(1 + 240\sum_{n=1}^{\infty}\sigma_3(n)q^n\right)^2,$$

which implies that

$$1 + 480\sum_{n=1}^{\infty}\sigma_7(n)q^n = 1 + 480\sum_{n=1}^{\infty}\sigma_3(n)q^n + 240^2\left(\sum_{n=1}^{\infty}\sigma_3(n)q^n\right)^2$$

so

$$\sum_{n=1}^{\infty}\sigma_7(n)q^n = \sum_{n=1}^{\infty}\sigma_3(n)q^n + 120\sum_{n=1}^{\infty}\sum_{j=1}^{n-1}\left(\sigma_3(n)\sigma_3(n-j)\right)q^n.$$

Therefore,

$$\sigma_7(n) = \sigma_3(n) + 120 \sum_{j=1}^{n-1} \sigma_3(n)\sigma_3(n-j),$$

as required.

10.17 By Definition 10.4,

$$j(z) = \frac{1728 \cdot 60^3 G_4(z)^3}{\Delta(z)} = \frac{1 + 720q + 179280q^2 + 16954560q^3 + \cdots}{q - 24q^2 + 252q^3 - 1472q^4 + \cdots}$$

$$= \frac{1}{q} + 744 + 196884q + 21493760q^4 + \cdots,$$

which shows that j is a modular form of weight 0, but not a cusp form.

Section 10.3

10.19 If we have isomorphic elliptic curves $E_1 \cong E_2$, then by Definition 10.12 on page 350,

$$j(E_2) = \frac{1728(g_2^{(2)})^3}{(g_2^{(2)})^3 - 27(g_3^{(2)})^2} = \frac{1728\alpha^{12}(g_2^{(1)})^3}{\alpha^{12}(g_2^{(1)})^3 - 27\alpha^{12}(g_3^{(1)})^2}$$

$$= \frac{1728(g_2^{(1)})^3}{(g_2^{(1)})^3 - 27(g_3^{(1)})^2} = j(E_1).$$

10.21 From the hint,

$$\sum_{n=0}^{\infty} x^n = (1-x)^{-1}.$$

Differentiating with respect to x, we get

$$(1-x)^{-2} = \sum_{n=1}^{\infty} nx^{n-1} = \sum_{n=0}^{\infty}(n+1)x^n = 1 + \sum_{n=1}^{\infty}(n+1)x^n.$$

10.23 Using the expansion in Exercise 10.22,

$$\wp'(z)^2 = 4z^{-6} - 24G_4z^{-2} - 80G_6 + \cdots,$$
$$\wp(z)^3 = z^{-6} + 9G_4z^{-2} + 15G_6 + \cdots,$$

and

$$\wp(z) = z^{-2} + 3G_4z^2 + \cdots,$$

so

$$f(z) = \wp'(z)^2 - 4\wp(z)^3 + 60G_4\wp(z) + 140G_6$$

is analytic around $z = 0$. Since $f(z)$ is an elliptic function with respect to L and $\wp$ is analytic on $\mathbb{C} - L$ by Remark 10.7 on page 349, then f is an analytic elliptic function, so by Liouville's theorem given in the hint, f is constant. But since $f(0) = 0$, then f is identically zero, so

$$\wp'(z)^2 = 4\wp(z)^3 - 60G_4\wp(z) - 140G_6 = 4\wp(z)^3 - g_2(L)\wp(z) - g_3(L),$$

as required.

10.25 From the definitions, if L_1 and L_2 are homothetic,

$$g_2(L_1) = g_2(\lambda L_2) = 60 \sum_{\lambda w \in \lambda L_2 - \{0\}} \frac{1}{w^4} = \lambda^{-4} 60 \sum_{w \in L_2 - \{0\}} \frac{1}{w^4} = \lambda^{-4} g_2(L_2),$$

and similarly,

$$g_3(L_1) = \lambda^{-6} g_3(L_2).$$

Therefore, by Equation (10.20) on page 349,

$$j(L_1) = \frac{1728 g_2(L_1)^3}{g_2(L_1)^3 - 27 g_3(L_1)^2} = \frac{1728 \lambda^{-12} g_2(L_2)^3}{\lambda^{-12} g_2(L_2)^3 - 27 \lambda^{-12} g_3(L_2)^2}$$

$$= \frac{1728 g_2(L_2)^3}{g_2(L_2)^3 - 27 g_3(L_2)^2} = j(L_2),$$

and by (10.24) on page 350, $j(E_1) = j(E_2)$, so by Exercise 10.19, $E_1 \cong E_2$. Conversely, assume that $E_1 \cong E_2$, so by Exercise 10.19, $j(E_1) = J(E_2)$, so by (10.24), $j(L_1) = j(L_2)$. If

$$g_2(L_2) \neq 0 \neq g_3(L_2),$$

then let

$$\lambda^4 = \frac{g_2(L_1)}{g_2(L_2)}.$$

Since

$$j(L_1) = \frac{1728 g_2(L_1)^3}{g_2(L_1)^3 - 27 g_3(L_1)^2} = \frac{1728 g_2(L_2)^3}{g_2(L_2)^3 - 27 g_3(L_2)^2} = j(L_2),$$

so by cross multiplying we get that

$$g_2(L_1)^3 (g_2(L_2)^3 - 27 g_3(L_2)^2) = g_2(L_2)^3 (g_2(L_1)^3 - 27 g_3(L_1)^2).$$

However, since

$$g_2(L_1)^3 = \lambda^{12} g_2(L_2)^3,$$

it follows that

$$\lambda^{12} (g_2(L_2)^3 - 27 g_3(L_2)^2) = g_2(L_1)^3 - 27 g_3(L_1)^2 = \lambda^{12} g_2(L_2)^3 - 27 g_3(L_1)^2,$$

so

$$\lambda^{12} g_3(L_2)^2 = g_3(L_1)^2,$$

or by rewriting

$$\lambda^{12} = \left(\frac{g_3(L_1)}{g_3(L_2)} \right)^2.$$

Taking square roots, we get

$$\lambda^6 = \pm \frac{g_3(L_1)}{g_3(L_2)}.$$

If the minus sign occurs, then replace λ by $\sqrt{-1}\lambda$, so without loss of generality, we may assume that the plus sign occurs. We have demonstrated that when $g_2(L_2)$ and $g_3(L_2)$ are nonzero, then there is a nonzero $\lambda \in \mathbb{C}$ such that

$$g_2(L_2) = \lambda^{-4} g_2(L_1) = g_2(\lambda L_1) \text{ and } g_3(L_2) = \lambda^{-6} g_3(L_1) = g_3(\lambda L_1). \quad \text{(S44)}$$

Now from (10.22) on page 350, since (S44) holds, then

$$\wp(z; L_2) = \wp(z; \lambda L),$$

namely, they have the same Laurent expansions about $z = 0$. Since they agree on a disk about $z = 0$, then

$$\wp(z; L_2) = \wp(z; \lambda L)$$

for all $z \in \mathbb{C}$. Since the underlying lattice is the set of poles of $\wp$, then $L_2 = \lambda L_1$. We have proved the result for all cases except where $g_2(L_2) = 0$ or $g_3(L_2) = 0$. Note that by Exercise 10.24, it is not possible to have

$$g_2(L_2) = 0 = g_3(L_2).$$

Suppose that

$$g_2(L_2) \neq 0 = g_3(L_2),$$

then as above, we get

$$g_2(L_2) = g_2(\lambda L_1),$$

and since $g_3(L_2) = 0$, then

$$0 = g_3(L_2) = g_3(\lambda L_1).$$

The other case is similar.

Section 10.4

10.27 We have from (10.30) and (10.32) that

$$\frac{c_4^3 - c_6^2}{1728} = \frac{1}{1728} \left[(b_2^2 - 24b_4)^3 - (-b_2^3 + 36b_2b_4 - 216b_6)^2 \right]$$

$$= \frac{b_2^2b_4^2 - b_2^3b_6}{4} - 8b_4^3 + 9b_2b_4b_6 - 27b_6^2. \tag{S45}$$

Now since

$$b_2^2b_4^2 - b_2^3b_6 = -b_2^2(b_2b_6 - b_4^2) = -b_2^2 \left[(a_1^2 + 4a_2)(a_3^2 + 4a_6) - (2a_4 + a_1a_3)^2 \right]$$

$$= -4b_2^2(a_1^2a_6 + 4a_2a_6 - a_1a_3a_4 + a_2a_3^2 - a_4^2) = -4b_2^2b_8,$$

then plugging this into (S45), we get

$$\frac{c_4^3 - c_6^2}{1728} = -b_2^2b_8 - 8b_4^3 + 9b_2b_4b_6 - 27b_6^2 = \Delta(E),$$

as we sought to prove.

10.29 By Exercise 10.28, the elliptic curve E given by $y^2 + y = x^3 - x^2 - 10x - 20$ has good reduction for all primes $p \neq 11$. Here is the good reduction table.

p	2	3	5	7	13	17	19	23	29	31	37	41	43
N_p	4	4	5	10	10	20	20	25	30	25	35	50	50
$a_p(E)$	-1	0	1	-2	4	-2	0	-1	0	7	3	-8	-6

10.31 According to the hint, we look at $\mathcal{F}(0,1,0) = Z$ for which $\partial\mathcal{F}/\partial Z(\mathfrak{o}) = 1 \neq 0$. Hence, no elliptic curve can be singular at infinity.

We know that E given by (10.26) is singular if and only if E given by (10.27) on page 353 is singular, the latter given by

$$f(x,y) = -y^2 + 4x^3 + b_2 x^2 + 2b_4 x + b_6 = 0.$$

Given that E is singular if and only if there is a point $P = (x_0, y_0)$ with

$$\partial f/\partial x(P) = 0 = \partial f/\partial y(P),$$

then

$$2y_0 = 12x_0^2 + 2b_2 x_0 + 2b_4 = 0,$$

so $y_0 = 0$, and $P = (x_0, 0)$, which is therefore a repeated root of $4x^3 + b_2 x^2 + 2b_4 x + b_6 = 0$, namely

$$4x^3 + b_2 x^2 + 2b_4 x + b_6 = (x-\alpha)^2(4x-\beta) = 4x^3 - (8\alpha+\beta)x^2 + 2(\alpha\beta+2\alpha^2)x - \alpha^2\beta = 0.$$

This implies that

$$b_2 = -8\alpha - \beta,$$
$$b_4 = \alpha\beta + 2\alpha^2,$$
$$b_6 = -\alpha^2\beta,$$

so

$$c_4 = b_2^2 - 24b_4 = 16\alpha^2 - 8\alpha\beta + \beta^2,$$

and

$$c_6 = -64\alpha^3 + 48\alpha^2\beta - 12\alpha\beta^2 + \beta^3.$$

Therefore,

$$c_4^3 = 4096\alpha^6 - 6144\alpha^5\beta + 3840\alpha^4\beta^2 - 1280\alpha^3\beta^3 + 240\alpha^2\beta^4 - 24\alpha\beta^5 + \beta^6,$$

and

$$c_6^2 = 4096\alpha^6 - 6144\alpha^5\beta + 3840\alpha^4\beta^2 - 1280\alpha^3\beta^3 + 240\alpha^2\beta^4 - 24\alpha\beta^5 + \beta^6.$$

Thus,

$$\Delta = (c_4^3 - c_6^2)/1728 = 0.$$

We have shown, by contrapositive, that E is nonsingular if and only if $\Delta \neq 0$.

10.33 By using the solution of Exercise 10.31 above, we see that we can transform $y^2 + y = x^3 - x^2$ via replacing y with $(y-1)/2$ to get

$$y^2 = 4x^3 - 4x^2 + 1,$$

which reduces, modulo 11, to

$$y^2 = 4x^3 - 4x^2 + 1 = (x-8)^2(4x-6),$$

where in the notation used above, $\alpha = 8$ and $\beta = 6$, observing that $\Delta(E(\mathbb{F}_{11})) = -11 \equiv 0\,(\mathrm{mod}\,11)$. Graphing the right-hand side we get

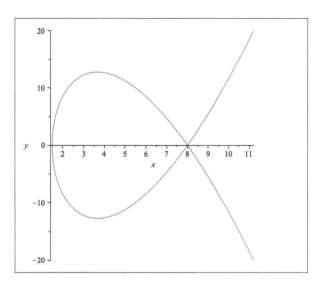

10.35 Since $a_1 = a_3 = a_4 = 0$, $a_2 = p$, and $a_6 = 1$, then $b_2 = 4p$, $b_4 = 0$, $b_6 = 4$, and $b_8 = p$. Therefore,

$$\Delta = -b_2^2 b_8 - 8b_4^3 - 27b_6^2 + 9b_2 b_4 b_6 = -16(p^3 + 27),$$

so since $p > 3$, then $p \nmid \Delta$ so E has good reduction at p by Exercise 10.31.

10.37 Reduced modulo p, we get $y^2 = x^3$ which is Figure 10.2 on page 359, so it has a cusp over $\mathbb{F}_p$.

10.39 From (10.38)–(10.39),

$$\frac{c_4^3}{\Delta} = -\frac{(2^4 \cdot 31)^3}{11^5},$$

so $11^5 \mid \Delta$, $2^4 \cdot 31 \mid c_4$, and $c_4 = 2^4 \cdot 31 \cdot \sqrt[3]{-\Delta/11^5}$. Also,

$$-\frac{(2^4 \cdot 31)^3}{11^5} = 1728 + \frac{c_6^2}{\Delta},$$

so

$$c_6^2 = \Delta \left(-\frac{(2^4 \cdot 31)^3}{11^5} - 1728 \right) = \frac{-\Delta}{11^5} \cdot 2^6 \cdot 41^2 \cdot 61^2,$$

which implies that $c_6 = 2^3 \cdot 41 \cdot 61 \cdot \sqrt{-\Delta/11^5}$. Hence, there is an integer $k \neq 0$ such that

$$\Delta = -11^5 k^6,$$
$$c_6 = 2^3 \cdot 41 \cdot 61 k^3,$$

and

$$c_4 = 2^4 \cdot 31 k^2,$$

as required.

10.41 For the Frey curve, in the notation of (10.26)–(10.34) on pages 353–354, we have that $a_1 = a_3 = 0$, $a_2 = a^p + c^p$, $a_4 = a^p c^p$, $a_6 = 0$, $b_2 = 4(a^p + c^p)$, $b_4 = 2a^p c^p$, and $b_6 = 0$, so

$$c_4 = 16(a^p + c^p)^2 - 48a^p c^p = 16a^{2p} + 32a^p c^p + 16c^{2p} - 48a^p c^p = 16(a^{2p} - a^p c^p + c^{2p}),$$

and

$$\Delta = 16(a^p + c^p)^2 a^{2p} c^{2p} - 8 \cdot 2^3 a^{3p} c^{3p} = 16a^{4p} c^{2p} - 32a^{3p} c^{3p} + 16a^{2p} c^{4p}$$

$$= 16(a^{2p} b^{2p} c^{2p}) \left(\frac{a^{2p} - 2a^p c^p + c^{2p}}{b^{2p}} \right) = 16(a^{2p} b^{2p} c^{2p}) \left(\frac{(a^p - c^p)^2}{b^{2p}} \right)$$

$$= 16(a^{2p} b^{2p} c^{2p}) \left(\frac{(b^p)^2}{b^{2p}} \right) = 16(a^{2p} b^{2p} c^{2p}),$$

which verifies (10.46)–(10.47).

If $p \mid \Delta$, then $p \mid abc$. Since a, b, c are pairwise relatively prime, $p \nmid c_4$, so by Exercise 10.32, (10.45) is minimal at p. Also, we see that if $p \mid ac$, then $\overline{E}((\bmod\ p))$ has a node at $(0, 0)$, whereas if $p \mid b$, $\overline{E}((\bmod\ p))$ has a node at $(a^p, 0)$. Therefore,

$$n = 2^\delta \prod_{p \mid abc} p$$

for some nonnegative integer δ, so it remains to check for $p = 2$. Given that an admissible change of variables "uses up" powers of 4 and $2^4 \| c_4$, then we may reduce only once. Without loss of generality, assume that c is even. Then

$$c^p \equiv 0 \pmod{32}, \tag{S46}$$

since $p \geq 5$. Also, we may assume, without loss of generality, that

$$a^p \equiv -1 \pmod 4, \tag{S47}$$

since if not then we interchange a and b to get (S47) given that $b^p \equiv -a^p \pmod 4$ and $p > 2$. Now, by setting $x = 4X$ and $y = 8Y + 4X$ as an admissible change of variables in (10.45), we get

$$Y^2 + XY = X^3 - \frac{1 + a^p + c^p}{4} X^2 + \frac{a^p c^p}{16} X, \tag{S48}$$

where the coefficients are integers via (S46)–(S47). Hence, (S48) is global minimal. Reducing modulo 2, the right-hand side of (S48) is either X^3 or $X^3 + X^2$. The sole singular point is at $(0, 0)$, then it must be a node since neither $Y^2 + XY$ nor $Y^2 + XY + X^2$ is a square–see Remark 10.14 on page 360. This proves that $\delta = 0$, so

$$n = \prod_{p \mid abc} p,$$

as we sought to show.

Index